"Houston, Tranquility Base here. The Eagle has landed."

WITH these words Astronaut Neil Armstrong announced to the world that the first manned space vehicle had touched down safely on the Moon. The time was 4:17 p.m. plus 42 seconds, Eastern Daylight Time, July 20, 1969. A new age for mankind had begun.

Behind that achievement—almost incredible in its accuracy and daring—lay more than eight years of effort involving hundreds of thousands of Americans and a carefully planned program of design, testing and space flight that led step by step to the final triumph.

In *Appointment on the Moon,* Richard S. Lewis traces the story of America's venture in space from its beginnings shortly after World War II, through the Mercury and Gemini projects, and on to the Apollo 11 flight which lifted man out of his terrestrial home and set him on the road to the stars.

About the Author

RICHARD S. LEWIS was born in Pittsburgh, Pennsylvania in 1916 and received his A.B. from Pennsylvania State College in 1937. Since then he has been a police reporter, war correspondent, drama and music critic, and city editor for several major newspapers scattered over the United States.

Mr. Lewis is the author of two previous books, *The Other Child,* a book for parents who have brain-injured children, and *A Continent For Science: The Antarctic Adventure,* and of numerous articles for national magazines. He is now managing editor of *The Bulletin of Atomic Scientists* and consultant on the space program for the Chicago *Sun-Times* of which he was formerly the science editor.

APPOINTMENT ON THE MOON

The full story of Americans in space from Explorer 1 to the Lunar landing and beyond

RICHARD S. LEWIS

BALLANTINE BOOKS • NEW YORK

Since the beginning of human history the expansion of the human race into new environments has inevitably taken a toll of the pioneers. So it has been in our time, when mankind has begun to expand into the solar system. To the four brave and skillful men who have perished in spacecraft, this book is respectfully dedicated:

ROGER B. CHAFFEE
VIRGIL I. GRISSOM
VLADIMIR KOMAROV
EDWARD H. WHITE II

First published in 1968 by The Viking Press, Inc.
625 Madison Avenue, New York, N.Y. 10022

Revised edition issued in 1969

Published simultaneously in Canada by
The Macmillan Company of Canada, Ltd.

Library of Congress Catalog Card Number: 68-22871

This edition published by arrangement with
The Viking Press, Inc.

First Printing: July, 1969

Printed in the United States

Ballantine Books, Inc.
101 Fifth Avenue, New York, N.Y. 10003

Contents

Prologue: Operation Backfire — *1*
1. The Phoenix — *5*
2. The World-Circling Spaceship — *25*
3. A Wind between the Worlds — *68*
4. Project Mercury — *100*
5. Fireball — *129*
6. Rendezvous — *155*
7. Men in Orbit — *198*
8. Mare Nubium — *230*
9. No Hiding Place — *263*
10. The Tall Towers — *300*
11. Gemini — *327*
12. Fire in the Cockpit — *373*
13. Saturn 501: The Big Shot — *408*
14. Genesis Revisited — *434*
15. Landfall — *463*

Epilogue — *540*

Reference Notes — *547*
Index — *553*
Illustrations following page — *280*

Acknowledgments

I wish to express my appreciation to Bailey K. Howard, President of the Newspaper Division of Field Enterprises, Inc., publisher of *The Chicago Sun-Times,* and to its editor, Emmett Dedmon, whose policy of assigning staff members to major national and international events led to my mission to cover the development of the United States space program.

For their helpful comments on Chapter 3, "A Wind between the Worlds," I am indebted to John F. Clark, director; George F. Pieper, assistant director; and Norman F. Ness, assistant head, Fields and Plasmas Branch, Space Science Directorate of the Goddard Space Flight Center, National Aeronautics and Space Administration.

In the five years this book has been in preparation, I have had the privilege of interviewing hundreds of scientists, engineers, and administrators. Among those whose insights and explanations have been invaluable are the following:

NASA: William H. Pickering, director of the Jet Propulsion Laboratory; Kurt H. Debus, director of the Kennedy Space Center; Robert R. Gilruth, director, and Christopher C. Kraft, Jr., flight operations director, Manned Spacecraft Center; Wernher von Braun, director, and Ernst Stuhlinger, Space Science Laboratory director, Marshall Space Flight Center; and Abe Silverstein, director, Lewis Research Center.

Universities: Eugene Parker, professor of physics, Anthony Turkevich, professor of chemistry, John A. Simpson, professor of physics, and James E. Lamport, technical services manager, of the Enrico Fermi Institute for Nuclear Studies, University of Chicago; Harold C. Urey, professor of chemistry, University of California, La Jolla; Sir Bernard Lovell, professor of radio astronomy, University of Manchester, and director of the Nuffield Radio Astronomy Laboratories, Jodrell Bank; J. Allen Hynek, director of the Lindheimer Astronomical Research Center and head of the Department of Astronomy, and Karl G. Henize, professor of astronomy, Northwestern University; John P. Hagen, professor of radio astronomy, Pennsylvania State University; and Homer J. Stewart, California Institute of Technology.

Industry: John R. Pierce, director of the Research-Communications Sciences Division, Bell Telephone Laboratories; Major General John B. Medaris (Retired), industry consultant; Grant L. Hansen, vice-president, Launch Vehicle Programs, General Dynamics, Convair Division; Joseph Gavin, vice-president, Grumman Aircraft Engineering Corporation; and Dale D. Myers, vice-president, Space and Information Systems Division, North American-Rockwell, Inc.

Prologue

Operation Backfire

The age of interplanetary exploration, like the world war from which it arose, has multiple origins. One of them was the development by Germany of the V-2, the world's first strategic rocket weapon. During the autumn of 1945, several months after the defeat of Germany, British Army engineers conducted a series of tests with the V-2 to complete their assessment of the weapon system from the launch end. (They already knew a great deal about it at the receiving end.) The tests were coded as "Operation Backfire" and they made a profound impression on United States military observers.

With the aid of German technicians, the British launched three of the rockets from the German Naval Artillery Range at Altenwald, near Cuxhaven on the North Sea. I was present at the third and final launching on October 12, 1945. It was a cold, icy, shadowless day, and the wind was harsh. Military observers stood behind a rope barrier nearly a half mile away from the rocket, but even at that distance the size of the missile was astonishing. Mounted on a low steel platform, the olive-green rocket reared up 46 feet. It seemed to be crouched on its enormous fins, which suggested the folded wings of some primordial beast of the upper air.

From time to time, a voice could be heard from a field

public-address apparatus, echoing unintelligibly through the pine trees. The voice stopped and there was a long silence, pierced only by the shrill cries of sea birds. On the launch stand, the crouching beast seemed to breathe. A plume of liquid-oxygen vapor emanated from it like frosty breath.

In the gray light, the missile gave the impression of a timeless dragon, quietly exhaling smoke as it prepared to spring upon humanity. When the sun flashed briefly through a break in the overcast, the rocket became silhouetted against the sky like a deep shadow. There was a signal from the public address system and figures began moving away from the launch platform. When they had fled, the beast was alone, wreathed in its white smoke. Suddenly the vapor vanished. A bright orange flame blossomed under the missile, changing quickly to a dazzling yellow. A growling roar came back on the wind. The beast rose on its haunches and began to lift. It seemed to hover at treetop height and then shot up with ever-increasing velocity. At a thousand feet, it entered the clouds. The bright candle flame of its exhaust remained visible but veiled for several seconds. It dimmed through successive cloud layers until it was gone. The range became silent, as though a storm had passed. The steel launch platform looked strangely empty.

That night, American observers at a United States fighter base near Cuxhaven were advised that a British destroyer had picked up the V-2's green-dye marker at the impact site where the missile had splashed down into the North Sea after a five-minute flight of 190 miles. The demonstration marked a transition point in the development of rocket technology in the West. That was the last V-2 fired in Germany. The engineering science which the weapon represented was carried off by the victors as spoils of war. In the United States, it was to evolve into an interplanetary spaceship technology.

It took only 22 years for the American industrial system to enlarge upon the technology sufficiently to develop a launch vehicle, the Saturn 5, powerful enough to propel three men in a 45-ton spaceship to the moon. The evolution of space technology in the postwar years was a resultant of powerful forces in human society. Most conspicuous was the rise of competition, at the outset of the International Geophysical Year 1957–58, between the United States and the Soviet

Union for supremacy in space, as a symbol of technological and military superiority.

As an outgrowth of World War II and the Cold War, it would appear, the American space experience has been essentially a response to competition and threat, moderated by scientific curiosity and a spirit of adventure. These are the characteristics of human expansion in every era. In these respects, the motives of the Soviet space program are the same as our own. Their program responds to ours. Both programs are continuously and mutually reactive.

Beyond the immediate stimulus of the competition of rival social systems, it may be that the venture into space is a product of biological determinism which impels us to explore a new environment when we are technologically ready. As Mark Twain is reputed to have said, "When it's steamboat time, you steam." The scientific colonization of Antarctica after World War II illustrates this process. Only an advanced technology could make that scenic but inhospitable icescape a "Continent for Science."

By professional chance, I have been able to observe many of the events which have shaped our space program. One of the earliest was that final V-2 test of "Operation Backfire" in 1945, which I attended as a correspondent for the United States Army newspaper *Stars & Stripes*. Here was a forerunner of things to come, for the victors certainly would exploit such a weapon. In the beast crouching on the launch ring was the beginning of an engineering science which could open up a new and limitless environment to man as well as rain terror and death from the skies.

With the development of space technology since then, the competition between the United States and Russia has resulted in a new imperative in America: a manned lunar landing by 1970. No enterprise in American history has been so daring and so demanding as our appointment on the moon. It has dominated the United States space program since President Kennedy created that goal in 1961. It is one of the most compelling, yet controversial, efforts to which any nation has ever been committed.

This is a history of how it came about and of its consequences.

Chapter 1

The Phoenix

The development of large rockets in the twentieth century began as an outgrowth of dreams. Although the first ballistic missile, the V-2, was a nightmarish military weapon, the engineering science it represented was not entirely a product of German rearmament and World War II. In origin, the V-2 was the offspring of an unlikely union of Middle European technological romanticism and the Treaty of Versailles.

Long-range rockets in modern times were the conception of men who dreamed of space travel. At the end of the nineteenth century, a Russian schoolteacher, Konstantin Ziolkovsky, proposed the development of rockets to explore the solar system. "The Earth is the cradle of mankind," he wrote in 1899, "but one cannot live forever in the cradle." [1] In 1903, he published an article, "Beyond the Planet Earth," in which he imagined the colonization of other planets. In America, Robert H. Goddard, whose rocket designs outran the technology of his time, suggested in 1907 that the heat of radioactive materials could be used to propel rockets in space.[2] A plan for space exploration by rocket was described in 1923 by Professor Hermann Oberth, a Rumanian engineer teaching mathematics in Mediash, Rumania. Oberth wrote a pamphlet, "The Rocket to Outer Space," in which he proposed the use of liquid propellants (instead of solids such as gunpowder) to make the rocket a truly long-range, steerable vehicle. Goddard had made the same proposal in 1919,[3] and had fired the world's first liquid-fuel rocket in 1926, but it was the Germans who developed it as a military weapon.

For 700 years, rockets propelled by gunpowder had been considered mainly as weapons. British rockets devised by an English inventor, Sir William Congreve, were used against Fort McHenry ("the rockets' red glare") in the War of 1812. From that time until Goddard began his experiments in 1914, there had been little development in rocketry. Goddard designed a solid-fuel missile in 1918, but could not persuade the War Department that it might be useful.

In the years following World War I, rockets were mainly vehicles of fantasy. They became associated firmly with space travel in a society which, for a time, sought escape from war and its aftermath. A new escape literature based on science fantasy arose in Europe and America. A German movie, *Frau im Mond* (Woman in the Moon), was shown throughout Central Europe. Fritz von Opel built the first rocket car, which roared to speeds of 60 miles an hour.

During the late 1920s, rocket societies sprouted like dandelions. The word "astronautique" was coined by Robert Esnault-Pelterie, an engineer, and André Hirsch, a Paris banker, in 1928, when an astronautics committee was established by the Société Astronomique Française. In Russia, Ziolkovsky, Friedrich A. Tsander, and Mihail K. Tikhonravov founded the Society for Studying Interplanetary Communication in 1924, and four years later a technical "Group for the Study of Reactive Motion" was organized in Moscow. The Soviet government also established a liquid-fuel rocket study group in Leningrad.

In Germany, Johann Winkler of Breslau, Willy Ley, and six others founded the German Society for Space Travel, or VfR (*Verein für Raumschiffahrt*). Oberth and Esnault-Pelterie joined the society, and so did Wernher von Braun, a student at the University of Berlin.

While the objectives of the VfR were similar to those of the American Rocket Society and the British Interplanetary Society, which arose at the same time, the German organization was destined to become considerably more influential than its American and British counterparts, for reasons having little to do with space travel.

The military potential of large, steerable, long-range rockets did not escape the attention of German military specialists. The long-distance strike had been an inherent feature of German military thinking since the Zeppelin and the Paris Gun of World War I. Technically, it would have been more feasible for German ordnance to develop long-range cannon to higher standards of accuracy and fire power, but that was forbidden by the Versailles Treaty. Artillery development, which is noisy, is difficult to carry on in secret. On the other hand, there was nothing in the treaty that prohibited the development of rockets for space travel or scientific purpose. So it was that by the end of the 1920s the interests of the romanticists who dreamed of going to the stars and of the militarists who dreamed of a new, long-distance weapon found a common expression in rocketry. And, as far as the technology was concerned, the military interest was critical, for it could gain access to the public funds required to develop big rockets.

The VfR had acquired a lease from the city of Berlin on a proving ground in the Berlin suburb of Reinickendorf in 1929, but the field proved to be too small for testing large rockets. In 1932, the Army Board of Ordnance opened a larger rocket-testing ground at Kummersdorf, 20 miles south of Berlin. Walter Dornberger, a captain of ordnance engineers, was placed in command by Major General Karl Becker, chief of ordnance development. Dornberger hired von Braun, then 26 years old, to assist him.

As the Army took over rocket development, the VfR quietly dissolved. The rocket for space travel became the rocket for war. In this century, rocketry got off the ground as a weapon.

ACTION AND REACTION

Although the gross mechanics of rocketry have been known for seven centuries, the conception of the rocket as a launch vehicle, that is, a means of propelling a payload from one place to another, arose in the first decade of the twentieth century, in America, in England, in Germany, and in Russia. One of the consequences of this idea was heightened interest in the problem of extending the range of rockets.

While the theory of rocketry had been understood for more than two centuries, there had been little incentive to develop the technology until the 1930s.

In 1687, Sir Isaac Newton described the principle of rocket propulsion as follows: "To every action there is an equal and opposite reaction. The mutual actions of any two bodies are always equal and opposite directly." The hand that rocks the cradle is rocked by the cradle. The equal and opposite reaction of a skyrocket's thrust in one direction causes it to move in another. That is why the reaction motor can function in space, where there is nothing, or relatively nothing, to push against.

By 1920, Goddard had discovered that the average velocity of the gases ejected from an ordinary fireworks skyrocket was about 1000 feet a second. In order to make such a rocket go farther and faster, he had to find a means of increasing the velocity of exhaust gases. By firing smokeless powder in steel chambers with smooth, tapered nozzles, he managed to boost the velocity of the expelled gases ten-fold, up to 10,000 feet a second. As the velocity of the gas jet increased, however, the heat it generated also increased. At 10,000 feet a second, a jet velocity produced a temperature of 12,000 degrees Fahrenheit. Since aluminum melts at one-tenth of that temperature and most steels at about one-fifth, the problem of finding material that could withstand the heat became a bottleneck to rocket improvement.

The development of rockets therefore became a technological problem at the outset involving the state of the art in metallurgy, chemistry, and gas dynamics. Goddard and his friends watched motors of the toughest steel burn out in the first fraction of a second after the fuel was ignited. What could they do about it?

Examination of the "burning" process showed that rather than being melted by the heat of the gases the metal was being eroded. Like an icicle in a blowtorch stream, it was being flaked away. As the surface of the metal wall softened in the heat, the blast of escaping gases carried off bits and pieces, exposing new surface to the erosion of the hot blast. Before the metal became hot all the way through, the walls of the motor were eroded away and the motor disintegrated.

The erosion problem seemed to put an upper limit on the velocity of the exhaust gases and therefore to the velocity any rocket could attain. Unless it was solved, the rocket would never amount to anything more than a short-range, military tactical projectile in one form and a fireworks display in another.

In Germany, the VfR experimenters had encountered this problem, too. They thought of placing the blast chamber in a tank of liquid oxygen, at 297 degrees below zero Fahrenheit. They reasoned that the supercold liquid oxygen would cool the thrust chamber sufficiently to keep the wall intact. But the heat of the exhaust caused the liquid oxygen to boil so rapidly that the expanding gas ruptured the oxygen tank and the motor blew up.

The VfR group then tried a water jacket to cool the blast chamber. Goddard tested the same approach in the autumn of 1932, but found, as the Germans did, that it was not the answer.

The answer was a rocket motor cooled by its own fuel. Such motors had been tested in the early 1930s in Vienna and Berlin, and by 1938 engineer James Wyld, who was president of the American Rocket Society, had designed a practical self-cooled rocket motor that made possible the development of high-velocity exhaust streams and, consequently, rockets that could attain high altitudes and long distances.

In its earliest form, the Wyld rocket motor was rather simple. The blast chamber consisted of an aluminum tube six inches long and two inches in diameter, with a nozzle of Monel metal at one end. A second tube, just a little larger, formed an outer jacket for the blast chamber. The fuel, gasoline or kerosene, was pumped into the space between the blast chamber and the outer jacket and forced upward and through a cluster of small holes which allowed it to spray into the blast chamber, where it was mixed with liquid oxygen and ignited. The heat of the burning fuel at the nozzle passed through the walls of the blast chamber and was carried away by the fuel flowing upward between the walls of the blast chamber and outer jacket.

So rapidly was the heat carried away that its only effect was to warm the fuel, which promoted combustion. Because of this feature, wherein the heat of combustion was returned to the engine to make combustion more efficient, Wyld's motor was called regenerative. It opened the era of large, liquid-fuel rocket engines, Once the problem of preventing the blast chamber and nozzle walls from melting was solved, it became theoretically possible to build intercontinental and interplanetary launch vehicles. The V-2, which applied the same principle used in Wyld's rocket motor, was a beginning.

It was at the Kummersdorf proving ground that the design for the V-2, or the A-4 as it was called by ordnance, was drawn in 1932. Dornberger, who was chief of rocket development for the German Army Board of Ordnance from 1931 until the end of World War II, described it as an extension of the Paris gun. That huge artillery piece, developed in World War I, would hurl 22 pounds of explosive a distance of 78 miles. What Dornberger and other artillerymen had in mind was a rocket that would hurl 100 times that payload, or 2200 pounds (a metric ton), twice as far, or 156 miles.

The rocket should be small enough to be hauled as a unit on conventional roads, through the narrow, medieval streets of small towns, and through European railroad tunnels. This transportation requirement established an over-all diameter at the guidance fins of 9 feet and a length of 46 feet.

In order to hurl a metric ton of warhead 156 miles, the rocket had to reach a speed of 3600 miles an hour. And to attain this speed, a thrust of 56,000 pounds was required at launch. At the time these specifications were drawn, the peak thrust obtained in liquid rockets was just short of 1000 pounds.

The first of the family of rockets with the code name "Aggregate" or "A," the A-1 was designed to develop 660 pounds of thrust, burning liquid oxygen and alcohol. It was a squat tank, 4½ feet long and a foot in diameter. Fully fueled, it weighed 330 pounds. The first test was planned for a few days before Christmas 1932. The rocket blew up on the stand. However, the explosion taught the Kummersdorf designers their first important lesson. They had been using compressed gas to force the fuel and oxidizer from their tanks into the mixing chamber, but that method was too slow. The system required a pump that would deliver 1100 gallons of 150-proof alcohol and 1000 gallons of liquid oxygen a minute to the combustion chamber. After a series of false starts and frustrations, a steam-driven pump was devised. The steam was generated from the decomposition of hydrogen peroxide by liquid potassium permanganate. The next year, a redesigned rocket, the A-2, was launched from the island of Borkum in the North Sea to an altitude of 6500 feet.

Now a workable ballistic missile was in sight. In 1936, Field Marshal Albert Kesselring of the Army was convinced by his rocket experts that a much more elaborate development center than Kummersdorf was needed to bring the missile to completion. The area should be away from densely populated areas, and near the sea, so that rockets could be tested over long distances. A site was chosen on the island of Usedom in the Baltic Sea, at the mouth of the river Peene. Acreage was leased from the town of Peenemünde. Then the Special Ordnance Development Section, under Dornberger's command, moved into the area. From then on, Peenemünde became the home of the German DSOD, the Department for Special Ordnance Devices. By the end of the war, the German investment in the Peenemünde development center totaled $75,000,000.

By 1937, the rocket program which had begun five years earlier with eight men had grown into a major development project employing 100 engineers and technicians. During that period there had been experiments to test the steering of rockets by external fins. They demonstrated that fins alone were not adequate because their steering reaction against the air stream was too slow to guide the vehicle accurately. Another solution to the problem was sought in the A-3, a vehicle which produced 3300 pounds of thrust for 45 seconds. It reached an altitude of 7½ miles and a range of 11 miles. Guidance was accomplished by movable vanes within the thrust stream. When turned from one side to the other, or up and down, the vanes would reflect the exhaust stream, causing the vehicle to yaw to the left or right, or pitch up or down. A combination of external fins and exhaust vanes seemed to solve the steering problem. The dual system was incorporated in a more powerful version, the A-4.

The first flight test of Fernrakete A-4 (long-range rocket A-4) was made on July 6, 1942. It reached an altitude of one yard and blew up. The second test vehicle was more successful. It flew three miles high before exploding. On October 3, 1942, the fourth rocket flew full range. The fifth also seemed to fly successfully, although it went off course and its impact point in the Baltic Sea was not located. In the next 13 launches, the A-4 failed every time. It required a hundred more launches to bring the vehicles up to 85-per-cent reliability.

According to Dornberger,[4] Hitler was uninterested in the rocket because he had dreamed that it would never reach England. Hence the program lacked high priority during the early war years. Engineers and technicians were not available to it because they were misclassified as riflemen, cooks, or ordnance mechanics, or because they were assigned to other technical tasks.

It was not until 1943 that the Peenemünde development program acquired significant priority in men and materials. The staggering losses of German bombing aircraft to the guns of Royal Air Force fighters were rapidly undermining the entire Nazi war machine. Hitler, the Oberkommando Wehrmacht, and the Luftwaffe began to look for miracle weapons. There were two possibilities: the Army's A-4 ballistic missile and the Luftwaffe's pilotless bomber, the Fi-103 "buzzbomb."

Magically, misclassified experts were pulled out of their units to serve in the rocket program. One of them, Dieter Huzel, an electrical engineer who was transferred to Peenemünde from an ordnance detachment on the Russian front, recalls: "Overnight, Ph.D.s were liberated from KP duty, masters of science were recalled from orderly service, mathematicians were hauled out of bakeries, and precision mechanics ceased to be truck drivers."[5]

The High Command ordered mass production of the A-4 even before the bugs had been taken out of the vehicle. Six weeks after the abortive attempt to assassinate Hitler on July 20, 1944, the A-4s began lifting from launch stands in the Netherlands against targets in liberated France and England. As an operational weapon, the A-4 was designated by Nazi propaganda as "Vergeltungswaffen Zwei" (Vengeance Weapon Two), or V-2. It followed by three months the V-1 flying bomb which had begun to harass London a week after D-day.

In the V-2, with its thrust of about 56,000 pounds, propulsion cutoff, or "Brennschluss," was critical. If the engine burned longer than 65 seconds, the rocket would overshoot its targets in England. At first, the guidance crew would cut off the engine by sending a radio signal which actuated an electrical circuit in the engine to close the fuel valve, and the missile would then coast on a ballistic trajectory just as an artillery shell would. However, the radio cut-off was susceptible to interference by the British, and the Peenemünde engineers devised another means of controlling the engine that would be immune to radio interference. It was an accelerometer, an inertial device which would react at a velocity of 3600 miles an hour to shut off the turbopump.

During the long period of A-4 experiments, British military intelligence became aware that long-range missiles were being developed at Peenemünde. The first intimation of the V weapons came in April 1943, and a subsequent investigation by Duncan Sandys, Joint Parliamentary Secretary to the Minister of Supply, turned up evidence of substantial progress in German development of long-range rockets. There was, however, no general agreement on the extent of the threat the rocket weapons posed. Lord Cherwell (F. A. Lindemann), the principal scientific adviser to Prime Minister Winston Churchill, believed that the long-range rocket story was a deliberate hoax to conceal the actual development of another weapon, such as a pilotless aircraft.

From several sources, British intelligence assembled this picture of the rocket weapon in the spring of 1943: a vehicle 38 feet long, 7 feet in diameter, weighing 60 tons, with a 5- to 10-ton explosive charge and a range of 90 to 130 miles. On the basis of these data, the Ministry of Home Security calculated that a single 10-ton warhead would cause demolition over a radius of 850 feet, damage over an area of 650 acres, with serious blast damage over a radius of 1700 feet. Casualties might amount to 600 killed, 1200 seriously injured, and 2400 slightly injured. One rocket an hour for 24 hours would kill 10,000 persons and seriously injure 20,000.

The Royal Air Force responded to the rocket threat by launching a series of bomber raids coded as "Operation Crossbow" on Peenemünde. In the first raid on August 17, 1943, about 600 aircraft dumped 2000 tons of bombs on the island of Usedom, according to the British War Office. The raid was repeated the following day. According to Dornberger's book, *V-2*, the raids, while damaging, did not knock out testing capability or interfere seriously with A-4 development. The following year, raids on July 18, August 4, and August 25 proved more destructive. Even so, the rocket development center never was put out of operation.

The pilotless aircraft which Lord Cherwell had foreseen

appeared on the night of June 12, 1944, in Greater London, killing six and injuring nine persons seriously at Bethnal Green. Called Vergeltungswaffen I (Vengeance Weapon 1) by the Germans, the flying bomb or "buzzbomb" was a 5000-pound airplane developed by the Luftwaffe. It was driven by an air-breathing, jet-propulsion unit in the rear and carried one ton of high explosives. It was launched at 245 miles an hour by a catapult and steered by a crude autopilot. A windmill-type cyclometer controlled the length of the flight by cutting off the engine after a given number of turns. The V-1 had a range of 160 to 190 miles and a speed of 370 miles an hour, which made it fair game for fighters and antiaircraft guns. Of 6725 flying bombs reported between June 12 and September 5, 1944, 3463 were destroyed and 2340 reached their targets.

The V-1 attacks were heavy enough to persuade the government to begin evacuating mothers and children from London, especially from the neighborhoods south of the Thames where the buzzbombs seemed to fall in high concentrations. Then, as Allied troops expanded their continental bridgeheads during the summer of 1944 and moved into the V-1 launching areas, the attacks faded away. On September 6, 1944, the Vice-Chiefs of Staff reported that all areas from which flying bombs or rockets might be launched against London were either occupied by Allied troops or evaculated by the enemy. "There should be no further danger to this country from either of these causes, except for the possibility of the airborne launching of flying bombs," an official statement said. On September 8, Duncan Sandys issued a statement saying that the "Battle of London was over except for the last few shots." [6] The statement was premature. The next day, Londoners traveling home from work were startled by a sharp clap of thunder at 6:40 p.m. The first A-4 rocket had fallen at Chiswick, killing three people and seriously wounding ten. Sixteen seconds later, a second rocket fell at Epping but did little damage. Here was Vergeltungswaffen 2 which had been promised—the V-2.

During the next ten days, rockets arrived at scattered places in southeast England at the rate of two a day. They were of the type which intelligence had forecast. The radar stations failed to detect them at first, partially because the radars were not covering portions of the Dutch coast where the V-2 rockets were being launched. Since the V-2s did not appear to be doing any more damage per vehicle than had the flying bombs, the Chiefs of Staff did not believe that elaborate countermeasures were called for.

Indeed, if countermeasures had been advocated, the Chiefs would have been hard pressed to devise any. The rockets came down from the stratosphere faster than sound, and the first inkling of their arrival was a shattering explosion. The Chiefs advised against any announcement that England was under a rocket attack. The London newspapers referred to the new weapon as a "flying gas main," a gibe at a War Office explanation that the mysterious new explosions were caused by exploding gas mains.

On September 17, 1944, the Allied Airborne operation against the lower Rhine at Arnhem, the Netherlands, forced German rocket troops to flee eastward. For a week, the V-2 attack on England stopped. Then it resumed on September 25, when Norwich was hit. London again became a V-2 target on October 3, 1944, and for the next four weeks received two or three hits a day. The intensity of the bombardment rose in November to four a day and in December to six. Unlike the flying bombs, the rockets could not be shot down or intercepted. On November 10, 1944, Churchill announced in Parliament the nature of the beast, the long-heralded long-range rocket.

The only defense against the V-2 that winter was a series of fighter-bomber sweeps across the Low Countries, and intensive reconnaissance by armed fighters to ferret out likely launch sites. Despite its size, the V-2 was so mobile it could be launched in a woods, from a portable steel launch platform placed on ground which could be hardened by being frozen with liquid oxygen. The launch sites, however, could not be picked at random because of the elaborate programming which was necessary to enable the rocket to hit its target. The precise distance between target and launch site had to be known and had to be within the range of the vehicle.

From a civil-defense point of view, the rocket attack was little different from the buzzbomb barrage during the summer. The rockets had greater penetrating power and caused more devastation in the immediate impact area, but the total area of blast damage was smaller than that affected by flying bombs. In spite of the increasing severity of the V-2 attacks during the winter of 1944–45, thousands of refugees who had evacuated London during the summer and early fall were returning to find work and shelter in the city. During January and February, both the frequency and accuracy of the V-2s increased. On January 26, 1945, there were 17 "incidents"—13 of them in Greater London. The weekly casualty list was twice as high as in December. While radar was beginning to pick up the rockets 60 seconds before impact, it was decided to wait until the warning time could be extended to four minutes before sounding air-raid alarms.

On March 8, 1945, one of the worst disasters of the rocket attack occurred at Southfield Market, where a V-2 fell into a crowd of shoppers, killing 110 and injuring 123. The German V-2 offensive ended March 27, 1945, when the last rocket to hit England fell at Orpington, Kent. During the seven months of the rocket bombardment, 2754 persons were killed and 6523 injured by V-2s in England. The flying bombs had killed 6184 and seriously injured 17,181, and aircraft bombing had killed 51,509 and seriously injured 61,423.[7]

According to Dornberger,[8] a total of 3745 rockets were launched by Germany between September 6, 1944, and March 27, 1945. Of these, 1115 fell on England and 2050 on targets on the continent. An additional 580 rockets were used in developing the missile system and in training launch crews. About ten per cent of all missiles launched blew up in the air. One cause of air bursting was found to be a fluttering of the outer skin which increased the heat of air friction to the point where the outer skin disintegrated. Air then rushed into the missile and ruptured the fuel tank.

Of all the V-2s launched against Allied targets, 74 per cent landed within 18 miles and 44 per cent of these within 6 miles, according to Dornberger. No rocket was intercepted nor were any attempts to deflect them by radio countermeasures detected, Dornberger said.

Mass production of the weapon was carried on at the rate of 650 flight articles a month at a gnomish underground factory called Mittelwerke near Nordhausen, in central Germany. After the V-2 was operational, Peenemünde continued to be a research and development center for more advanced rockets.

Against military installations in the field, the V-2 was singularly ineffective. There were few American military casualties from it. In the light of London's experience, the V-2 attack was the least serious of the three types of German assaults from the air, the others being the Luftwaffe and the flying bombs. The rocket hardly lived up to Germany's advance billing of its horrendous destructive power or to early British estimates of its lethal effects, based on a 10-ton warhead.

Yet this piece of ordnance represented the most important technological development in warfare since the tank and the aircraft bomber. It was important in its implications rather than its effects, which, while impressive and frightening, were never strategically significant.

Larger and longer-range weapons were implied by the operating successes of the A-4. These and discoveries about other secret weapons German ordnance was working on were disclosed at a United States Ordnance Service Intelligence briefing in Paris in May 1945.

The briefing officer, Lieutenant Colonel John A. Keck, a retired engineer from Pittsburgh, Pennsylvania, disclosed that Nazi rocketeers had been working on a V-3, as he called it. It was an intercontinental ballistic missile which could hit New York City from a launch site on the French coast.

Beyond that, Keck described an array of bizarre weapons the Germans were developing which seemed to his audience to have been inspired by the comic-strip pseudoscience of the 1930s. The most formidable of these was a "sun gun," which would annihilate cities. The sun gun was a giant mirror which would focus the sun's rays on any point on earth. The mirror was to be placed in an orbit 5100 miles high by one of the super-rockets the Germans were also working on and controlled by radio. No one at the time knew why an altitude of 5100 miles had been specified for the sun gun, but it was speculated that perhaps this was the minimum altitude at which an orbiting space vehicle would be free of the drag effects of the atmosphere.

Keck also described plans for a submarine V-2, which would be launched from a U-boat. Such a weapon was being developed near a lake in the Austrian Alps. Also, a new German antiaircraft rocket that would seek its target was on the list of superweapons. This reference may have been to the "Wasserfall," a supersonic antiaircraft rocket. Its existence was discovered by the British when one went out of control during a test and landed in Sweden.

There was an array of lesser weapons in various stages of

evolution. One was a curved gun barrel with mirror sights that would enable the trooper to shoot around corners. There was a small self-propelled bomb, like a toy tank. It would scoot across the terrain for several thousand yards and blow up on radio command.

Except for the V-2, none of these schemes came to fruition. Looking back, it seems obvious that the Peenemünde engineers had conceived of the ICBM, the Polaris, and an orbiting space station. They were mistaken about one point. The final shape of the sun gun would not be a space mirror to focus sunlight on hapless victims down below, but rather a means of replicating the energy that drives the sun in a device called a hydrogen-fusion bomb, and Nazi science was not to achieve that.

The German ICBM was designated the A-10. It had a projected range of 2600 miles, or about ten times the range of the V-2. This was Colonel Keck's V-3. The A-10 would develop 440,000 pounds of thrust for 50 seconds in the first stage—almost as much as the first stage of the Titan 2 ICBM of 1961. The second stage was a winged version of the A-4 called the A-9. The over-all length of the rocket was to be 72 feet, and the diameter 12 feet. Liftoff weight was calculated at about 192,000 pounds. Generally, the A-10 design was similar in size to the Titan missile developed 15 years later, except for the fact that the German ICBM of 1945 was two feet larger in diameter.

Although the A-10 never got off the drawing boards, its fame spread far and wide. After the Battle of the Bulge, there were rumors in Paris that it would be launched from the Bavarian Redoubt against New York as a final gesture of defiance. The Redoubt, it will be remembered, was to be the last stand, the Nazi Gotterdämmerung, wherein the diehards would hole up in the mountains of Bavaria and fight on until the last man was dead. There was no Bavarian Redoubt and there was no A-10. But in the gloomy winter of 1944–45, when the war seemed to be an interminable punishment from which mankind could never extricate itself, both of these rumors appeared to be uncomfortably plausible.

After the A-10 plan was disclosed by United States Ordnance Intelligence, a letter to the B-Bag column in the Paris edition of the *Stars and Stripes*, the American Army newspaper, summed up a typical GI reaction to it. "At least if a rocket lands in Manhattan," the writer observed, "they'll find out there's a war on."

During January and February of 1945, Peenemünde was evacuated in stages as the Russian armies came nearer week by week. "By January 1945," von Braun said, "the situation in Peenemünde was the following: The Russian Army was approaching from the east. It was about 100 or 80 miles from Peenemünde, so close that we could already hear the artillery fire at night. It was very obvious to me and my associates that the war was lost, and the decision whether we wanted to wind up on the East side or on the West side had to be made before the Russian Army arrived. So I held a meeting and we took a vote and the vote was unanimous to the effect that we should go West." [9]

Soviet infantry in trucks rolled into Peenemünde early in March, but by that time the German technical staff had fled to the south. Most of the engineering staff, headed by von Braun, moved to Magdeburg. As United States tanks of the Fourth Armored Division approached Mühlhausen, von Braun ordered Dieter Huzel, the electrical engineer, to find a cave or a mine where the V-2 technical documents could be hidden. These amounted to priceless booty, for they contained the whole story of German rocket development. As Huzel put it: "Whoever inherited them would be able to start in rocketry at that point at which we had left off, with the benefit not only of our accomplishment but of our mistakes as well—the real ingredient of experience. They represented years of intensive effort in a brand new technology, one which would play a profound role in the future course of human events." [10]

Huzel loaded the documents into a truck. Dodging American fighter bombers, he and his driver managed to reach the town of Dorten in the Harz Mountains. Outside the town he stored the crated documents in a potassium mine. The engineering staff then moved to Oberammergau. Huzel said von Braun was uneasy and told him, "Regardless of what is happening to Germany, our personal, technical knowledge, our knowledge of the hidden location of our research results—in these things we are the bearers of an entire engineering science."

The technical group was apprehensive because Oberammergau was filled with tough and morose SS men. Von Braun expressed the fear that the SS would develop an "Attila complex and try to destroy us and everything we've done" rather than let the engineering science represented by the V-2 fall into other hands.

Presumably, von Braun was drawing for analogy on the legendary interment of Attila, the Hun. The body of the barbarian chieftain was buried in a river bed, after the river was diverted just long enough to dig the grave. The workmen on this project were massacred by Attila's bodyguard to protect the secret of the burial place. Von Braun feared that he and his engineering staff might suffer a similar fate at the hands of SS fanatics seeking to bury the secrets of rocket technology. Hoping for safety in numbers, the von Braun group joined forces with Dornberger at Haus Ingeborg, an inn at the town of Oberjoch near the Adolf Hitler Pass.

On the dull, rainy afternoon of May 2, 1945, seven men assembled in front of the inn. They were Wernher von Braun; his brother Magnus, an engineer who had been in charge of gyroscope production at Mittelwerke; Dornberger; Hans Lindenberg, the V-2 combustion-chamber engineer; Bernhard Tessman, chief designer of the test facility at Peenemünde; Dr. Herbert Axtel, an officer on Dornberger's staff; and Dieter Huzel. The group descended the Adolf Hitler Pass in three cars, Huzel relates, toward the Austrian village of Schattwald. As the caravan rounded a curve, an American soldier stood in the road. He waved it to a stop.

OPERATION PAPERCLIP

It had become clear to United States Army Ordnance in 1944 that the Germans had created spectacular innovations in the development of the V-1 and V-2 weapons systems. There was a general belief then that if the V-2 had been deployed earlier and in greater numbers, it alone could have crushed British power to continue the war. Assessments of bomb damage after the war, however, did not support that view.

Major General H. J. Knerr of the United States Strategic Air Force made a recommendation to Lieutenant General Carl Spaatz, USSTAF commander, that the Army Air Force "make full use of established German technical facilities and personnel before they are destroyed and disorganized."[11] Specifically, Knerr wanted to bring the German scientists and engineers and their families to the United States, where they could continue their work in rocketry without fear that their families would be taken as hostages by the Russians.

On April 26, 1945, the Joint Chiefs of Staff issued Order 1067 directing General Dwight D. Eisenhower to "preserve from destruction and take under your control records, plans, documents, papers, files and scientific, industrial and other information and data belonging to . . . German organizations engaged in military research."

This roundup of German technology was known as "Operation Paperclip." The rocket and missile aspects of it were supervised by Colonel Holger N. Toftoy, who called Washington from Paris in May 1945 to request the transfer to the United States of 300 German rocket scientists and technical men.[12] When he received no reply, Toftoy flew to Washington to get the transfer on the road. He succeeded in persuading the War Department to authorize the emigration of 127 German scientists and engineers. This group was held in special camps for several months, under varying conditions of uncertainty and confusion. The first seven Peenemünde émigrés, headed by von Braun, arrived at Fort Strong, Massachusetts, on September 20, 1945. The other six were Erich W. Neubert, Theodor A. Poppel, August Schulze, Eberhard Rees, Wilhelm Jungert, and Walter Schwidetzky.

They were initially put to work at the Aberdeen Proving Ground in Maryland processing captured German missile documents, many of which they themselves had written, edited, or approved. Tons of reports, including five truckloads of Dornberger's notes and all the documents hidden in the potassium mine in the Harz Mountains, were organized, translated, and filed. The original seven were then transferred to Fort Bliss, Texas, where they were joined by 55 more V-2 specialists in December 1945. All by that time were under contract to the Army.

The Army moved promptly to create a rocket development group of its own, using the Germans from Peenemünde as a nucleus. Rocket-test facilities, including a blockhouse with walls 10 feet thick, were built at the White Sands Proving Ground in New Mexico, a flat desert 125 miles long and 40 miles wide.

By the end of 1945, all that was needed was rockets. These, happily, were provided by the energetic Colonal Toftoy and his associate, Major James J. Hamill of Ordnance/Technical Intelligence. They let nothing escape them that they could get their hands on, including vast stores of technological loot in the Mittelwerke plant at Nordhausen. The plant was in the Russian zone, but Toftoy's crew got there before Soviet troops took possession. From Mittelwerke storage depots, Operation Paperclip collected 100 nearly completed V-2s, machine tools specially designed for rocket production, plans, and manuals. These were loaded in 300 boxcars and hauled to Antwerp, where they were shipped to the United States.

In addition, American forces also picked up a big supersonic wind tunnel which had been moved from Peenemünde to Bavaria in 1944, where it was less vulnerable to bombing. They dismantled it and shipped the parts to the Naval Ordnance Laboratory at White Oak, Maryland, where it was reassembled and used for Navy rocket development.

When the Russians arrived at Mittelwerke, they took into custody the production men who had been mass-producing V-2s there and carted off the tools which the Americans had left. Thus, Soviet rocket technology too had access to the German development. In later years, however, Germans returning from Russia reported that they had been kept isolated and were never integrated with Russian engineering groups in Soviet missile development. For the most part, the Peenemünde émigrés in America remained together at Fort Bliss, but their role in American rocket development was considerably more open than that of their colleagues in Russia.

The contract under which the Germans worked for the United States Army—which gave them the status of employees rather than prisoners of war—was one of the most unusual documents of the space age. It consisted of a single mimeographed sheet of paper and two carbon copies, signed by an Army captain. The original copy was sent to Headquarters, United States Forces, Europe. The new employee retained the second copy, and the third was sent, ironically, to the "Discontinued Projects Branch," War Department, Omaha, Nebraska.

Twenty years after the first seven arrived, the Peenemünde

émigrés held a reunion at the Marshall Space Flight Center at Huntsville, Alabama, where von Braun now is director for the National Aeronautics and Space Administration. All had become United States citizens and many of them were still on the team, deeply involved in developing the giant Saturn 5 rocket that is scheduled to boost three American astronauts to the moon by the end of this decade.

Of 118 who had signed the mimeographed contracts to work for the Army in 1945, eight had died, 15 had returned to Europe, and 31 were working for other government agencies or for private industry. The remaining 64 were engaged in rocket development and related technology for NASA in Huntsville.

The German rocket developers quickly became a technical elite in America, but their image as a group has been ambiguous in the minds of many Americans. Huzel, perhaps the most articulate of the émigrés, has attempted to rationalize their role as Promethean bearers of a great technological gift for mankind, forged in the fires of war but of greater import as a means of exploring space than as a weapon. In the chaos of Germany's collapse, Huzel saw himself and his colleagues as men with a mission to perpetuate the engineering science they had created. These men, Huzel believes, tend to think of themselves as a group apart from the Nazi war machine, with ambitions transcending its military and political objectives. Whether or not this image of the V-2 creators is credible to Americans, many of the émigrés firmly believe it and have adjusted readily to American communities and technical societies.

Irrespective of what inner motives these engineers harbored, they had succeeded in building the first big ballistic missile, the V-2. When developed further and mated with a nuclear warhead, such a missile was to revolutionize strategic war and to challenge air power. It was also to provide a basic design for the evolution of interplanetary launch vehicles. To that extent, the V-2 became an element in the original space-travel dream.

Once assigned to Peenemünde, the technical men displayed attitudes and motives no different from those of their American counterparts working in wartime bomber plants or in missile factories today. Rocket development at Peenemünde or production at Mittelwerke were preferable to duty at the front.

It was apparent, however, that the prospect of space travel had a strong appeal to this group. Von Braun became over-enthusiastic in discussing it at a party in 1944, and was arrested by agents of Heinrich Himmler who threatened to try him on charges of treason and sabotage. It took determined intervention by Dornberger to get von Braun out of the Gestapo's clutches.

Whatever the dreams of this group were, their talents were utilized by the Nazi war machine to serve its own dream of conquest, and they "went along," as most Germans did. Their mission was to build a rocket that would dump a ton of high explosives on civilian populations, and they carried it out. If they could have completed the A-10 "America Missile" in time, they would have fired it at New York City.

Yet the leitmotif of space exploration cannot be dismissed simply as a postwar rationale, designed to make the German émigrés more acceptable to the American technical and scientific community. Von Braun, Dornberger, Eberhard Rees, Kurt Debus, and others working in the American program now are totally committed to space faring. It is their life's work.

What is unique about these men is that they gained access to the resources of two of the world's most advanced nations to help develop an engineering science that is taking men to the moon. And it took them only 30 years to do it.

Chapter 2

The World-Circling Spaceship

Twelve years after the last V-2 was launched from German soil, another rocket stood on another beach beside another sea.

It was tended by many of the same men who had developed and launched the old ballistic vengeance weapon. Although it was a lineal descendant of the V-2, the new machine had a different mission. Instead of a metric ton of high explosive, its payload consisted of a steel cylinder 80 inches long, 6 inches in diameter, and weighing only 18.13 pounds. Among other instruments, the cylinder contained a Geiger-Müller radiation counter, which was to make one of the great scientific discoveries of the century.

The mission of Missile No. 29, a Jupiter-C, was to insert the 80-inch tube into an orbit around the earth so that its instruments could sense and report back by radio the nature of the interplanetary medium beyond the atmosphere. Here was the first American earth satellite, Explorer 1. On the night of January 31, 1958, it was poised for flight atop the missile on Pad 26-A at Cape Canaveral, Florida.

At X minus 50 minutes, the red steel service tower enfolding the missile was rolled back and the vehicle became visible for miles along the Atlantic beaches, gleaming whitely in floodlights. Standing 68.6 feet high, Missile No. 29 was encircled by halos of liquid-oxygen vapor which lent an ethereal aura to the brilliant scene. Ice had formed on the flanks of the Redstone first stage, chilled by the oxidizer tank at 297 degrees below zero Fahrenheit.

A humid breeze came in from the ocean, blowing away the oxygen smoke The upper stages of Missile 29 were hidden by a tublike aluminum cylinder which rested on the big Redstone first stage Only the satellite and a four-foot rocket attached to it were visible above the tub, looking insignificant as a burden for the 62,741-pound rocket.

The tub began spinning, a means of stabilizing the upper-stage rocket clusters in flight. The countdown was nearing X, the time of the firing command. In the damp Florida night, a moment in history was materializing out of many past events. A new pattern of human effort was taking shape. Explorer 1 marked the beginning of the American exploration in space. It opened up a new dimension for science, technology, industry, and national purpose.

Yet the countdown for this new American enterprise had begun more than a decade earlier, at the end of the war. In the fall of 1945, the Army, with the aid of the émigrés from Peenemünde, began testing the V-2 spoils of war at White Sands, New Mexico. The V-2 phase in American rocketry was to last six years. The first of 63 German rockets assembled for flight from 100 hauled to America was launched June 7, 1946, and the last in September 1952.

During this time, the V-2 served two purposes. It formed the template for the development of American liquid-fuel rockets and it also enabled scientists to make the first soundings of the upper atmosphere beyond altitudes that could be reached by balloons. An Upper Atmospheric Research Panel was formed by scientists to exploit the V-2 missiles for science. Members were drawn from the Army Signal Corps, the Applied Physics Laboratory of Johns Hopkins University, the Air Force, the Naval Research Laboratory, and universities such as Harvard, Princeton, the California Institute of Technology, and the University of Michigan.

As instrument carriers, the V-2s enabled research men to obtain data on the structure and density of the atmosphere to an altitude of 100 miles. For several years, this altitude was novel enough to satisfy the panel members. But then the need to go to even higher altitudes and the dwindling pile of surplus war rockets demanded new and better launch vehicles.

First of the American rockets to "descend" from the V-2 was the Aerobee, developed by the Johns Hopkins Applied Physics Laboratory. Early models could reach an altitude of only 80 miles, however, and a more powerful vehicle was required. Working with the Naval Research Laboratory, the Martin Company of Baltimore developed a more powerful rocket called the Viking. It could lift a 100-pound payload to an altitude of 150 miles.

Neither the Aerobee nor the Viking was a war rocket, and in a curious way the American experience with rockets in the immediate postwar period recapitulated the German experience in the 1930s. Rockets were being developed initially to explore the environment.

X MINUS 13 YEARS: "HATV"

It was the Navy, perhaps logically enough, that first revealed an interest in exploring the new ocean of space beyond the atmosphere. The Navy's Bureau of Aeronautics in 1945 had established a Committee for Evaluating the Feasibility of Space Rocketry (CEFSR). The committee promptly presented a proposal unique for its time. It called for the development of a single-stage rocket using liquid hydrogen as the propellant to boost a satellite into orbit.

Both the mission and the means of accomplishing it were quite advanced. While liquid hydrogen had been proposed as a high-energy fuel as early as 1895 by Ziolkovsky, the difficulty of handling liquid hydrogen was still unknown at that time. As engineers discovered many years later, the art of welding had to be improved in order to build tanks that would hold liquid hydrogen without leaking. At 423 degrees below zero Fahrenheit, liquid hydrogen leaks through apparently impermeable seams.

The Navy's CEFSR called its satellite launcher the High Altitude Test Vehicle, or HATV. The Bureau of Aeronautics contracted with the Guggenheim Aeronautical Laboratory of the California Institute of Technology to make a feasibility study of the single-stage hydrogen-rocket concept. As part of the Guggenheim facility there existed a unit called the Jet

Propulsion Laboratory (JPL). It had been working toward a high-altitude rocket since 1936, under the direction of the Hungarian-born aerodynamicist Theodor von Kármán. The JPL project leader William H. Pickering, a New Zealand-born physicist, reported that the one-stage hydrogen vehicle was feasible for placing a satellite in orbit. The Navy then let a contract to a small industrial organization which had been formed by former JPL engineers, the Aerojet Engineering Corporation, to build a plant to liquefy hydrogen and a stand on which to test a hydrogen engine. Preliminary engine tests verified the theoretical performance which JPL had predicted. All that was now required for development was money.

The cost of developing the engine alone was estimated at $8,000,000. Upper echelons of the Navy command balked at the expense, since there was no military justification for it. (The cost of developing the first hydrogen engine, the RL-10 for the Centaur rocket, exceeded $100,000,000 between 1961 and 1964.) Undiscouraged, the CEFSR approached the Army Air Force with a proposal to set up a joint satellite project. Navy and Army representatives met in Washington on March 7, 1946, and agreed that ". . . the general advantages to be derived from pursuing satellite development appear to be sufficient to justify a major program in spite of the fact that the obvious military or purely naval applications in themselves may not appear at this time to warrant the expenditure."[1]

In essence, this farsighted document expressed the basic optimism of the space program since its earliest beginnings. No one was sure what the penetration of the unexplored domain beyond the earth would lead to, but many scientists were convinced it would lead to something mankind needed to know.

This instance of interservice effort demonstrated that the lower echelons of the service commands were inclined to cooperate. It was the top brass that perpetuated rivalry and disrupted cooperative and economical development of rocket technology. Had the joint satellite project been pursued, the history of the United States in the last twenty years might have been quite different. Americans might have been on the moon in the first half of this decade. At least, the disastrous consequences of Sputnik, which marked the beginning of the eclipse of American prestige in Europe, would have been averted. But interservice cooperation in space was not in the cards. It was thwarted at the top. Interservice competition became the order of battle.

On March 15, 1946, General Curtis E. LeMay, Chief of the Army Air Force, advised Navy representative Dr. Harvey Hall that the Army would not support the Navy's satellite proposal. The Army Air Force then turned to the RAND organization, which it had created in 1943 as a thinking agency to solve tough problems. RAND was instructed to make a separate feasibility study of an Army Air Force satellite launcher. RAND, an acronym for Research and Development, was operated for the Air Force by the Douglas Aircraft Company. Its mission to solve technological problems and make difficult feasibility studies had attracted some of the brightest young mathematicians, scientists, and engineers in the country.

Undismayed by the advent of Army Air Force competition, the Navy CEFSR gave a contract to North American Aviation, Inc., to make preliminary studies of a rocket using the hydrogen engine. As the Navy group designed it, the HATV had nine motors burning hydrogen and oxygen. It stood 86 feet high and had a diameter of 16 feet. It weighed 101,440 pounds, with 89,000 pounds of propellant (fuel and oxidizer) and a 1000-pound satellite. The launch weight required a thrust of 233,000 pounds at liftoff, nearly five times that of V-2. HATV had to be capable of accelerating the payload to 25,400 feet per second at an altitude of 150 miles.

North American Aviation engineers confirmed JPL's finding that the HATV was feasible. However, they believed the same result could be achieved more cheaply by improving the old V-2 and adding upper stages to it.

X MINUS 12 YEARS: THE RAND SPACESHIP

Spurred by the Navy's persistence, the Army Air Force accelerated its own plans to place a satellite in orbit. In the spring of 1946, RAND issued a startling study entitled "Preliminary Design for an Experimental World-Circling Space Ship." The report dispatched to the Air Matériel Command at Weight Field in Dayton, Ohio, contained prophetic exposition of what a satellite might do. It said in part:

1. A satellite with appropriate instrumentation can be expected to be one of the most potent scientific tools of the twentieth century.

2. The achievement of a satellite by the United States would "inflame the imagination of mankind and would probably produce repercussions in the world comparable to the explosion of the atomic bomb."

3. "Since mastery of the elements is a reliable index of material progress, the nation which first makes significant achievements in space travel will be acknowledged the world leader in both military and scientific techniques." To this the report added an amazingly clairvoyant warning: "To visualize the impact on the world, one can imagine the consternation and admiration that would be felt if the United States were to discover suddenly that some other nation had already put up a successful satellite."

RAND considered the potential of a satellite as an orbital bomb but rejected the idea as impractical. Only an atomic bomb made such a weapon worth considering and the A-bombs available at that time were too heavy to be boosted into orbit by any propulsion system RAND knew of. However, the RAND report pointed out that a booster powerful enough to orbit a satellite was capable of serving as an intercontinental ballistic missile. It said: "Such vehicles can be used either as long-range missiles or for carrying human beings."

In contrast to the Navy's single-stage HATV, RAND proposed a four-stage rocket with V-2 propulsion, burning alcohol and liquid oxygen, weighing 233,669 pounds. As an alternate, the study considered a three-stage, hydrogen-powered vehicle. RAND was not impressed with the Navy's single-stage HATV concept as a feasible satellite launcher. "If a vehicle can be accelerated to a speed of 18,000 miles an hour and aimed properly," the study said, "it will revolve on a ground-circle path above the atmosphere as a new satellite. The centrifugal force will just balance the pull of gravity. Such a vehicle will make a complete circuit of earth every one and a half hours."

The proposal called for a 500-pound satellite circling the earth in a 300-mile orbit. However, RAND's explanation of why a satellite would remain in orbit would not be acceptable in today's high-school physics classes. The notion of centrifugal force has been widely rejected in current science teaching. Only one force acts upon an orbiting satellite in Newtonian physics, the newer textbooks inform us, and that is gravity.

Newton's first law says: "A body remains at rest or travels with constant velocity in a straight line unless an unbalanced external force acts upon it."

Imagine a super-baseball pitcher in a cosmic baseball game astride an extravagant pitcher's box 100 miles high. He hurls a fast, straight ball parallel to the ground at a speed of 18,000 miles an hour. In the first second after the ball leaves his hand, the earth's gravity causes it to fall 16 feet. This is the "unbalanced external force" Newton talked about. But while that super-baseball is falling 16 feet, it is also traveling 5 miles in the same second. In a rough way, this happens to match the curvature of the earth, which drops 16 feet every 5 miles. When the path of a falling object matches the curvature of the earth, the object will not hit the surface of the earth but will fall around the earth—in orbit. There it will remain forever unless something decreases its velocity. At altitudes of 100 miles or so, something does slow it down: the topmost fraction of the atmosphere. There is just enough air 100 miles up to offer some resistance to an orbiting object and gradually slow it down until its flight path intersects the ground. There is no centrifugal force in this picture, nor is there any "zero *g*'—another myth. Gravity is the string that ties a satellite to the earth. If gravity suddenly vanished, the vehicle would fly off at a tangent into infinity. A man in an orbiting satellite feels as though he is weightless because he is in free fall.

While these elementary principles of orbital mechanics have been known to astronomers and physicists since Newton's time, they were inner-sanctum mysteries to the average American in 1946. Newton's laws then were involved sentences one memorized with no realization of their application to current history and national survival.

RAND suggested that the satellite rocket be launched from a Pacific island near the equator, in an easterly direction in order to take advantage of the earth's rotation, which would add about 1400 feet a second to the satellite's velocity. One portion of the RAND report remarked that there was little difference in the design and performance between a satellite launcher and an intercontinental missile. A rocket with a 6000-mile trajectory required a velocity of 4.4 miles a second, while a satellite required 5.

Eleven years later, at the end of a Senate investigation into America's space effort, Senator Lyndon B. Johnson said of Sputnik: "It demonstrates beyond question that the Soviet Union has the propulsive force to hurl a missile from one continent to another."

At the end of 1946, the Army and the Navy presented their satellite plans to the War Department Aeronautical Board. The Board merely suggested that each service pursue its studies independently. The projects were again reviewed after the Department of Defense was created by the National Security Act of 1947 to supplant the War Department.

This time they were subjected to the scrutiny of a new jury, the Technical Evaluation Group of the Department of Defense Research and Development Board. The name had changed, but it was still the same game. The Board took a negative view of both Navy and Air Force projects. Neither "has yet established a military or scientific utility commensurate with the presented expected cost of a satellite vehicle," the Board ruled. The expected cost of the RAND proposal was $82,000,000.

During this postward period, the development of expensive new weapons systems, whose effects were speculative, was deemed to be an unjustified gamble. With the sole proprietorship of the atomic bomb, in addition to the most powerful air and ocean fleets in the world, the United States seemed invulnerable. The Department of Defense could not accept an intercontinental missile as a military requirement. It was entirely too expensive and exotic.

X MINUS 10 YEARS: BUMPER-WAC

Nevertheless, it was consistent within this frame of reference to continue the low-level development of tactical missiles which could be used by troops in the field and to pursue studies of a satellite launch vehicle, short of fabricating the hardware. In September 1948, the Research and Development Board's Committee on Guided Missiles approved an

Army project called "Hermes" which was created to continue missile development on a bargain-basement level.

Until this time, the satellite proposals had been labeled "secret." Suddenly the lid was lifted. In his first annual report in 1948, none other than Defense Secretary James E. Forrestal let the satellite yearnings of the military out of the bag. The report stated: "The earth satellite vehicle program which was being carried out independently by each military service was assigned to the Committee on Guided Missiles for coordination. . . . The committee recommended that current efforts in this field be limited to studies and component designs."

While the essential message here could be paraphrased as "We are not going to build a satellite but will continue to think about one," Russian reaction to the disclosure was as violent as though a launching was imminent. The Soviet journal *New Times* referred to "madman Forrestal's idea of an earth satellite" as an "instrument of blackmail." It ridiculed America's "Hitlerite ideas" in developing rockets, and particularly the "fantastic idea of reconnaissance satellites."

By mid-1948, the Navy had gone as far as it could with its HATV studies without developing hardware to try them out. But the Bureau of Aeronautics estimated that a 12-vehicle program would cost $150,000,000, and this sum was regarded as fantastically high for a project without a definable military purpose. Once more, the Navy went to the Air Force with a proposal for a joint program, but the Air Force would agree on only one point: the cost was fantastically high.

After that, the high-altitude test vehicle faded away, and thousands of man-hours of planning and calculating seemed to go down the drain. But there is an odd little rule that persists in research and development. It states that research is never really wasted whether it comes to term or not. The Navy's HATV had laid the groundwork for the hydrogen engine, the first new advance in rocketry since the V-2.

Outside the Pentagon, there was little support for long-range rockets. Dr. Vannevar Bush, America's chief military scientist during World War II, had stoutly rejected rockets as strategic weapons. He had insisted that guidance posed an insurmountable problem which made the whole rocket scheme a wild one. The historian for the Air Force Systems Division, Robert L. Perry, has observed that "by 1947 it was a precept of American folklore that the only possible opponent of the United States, the Soviet Union, was incapable of developing an advanced technology and was in a near comatose condition as a result of war damage." He added, "To have undertaken a serious ballistic missile program in the immediately postwar years would have required a very substantial investment in dollars and in skilled manpower." [2] The relaxed mood of the country would not support such an outlay.

Work on aerodynamic missiles was continued by the Jet Propulsion Laboratory group at the California Institute of Technology with Army Ordnance funds. One of the most successful missiles the group had developed was the Wac Corporal, which had been fired to an altitude of 235,000 feet from White Sands on October 11, 1945. It was the basis for the more sophisticated Aerobee sounding rocket.

By the end of 1946, JPL technical people were considering the possibility of using the Wac Corporal as a second stage of the V-2. Staging, or putting one rocket atop another to fire in sequence, was the next move in the game.

This kind of experimentation, with off-the-shelf rockets, appealed to the Ordnance Corps because it was feasible and relatively cheap. Accordingly, a project called "Bumper" was created, using the V-2 as a first stage to "bump" the Wac Corporal to high altitude, whereupon the smaller missile would fire.

The first successful Bumper-Wac was launched to an altitude of 244 miles from White Sands on February 24, 1949. After that, the action in this primitive operation shifted to a new scene, a vast, empty expanse of sand and scrub palmetto on the east coast of central Florida called Cape Canaveral. Ponce de Leon had given it its name, which means the cape of reeds, as he sailed by in 1547 to found St. Augustine. Aside from a lighthouse to warn mariners away and a few residents struggling to grow oranges, the Cape was principally a subtropical wilderness of snakes and alligators when it was activated October 1, 1949, as the Joint Long Range Missile Proving Ground by order of President Harry S. Truman.

Until this event, the only military installation in the vicinity was the Banana River Naval Air Station, a hard-luck post where personnel were not required to wear neckties between 1 March and 15 October. In the milieu of Naval Aviation, the region was known as the bug capital of the world.

Shortly after the first concrete pads were poured on the Cape, hard by the lighthouse, in the summer of 1950, the first rocket was erected beside a spindly scaffold for launching. It was Bumper No. 8, the shotgun marriage of V-2 and Wac Corporal.

On the morning of July 24, the countdown proceeded more or less without incident. There was a momentary flurry when an alligator slithered into one of the bunkers which served in lieu of a blockhouse to protect the launch crew against a premature explosion, but this incident did not interfere with the count. At X minus 30 minutes, the V-2 first stage was fueled. A red smoke grenade was detonated at X minus 15 minutes. Everyone on the pad ran for the sandbagged bunkers or took cover in a war-surplus Sherman tank.

The 14-ton rocket thundered into the overcast at 9:29 a.m., and 63 seconds later, 15 miles downrange, the V-2 first stage "bumped" off the Wac Corporal second stage. Two elements in the United States space program were inaugurated that day: Cape Canaveral and the concept of multistage rockets, the key to exploring space.

X MINUS SEVEN YEARS: REDSTONE

In mid-1950, the Army turned to the development of long-range rockets. Ordnance called upon the Peenemünde émigrés, most of whom by this time had become naturalized American citizens, to make a feasibility study of a mobile, 500-mile rocket, which troops could launch in the field. The mission reflected developments in atomic-bomb technology, which suggested the practicality of mating a nuclear warhead to a rocket. The von Braun team was back in the war-rocket business.

The Ordnance Corps had taken over the Redstone Arsenal at Huntsville, Alabama, as its headquarters for missile development. Sprawling over 40,000 acres, the arsenal had been used by the Chemical Corps in World War II, and virtually abandoned after the war. Early in 1950, Ordnance moved in and transferred 120 members of the von Braun group from White Sands, New Mexico, to northern Alabama. The German émigrés settled happily in Huntsville, a fairly prosperous trading center with a population of 16,000 in an Appalachian valley watered by the wide Tennessee River.

Except for the wooded mountains, the most conspicuous feature of the region is its ubiquitous red clay, which inspired the name of the arsenal. It seemed appropriate to the von Braun team to call its 500-mile missile "Redstone."

Established at a headquarters with good industrial space and an open-end potential for more, the German-born team began to create a rocket-development center where its own vehicles could be evolved "in house" rather than by outside contractors. The team contracted for the manufacture of the major components including the engine, but assembled and tested them in its own shop.

Von Braun liked the liquid-fuel engine which North American Aviation, Inc., had developed from the V-2 for an unmanned bomber called the Navaho. This project had been started by the Air Force and then abandoned after more than $700,000,000 had been invested in it. The Huntsville engineers gave North American's Rocketdyne Division a contract to modify the Navaho engine for the Redstone. The result, with later improvements, became Rocketdyne's H-1 engine, which was to be the basic power-plant design for the Saturn family of true space rockets.

As development of Redstone proceeded, it became apparent that the 500-mile range was beyond its capability if the rocket was to carry a nuclear warhead. The Army settled for 200 miles and Redstone became officially a medium-range ballistic missile. Development was speeded up during the Korean war, but that conflict ended before Redstone was operational. In fact, the missile failed its first flight test at Cape Canaveral on August 20, 1953. The inertial-guidance system went awry and the rocket rose only 8000 yards when the range safety officer detonated the "destruct package" of dynamite on the missile to blow it up before it hit something. However, enough telemetry had been received from the bird to enable the engineers to identify the trouble. The second Redstone flew.

Redstone development extended over the next five years. The Chrysler Corporation received the contract to build the missiles in Detroit after the first models were developed in the ordnance shops at Huntsville.

Redstone basically was a single-stager, like the old V-2, but it was larger and more powerful. The tank was 70 inches in diameter and, in later models, 56 feet long. The fuel consisted of ethyl alcohol and water, which burned when mixed with oxygen. The engine put out 75,000 pounds of thrust at sea level and burned for 2 minutes and 1 second. At launch, the rocket weighed 61,000 pounds. It could carry chemical or nuclear warheads. After it was checked out in the field in 1958, it was shipped for deployment in Europe by NATO forces.

X MINUS THREE YEARS: INTERSERVICE SWEEPSTAKES

Redstone was deployed extensively as a missile, but, like the Model-A Ford, its chief significance was ancestral. It began several generations of descendants. The first was a vehicle called Jupiter C, which was a Redstone with an elongated tank to increase propellant-burning time and total thrust. On top of it, like Bumper-Wac, were fastened two upper stages consisting of clusters of solid-propellant "Loki" missiles. The Loki was a small, solid-fuel rocket.

This lash-up was evolved at Huntsville in respone to an Ordnance directive to develop vehicles to test nose cones. The Army had a long-range missile in mind, one that would travel 1500 to 2500 miles. In order to reach those distances, the warhead had to be lobbed on a high trajectory that would take it out of the atmosphere and plunge it back in again at nearly four miles a second. During reentry into the atmosphere at such a speed, atmospheric friction would heat the warhead to 3500 degreed Fahrenheit. Unless some means was found to protect it from the high heat, the missile would disintegrate aloft.

Jupiter C was an off-the-shelf product, and to a number of experts it looked as though it had been designed on a tablecloth during lunch hour. It seemed the least likely vehicle in the American inventory to get off the pad. The clusters of Loki rockets stood in a metal cylinder, the size of a galvanized washtub, which sat atop the Redstone. In order to stabilize the Lokis in flight so they would fly true, the washtub assembly had to be rotated rapidly by an electric motor. Jupiter C (the "C" meant "composite" vehicle) looked impossible, but only to outsiders. Insiders at Huntsville knew it would work.

In fact, they were sure it could not only propel a nose cone into space, but also launch a satellite. That idea occurred to the Navy, too.

A new proposal for an interservice satellite project was broached to von Braun by Commander George Hoover of the Office of Naval Research (ONR). The Navy would provide the satellite and the Army the launch vehicle. The plan was approved by the ONR top brass and by Toftoy, now a major general, who commanded the missile group at Huntsville.

"There is only one method for getting ahead in this game," Toftoy observed to Navy Captain W. C. Fortune, chief of ONR. "Action." [8]

But this demonstration of interservice cooperation was also destined to collapse. Another Navy development group, the Naval Research Laboratory, had an idea it could develop its Viking rocket, which had achieved a record altitude of 158.4 miles on May 24, 1954, into a satellite launcher. The launch vehicle had been used only for ballistic probes.

New interest in satellites as scientific instruments was generated early in 1954 by the prospect of an International Geophysical Year (IGY), to be held in 1957–58. The geophysicist S. Fred Singer, then at Johns Hopkins University, had set the stage for consideration of a satellite in the IGY at the Fourth International Congress on Astronautics in Zurich in 1953. He had proposed a Minimum Orbital Unmanned Satellite Experiment (MOUSE). The proposal impressed a number of scientists who were active in promoting the IGY, namely Harry Wexler, chief scientist of the United States Weather Bureau; Joseph Kaplan, professor of physics at the University of California at Los Angeles; Athelstan Spilhaus of the University of Minnesota; James A. Van Allen of Johns Hopkins; and Homer E. Newell, head of the atmosphere and astrophysics division of the Naval Research Laboratory.

During 1954, the International Scientific Radio Union and the International Union of Geodesy and Geophysics adopted resolutions calling for satellites during the IGY. At its third general assembly in Rome in October 1954, the Special Committee for the IGY (CSAGI) recommended that satellites be considered "in view of the great importance of observations of extraterrestrial radiations and geophysical phenomena and in view of the advanced state of rocket techniques."[4] This recommendation was given to President Eisenhower by Alan T. Waterman, director of the National Science Foundation, but the President took no action on it until the following year.

In the autumn of 1954, the peak of American rocket development was represented by the Jupiter C, with its washtub-emplaced upper stages. Inspired by the CSAGI recommendation, von Braun proposed "A Minimum Satellite Vehicle Based on Components Available from Missile Development of the Army Ordnance Corps." The proposal offered the Jupiter C as the launch vehicle for an American IGY satellite. It was formally presented to the Department of Defense by the Army as "Project Orbiter."

Meanwhile, American scientists and missile experts had become convinced that the Russians were planning a satellite. The first open indication of Soviet interest in satellites had appeared in Russian magazine articles in 1951. Early in 1954, Alexander N. Nesmeyanov, the president of the Soviet Academy of Sciences, issued a statement that Soviet science and technology had reached a stage at which launching earth satellites and sending a spaceship to the moon were feasible.[5] On April 15, 1955, the Soviets announced the formation of the Permanent Interdepartmental Commission for Interplanetary Travel as an arm of the Astronomical Council of the Soviet Academy of Sciences. Moscow Radio reported on April 26, 1955, that Russian scientists planned not only to put up a satellite but also to explore the moon by means of a tank remotely controlled by radio.

Early in July 1955, the United States National Committee for the IGY received confidential assurances from the White House that the President had agreed to support a satellite project.

On July 26, 1955, Kaplan, then chairman of the United States National Committee for the IGY, advised CSAGI that the American participation in the IGY "now includes definite plans for launching small satellites."

Plans for the orbiting of a satellite were formally announced on July 29, 1955, by the National Science Foundation, the National Academy of Sciences, and the Department of Defense. The following day, the Soviet Union announced that it, too, would raise a satellite or two for the IGY. The race for space was on.

The arrangement approved by the President was that the Department of Defense would launch the satellite, the National Academy of Sciences would determine the scientific experiments to be lofted in the capsule, and the National Science Foundation would finance the venture. This last assignment proved to be wholly unrealistic, since NSF's budget could barely carry the cost of the paper work. The Department of Defense would have to pay the tab.

Russian intentions were spelled out more clearly by Leonid I. Sedov, chairman of the Soviet Academy of Sciences' Interdepartmental Commission in Interplanetary Communications. On August 2, 1955, he told a press conference at the International Congress of Astronautics at Copenhagen that "In my opinion, it will be possible to launch an artificial satellite of the earth within the next two years and there is the technical possibility of creating artificial satellites of various sizes and weights." Sedov also predicted that two- or three-stage launch vehicles would be used.[6]

With United States Army and Navy satellite proposals waiting in the wings, the stage was set for a grand competition between the services for the prestige of carrying the American flag into space. Only a nation rich enough to afford chronic rivalry among its military services could have afforded such a contest. It was not long before the Air Force announced that it, too, had a launch vehicle to enter into the satellite sweepstake: the Atlas missile, then in development.

In January 1955, the Department of Defense asked eight civilian consultants to serve on an *ad hoc* committee to examine the satellite projects. Two members were picked by each of the armed services (Army, Navy, and Air Force) and two by Donald A. Quarles, then Department of Defense Assistant Secretary for Research and Development. Homer J. Stewart of the California Institute of Technology, a physicist active in the Jet Propulsion Laboratory, became chairman. The consultant group was called the Ad Hoc Committee on Special Capabilities. Its decision on the IGY satellite was to have a profound influence on the directions taken by American science, technology, and education over the next decade. Other members of the advisory group at the outset were Kaplan; Robert W. Buchheim and G. H. Clement of RAND; Charles C.

Lauritsen, California Institute of Technology; Robert R. McMath, University of Michigan; Richard W. Porter of the General Electric Company, the only contractor executive on the committee; and J. Barkley Rosser of Cornell University. Later, Clifford C. Furnas, former director of the Cornell Aeronautical Laboratory and Chancellor of the University of Buffalo, was added to the Committee. When the committee's work was done, Furnas was appointed Assistant Secretary for Research and Development to succeed Quarles, who then became Secretary of the Air Force.

The committee was instructed to recommend a satellite-equipment program by August 1, 1955. The Navy presented its proposal to develop a three-stage rocket, using an enlarged Viking as the first stage, the Aerobee as the second stage, and a solid rocket as the third. The launch vehicle would be called "Vanguard." It would place into orbit a payload weighing 20 pounds. Von Braun presented Project Orbiter for the Army. The Jupiter C would place at least 10 pounds in orbit. The composite vehicle would use the Redstone as the first stage and spin-stabilized clusters of solid rockets in the second and third stages. The fourth stage would be a single solid-fuel rocket to which the payload would be joined. The Air Force proposal to prepare the Atlas as an IGY satellite launcher was dismissed lest it interfere with the intercontinental missile's development as a weapon.

As Dr. Stewart recalls it, Quarles appeared to favor von Braun's Project Orbiter approach. The Ad Hoc Committee was advisory to Quarles as Assistant Secretary of Defense, but there was no doubt that it would be influenced by preferences he indicated in briefings for the members. However, before the committee took a vote, Quarles was named Secretary of the Air Force and became immersed in his new duties. The Ad Hoc Committee then voted for Vanguard, seven to two on August 4, 1955,[7] mainly, in Stewart's opinion, because that project was believed to be free of military taint. Even so, Vanguard carried a military security classification, and Stewart recalls that the same security checklist applied to it as to Orbiter or any military rocket.

Stewart and Furnas held out for Orbiter, but were unable to convince anyone else that it was the logical choice to launch a satellite into orbit promptly. Stewart wrote a minority report asserting that Redstone was big and powerful enough to compensate for less efficient upper stages.

"The Jupiter C was a messy design," he said, "but it was quick enough and efficient enough to do the job."

The committee's action was then considered by the Defense Department's Research and Development Policy Council, which allowed the Army to amend its proposal by increasing the payload from 10 to 18 pounds. This could be done by substituting a scaled-down version of JPL's Sergeant, solid-fuel rocket for the less powerful Lokis which the Army initially planned to use in the upper stages. Air Force, Navy, and Marine Corps members of the Policy Council then endorsed the Ad Hoc Committee's recommendation, with Army members dissenting. Quarles approved the Navy proposal and issued a directive to carry it out on September 9, 1955. However, in Defense Department inner circles it was understood that the final decision on Vanguard was made by President Eisenhower.

According to the Department of Defense, the decision to use Vanguard was based on "technical recommendations of the Advisory Group," [8] and these were founded on the belief that a three-stage launch vehicle appeared "less inherently complex" than a four-stage vehicle. Further, the spin-stabilized upper stages in the Army's proposal had not been adequately flight-tested. In addition, the Redstone proposal involved reliance on the Redstone weapons program and "possibly some competition for diversion of Redstone engines intended for weapons." The fact that the three-stage configuration of Vanguard had never been flight-tested either, though not mentioned in the Defense Department statement, was aired during the hearings in 1958 on the National Aeronautics and Space Act by the Select Committee of Astronautics and Space Exploration.

There were other beliefs which worked against the Jupiter C. One was that Vanguard was a more sophisticated and potentially useful vehicle for science than the lashed-up Jupiter C appeared to be. Another was that American industry—such firms as Martin Company, which built the Viking rocket, and General Electric Company, which developed the engine—represented a more substantial industrial know-how than the German émigrés demonstrated in Huntsville, with their cluster complex. Navy technicians referred to the Army's

Jupiter C as "cluster's last stand," a term later applied to the Saturn family of rockets.

During this period, aerospace industry leaders were concerned that if the Army put up the first satellite the feat would expand the practice of "in-house" technological development by the government. The aerospace industry tended to regard this traditional practice by the Army as an usurpation of the development role which industry had acquired in aviation. The Army followed the "arsenal system" in which it developed its own weapons before putting production into the hands of a contractor. At the Redstone Arsenal, the Guided Missile Development Division of the Army Ordnance Corps maintained its own shops, guidance and control laboratories, and static test stands for rocket development. Yet the Army division worked closely with contractors, buying engines they had developed for other programs, and turning production over to them once development had been completed. For example, it was mentioned earlier that the Chrysler Corporation manufactured the Redstone rockets after the development phase had been accomplished at Huntsville.

Army commanders favored the arsenal system because it kept the contractors honest. When industry did the development of a weapon system, it managed to control production, too. The contractor was supposed to produce complete manufacturing drawings, but it was the experience of the Army Ordnance officers that no one could produce the weapon from those drawings but the development contractor. When the government produced the blueprints, however, any qualified manufacturer could fabricate the weapon from them. In addition to keeping the contractors honest, the arsenal system kept them competitive.

So far as America's first satellite was concerned, Project Vanguard brought major contractors into the development picture, while Project Orbiter kept them out.

A month after the Defense Department had announced that Vanguard would be the official satellite launcher, it made public the letting of an initial contract for $2,035,033 to the Glenn L. Martin Company of Baltimore to commence first-stage development. In addition, Defense said it had authorized the negotiation of a contract with General Electric Company for a first-stage rocket engine.

The fact that Vanguard, even though a Navy project, had

been conceived as a nonmilitary vehicle carried a good deal of favor with American scientists, who could argue that it was a more appropriate symbol of the spirit of the IGY—peaceful cooperation—than the Army's Jupiter C.

Behind that lofty sentiment, however, lurked the specter of military advantage and prestige which speedily came to dominate the space adventure in Russia as well as America. Yet there was no feeling of urgency, or at least there appeared to be none, in the United States government about the timing of the satellite program. Sedov's forecast during the summer of 1955 that the Russians would have a satellite up in two years made no impression on the Defense Department or the White House. Sherman Adams, the President's first assistant, compared United States–Soviet competition in satellites to an international basketball game.

Army Ordnance missile men and the team at JPL took Sedov's words seriously. There were suggestions in technical publications in Eastern Europe and other strong rumors that the Russians were developing very large and powerful boosters.

X MINUS 18 MONTHS: ARMY'S JUPITER C

So far as Project Orbiter was concerned, however, the von Braun team was advised by the Pentagon to retire to Huntsville and forget about satellites. The Navy would blaze the trail into space. Immediately after the Ad Hoc Committee decision, Sewart and Pickering flew down to Huntsville and urged von Braun to find means of keeping Project Orbiter alive as a backup to Vanguard.

The Army needed no urging. Von Braun and his associates knew they could go into orbit with Jupiter C, and they continued to work on it. Night after night, experts such as Ernst Geissler, the team's aerodynamics expert, put in hours of their own time solving the new problems posed by multistage rockets. In November 1955, the Defense Department decided to have the Army develop an intermediate-range ballistic missile for land or ship launching, complementary to the Thor IRBM which the Air Force was developing. The 1500- to 2500-mile rocket was designated as the Jupiter IRBM.

Jupiter C is frequently confused with the Jupiter IRBM,

and this was intentional. Early in the program, Redstones were used to test IRBM components in flight. The von Braun team availed itself of the Jupiter IRBM firing priority at Cape Canaveral by calling the modified Redstones "Jupiter A" vehicles. When the upper-stage clusters of Sergeant solid-fuel rockets were grafted onto the nose of the "Jupiter A," it became "Jupiter C"—the composite-rocket vehicle.

The Guided Missile Development Division of the Army Ordnance Corps was reorganized early in 1956 as the Army Ballistic Missile Agency, under the command of a plain-spoken artilleryman, Major General John B. Medaris. Von Braun was appointed technical director of ABMA.

The first Jupiter C, known as missile No. 27, was ready for flight test in September 1956. As it was being hauled in an airplane from Huntsville to the Cape, however, its outer structure buckled because of changes in air pressure during the trip. The rocket's hull was constructed of thin aluminum alloy no thicker than a dime and would not bear its own weight unless it was pressurized with gas, like a balloon. Somehow, the ABMA team had failed to pressurize the rocket for the flight to the missile test center. The buckling of the skin threatened to prevent the launch until William Mrazek, the Huntsville team's director of structures and mechanics, pumped compressed air into the thin-skinned rocket and inflated it back into its proper shape.

Standing 70 feet high on Pad 26-A overlooking the gray Atlantic Ocean, Jupiter C was the most impressive piece of rocket machinery in the Western world at the time. Newspaper reporters, who were barred from the Cape in those days, peered at it through binoculars and telescopes from the sands of Cocoa Beach to the south. Missile No. 27 was to be tested in its satellite configuration, that is, with all four stages and a dummy satellite. The fourth stage, however, was not to be live, so that a weight simulating a satellite would not go into orbit. While the Jupiter C was being checked out for the test on its launch pad, concern developed in the Pentagon that von Braun might try to pull a fast one and orbit a satellite without authorization. General Medaris was ordered to make certain that the fourth stage was loaded with sand. He also received a teletype from Lieutenant General James M. Gavin, deputy chief of the Office of Research and Development, stating, "I'm holding you personally responsible to see that there are no accidents."

Missile No. 27 was launched September 20, 1956. The three stages fired with marvelous precision, hurling the payload to an altitude of 700 miles and a distance of 3300 miles down the Atlantic Missile Range. This first test of the cluster system established a long-distance record in American rocketry that stood until the first Atlas ICBM flew 4000 miles three years later. Von Braun was delighted. "The old cucumber did it," he whooped. He recalls: "On the morning of September 21, 1956, we knew that with a little bit of luck we could put a satellite into space. Unfortunately, no one asked us to do it."

Though Missile No. 27 was not used as a satellite launcher, one "experiment" was placed aboard. It was a letter addressed to Medaris, from Kurt H. Debus, director of the Missile Firing Laboratory at Cape Canaveral, reading:

> If this letter reaches your desk, it has been transported by a rocket over a distance of approximately 1500 nautical miles to an intermediate delivery point in less than 20 minutes, floated in the ocean for the next several hours, was picked up by a U.S. Navy vessel, and then sent by more conventional means to ABMA, Huntsville, Ala. This letter is, therefore, in commemoration of the successful accomplishment of a significant step in the development of your Jupiter IRBM project as well as the first "rocket mail" delivered over a practical distance.

The second Jupiter C was launched May 15, 1957, carrying a nose cone on a reentry test. The terminal guidance failed, however, and the nose cone fell in the ocean so far from the target area that ships standing by could not find it. However, telemetry signals from the cone indicated that its phenolic resin coating had enabled it to withstand the heat of reentering the atmosphere at 12,000 miles an hour. The coating worked by ablating, or eroding away, carrying the heat with it. This principle applied to heat shields was to make possible a manned space-flight program by assuring a safe reentry.

A third Jupiter C was launched August 7, 1957, to test a one-third scale model of a Jupiter IRBM nose cone. It reached an altitude of 600 miles and was recovered by the Navy 1200 miles downrange. It had successfully survived reentry heating and was the first man-made object recovered from space. This was the nose cone shown by President Eisenhower in his memorable television address of November 8, 1957, on the state of American defenses after the nation had been shaken by the Sputniks.

During the early testing phase, the Jupiter C program had been kept under wraps. One way or another, however, "everybody found out about it," General Medaris told a Senate investigating committee, referring to the spectacular success of Missile 27.[9]

"We had on hand a backup missle still in the original satellite configuration," he told the committee. "At various times during this period [1956–57] we suggested informally and verbally that if they really wanted a satellite we could use the backup missile. In various languages, our fingers were slapped and we were told to mind our own business, that Vanguard was going to take care of the satellite program."

Medaris related that as early as April 1956, the ABMA had made a presentation to the Defense Department on the use of Jupiter C as a backup to Vanguard. The suggestion was turned down. There was no indication then that Vanguard was in any difficulties. In fact, Medaris said, ABMA was pointedly advised to mind its own business. William M. Holaday, director of guided missiles in the Defense Department, issued a memo dated May 15, 1956, to General Gavin, stating that "Without any indications of serious difficulties in the Vanguard program, no plans nor presentations should be initiated for using any part of the Jupiter or Redstone programs for scientific satellites."

"We again presented the proposal in November 1956, after we had fired Missile No. 27," Medaris continued. "We had the hardware now and some could be spared and we figured we would still take another swing at it and the proposal was taken under consideration. I know it was sent up by the Secretary of the Army officially to Defense and was under consideration during May and June. On June 21, 1957, we were again reminded of the directive of the Department of the Army of May 15, 1956, to stay out of the satellite business, which we did. It was at the direction of the Department of Defense."

Senator Estes Kevauver asked Medaris, "Do I understand, then, that in September 1956, you had the hardware, the capability, and you proved you had it, of firing a satellite?"

"That is correct, sir," replied Medaris.

In April 1957, the ABMA proposed to the Chief of Research and Development, Department of the Army, that Jupiter C launch vehicles be used to place in orbit satellites as a backup for Vanguard. Each satellite would weigh about 17 pounds. The plan called for lofting the first satellite in September 1957 and the second one by the end of the year. It was estimated the program would cost $1,000,000.[10] The response was emphatic General A. P. O'Meara flew to Huntsville on June 21, 1957, with a message from the Department of Defense. It said that ABMA's mission was not concerned with earth satellites.

On June 1, 1957, Soviet Academy President Nesmeyanov was quoted in *Pravada:* "As a result of many years of work by Soviet scientists, rockets and all necessary equipment and apparatus have been created by means of which the problem of an artificial earth satellite for scientific research purposes can be solved. Soon, literally within the next month, our planet earth will acquire another satellite."

X MINUS 12 MONTHS: VANGUARD

The saga of Project Vanguard, the authorized satellite booster for America in the IGY, was a hard-luck story from the beginning. The government underestimated the difficulties of developing a new rocket, the cost of the development, and the amount of damage that losing the race to orbit could inflict on the American position in the world. For example, the cost of Vanguard was estimated at $11,000,000. The final bill came to ten times that much, $110,000,000.

As soon as the Defense Department gave the green light, Project Vanguard was started under the direction of John P. Hagen, a Naval Research Laboratory physicist. He and other project leaders had been counting on the Martin Company to put its Viking engineering team to work on Vanguard. However, this talented group of men had been broken up when several of the team members had been transferred in the company to the Air Force's Titan ICBM program, then in its initial stage. According to Hagen, Vanguard development was relegated to a stuffy loft in the Martin plant, where workrooms were not air-conditioned against the heat of Baltimore's summer and work space was overcrowded. "We were dealt a hard blow right at the start," said Hagen.[11]

That was only the beginning. Originally, Vanguard design had specified a missile-style nose cone in which the instruments were to be housed. Instead of a cone, the United States National Committee for the IGY wanted a 30-inch sphere, and meeting that requirement involved the redesign of the third stage of the rocket. Hagen and his team found they could not alter the design sufficiently to accommodate a 30-inch sphere and the IGY committee compromised on a 20-inch sphere. Even this size sphere, however, caused difficult problems.

"To meet the requirement of a sphere," Hagan said, "we needed a larger second stage than the Aerobee as well as a larger third stage. We developed a new second stage with its own guidance and control."[12]

No one can assess how much delay the requirement of the IGY National Committee for a sphere, instead of a nose cone, caused in Vanguard development, Hagen said. "In retrospect," he added, "it was serious."

As work progressed, it was found that the guidance system would weigh more than the blueprint estimate. One of the most familiar problems of rocket development is the tendency of components to gain weight from the time they are designed until they are manufactured, as development proceeds. The von Braun team had learned this from twenty years of experience. The Vanguard team was just learning it and was struggling futilely against the necessity of reducing the weight of the payload. By late autumn of 1955, the entire Vanguard vehicle had to be redesigned to correct for underestimates and bad guesses. The new design was not completed until March 1956. At that time, six test launchings were scheduled before the satellite was to be flown. The next question was where to test the rocket. The Navy had been using the White Sands Missile Range for tests of both the Viking and Aerobee rockets before they were modified to become components of Vanguard, but the range was too small for Vanguard. The second stage would drop some 1500 miles from the launch site.

Accordingly, the project was forced to turn to Cape Canaveral. There was an advantage here. The eastward rotation of the earth would add a velocity of about 1400 feet a second to the satellite on an easterly launch.

The Navy, however, had no launch facilities at the Cape, which at that time was mostly covered by palmetto scrub and infested with rattlesnakes. "Civilization" consisted of the Redstone pads where the ABMA was testing Jupiters. The Army turned down a request by Vanguard officials to borrow the Jupiter pads on the grounds that this would interfere with the missile test program. The project had to build its own launch complex and blockhouse. No funds had been provided for this, and the money had to be obtained from an emergency fund in the Department of Defense. It took eighteen months to construct a blockhouse control center.

From the point of view of the Vanguard directorate, Cape Canaveral was not equipped to handle a satellite launching. It lacked downrange stations from which to radio firing and steering commands to the upper stage of the rocket, and to track it. Vanguard required radio control during second- and third-stage burning, when it would be hundreds of miles down the Atlantic Missile Range. The Navy's minitrack network, established to track guided missiles, was inadequate. Additional radio stations had to be built on the islands of Grand Turk, Mayaguana, and Antigua. Large radar stations were set up on the Cape and on Grand Bahama Island to track the vehicle during early phases of flight.

Such an array of installations had not been deemed essential by the émigrés from Peenemünde, who had been accustomed early in the game to operating with meager facilities. In fact, when some of von Braun's engineers first saw concrete launch pads at the Cape, they shook their heads with wonder, since in Germany they had launched rockets on portable steel platforms set on ground which had been hardened by freezing it with liquid oxygen.

Vanguard also found it necessary to build its own hangar on the Cape. Hangar S was erected with project funds. It was later to become better known as the ready room for astronauts in Project Mercury. Since the Cape had no service tower from which to work on the rocket, once it was erected on the pad, Vanguard engineers disassembled the Viking gantry at White Sands and shipped it to Florida.

With these installations, the Vanguard Project converted the Cape from a rather primitive missile-test site to a well-equipped base for launching space satellites. All of this was accomplished with a total force of 180 persons, including clerks and shop mechanics.

In order to track the satellite when it achieved orbit, the project set up a line of seven north-to-south tracking stations, equipped to receive signals from the satellite on a frequency of 108 megacycles. The line began at Blossom Point, Maryland, and ran south through Central and South America to Santiago, Chile.

Vanguard also built radio stations in Woomera, Australia, and Pretoria, South Africa, modeled after a Naval Research Station at San Diego. These facilities became the foundation for global tracking networks established later. As a backup to the radio-tracking network, a system of optical-tracking stations was set up by the Smithsonian Astrophysical Observatory at Cambridge, Massachusetts. The stations had telescopic cameras, capable of photographing small satellites as they passed over. They were built around the world within 30 degrees of the equator. In the spring of 1957, the down-range tracking system was tested by launching a Viking from the Cape. On another flight test, the ability of the second or Aerobee stage to separate from the first-stage Viking, was demonstrated. During that period of the program, all seemed to be going well.

Then, on August 26, 1957, Nikita Khrushchev announced on Moscow Radio that "a super-long-distance intercontinental multi-stage ballistic missile was launched a few days ago. Tests of the rocket were successful. They fully confirmed the correctness of the calculations and selected design. The missle flew at very high, unprecedented altitude, covering a huge distance in a brief time, and landed in the target area. The results show it is possible to direct missiles to any part of the world." The statement generated no official concern.

However, when General Gavin met with the Army Scientific Advisory Panel on September 12, 1957, he warned that the Soviet Union would launch a satellite in 30 days.

X MINUS 119 DAYS: SPUTNIK

Early in the morning of October 5, 1957, the Russians launched one of the intercontinental missiles Nikita Khrushchev had described. It carried into orbit an aluminum alloy sphere 22.8 inches in diameter weighing 184 pounds. The sphere was equipped with two radio transmitters and four spring-loaded whip antennas, which made it look like a floating mine. The Russians called this strange-looking object "Sputnik," meaning something that travels along with a traveler, Earth being the traveler. In other words, a satellite.

In Washington, it was late evening local time, October 4 that men in science and government became aware that the Soviets had put up the first earth satellite. Scientists from thirteen nations attending a conference on rockets and satellites, called by the Special Committee for the International Geophysical Year, were apprised of the event at a cocktail party at the Russian Embassy. They had spent most of the day at the National Academy of Sciences discussing the satellite plans of the United States and other countries.

The announcement was made not by the Russians but by an American scientist, Dr. Lloyd V. Berkner, a member of the President's Special Committee for the IGY. Berkner had not been told anything by the Russians. He had been informed of the satellite by the Washington Bureau of the Associated Press and asked for a comment.

News wires hummed with the story around the world.

In Chicago, Jerome Tannenbaum, an amateur radio enthusiast, picked up Sputnik's radio signals on a short-wave frequency at 12:03 a.m. He described the satellite's signals as a series of dashes one-third of a second long, spaced by an equal period of silence. They became stronger after he first picked them up and then they faded out at 12:16 a.m.

At Urbana, Illinois, a party was in progress that Friday night at the home of Professor George McVittie, head of the astronomy department of the University of Illinois, when the flash came over the radio. McVittie and Professor George W. Swenson of the electrical engineering department jumped into Swenson's car and sped to the Electrical Engineering building, where they hastily rigged up an antenna, just in time to hear Sputnik's midnight pass.

In the South Pacific, Professor James A. Van Allen, an upper-atmosphere physicist of the State University of Iowa, heard the news of Sputnik aboard the U.S.S. *Glacier*, an icebreaker bound for Antarctica.

"We got the radioman out of bed and we got on it right

away," he recalled. "We were receiving it on the *Glacier* two hours after it was fired."

One of the earliest visual sightings of Sputnik was made by an amateur observer, Larry Ochs, at an IGY moonwatch station at Columbus, Ohio. He spotted the satellite at 10:58 p.m. It took 24 seconds to cross his telescope's field of view. He picked up the satellite again at 12:06 a.m. Saturday on its next pass over the United States.

North of Boulder, Colorado, research men of the University of Colorado and the High Altitude Observatory of the National Bureau of Standards raised antennas quickly enough to pick up Sputnik in its first pass over the United States.

Sputnik was broadcasting on two frequencies, which made it easy for amateur radio operators to tune in the satellite's beeps. The Naval Research Laboratory recorded the signals of Sputnik's first pass over the United States. RCA Communications Division heard them, too. In London, the British Broadcasting Company picked them up.

Coincidentally, the new Secretary of Defense, Neil H. McElroy, was dining in the officers' mess at the Redstone Arsenal that fateful evening, along with Army Secretary Wilbur M. Brucker, General Lyman Lemnitzer, chairman of the Joint Chiefs of Staff, Gavin, and other dignitaries.

According to Medaris, the group had completed a tour of the Arsenal, had been entertained at a cocktail party, and was finishing dinner when Medaris's public relations officer, Gordon Harris, dashed into the mess hall with news of Sputnik.[13]

Medaris relates: "Suddenly, von Braun began to talk as if he had been vaccinated with a phonograph needle: 'We knew they were going to do it. Vanguard will never make it. We have the hardware on the shelf. We can put up a satellite in sixty days.' "

As von Braun tells the story, General Medaris gulped at this point and said, "No, Wernher. Make it ninety days."

"That turned out to be more accurate," Medaris recalls.

By Saturday morning, a great many people in this country and the world had become aware that the space age had dawned, ushered in by a supposedly second-rate technological power. In the United States, the reactions of political figures tended to diverge along party lines. By Sunday morning, October 6, 1957, spokesmen whose views normally are sought on major events began to collect their wits and issue statements.

From Calhoun, Georgia, Senator Richard Russell declared that "Sputnik confronts America with a new and terrifying military danger and a disastrous blow to our prestige." [14]

"A spectacular event," said Vice-President Richard M. Nixon. "No more dramatic incident could have occurred to remind both the Communists and the free world of the increasingly terrifying aspects of modern warfare." [15]

"The rueful reaction in Washington is that Russia's earth satellite launching has beaten the daylights out of the United States in an epic contest of the 20th century," said a compilation of wire-service reports in the Chicago *Sun-Times,* the major liberal-Republican, pro-Eisenhower newspaper in mid-America.

Congressional and military leaders observed in unison that a nation which could place a satellite in orbit had the rocket capability of delivering a nuclear bomb over intercontinental distances.

Perhaps the only analytical voice in the crowd came from Vanguard's Hagen, who was still struggling to get his rocket off the ground with a satellite ever-diminishing in weight as new technical problems appeared. He noted that the Russian satellite was not a scientific package at all and did not compare with the one the United States was planning to launch. All Sputnik carried was a pair of radios to announce its presence. It contained no instrumentation to measure radiation or any other physical attribute of the region beyond the atmosphere. In this respect, it was a technological demonstration, not a scientific contribution to the IGY.

Not everyone, however, viewed the event with alarm and chagrin. Clarence Randall of Chicago, a special assistant to President Eisenhower for foreign economic matters, was quoted as calling Sputnik "a silly bauble in the sky." [16] He was gratified, he said, that the United States was not the first to put one up. Of greater significance than Sputnik, he said, was the United States supermarket exhibit in Zagreb, Yugoslavia, sponsored by the National Association of Food Chains and the United States Department of Commerce. The Russians would have to go some to beat that.

Percival Brundage, director of the budget, remarked at a

dinner party in Washington that Sputnik would be forgotten in six months.

"Yes, dear," replied his dinner companion, Perle Mesta, "and in six months we may all be dead."[17]

Army Secretary Brucker regarded Sputnik as a benchmark in Russian missile technology. He spoke of the "urgency" to get moving "in this critical hour." He said that "complacency is a luxury we cannot afford."[18]

Defense Secretary McElroy had returned to Washington from Huntsville impressed with the intensity of the feelings of urgency and frustration which pervaded the Army's missile establishment. Yet it was going to take time to overcome the inertia of the complacent mood which still gripped official Washington.

At the Eighth International Astronautical Congress in Barcelona, scientists and engineers from 18 nations were electrified by Sputnik. A French scientist contrasted the 23 papers presented by Americans with the 5 presented by Russians and commented: "The Americans talk about it and the Russians do it."[19]

The Monday after Sputnik, aircraft and missile shares boomed, but bad reports from the steel industry and general business uneasiness sent the stock market as a whole down to a 1957 low.

A. A. Blagonravov, the Russian rocket expert, announced that a second Russian satellite was being prepared for launching when enough information was collected from the first. The main intelligence the Soviets needed from Sputnik I was the rapidity with which its orbit decayed. This would indicate the extent of air drag on the vehicle at perigee, the lowest point in the satellite's orbit.

Sputnik's initial orbit was 142 miles high at perigee to 588 miles at apogee, the highest point in its looping path around the earth. No one could foretell how long the vehicle would remain in orbit until there was some information about the density of the residual atmosphere at very high altitudes. To the extent that changes in Sputnik's orbit would indicate this, it was serving a scientific purpose, but the feat essentially was an effective proof of the Soviets' astonishing technological proficiency in rockets.

Conflicting reports came from overseas on European reaction. Roscoe Drummond, the newspaper columnist, reported from Paris: "Only a few days have passed since the launching of the Soviet earth satellite. But these few days have been enough to show that this is a disastrous event." It represented a climax, he said, to a long, complex, and ominous process of deterioration of the Western Alliance, which began when President Eisenhower first took office.

On November 3, Sputnik II, carrying the dog Laika, was launched into orbit. It weighed 1120 pounds. Walter Dornberger, the ex-German general who had commanded V-2 development at Peenemünde, observed, "There can be absolutely no doubt now that the USSR has the means for sending an atomic or hydrogen warhead anywhere in the world." At Huntsville, von Braun wholeheartedly agreed with his former chief.

During November, a wave of consternation washed over America. It was becoming clearer that Eisenhower and McElroy had underestimated not only Russian rocket prowess but also the deleterious effect on the American image in the world of the Russian breakthrough into space. Reassuring statements by the President and the Secretary of Defense merely enhanced a widening feeling of alarm and insecurity.

"Now, so far as the satellite itself is concerned, that does not raise my apprehension one iota," Eisenhower had observed at this post-Sputnik press conference. But it raised the apprehensions of American rocket experts who knew that the United States had no launch vehicle comparable to the Russian ones that had orbited the Sputniks.

In the small, thoughtful journals of the nation, open anxiety was displayed. "Those few who scoffed at the importance of the earth satellite, including the President himself, failed to convince anyone," said *Commonweal*. "Indeed, the tone and the nature of their protests suggested that they were not convinced themselves."

The TRB column in the *New Republic* observed that Eisenhower was "terribly mauled" by the press October 9, 1957, when he said that Sputnik hadn't increased apprehension "one iota." "We think we have never seen a time when the Washington press so universally belittled a president privately, and some of it is getting into the papers," the column said.

In Spain, Generalissimo Franco, a sage infrequently heard from, drew a singular moral from the Soviet accomplishment.

"The achievement of great exploits demands political unity and discipline," he noted. He explained that, like his own regime, the Russian one was authoritarian. It was the only point of similarity, he added.[20]

The national Catholic weekly *America* called for an "agonizing reappraisal in the land."

X MINUS 84 DAYS: THE RISE AND FALL OF VANGUARD

On November 8, 1957, two events occurred. One was a nation-wide radio and television report by Eisenhower on the state of the nation's defenses. He admitted that the Sputniks represented "an achievement of the first importance." But although the Soviets "are quite likely ahead in some missile and special areas . . . as of today, the over-all military strength of the free world is distinctly greater than that of the Communist countries." The President then showed his television audience the scaled-down nose cone of the Jupiter C which had been launched to an altitude of 600 miles from Cape Canaveral on August 8, 1957, and recovered 1200 miles downrange. To avoid being taken unawares on the technological front again, Eisenhower announced he had created the new office of Special Assistant to the President for Science and Technology. The post had been accepted, he said, by Dr. James R. Killian, president of the Massachusetts Institute of Technology.

The second event of November 8, 1957, was a telegram from Defense Secretary McElroy instructing General Medaris to prepare vehicles to launch two satellites during March 1958. The President wanted the first launched March 6, if possible. The joy at Huntsville was unbounded. Missile No. 29, the Jupiter C backup for the rocket which had set an altitude and distance record in 1956, was in the hangar. At the suggestion of Medaris, Army Secretary Brucker recommended to McElroy that the launch date for the first satellite be advanced to January 30, 1958. Medaris explained that the ABMA could do it by then and a flight on that date would give assurance of success on March 6. McElroy agreed, and the ABMA went into action to prepare Missile No. 29 for air shipment to Cape Canaveral.

In the meantime, the Navy stepped up its preparations for

the first test of Vanguard in full-flight configuration, with a satellite. Although President Eisenhower had been briefed that "it was only a test, which had a very remote bonus—a satellite," White House Press Secretary James C. Hagerty had released an announcement on October 11, 1957, that Project Vanguard was getting ready to launch a satellite. According to Hagen,[21] the announcement had come as a rude shock to the Navy, as "Our first, live, three-stage launching was billed as a satellite-launching success in advance and committed us to a public deadline with an untried vehicle."

The launch test was attempted Friday, December 6, 1957. The first-stage engine ignited with a comforting roar and the pencil-thin vehicle began to lift. Then the observers watched in growing dismay as the rocket faltered and fell back on the pad, exploding in a billow of smoke and flame. It was Black Friday for the United States satellite program. In the light of the advance build-up in the press, the test failure was a disaster. The blow-up evoked expressions of chagrin, sympathy, consternation, and ridicule all over the world. The Communist press hooted. In London, the *Daily Express* headlined, "U.S. Calls It Kaputnik;" the *Daily Herald* bannered, "Oh, what a Flopnik!" and the *Daily Mail* observed, "Phut Goes U.S. Satellite." The *Daily Mail* commented that the failure was a blow to "Mr. Eisenhower's declining prestige." In New York, the *Times* warned that Moscow would exploit this propaganda defeat "in its current drive to convince the world it is now the rocket and therefore the military master of this planet." In Tokyo, the newspaper *Yomiuri* summed up the experience as a "Pearl Harbor for American science." One American reaction was expressed vividly by the columnist Joseph Alsop in a dispatch from Beirut, Lebanon: "Language that can be printed in a family newspaper cannot possibly convey the emotions of an American traveler abroad on receiving the news of the failure of the first American attempt to launch an earth satellite."

X MINUS 41 DAYS: MISSILE NO. 29

The Redstone first stage of Missile No. 29 arrived at the Cape December 20, 1957, aboard a C-124 Globemaster aircraft. The stage was 56 feet long with instrument section, and

70 inches in diameter. The engine had been modified to burn a new, high-energy fuel called Hydyne, which at that time was secret. It was a blend of UDMH (*u*nsymmetrical *dim*e-thyl*h*ydrazine) and diethylene triamine. It increased Redstone's thrust from 65,000 to 83,000 pounds at liftoff.

The upper stages of the missile had been checked out at the Jet Propulsion Laboratory's spin-test facility on the Cape. Eleven of the scaled-down Sergeant rockets, each 4 feet tall and 6 inches in diameter, were clustered in one ring, constituting the second stage. Their total thrust was 733 pounds. Three more within the ring formed the third stage, with a thrust of 200 pounds. A single Sergeant rocket was the fourth stage, with a thrust of 67 pounds to "kick" the satellite attached to it into orbit.

The Redstone was sent to the Army's missile firing laboratory in Hangar D on the Cape for checkout under the supervision of Kurt Debus, a Peenemünde veteran. By January 13, 1958, the power plant had passed inspection. The gyroscopes were tested and found in good order. The "destruct'" system, which enables the range safety officer to blow up the rocket if it veers off course, was checked out.

The Air Force, which operated the Cape, assigned a launch date of January 29, 1958, to Missile No. 29 and granted two days' leeway in the event of delays. There was other traffic on the missile range. A mile away, a second Vanguard in Hangar S was being readied for launching early in February.

The launch of Missile No. 29 was prepared under two restraints. Safety requirements dictated that the launch azimuth or direction would be just south of east, down the Atlantic Missile Range. This meant that the flight path of the satellite would not go far enough north to pass over Europe where its radio voice might reassure American allies that the West, too, was on the move in space. Instead, the "new American moon" would be confined mainly to sub-tropical latitudes over the United States, Africa, Asia, Australia, and South America. It would not fly over any part of Russia.

A second restraint imposed by the Army was the avoidance of any premature publicity about the prospective launch, and this was carried out by keeping it secret. When Missile No. 29 had been checked out at Hangar D, it was erected on Pad 26-A under cover of darkness. This was done to avoid alerting the "bird watchers"—newsmen who tried to observe what was going on at the Cape through field glasses from the beaches to the south—that something was going on.

Once the missile was set up on the pad, the peculiar washtub configuration of the upper stages, which would have given the show away, was concealed by the service tower enclosing the missile. The lower part of the first stage, an ordinary Redstone, was all that could be seen. Only one reporter, the late Chris Butler of the Orlando, Florida, *Sentinel* noted that a large Redstone had been erected.

Then on January 24, 1958, the Army decided it was time to let the mass communications media in on the game, and Major General Don Yates, commander of the Air Force Missile Test Center, agreed. However, the Army insisted on briefing the press in confidence, with no release until the flight. Both the Army and the press kept the pact. Day by day, the reporters were briefed on the progress of the vehicle checkout. As the January 29, 1958 launch date approached, the missile was pronounced ready to go. A night launching between 10:30 p.m. and 2:30 a.m. Eastern Standard Time had been decided on in order to equalize the time the satellite would be in daylight and darkness on its first orbit. It was believed this would be a safeguard against damage to the storage batteries and radios from extremes of temperature. It was estimated that the outer skin of the satellite would be subjected to temperatures ranging from 100 degrees to about 70 degrees below zero Fahrenheit. Insulation and the use of reflecting paint to reduce solar heating were expected to keep the instruments and electronic gear fairly close to "room" temperature. The proper functioning of the electrical system and the radios was critical for the first revolution, so that ground observers could determine by radio signals from the satellite that it was indeed in orbit. Solar cells were not used.

The satellite was a tube 6 feet, 8 inches long and 6 inches in diameter, with a rocket nozzle at one end and the nose cone at the other. The rear section of it was one of the scaled-down Sergeant rockets, designed to kick the entire assembly into orbit with a final thrust of 67 pounds for one second. When the fuel was burned, the tube weighed 30.8 pounds, including its 18.13-pound payload, consisting of two radios, two antennas, the batteries, and 11 pounds of scientific instruments. The tube itself was made of steel. Eight aluminum oxide stripes were painted lengthwise on its surface to control temperature inside by reflecting part of the sunlight falling on the metal skin.

One antenna had four whip elements, each 22.5 inches long, extending out from the cylinder like tentacles, and the other was a dipole antenna, using the skin of the satellite itself. One radio operated on a frequency of 10 megacyles with a power of 10 milliwatts, designed to transmit a signal so faint that only powerful military stations could receive it. The second radio, operating on 108.03 megacycles at 60 milliwatts, could be received by amateur radio operations. Both radios would carry essentially the same telemetry information, however.

The scientific package consisted of instruments to measure cosmic radiation, micrometeors, and temperatures inside the tube and on the outside metal surface. The cosmic-ray experiment was designed by James A. Van Allen and his physics students at the State University of Iowa, where Van Allen was chairman of the physics department. Similar to the equipment which Van Allen had flown in sounding rockets for 12 years, it consisted of a single Geiger counter, an instrument invented in Germany 40 years before. From the Geiger tube, Van Allen hoped to receive data on the flux and intensities of cosmic rays at a number of points in the orbit.

The experiment to measure the density of micrometeors, or space dust, was designed by the Air Force Cambridge Research Center's Geophysics Research scientists, and consisted of a microphone which would pick up the sound of a dust particle striking the tube and broadcast it over the radio. The Cambridge group also had an erosion gauge on the satellite, consisting of coils of fine wire electrically connected in parallel. When one of the wires was struck by a particle large enough to break it, the electrical resistance in the coils would change. The erosion gauge also had a thin film of electrically resistant material. If high-velocity cosmic dust struck the satellite, the sand-blast effect would so erode the film that its resistance would change. The resistance changes in both the wire coils and the film would be telemetered to ground stations by the radios. Temperature readings would be telemetered by JPL thermometers placed inside the tube and between the exterior metal shell and insulation.

X MINUS TWO DAYS: COUNTDOWN

The flight-simulation test of Missile No. 29 was completed January 28, 1958, and the next morning Medaris and Debus awaited only the weather report to prepare to launch. But that morning, Lieutenant John L. Meisenheimer, the Air Force weather officer, came into the launch control room with bad news. A jet stream of 146 knots (168 miles) per hour was predicted out of the west at altitudes of 36,000 to 40,000 feet. With its spinning upper stages, Jupiter C was vulnerable to being buffeted off course by high winds. Also, the igniters used to fire the Sergeant upper-stage rocket clusters could be set off prematurely by lightning, even while the vehicle stood on the pad.

General Medaris went into a huddle with Debus and Walter Hauessermann, the guidance and control expert on the von Braun team. In view of unfavorable winds aloft, they decided to postpone the final countdown for 24 hours. Von Braun, who had been summoned to Washington to stand by in the Pentagon, agreed.

The next day, January 30, found upper-air conditions scarcely improved. Nevertheless, the final countdown was started at 11:30 a.m. at X minus 11 hours and the Redstone tank was filled with Hydyne. During the day, which was cool and cloudy, the Weather Detachment at Patrick Air Force Base sent up balloons to measure the force of high-altitude winds. At 40,000 feet, the winds were blowing at 217 miles an hour. When they continued to blow that hard after sundown, a "scrub" appeared inevitable. At 9 p.m., the countdown was halted.

Overnight, the velocity of the jet stream slackened somewhat, but not enough to satisfy Medaris. Debus was worried. He feared the corrosive Hydyne in the tank would open the seals. Lieutenant Meisenheimer forecast that the high-altitide winds would die down on January 31, and Debus pleaded with Medaris to launch. The general reluctantly agreed, and once more the final countdown began. As the launch directors and crew laboriously went through hundreds of checks and tests in the countdown manual, predictions of the chance of success varied considerably. A JPL engineer gave the satellite one chance in ten of achieving orbit. General Yates, the Cape commander, gave it one in three. Hans Gruene, the deputy director of the missile firing laboratory, gave it one in two. General Medaris was the optimist, giving the satellite a 90 per cent chance of success.

By early evening most of the launch crew and technicians had been called into the blockhouse, which contained sixty persons. Robert Moser, who was conducting the countdown, sat at a bank of consoles with his crew of five men. Debus stationed himself near a window overlooking the launch pad, with his deputy, Gruene, beside him. Captain Ballard B. Small in charge of missile-range support was stationed nearby. General Medaris moved restlessly from one group to another, exuding confidence.

In the telemetry room of Hangar D, Ernst Stuhlinger, von Braun's chief of research, was checking out an instrument called an Apex Predictor, which looked like an overgrown slide rule. When the missile would leave the pad, a microswitch in the instrument would begin traveling down a wire at a rate proportional to the time it would take the Redstone first stage to reach the apex of its ballistic trajectory. At this precise moment, the traveling microswitch would be halted by a bar. And at this contact, the instrument would activate a radio signal commanding the second stage to separate from the Redstone and fire. The second stage would boost the satellite's velocity to 9000 miles an hour. Ignited by its own timing device, the third stage would fire to increase the satellite's velocity to 15,000 miles an hour. Then the kick stage would boost final acceleration to 18,000 miles an hour—orbital speed. Stuhlinger was in high spirits. A few days before, his wife had given birth to a boy. She had wired him: "I've had my little satellite; now you have yours."

In the blockhouse, Captain Henry C. Paul was prepared to send a backup signal via the Air Force tracking station on Antigua. It would activate the upper stage if the firing command from the Cape failed to reach the flying washtub, which by that time would be spinning over the Atlantic Ocean several hundred miles downrange. When Stuhlinger's Apex Predictor microswitch hit the bar, a light would glow on a panel in front of Captain Paul. This was his cue to count off two seconds—"one thousand and one, one thousand and two"—and press the button alerting Antigua to send the firing signal. The officer was so keyed up about the operation —as was everyone in the blockhouse—that the night before the shot, his wife Judy related, he had said in his sleep, "One thousand and one, one thousand and two—oh my gosh, the power failed!"

Next door to Hangar D, JPL engineers were busy in their laboratory building. They had a direct line to their tracking facility on Antigua and also a computer at Pasadena. Tracking data from the Cape and Antigua were to be fed to the computer in California to indicate whether the payload would achieve orbit. When the computer said that orbit was achieved, General Medaris of ABMA and Dr. Jack Froelich of JPL would make a joint announcement.

Outside, about a hundred newspaper, radio, and television reporters had assembled during the evening at a hastily erected wooden grandstand near Hangar D—press site No. 1. It was 7500 feet from the launch pad, the closest the range-safety officer would allow correspondents to stand.

X MINUS ZERO: OPEN VALVES

Missile 29 stood gleaming whitely in the glare of flood-lights. Hundreds of men, women, and children had begun to gather on nearby beaches, alerted by the activity and bright lights on the Cape and widening rumors that "this was it." The launch pad and gantry appeared to be a vague blaze of light from Cocoa Beach. From the press grandstand, fumes of liquid oxygen could be seen wreathing the upper tub stages.

The countdown went quickly. All personnel had been cleared from the pad at X minus 15 minutes. At X minus 12 minutes, the aluminum tub containing the upper stages began to spin at 550 revolutions per minute as power was fed to its electric motors. At X minus 8 minutes, final checks were made to see that the liquid-oxygen tank had been "topped off" —that is, filled to the brim.

At X minus 6 minutes, there was a "nose count" in the blockhouse to make sure all members of the launch crew had left the pad. At X minus 3½ minutes, the final voltage checks were made.

X minus 3 minutes: Firing key to launch.

X minus 2½ minutes: Rudder drive check, gyroscope check, power check.

X minus 1½ minutes: Clear signal to launch.

X minus 45 seconds: Rudder drive on, arm destruct package, telemeter recording on.

X minus 3 seconds: Record sequence at fast speed. Suddenly, Moser called out that one of the steering vanes was improperly deflected. It might make a difference; it might not. Debus said to forget it and General Medaris gave a thumbs-up signal to go ahead.

X minus 0: Firing command.

X plus 0: Open fuel prevalves (to let the fuel start flowing); close liquid-oxygen vent. The halo of white smoke around the missile vanished.

X plus 3 seconds: Pressurizing of fuel and liquid oxygen in tanks.

X plus 10 seconds: Missile on internal power.

X plus 12 seconds: Drop boom plug, leaving missile free.

X plus 14 seconds: Ignition.

X plus 15.75 seconds: Liftoff. The time: 10:47:56 p.m. EST.

Missile No. 29 leaped off the launching ring in a gout of bright flame and raced upward with astonishing speed. It shed flakes of frost as it climbed vertically for an unusually long time compared to the ballistic missiles, which arced over to fly parallel to the ocean early in their flights. A long candle flame burned from the tail, leaving a shower of bright sparks. These were fragments of graphite from the erosion of the carbon vanes that extended into the jet blast.

The jet vanes guided the missile by deflecting the exhaust jet in response to the missile's autopilot commands. The premature deflection of one vane apparently was corrected by the initial jet blast when the Redstone engine ignited.

In the blockhouse, the tense crew heard a high-pitched scream which fell down the scale as the missile rose, like the whistle of a departing locomotive. It was a Doppler signal from the rocket's telemetry system, made audible in the blockhouse. One observer remarked, "They knew right away they had a good operation. Then everybody made a beeline to the hangar where Stuhlinger had the Apex Predictor, which was to fire the second stage."

Seventy seconds after liftoff, the launch vehicle was turned by its autopilot on a long arc. In its instrument section, a tape unwinding slowly on a reel triggered a command to the electric motor to increase the spin rate of the aluminum tub to 650 revolutions a minute. At 115 seconds into the flight, the motor was ordered to speed up the tub spin to 750. It had been calculated in advance that this step-up in the tub's spin rate would avoid resonant vibration with the shaking of the Redstone as it hurtled through the atmosphere. Resonant vibration could shake apart the upper-stage assembly.

The first-stage propellant burned out 2 minutes and 36.7 seconds after liftoff, a small fraction of a second earlier than predicted. At this time, the rocket was tilted at an angle of 40 degrees to the horizon and was coasting upward. Six seconds later, the instrument-section clock sent an electrical signal causing explosive bolts to fire and free the tub stages and satellite from the missile.

This assembly coasted upward for the next four minutes. Small jets on the tub, powered by compressed air, then fired automatically, tilting the tub assembly and satellite until they were parallel to the earth's surface. With sufficient velocity, the satellite would then fall around the earth in an orbit.

When the tub stages and satellite reached an altitude of 225 miles, Stuhlinger's Apex Predictor microswitch hit its bar and sent the radio signal to ignite the eleven Sergeant rockets of the second stage. The light flashed on Captain Paul's console. He counted two seconds and signaled Antigua to repeat the radio firing command.

Stage-two rockets ignited 6 minutes and 43 seconds after liftoff and burned 6.5 seconds. Then the third-stage cluster of three Sergeant rockets ignited, burning another 6.5 seconds. At 6 minutes, 50 seconds in the flight, the "kick" stage ignited and burned for 6 seconds.

The payload, known as Explorer I, was on its own. In the control room at the Cape, telemetry indications said that the satellite had reached orbital velocity. All that remained was confirmation.

EXPLORER 1: 1958 ALPHA

Meanwhile, at the Pentagon, von Braun, Pickering, and Brucker were awaiting word from the Cape. As soon as von

Braun was notified that all four stages had fired properly, he advised Brucker that Explorer 1 was certainly in orbit.

However, Brucker wanted confirmation before notifying President Eisenhower, who was playing bridge in Augusta, Georgia. The Naval Research Laboratory had a radio station at San Diego and an IGY moonwatch team had been alerted to listen for Explorer 1's radio beep. In addition, JPL's antenna at Earthquake Valley, California, had just been completed and would pick up Explorer 1 as the satellite came over the West Coast on its first revolution.

Von Braun calculated that the satellite should pass over California at 12:41 a.m. Eastern Standard Time. Pickering opened a telephone line to Pasadena in California to alert observers at the Jet Propulsion Laboratory to watch for the radio signal from the tiny satellite.

At 12:40 a.m., Pickering queried Pasadena, but the answer was, "Not yet."

At 12:41 a.m., he asked, "Do you hear her?"

"Negative," came the reply.

At 12:43 a.m., Pickering again asked if they heard anything.

"Negative, sir," replied JPL.

"Well, why the hell don't you hear anything?" demanded Pickering.

Brucker turned anxiously to von Braun: "Wernher, what happened?"

Von Braun remained silent, staring at his watch. The minutes were moving like centuries.

At the Cape, where he had been on the line to JPL in Pasadena, Medaris put down the phone and said, "Earthquake Valley has the bird."

At the same moment in the Pentagon, Pickering yelled, "Wernher, they hear her."

Von Braun took a deep breath. Perspiration beaded his forehead and upper lip. Still staring at his watch, which he held in a trembling hand, he said in a flat voice, "She is eight minutes late. Interesting."

Chapter 3

A Wind between the Worlds

Explorer 1 not only opened the space age in the United States, but inaugurated a revolution in the conception of the interplanetary medium. The first soundings of this new sea were soon to reveal the shoals of forces and fields which lay immediately beyond the planet Earth, and to provide a new perspective of the relationship between the earth and the sun.

As Missile No. 29 bore the satellite upward toward orbit, the early returns from Van Allen's Geiger tube showed that radiation increased with altitude, as expected. When Explorer 1 passed beyond range of the Cape radio receiver, the scientific data were picked up by Naval Research Laboratory and JPL stations in Ecuador, Peru, Chile, and Australia, where they were recorded on tape. Van Allen had to wait until the tapes reached him by air mail at Iowa City three weeks later.

The tapes showed increases in radiation intensity with altitude at about the rates which previous balloon, rockoon,* and rocket probes had shown. But then, to the investigators' dismay, an inexplicable event appeared. At an altitude of 300 miles, the radiation intensity dropped abruptly, as though the

* A rockoon is a combination of a rocket and balloon, the rocket firing after the balloon lifts it to a high altitude.

Geiger tube had stopped counting. Van Allen and his associates were mystified. Had the Geiger tube failed? Had the telemetry failed? Since Explorer's radio continued to send other data, it was apparent that something had affected the radiation counter. There was no way of telling what it was.

Explorer 2 was coming up, the second of the four satellites which the Army had been authorized to launch as the backup for Vanguard. Van Allen installed another Geiger radiation counter on the vehicle. But the "kick" stage failed to fire and Explorer 2 did not reach orbit. There were two more Jupiter C satellite launchings in Project 416, as the quadruple launch program was called in Pentagonese, meaning four shots for $16,000,000.

Explorer 3 was boosted into orbit on March 26, 1958. The Geiger tube repeated the radiation data which had been returned by its predecessor on Explorer 1. Again, the counting rate rose rapidly. Above 300 miles altitude, the instrument cut off its reports, and nothing further was heard from it. Again, Van Allen and his staff pondered the mystery. What was happening up there? It was Carl McIlwain, one of Van Allen's assistants, who hit upon the amazing answer: The counters on Explorers 1 and 3 hadn't failed. They had been silenced—jammed—by radiation so intense that they could not record it. "We had discovered an enormously high rate of radiation—not the lack of it," Van Allen said.[1]

When the counter of Explorer 3 jammed, it was showing radiation counts more than a thousand times those which had been theorized for cosmic rays. The actual counts must have been many times higher than the instrument had been able to record, Van Allen surmised. What did it mean? What was up there?

There was a significant clue in the cosmic-ray work that Van Allen and his students had done in previous years, and now the earlier balloon, rockoon, and rocket probes, when considered together with the Explorer satellite revelations, suggested a fantastic—but not altogether unexpected—phenomenon above the atmosphere.

The earlier probes had detected a "hot zone" of intense radiation above 30 miles altitude in both the northern and southern auroral zones, in polar regions. It was while he was sailing to Antarctica on the *Glacier* in October 1957 to make auroral observations that Van Allen had heard the fateful beeps of Sputnik I.

Although it was tempting to assume that the radiation was connected with the auroras, the probes showed that each phenomenon operated independently of the other. The radiation intensity persisted whether the auroras were "on" or "off." It did not flicker out when the auroras did, but remained steady in the dark, empty heavens.

Early in 1958, McIlwain launched sounding rockets through the bright auroras over Fort Churchill, Manitoba. The data he got back showed that the "hot zone" contained both protons and electrons and was independent of the auroras.

These findings did not support the theory that the auroras were the products of solar particles interacting with the topmost layer of the atmosphere, since they failed to detect any correlation between radiation intensity and auroral activity. In fact, the evidence of rockoon probes strongly suggested there was no correlation at all. Yet the radiation zones in polar regions and the auroral displays seemed to represent some kind of process in the earth-sun relationship involving the magnetic field.

Fitted together, the data from the rockoon research and from Explorers 1 and 3 suggested a new and exciting possibility that a zone of intense radiation existed hundreds of miles above the earth. Perhaps it followed the contour of the magnetic field, in which case it would dip down fairly close to the ground in polar regions.

On May 1, 1958, Van Allen put the puzzle together in an historic report to the joint meeting of the National Academy of Sciences and the American Physical Society in Washington. Explorers 1 and 3, he said, had found a new phenomenon: a belt of intense radiation above the earth, starting at 300 miles. The region had been "sounded" by the satellite Geiger tubes across 68 degrees of terrestrial latitude, 34 degrees north and 34 degrees south of the equator. It was filled with a higher radiation flux than had ever been detected above the atmosphere.

Van Allen did not believe that the radiation had the same origin as the high-energy protons of cosmic rays. The "belt" of radiation near the earth was much "softer" than cosmic rays. Its particles were less energetic. It consisted of protons and electrons which were trapped in the earth's magnetic field. The radiation zone seemed to be a permanent appurtenance of the planet.

The revelation of this "belt" of radiation around the planet—the Van Allen radiation belt—electrified the world scientific community. It was the most spectacular discovery of the International Geophysical Year. And it offered mankind its first glimpse into the new and strange reality of the interplanetary medium, a region once thought "empty" or filled with "ether." The first American satellite, Explorer 1, had opened up a new vista to mankind, and it was only the beginning.

VANGUARD

The success of Explorer 1 was offset a few days later by another Vanguard launching failure, the second in a row for that vehicle. On its third flight test, however, Vanguard at last vaulted into space March 17, 1958, lofting its famous 3.25-pound satellite, a 6-inch sphere, into a looping orbit 405 to 2462 miles in altitude. Compared to the monstrous Sputniks, Vanguard's grapefruit-sized payload appeared inglorious. Yet it became one of the most successful and scientifically useful satellites of the space age.

For more than six years, its tiny radio, powered only by six quartz-covered arrays of solar-energy cells, continued to transmit signals on 108 megacycles. Vanguard 1 (1958 Beta 2) achieved such a stable orbit that it enabled scientists to observe changes in atmospheric pressure during solar activity, the influence of the sun's radiation pressure on the orbit of a satellite, and the atmospheric density at several altitudes. The most notable contribution of Vanguard 1, however, was the evidence it provided that the earth is slightly pearshaped, with the stem of the pear pointed toward the north pole. While this distortion in the planet's calculated ellipsoidal shape is very slight, it has important implications in geophysics. For example, it sheds new light on the long-standing controversy of continental drift, and supports the convection-cell theory of sea-floor spreading. How did Vanguard provide information on the shape of the planet?

Because the equatorial bulge which produces the ellipsoid contour and the pear-shaped component are relatively small compared to the size of the planet, the earth can be considered practically a sphere. It attracts a satellite with the gravitational force of a single point of mass at the planet's center of gravity. It has been known for a long time that the earth is not a perfect sphere but flattened slightly at the poles and bulges at the equator under the influence of the centrifugal "force" of the planet's rotation. The equatorial bulge, which has been calculated at 12.6 miles, exerts additional gravitational attraction on a satellite when it passes over the equator, pulling the trajectory slightly out of line at every revolution, so that the plane of the orbit rotates.

It was from the analysis of a second, and unexpected, perturbation in Vanguard's orbit that John A. O'Keefe, Ann Eckels, and R. K. Squires of the Goddard Space Flight center deduced the pear-shaped component, another region of high gravitational attraction. Its presence indicated "a very substantial load on the surface of the earth," the Goddard scientists said.[2] Such a load would produce stress that must be supported either by a greater mechanical strength than was generally assumed to exist in the interior of the planet, or by large-scale convection currents in the mantle.

The idea of convection currents in the mantle rock which lies between the crust of the earth and its molten-iron core is not a new one. The theory says that as rock at the boundary of the molten core is heated, it becomes lighter and tends to be displaced by heavier, cooler rock from above. A convection flow or current thus begins in the heated rock and moves throughout the mantle.

Convection in the mantle is believed by a number of geophysicists to be the source of energy that moves the continents about and rips the crust asunder to form the undersea rifts in the Atlantic, Indian, and Pacific Oceans. Continued analysis of the orbits of later satellites, as well as of Vanguard's, disclosed another curious fact about the shape of the earth. It was slightly more flattened than the theory accounted for, based on the shape a rotating body would assume if it was plastic inside, as a result of internal heating. However, the orbit analyses showed that the equatorial bulge was some 230 feet greater than the theoretical estimate—as though the earth actually were rotating faster than it does.

This discrepancy could be explained by assuming two facts about the planet. First, the mantle is not as plastic as it was believed to be but has relatively high mechanical strength. Second, the earth today is rotating more slowly on its axis than it did many millions of years ago, so that the day, in terms of time as measured by an atomic clock, is getting longer.

Studies by investigators at the National Bureau of Standards have shown that the earth's rotation at present is decreasing at a rate of about one-thousandth of a second a century. At the current rate of about 1 second per 100,000 years, the length of the day would have increased by 30 minutes in the last 180,000,000 years. The slowing down is attributed to friction caused by the tides, which the moon raises not only in the oceans but on land as well. Lunar tides in the mantle amount to several inches.[3]

Robert Jastrow, director of the Goddard Institute for Space Studies, suggested that the flattening deduced from the Vanguard 1 orbit data would correspond to the shape a plastic earth would have had when the day was only 23 hours and 30 minutes long. It appeared, he said, that while the mantle is sufficiently warm and plastic to respond to the changing stresses of the slowing down of the earth's rotation, it has enough internal strength to cause the response to lag over long periods of time.[4]

PARTICLES IN SPACE

Van Allen's discovery of the radiation belt stimulated more public interest in the strange environment above the atmosphere than any astronomical event since Halley's comet. Interplanetary space was an unimaginable region which seemed to those who thought about it, whether educated or not, to be the abode of nothingness, except for meteors, comets, and dust. Electromagnetic radiation in the form of light and radio waves came from the sun and the stars. So, too, did a rain of energetic atomic particles, also called radiation, to the confusion of the uninitiated. However, these particles also were forms of energy radiating from the sun or the distant stars. Far from being empty, the interplanetary medium—space—was alive with matter and energy, swirling in unfathomable patterns in response to indefinable forces and fields.

Polar auroras had been thought of as the visible evidence of the interaction between the sun's particle radiation and molecules and atoms in the high atmosphere. But the precise mechanism that produced these flaming banners in polar skies eluded a generally acceptable definition.

Since 1911, when mysterious radiation from space was detected by the Austrian physicists Victor F. Hess and Wernher Kolhorster in high-altitude balloon experiments, physicists had known of a class of particles, or corpuscular radiation, that seemed to come from stars. This class had been labeled "cosmic rays" by Robert A. Millikan, who established the tradition of cosmic-ray investigation at the University of Chicago. Millikan believed the "rays" were a form of electromagnetic radiation. However, Arthur Holly Compton at the University of Chicago found that cosmic rays were deflected around the curve of the earth's magnetic field and were more intense in polar regions where the lines of magnetic force dip down to the ground. This observation proved that cosmic rays were actually electrified particles, principally protons comprising the nuclei of hydrogen atoms, but also the nuclei of helium, carbon, nitrogen, oxygen, and the heavier elements up to and including iron. These atoms, it was theorized, are stripped of their electrons and the nuclei are accelerated to speeds approaching that of light in the explosion of stars. Much of this stellar debris collides with atoms of oxygen, nitrogen, or other gases in the atmosphere, smashing the nuclei and producing additional high-energy particles which are called "secondary" cosmic rays. Most of the particles detected near the surface of the earth are believed to be secondary cosmic rays.

The mystery of the origin and distribution of cosmic rays challenged the imagination of physicists for nearly half a century before the first Sputniks and Explorers inaugurated the space age. The radiation represents a major form of energy in the universe—how major is a subject of continuing speculation. Are cosmic rays confined to our galaxy—the Milky Way—or do they fill all of intergalactic space? If they pervade intergalactic space, with the same energy-density as they have within our galaxy, then they would contain energy equal to one per cent of the mass energy of all the stars, according to George F. Pieper, assistant director of the Space Science

Directorate at the Goddard Space Flight Center.[5] Cosmic radiation would be a major repository of energy in the universe.

A second type of radiation, "softer" than cosmic rays, was deduced by a Norwegian physicist, Carl Störmer, about the time that Hess and Kolhorster were doing their balloon experiments. Störmer's laboratory experiments later indicated that when high-energy primary cosmic-ray particles zipped into the atmosphere and struck atoms of gas there, the initial products of the collisions were neutrons, which then decayed into protons and electrons. By observing the behavior of charged particles in the magnetic field of a magnetized sphere representing the earth, Störmer concluded that charged particles can be trapped in certain regions of space around a magnetic dipole. In 1957, S. Fred Singer of the University of Maryland suggested that trapped radiation may exist in the earth's dipole field where charged particles would race back and forth between magnetic poles.[6]

Cosmic rays could be distinguished from Van Allen's trapped radiation because of their higher energy level. They were in the range of tens of millions of electron volts, while the energy range of the trapped particles appeared to be in the tens of thousands of electron volts. The source of the trapped particles was not clear. At first, Van Allen believed they came from the sun, but other physicists supported a theory that the trapped radiation was the product of neutron decay from the collision of cosmic rays with the atmosphere.

Once caught in the magnetic field, the particles spiraled back and forth between the magnetic poles like electrified shuttlecocks. Each particle followed a corkscrew path along a line of magnetic force. At the lower end of the path, the convergence of magnetic-force lines flattened the spiral and reversed it, causing the particle to zip back the way it had come, until it neared the force-line convergence, or mirror point, in the opposite hemisphere. Then it would again reverse its direction. As they spiraled in longitude, the particles would drift in latitude. An electron tended to drift to the east and a proton to the west. Now and then, a trapped particle collided with an atom or molecule of atmospheric gas and vanished from the force field.

The earliest analysis of trapped-particle behavior in a magnetic field had been attempted in the laboratory. Now, with the advent of the satellites, the laboratory could be extended into the vastness of the solar system. The earth was a convenient place to stand.

EXPLORER 4

The data from Explorers 1 and 3 had indicated the existence of the radiation zone in a negative way: the zone was presumed to exist not because it was fully observed but because it could not be. That led Van Allen and his associates to conclude that the radiation flux (the number of particles going by a given point in a second) was considerably higher than the instrument could measure. As negative evidence goes, it was good evidence, and Van Allen's fellow scientists accepted it.

What was needed, however, was a counting system which could measure the full flux of the radiation at an altitude of 500 miles. Van Allen and his associates adjusted one Geiger tube so that it would give a greatly scaled-down particle count, and hooked it up to a circuit that would scale the count down further, in order to simplify transmission of the gross count by radio. They shielded another Geiger tube with one millimeter of lead so that it would record only the most energetic particles. Two more devices were added to the system. One was a plastic scintillator, or radiation counter, attached to a photomultiplier tube. It was designed to measure the flux of medium-energy electrons, such as those Van Allen expected to find in the geomagnetic field. The other was a photomultiplier tube with an iodide-crystal barrier glued over the opening. When radiation hit the crystal, light would be emitted that could be measured by the tube. The intensity of the light would indicate the energy of the particle striking the crystal. This device would show the energy range of trapped radiation, as well as count particles.

With this array of detectors, Van Allen was prepared to analyze the flux and energy range of particles in the belt. The entire package was installed on Explorer 4. The last of the Jupiter Cs to fly in Project 416 boosted Explorer 4 into orbit on July 26, 1958, and the data that flowed from the radiation detectors were voluminous. They enabled the Iowa investigators to plot radiation intensities to an altitude of 1300 miles, the apogee (peak altitude) of Explorer's orbit. A clearer picture of the radiation zone began to emerge. The zone seemed to bracket the earth like parentheses, open over the poles. The term "radiation belt" was applied quite early in press reports of the discovery. Although it did not depict the parenthetical image of the zone, it stuck.

The contour of the radiation belt seemed to follow that of the magnetic field, which had been sounded roughly during the IGY by whistlers: attenuated lightning discharges which traveled from one hemisphere to the other along magnetic force lines. These discharges could be heard on a radio as whistles descending in pitch. Dipping low over the polar "openings" in the magnetic field, the radiation zone seemed to billow out to 6 or more earth radii (24,000 miles) in equatorial regions. In the polar auroral zones, a curious contour appeared. There the belt grew a horn pointing groundward. It was the horn, apparently, which rockoon-borne radiation counters had detected years before.

How far out did the radiation zone extend? To find out, it was necessary to fly a radiation counter higher than 1300 miles. This was done on Pioneer 1, launched by Thor-Able rocket October 11, 1958. It was an Air Force shot, America's first attempt to dump a payload on the moon. The lunar mission failed because the launch vehicle did not generate enough push to boost the payload more than 70,000 miles, but it was far enough to show Van Allen and his colleagues that the radiation region extended outward for thousands of miles. The highest intensities appeared to be at an altitude of 10,000 miles.

On November 8, the Air Force again tried to hit the moon with Pioneer 2, launched by a Thor-Able rocket. The shot fizzled. Pioneer 3, the nation's third attempt to hit the moon, was launched December 6 by the Army Ballistic Missile Agency's Juno 2, a modified Jupiter intermediate range ballistic missile. While it failed to reach the target, the radiation counter reported data to an altitude of 63,580 miles. The counting showed two peaks of radiation intensity, the first at about 2000 miles and the second at 10,000 miles. Beyond 10,000 miles, the radiation diminished and finally faded to assumed interplanetary levels at 40,000 miles.

Now there appeared to be not one but two Van Allen radiation belts, an inner and an outer one. The data from the Pioneers suggested that most of the particles of the inner belts were protons and most in the outer zone were electrons.

While the image of inner and outer belts became widely accepted, Van Allen had reservations about the accuracy of that picture. There were many questions to answer. How did the flux of protons and electrons in the belts change from hour to hour, day to day, month to month? What was the relationship of the radiation flux to activity on the sun?

By the end of 1958, after nearly one year in space, Americans discovered that there were zones of trapped radiation encompassing the earth to a distance ten times the planet's radius, with a maximum proton flux of 40,000 particles per square centimeter per second. The discovery illuminated the shape and extent of the earth's magnetic field, the cage which trapped the particles. Van Allen and his associates at the University of Iowa were convinced that most of the particles in the belts came from the sun, even though some in the inner belt might have been injected into the magnetic trap by decaying neutrons from the atom-smashing of cosmic-ray collisions with the atmosphere.

THE OUTER REEF

Once in the magnetic field, the particles were held captive by their electrical charge, positive in the case of protons, negative in the case of electrons. The radiation belts were a new geophysical phenomenon for mankind to contemplate. So far as spacefaring by man was concerned, they formed an outer reef which had to be crossed. The energy and intensity of the radiation posed a hazard to manned space flight above 300 miles altitude. On a flight to the moon, at 7 miles a second, astronauts would be in the belts for 95 minutes. It was a danger which had not been anticipated.

By the end of the International Geophysical Year, December 31, 1958, analyses of the radition belts had suggested to Van Allen and others that these phenomena had something to do with auroras and with solar activity. Van Allen hypothesized that solar particles captured in the magnetic field eventually leaked out in polar regions to produce the auroras by colliding with atoms of gas in the atmosphere. He compared the belts to a leaky bucket from which streams of atomic particles dribbled out. The bucket would be constantly refilled by new supplies from the sun. An unusually heavy influx of particles emitted by a flare would cascade into the bucket and cause it to overflow into the polar zones, generating unusually widespread and spectacular auroras. Possibly, slow leakage of particles from the belts might account for the upper-air radiance called airglow, a dimmer illumination than auroras, believed to be produced by the same process. The "leaky bucket" theory also offered an explanation for high temperatures which rocket- and balloon-lofted instruments reported in the upper atmosphere over the poles.

The idea that the particles which kept filling up the "bucket" came from the sun raised several questions, however. Their energy was considerably higher than that of slower-moving streams of solar particles which had been detected beyond the Van Allen belts by Pioneer 3. What accelerated them in the magnetic field, when they were captured?

The cosmic-ray-neutron theory of the origin of the trapped particles was supported by Singer; Nicholas Christofilos of the University of California's Lawrence Radiation Laboratory; and Sergei N. Vernov of the Soviet Union. Van Allen was skeptical. While the neutron-decay process undoubtedly goes on, he said, it did not seem likely that it could make any significant contribution to the belts. No, the solar mechanism by which particles emitted by the sun were captured directly by the belts appeared more reasonable. Furthermore, it tied the belts, the auroras, and solar-particle streams into a neat package of cause and effect.

ARGUS

Whether the Van Allen belt particles came from the sun, from cosmic-ray-produced neutrons, or from both was one problem. Whether the particles in the belts produced the auroras was another. The first required much more study in space, but the second could be proved by a test proposed by Christofilos. He suggested artificial injection of radiation particles into the high atmosphere to see if their interaction with the atmosphere would produce an aurora. The source of the particles could be an electron accelerator, a machine for speeding up and firing electrons like bullets, or a rocket-lofted nuclear bomb. Christofilos theorized that charged particles, principally electrons, would become trapped in the magnetic field. As they spiraled down into the atmosphere to collide with molecules of oxygen or nitrogen, the collision should produce auroral effects.

The Department of Defense agreed to conduct the test by launching nuclear rockets from a ship in the South Atlantic Ocean at the end of August 1958. The test was coded "Argus." Aside from auroral theory, the Defense Department was interested in other questions too: Would the electrons produced by nuclear fission cause a radio blackout if the bomb was exploded at very high altitudes? Would they interfere with radar detection of aircraft and missiles? How long would hypothetical blackout effects last?

Some inkling of the answers had been provided by high-altitude hydrogen-bomb tests July 31 and August 10, 1958, from Johnston Island in the Pacific Ocean. On the first test, coded "Teak," a bomb had been rocketed to an altitude of 50 miles and detonated, resulting in a radio blackout over the Central Pacific lasting several hours. The second test, coded "Orange," in which a bomb was exploded at an altitude of 28 miles, produced a flash seen in Hawaii and a strange "aurora" which was observed by a group of New Zealand weathermen in Western Samoa.

In "Argus," the Defense Department was intent on investigating the "side effects" of the bomb explosion on the environment at 300 miles altitude. The first event linked to the Argus tests was the launching of Explorer 4, which was instrumented to monitor the effects of the Argus test series as well as to perform its scientific mission of measuring Van Allen belt radiation. Data from the satellite might tell whether it was possible to detect the "signature" of a nuclear explosion in space by sensing a sudden increase in electrons with energies suggesting a bomb origin. This portion of Explorer 4's mission was suppressed when other data from the satellite were published.

The first of three atomic bombs was launched by a three-stage rocket with a Polaris first stage from the deck of the missile ship U.S.S. *Norton Sound* on August 27, 1958. It produced an aurora above the launching site, about 75 miles southeast of the South Atlantic island of Tristan da Cunha.

The question was: would auroral effects be observed at the opposite end of an arc of magnetic force extending from this region in the southern hemisphere to a point 5000 miles away in the northern hemisphere? Observers were ready aboard another Navy ship, the *Albemarle,* in the North Atlantic. They had not long to wait. Within seconds after the flash in the southern hemisphere, the ship's radar detected auroral effects overhead. Two more bombs were detonated aloft on August 30 and September 6. The third and last shot produced a visible aurora in the northern hemisphere.[7]

These high-altitude nuclear tests over the Pacific and Atlantic Oceans confirmed the mechanism of the trapping of charged particles by a magnetic field, which had been proposed to account for the existence of the radiation belts. But there was more to come. Four years later, the United States exploded a 1.5-metagon bomb at an altitude of 250 miles above Johnston Island. The results were more spectacular than the planners of the experiment had imagined. The detonation increased the radiation of the lower portion of the Van Allen belts to such intensity that it damaged the power supply of two Navy satellites, Transit 4-B and Traac, and the British satellite Ariel, and along with radiation from similar Russian bomb tests, reduced the longevity of the Bell System's Telstar 1, the first communications satellite.

THE MAGNETOSPHERE

During the first year of the space age, a new conception of the interaction between the earth and the sun had made its debut, as a result of man's new ability to probe the surf of an interplanetary sea he might some day cross. In the second year of the space age, 1959, the new ideas matured rapidly with additional probes of the interplanetary medium.

The radiation belt was not, it became apparent, an intermittent phenomenon. The belt—or belts, if there were actually two—were found to be permanent fixtures of the planet. Moreover, scientists were forced to think of the magnetic field and the radiant energy it contained as another "sphere" to be added to the lithosphere (the solid parts of the earth), the hydrosphere (the seas, rivers, lakes, and water vapor), and the atmosphere. The new sphere was the "magnetosphere," the region beyond the earth encompassed by a field of magnetic force, generated deep within the molten core. In the magnetosphere, the disciplines of astrophysics and geophysics merged. The newly discovered phenomena of the radiation belts gave physics a new dimension. There was a wonderful unity in the conception that a force generated by a rotating molten-iron core inside the earth created a zone of energetic particles around the planet to a distance ten times its radius, that the particles seemed to originate in the sun, and that they, perhaps, fired the top of the atmosphere with auroral lights. What other effects and connections might there be?

The mechanism by which the auroras were ignited continued to elude researchers. New evidence supporting the leaky-bucket (or magnetospheric-dumping) theory was interpreted in the reports of Explorer 7, which was launched into a elliptical orbit 345 to 675 miles high on October 13, 1959, by a Juno 2 rocket. Its radiation detector reported changes in radiation-intensity patterns of the Van Allen belts during a severe magnetic storm on November 28, when brilliant auroras were observed over North America. The term "magnetic storm" refers to the supposed disruption of the magnetosphere by a barrage of energetic solar particles which reach the earth a few days after a flare is observed on the sun. Such a storm cannot be detected by human sensory organs, but its effects can be. Particle bombardment from the sun disrupts the ionosphere, the electrified region of the high atmosphere which reflects radio waves around the curvature of the earth. Instead of being reflected back to the ground, where they can be received, long-distance-radio signals are absorbed in the ionosphere, causing a blackout in radio communications. The magnetic storm of November 28 silenced radio communications between New Zealand, Australia, and IGY science stations on the icecap of Antarctica. It signaled that a cloud of fast-moving atomic bits and pieces had struck the magnetosphere. It was then supposed that myriads of the particles had cascaded into the outer radiation belt, causing a dumping effect of trapped particles into the atmosphere. These in turn caused an intensification of the illumination and extent of the aurora borealis, so that it was seen as far south as Chicago and Des Moines.

The correlation of Explorer 7 data with observation of the

auroras from the ground amounted to "direct evidence of dumping into the upper atmosphere during times of geomagnetic disturbances," according to the Explorer 7 experimenters.[8] They were Van Allen and Brian J. O'Brien of the State University of Iowa, F. E. Roche of the National Bureau of Standards, and C. W. Gartlein of Cornell University.

Two years later, O'Brien revised this conclusion. He said that the number of charged particles moving into the auroral zones after a solar flare was too great to be accounted for solely by magnetospheric dumping. If the aurora was caused by dumping from the belts alone, he estimated that the entire magnetosphere would be drained of radiation in one hour. It appeared to him that when a solar disturbance occurred, only a part of the cloud of particles reaching the earth became trapped in the magnetosphere. Most of them passed directly into the atmosphere without spending any appreciable time in the belts, O'Brien theorized. In this vew, the outer belt became a slope down which most of the inrushing solar particles slid, like rain off an umbrella, into polar regions. O'Brien based his observations on reports from State University of Iowa detectors on the satellite Injun 1. He finally concluded that charged particles in the auroral zone are not fallout from the magnetosphere, but descend directly down the force-field lines into the atmosphere, nonstop from the sun.

With continued probing, scientists modified their ideas about the great radiation zones. An important new discovery in the outer belt was reported by Explorer 12. This 83-pound satellite, equipped to measure the flux of particles in several regions of the magnetosphere, was launched August 15, 1961, by one of the new Thor-Delta rockets. The Delta, a composite of a Thor booster and Vanguard second and third stages, was becoming one of the most reliable launch vehicles in the American inventory. It made the instrumented exploration of space a considerably more efficient enterprise than it had been in the beginning of the space age, when balky rockets sent 20 per cent of their scientific payloads to the bottom of the sea.

The orbit of Explorer 12 was a long ellipse ranging in altitude from 182 to 48,000 miles. The spacecraft traversed both radiation zones in 26½ hours. The satellite ceased broadcasting data after 112 days, but during the time it was on the air it returned more information on the magnetosphere than did all of its predecessors combined.

The craft's 2-watt transmitter reported that there were a thousand times fewer electrons in the outer belt than had been indicated by the Pioneer probes, and that the region contained many low-energy protons. This finding strengthened an idea that the outer belt was trapping protons emitted by the sun. In 102 revolutions of the earth in its elongated orbit, Explorer 12 crossed the outer belt 204 times. Electron concentration was measured by instruments supplied by Van Allen and O'Brien. The protons were reported by Goddard Space Flight Center counters.

The National Aeronautics and Space Administration reported that the satellite gave a particularly vivid demonstration of the perturbing effects of the sun and the moon on a satellite's orbit, especially one as elongated as that of Explorer 12. During the 112 days the satellite's radio was working, tracking stations noticed that the elliptical flight path of the vehicle became more circular. Twenty-seven days after it was launched, Explorer 12's apogee had dropped from 48,000 to 47,683 miles and its perigee had risen in altitude from 182 to 278 miles. By 112 days after the launching, the space agency reported, the perigee had risen to 497 miles, and was expected to continue rising to 620 miles before descending again toward the earth.[9] Further changes in apogee after 27 days were not reported.

The discovery of the extent of the magnetosphere had great geophysical implications. It enlarged man's conception of his environment in a manner unique in human experience.

Above the surface of the earth, there was first the sensible atmosphere, the gaseous mixture consisting principally of oxygen and nitrogen which man could sense. Above it, there was a region of helium gas, first predicted by a Belgian scientist, Marcel Nicolet, in 1957 and confirmed in 1960 by Explorer 8 data, as analyzed by the Goddard Space Flight Center. Above the helium was a layer of hydrogen gas, first recognized by Francis S. Johnson, director of the Earth and Planetary Sciences Laboratory at the Southwest Center for Advanced Studies.

Beyond the hydrogen, there was the magnetosphere. And beyond that—what? Where was the boundary between the planet and the interplanetary medium? Where was the true terrestrial frontier?

At the edge of the magnetosphere, the satellites had found a region of strange turbulence, of shock fronts. Here a stream of solar particles washed against the magnetic force field like waves against a beach. Was this the boundary between heaven and earth?

Investigators called this region the "magnetopause." On the sunward side of the planet, it extended outward about 40,000 miles. During the early 1960s, space probes flew in and out of the magnetopause, reporting the particles and fields of the region in their dissonant, binary language. The sing-song of satellite telemetry in the radio receivers told of a continuous rushing of particulate or corpuscular matter which blew against the magnetosphere. It was like a wind blowing from the sun.

THE SOLAR WIND

While the discovery of the Van Allen belts and the definition of the magnetosphere are the lasting monuments to the early years of the space age, the soundings which disclosed the undiscovered realm and its nature were not the primary motives for the flights. Explorer 1 was essentially a scientific rationale to the astropolitical requirement that America get something into orbit as quickly as possible. The first three Pioneers, carried by the newly evolved Air Force Thor with a second stage called Able, had primary missions of impacting the moon. It transpired that the data received from the Pioneer series about the nature of space this side of the moon were of considerably greater consequence to science than hitting the target would have been.

Pioneer 4 missed the moon by 37,300 miles, but it showed how a solar flare enhanced the outer radiation belt. The first three years of America's venture into space was characterized by this kind of serendipity. In those years, science was a hitchhiker on missions which were primarily engineering tests. Not for years would science and exploration displace technological muscle-flexing and national prestige as the prime motives for rocket development.

Despite their many successes, the Pioneers demonstrated

the technological point that the Jupiter C, based on the Redstone, and the Juno 2, based on the Jupiter IRBM, were not powerful enough for deep-space cruising. Better boosters had to be developed. These began to appear in the Thor intermediate and Atlas intercontinental ballistic missiles.

The Russians, too, were shooting at the moon. Their first shot, Lunik I, or Mechta, launched January 2, 1959, missed the target by 4660 miles. It sailed on into solar orbit with a 15-month period of revolution—the first artificial planetoid.

Then, at 4:02 p.m. Eastern Standard Time on September 13, 1959, the Russians hit the moon with Lunik II to win the lunar turkey shoot. They demonstrated by this feat that they had more powerful rockets than America had been able to develop up to this time, and the military significance of that advantage provided the stimulus for accelerating rocket development in this country.

In the meantime, however, the radiation detector aboard Lunik II told a fascinating story to scientists. It reported a stream of ions with a positive electrical charge between 400 earth radii (160,000 miles) and the moon. The particles were low-energy protons with a flux of 200 million per square centimeter per second.

A similar particle stream beyond 40 earth radii was reported by Lunik III, launched around the entire earth-moon system in an orbit of 25,257 to 291,439 miles October 4. This was the Russian vehicle that photographed the far side of the moon from distances of 37,000 and 43,000 miles and radioed the photos back to receiving stations in the Soviet Union on command when once more in line of sight of the earth.

The next vehicle to encounter this particle stream far out from earth was the Russian Venus probe. The spacecraft was launched piggyback from the 7-ton Sputnik VIII, boosted into low earth orbit February 12, 1961. It would be three more years before the United States would attempt to lift a comparable payload.

Technologically impressive and militarily threatening, the big payloads the Russians were putting into space in 1961 were also scientifically interesting. The Venus probe's electronics failed long before it reached the vicinity of Venus. But while its radio was still working, the big spacecraft's radiation detectors reported a flux of a billion protons per square centimeter a second at a distance of one million miles from earth.

There was no longer much doubt about what was out there. It was a wind blowing off the sun and it filled interplanetary space, at least as far as a point some 40,000 miles from the surface of the earth. At the magnetopause, the solar wind divided and flowed around the magnetic shell of force, as a stream of water flows around a boulder.

Like the trapped particles in the magnetosphere, the solar wind had been imagined long before it was actually detected. In 1919, the British physicist F. A. Lindemann, later to become Lord Cherwell, Britain's prime science adviser during World War II, suggested that streams of ionized gas from the sun caused geomagnetic storms and auroras on earth.[10] In 1951, the astrophysicist Ludwig F. Biermann of the Max Planck Institute in Munich speculated about the effects such a solar particle stream would have on the motion of comet tails.[11] As they approach the sun in their elongated orbits, comets develop long tails, which are formed by gases boiled by solar heat out of the coma, the cloudy mass of the comet's head that surrounds a nucleus of ice and rocks. The tail itself was believed to be quite similar to the vapor trail of a jet aircraft or of a rocket. Because it reflected light, its visibiliy was high and might extend for millions of miles. Biermann and others wanted to account for the fact that the comet tail always pointed away from the sun. As the comet approached the sun, the tail extended behind it. But as the comet began moving away from the sun on the outward-bound leg of its orbit, the tail preceded the comet head. It was apparent that something was pushing against this gaseous material, something offering resistance to the passage of the comet, like air resistance.

Biermann developed a hypothesis that streams of particles move out from the sun at about 300 miles a second, with a density of 100 particles per cubic centimeter, when the sun is quiet. When the sun is active, the motion of comet tails indicated that the solar particle stream moves more rapidly, at about 900 miles a second, and may have a density of 10,000 particles per cubic centimeter. Here was an atmosphere—the atmosphere of the sun.

In 1957, Sydney Chapman, the British physicist who headed the special international committee for the IGY, proposed that the solar atmosphere was an extension of the sun's corona, an ionized gas extending beyond the earth to at least 215 solar radii (93,029,425 miles). At the University of Chicago, Eugene N. Parker supported this idea by suggesting that because the outer surface of the sun was very hot, it would expand at high speed into interplanetary space.[12] Parker called the expanding corona gas the "solar wind." He visualized it as the atmosphere of the sun, blowing outward into interplanetary space and around the inner planets. In the case of the earth, the solar wind flowed around the reef of the magnetosphere at 40,000 miles out. For that reason, it was not detected until instrumented satellites emerged from the cocoon of the magnetosphere into "true" interplanetary space beyond the turbulent region called the magnetopause.

Ionized gas of high energy is called a "plasma." It is regarded by physicists as a fourth state of matter—not a liquid, nor a solid, nor a gas, but a supergas, with electrons torn away from the nuclei (mainly of hydrogen atoms), leaving a neutral stream of separated protons and electrons.

The image of the solar wind intrigued the international scientific community. Like the Van Allen radiation belts and the magnetosphere, its positive identification and definition represented a major advance in understanding the physics of the solar system.

A Russian group headed by the Soviet physicist K. I. Gringauz designed a detector to pick up low-energy solar protons which were believed to be a component of the solar wind. The instrument was flown on the Sputnik VIII Venus probe (launched February 12, 1961). In summarizing the data, E. R. Mustel of the Astronomical Council of the Soviet Union said that "the outflow of gas from the sun was largely sporadic during solar storms and if there is any persistent solar wind it must be extremely slight."[13]

American physicists Bruno Rossi and H. S. Bridge and their associates at the Massachusetts Institute of Technology built a solar plasma detector which was flown on Explorer 10, launched from Cape Canaveral March 25, 1961. The 79-pound satellite entered a long, elliptical orbit ranging from 100 to 145,000 miles. The detector weighing 2.5 pounds was designed to measure the density, direction, and velocity of the plasma. Particles entered the instrument through a 6-inch hole and passed through a series of grids to a plate sensitive only to protons with velocities of 6 to 1000 miles per second. In addition, the satellite carried three sensitive magnetometers to measure the strength and direction of weak magnetic fields in conjunction with the plasma.*

Unfortunately, Explorer 10's lifetime was short. It sent back data for only 58 hours as it sped outward to 42 earth radii (168,000 miles). The Massachusetts Institute of Technology plasma probe picked up the solar wind at 22 earth radii (88,000 miles). But then, at intervals of a few minutes to a few hours, the data showed that the plasma kept disappearing and reappearing. When the plasma was present, the magnetic field appeared to be weak and fluctuating, but when the plasma was absent, the field was strong and steady. This effect indicated that the magnetic field was keeping out the plasma, except where there were gaps or fluctuations in the field allowing some of the ionized gas to reach the satellite. The investigators then realized that the path of Explorer 10 during a portion of the flight followed the flapping boundary of the magnetosphere. Now and then, the vehicle seemed to cross the boundary; or, rather, the boundary of the magnetosphere seemed to envelop and then recede from the flight path of the vehicle. When the little probe was outside the boundary, it sensed the solar wind, and when it was inside, the wind was gone, as though a window had been shut. The combined data from the plasma probe and the magnetometers showed that the magnetospheric boundary was being pushed earthward from time to time by gusts of wind from the sun, blowing at speeds of 500 miles a second.

With these reports from Explorer 10, the geography of the magnetosphere became clearer. On the sunward side of the earth, the magnetopause, at 40,000 miles, appeared to be an elastic boundary, moving inward and outward with pressure from the solar coronal winds. Beyond the magnetopause, the solar wind appeared to blow more steadily than the Russians' observations had indicated.

Explorer 12 also contributed to the measurement of the magnetosphere and the magnetopause boundary. It returned data indicating that the magnetopause was about 60 miles thick and formed a boundary between the Van Allen radiation belt inside the magnetosphere and the solar wind outside.

* They were a rubidium-vapor and two fluxgate magnetometers.

Further, Explorer 12's magnetometer confirmed that the magnetopause was compressed toward the earth when a solar flare accelerated the plasma and caused the solar wind to blow harder against the earth's magnetic frontier.

The next major opportunty American scientists had to look at the solar wind came in 1962 when Mariner 2 roared off Cape Canaveral, bound for the planet Venus. Among its instruments, it carried a plasma probe built under the direction of Marcia Neugebauer and Conway W. Snyder of the Jet Propulsion Laboratory. After Mariner 2 had cleared the magnetosphere, the plasma data came in loud and clear. The JPL investigators received 23,500 solar-wind measurements along the spiral flight path Mariner followed to Venus. The velocity of the solar particles consistently was reported at 228 to 414 miles a second. Mariner 2 expanded on the message from Explorer 10: the solar wind blew continuously throughout the entire volume of interplanetary space. Beyond the cavity formed by the earth and its magnetosphere, Mariner 4 in 1965 reported the solar wind as far out as the orbit of Mars.

During the period of the quiet sun (1963–65), Pioneer 6, launched on December 16, 1965, as an interplanetary probe between the earth and Venus, reported that the wind was blowing relatively mildly at a velocity of 190 miles a second (670,000 miles an hour), compared with a speed of 555 miles a second (2,000,000 miles an hour) when the sun was turbulent.

In addition to hydrogen-nuclei protons, the solar wind contained ions of helium with one or both electrons stripped off. A plasma analyzer aboard Pioneer 6, designed by NASA's Ames Research Center, Moffett Field, California, showed that, instead of moving straight out from the sun, the plasma stream flowed at an angle to a line from the spacecraft to the sun. This finding, when tied to other reports from the satellite on the behavior of solar cosmic rays, became an important clue to the structure of the interplanetary magnetic field.

While most particles emanating from the sun are composed of fragments of hydrogen and helium atoms, they are classified by physicists according to their energy, or velocity. The plasma consists of ions moving about 300 miles per second, with energies in the thousands of electron volts. Solar cosmic rays also are ions, principally of hydrogen, as are most of the plasma particles, but they have higher energy, in the range of millions of electron volts. Apparently their source is different. Instead of being a part of the solar corona, the solar cosmic particles originate in solar flares, titanic eruptions on the face of the sun. The galactic cosmic rays are another story, as we have seen. They appear to originate in the explosion of stars far beyond the sun, but acceleration processes in stellar magnetic fields, like those used in atom smashers, boost their energies to billions of electron volts.

Instruments aboard Pioneer 6 showed that solar cosmic rays came from all directions, and that there were sudden changes in the direction of the interplanetary magnetic field. These indications could be explained by imagining that the magnetic lines of force were spiraling through space with the rotation of the sun and the solar cosmic-ray particles spiraled along with them. The figure that the magnetic force lines would describe as they spread out through interplanetary space from the rotating sun was the classical one called an Archimedes spiral, like the figure made by the water spray from a rotating lawn sprinkler.

If a spiral structure was assumed, it would account for the satellite's observations that solar cosmic rays from a particular flare would appear to be coming from all directions at once by the time they reached the satellite at a distance of 1,680,000 miles from earth. Norman F. Ness of the Goddard Space Flight Center and Kenneth G. McCracken of the Southwest Center for Advanced Studies reported that the detailed structure of the magnetic field was filamentary. Some spiraling filaments had kinks or "regressions" in them, they said.[14]

Later analysis of the data from Pioneer 6 and other satellites showed that the magnetic-force lines were swinging around the solar system with the rotation of the sun, which turns on its axis once every 27 days. Something that moved with the rotation of the sun was dragging or carrying the magnetic field along with it. And that something was the solar wind, which pulled the solar magnetic field out as it expanded into interplanetary space.

There was a relationship between the solar particles and the magnetic field. Physicists at Goddard and other institutions expressed it in terms of the ratio of the particle energy per unit of volume to the magnetic-field energy per unit of volume. To compute it, all one had to do was to add up the kinetic energies of all the particles in a unit of volume, such as a cubic centimenter, and divide the sum by the magnetic-field energy in that volume. If the answer was greater than 1, it meant that the particles had greater energy than the magneic field to which they were electrically attached. Consequently, the field followed the particles. If the answer was less than 1, the particles had less energy and followed the field. The ratio of particle energy to field energy became known as the "Beta" of the particles. Because of their density, the low-energy particles of the solar wind had a Beta greater than 1. Their total energy per unit volume in interplanetary space was more than the magnetic-field energy in that volume. Therefore it was apparent that the plasma of the solar corona carries the magnetic field with it.

Even though they have higher energy than the plasma, solar cosmic rays, being less concentrated, have a Beta of less than 1. Consequently, they are prisoners of the solar magnetic lines of force, and must follow wherever the plasma drags the field. Lines of magnetic force from the sun swing around the solar system with the sun's rotation. To an observer on earth, they execute spirals, which solar cosmic rays faithfully follow since they are tied to the force lines like bits of paper moving along the string of a kite.

This picture of interplanetary space, alive with fields and forces, matter and energy, supplants the primitive one of emptiness, nothingness, or void. It presents man with the idea that he is after all a solarian; he lives within a cavity of the corona or "heliosphere" of the sun, a cavity formed by the magnetosphere of his planet. With the advent of the satellites, the wind between the worlds has been translated from a hypothesis to a reality.

Is it really a wind? Colin O. Hines, professor of aeronomy at the University of Chicago, compared it to a terrestrial wind blowing at one-fifth of a mile an hour.[15] By terrestrial standards, that is not much of a breeze. However, the winds that blow from the sun are not always so mild. When the sun is turbulent, it whips up a hurricane in space, a hurricane of plasma containing protons with energies of 300 to 600 million electron volts.

THE SOLAR BROOM

The origin of solar "hurricanes" can be seen in regions of turbulence on the sun, but the mechanism which accelerates the protons to high energies and the origin of the turbulence remain matters of speculation. Of primary interest to physicists are the effects of the plasma cloud and streams of energetic particles emitted during solar flares on the structure of space and the environment of man. The plasma hurricanes are associated with a brightening of the auroras, magnetic storms, and the blackout of long-distance-radio communications. Obviously, there is an impact of the solar-particle stream on the magnetosphere and the atmosphere; yet its exact nature eluded researchers. While a link between a solar storm and terrestrial weather seemed difficult to establish, there was some observational data in polar upper-atmosphere studies suggesting an energy transfer between the solar cloud and the high atmosphere. How that worked was not at all clear.

One of the earliest solar-hurricane effects observed was on cosmic rays. More than thirty years ago, Scott E. Forbush, a physicist in the Department of Terrestrial Magnetism of the Carnegie Institution of Washington, proved that galactic cosmic rays diminished near the earth a few days after flares were observed on the sun. Magnetic storms then appeared on the earth. It was evident that a single cause linked these effects, but the nature of it eluded definition, and the phenomenon was simply called the "Forbush effect." It was as though the solar flare loosed a physical force that acted as a broom, sweeping space clear of galactic cosmic-ray particles for a time.

Whether the broom was being wielded by the sun or in some mechanism near the earth became a matter of considerable interest. Both possibilities were debated at a symposium on space exploration in April 1959 in Washington, D.C. An experiment was proposed to attempt to find the answer, calling for a satellite equipped with a cosmic-ray detector to be orbiting far out in space when a solar flare occurred. If there was a Forbush decrease at some distance from the earth, the satellite would report it. This would be proof that the Forbush effect originated on the sun, not in the vicinity of the earth.

The space probe selected for the test was Pioneer 5—launched March 11, 1960, from Cape Canaveral—whose primary mission was to reach Venus. The Thor-Able rocket boosting the 95-pound probe failed to develop enough thrust to put the shot as far as Venus, and Pioneer 5 went into a solar orbit between the earth and Venus. There it began to radio back some interesting information. Pioneer 5 carried two instruments for settling the question of the origin of the solar broom. One was a magnetometer built by C. P. Sonnett and his associates at Space Technology Laboratories. The other was a cosmic-ray telescope built by the University of Chicago's Laboratory for Applied Science under the direction of John A. Simpson.

On March 30, 1960, when Pioneer 5 was 3,000,000 miles from earth a solar flare was observed at 9:55 a.m. Eastern Standard Time. A magnetic storm began on earth 21 hours later, attended by the usual effects of fading radio communications and heightened auroras. The Forbush decrease then appeared, as the vicinity of the earth was swept clean of cosmic rays. This event was reported on the ground by a neutron monitor at Deep River, Canada. At the same time, 3,000,000 miles away from earth, a Forbush decrease was reported by Pioneer 5.

The solar broom *was* wielded by the sun, and it was related to the flare and to terrestrial effects as well. At the time that Pioneer's cosmic-ray detector showed that the cosmic rays had vanished, the spacecraft's magnetometer reported a tenfold increase in the strength of the interplanetary magnetic field!

All these reports depicted a gigantic solar broom of plasma and magnetic-force lines sweeping out the cosmic rays from Sol's front yard. The rise in magnetic-field strength showed that the plasma cloud launched by the sun was pulling the solar magnetic field out with it, like taffy. As the cloud and its magnetic strands raced earthward, it filled the entire volume of space between the earth and the sun. Presently, it enveloped the earth itself. It was as though a great net of force had been cast out by the sun, over Mercury, Venus, the earth, and the moon. The earth was now enclosed in this magnetic "bottle." And the bottle shut the cosmic rays out.

That was the way the Forbush decrease worked. It was as though a tarpaulin had been thrown over the planet to keep out the cosmic rain.

Two days later, at 3:45 a.m. on April 1, a second big flare was seen on the sun. It lasted nearly four hours. Unlike the first, the second flare emitted a burst of fast protons in the cosmic-ray range of 100 million electron volts. The fast protons of flare No. 2 sped earthward inside the magnetic bottle created by flare No. 1. They were detected by Simpson's cosmic-ray counter on Pioneer 5, which, though 3,000,000 miles from earth, was well within the bottle. The solar protons were also inside the bottle. The higher-energy galactic cosmic rays from other stars were outside the bottle.

University of Alaska detectors reported that the protons from flare No. 2 arrived one hour after the flare had been observed. It had taken 8 minutes for the light of the flare to reach earth. Consequently, the protons appeared to be traveling about one-eighth the speed of light. Their energies, however, suggested that they were actually moving a great deal faster than that, at speeds that would have brought them to the earth's magnetosphere in 18 minutes. It was apparent that the fast protons, with a Beta of less than 1, were following the spirals and kinks in the solar magnetic-field lines.*

Correlated with ground monitors, the Pioneer 5 data showed that when earth is inside the magnetic bottle it is linked directly with the site of the flare on the sun by the solar magnetic field.

When two flares come close together, fast-moving protons from the second one may reach the earth before the slower-moving plasma cloud from the first flare has had time to drag out the magnetic field to encircle the planet. This situation was seen during a series of solar flares from November 10 to 15. The first flare appeared at 10:15 a.m. Eastern Standard Time November 10, ejecting a slow-moving cloud of plasma at a velocity of 500 miles a second. At 8:25 a.m. November 12, the second flare began and expelled a shower of fast protons. Just 25 minutes later, the Deep River neutron monitor reported the arrival of the protons. They reached earth 5 hours and 40 minutes ahead of the plasma cloud from flare No. 1.

* Theorized by Robert Jastrow and others at the Goddard Space Flight Center.

There were no taffy-like solar magnetic fields to detour them during the morning of November 12 because the bottle had not yet been formed. At 2:30 p.m., the slow-moving plasma from the November 10 flare arrived at earth, dragging solar magnetic fields with it.

The earth was then enclosed in a magnetic bottle. A second burst of protons was immediately slowed as the particles were captured by the force lines of the bottle. Cosmic rays vanished. Northern auroras lit up like neon signs. A radio blackout descended in polar regions. It cut off Antartica entirely, just as the men of Operation Deep Freeze were awaiting the final tallies of the 1960 presidential election.

THE TAIL THAT WAGS

While these findings broadly depicted an interaction between the sun and earth, the manner in which energy was transferred from the solar winds to the atmosphere to cause auroras and affect the ionosphere remained uncertain. Van Allen's earlier theory of particle-dumping from the belts seemed to be refuted by O'Brien's data from Injun 1, which indicated there were not enough particles in the belts to produce auroras. No other explanation commanded general support. However, there was a major clue.

Colin Hines and others hypothesized that the solar wind would behave like terrestrial wind in a wind tunnel when it flows past an obstacle at supersonic speed. A shock wave is set up and it slows the supersonic flow into a turbulent, slower flow around the obstacle. After the obstacle is passed, the supersonic flow resumes. In the case of the solar wind, the analogous obstacle would be the magneto-pause. The shock wave and turbulence propagated through the magnetosphere and transferred energy to the atmosphere in the form of heat, ionization, and auroral illumination. This model of energy transfer needed proof, however, that there was a shock wave and turbulence as predicted.

That proof was provided by Explorer 18, the Interplanetary Monitoring Platform, launched November 27, 1963, from Cape Canaveral. Its magnetometers and radiation detectors confirmed the existence of a shock wave from the collision of the solar wind with the magnetic-field boundary. Turbulence in the region was confirmed also. The shock wave and a turbulent transition zone trailed along the boundary of the magnetopause for a quarter of a million miles. It might be thought of as resembling the wake of a ship at sea, or the diagram of a shock wave in the atmosphere created by a supersonic jet aircraft. Explorer 18, better known as IMP 1, also observed that the force lines of the interplanetary magnetic field mimic the field of the sun. Changes in the direction of the main solar field were followed four days later by similar changes in the interplanetary field where IMP 1 was in orbit. This information implied a direct connection of magnetic-field lines across one astronomical unit, the distance of the earth from the sun. Magnetic-field strands stretched between the sun and its inner planets like strings, and this linkage suggested a mechanism by which events on the sun in-

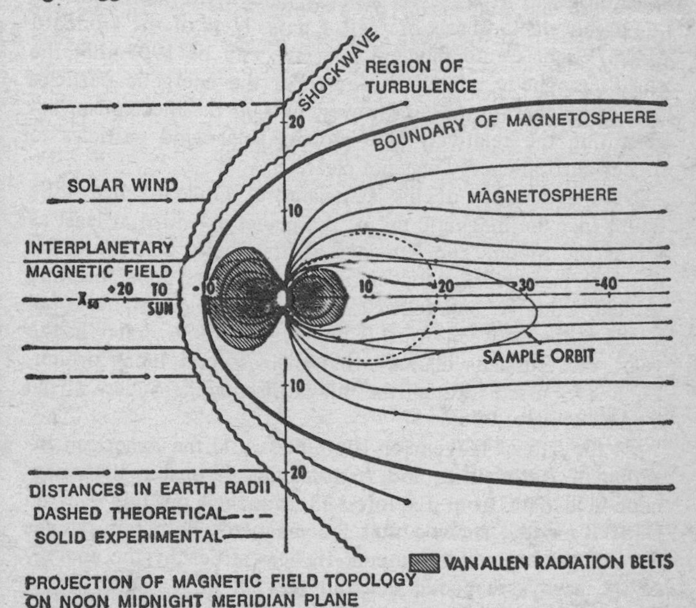

Sample orbit depicts the highly elliptical path of the IMP satellite, Explorer 18, with an apogee of 120,000 miles from earth.

fluenced the magnetic environment of the earth. The activity of the solar magnetic field was observed by a magnetograph at the Mt. Wilson Observatory while IMP 1 was sounding the field in space.

What became of the solar wind once it passed the magnetopause? The picture of the plasma, dividing at the magnetopause and flowing around it like a stream around a boulder, was incomplete. How far beyond the earth did the great magnetic cavity in the solar wind close?

In 1961, data from Explorer 10, obtained by J. H. Piddington of the Goddard Space Flight Center, confirmed his theory that the earth's magnetic-field lines were blown out on its night side by the solar wind. Explorer 14, launched October 2, 1962, reported that the magnetic field extended in the direction away from the sun to a distance of 64,000 miles. Finally, IMP 1 showed that the magnetic field tailed out behind the earth to at least a distance halfway to the moon.

Largely on the basis of IMP 1 data, Ness of the Goddard Space Flight Center suggested at the end of 1965 that the earth's magnetic tail is the source of the energetic particles that cause the aurora.[16] However, he said the mechanism accelerating the relatively slow-moving solar-wind particles to higher auroral energy was not clear.

Ness and some of his associates at Goddard had little doubt that the magnetic tail of the planet extended at least as far as the moon. The Russians at first threw cold water on this idea by reporting that data from their first lunar orbiting vehicle, Luna X, had failed to detect the magnetospheric tail at the moon between April 3 and May 4, 1966. Later in the year, the Russians claimed that their second lunar orbiter, Luna XI, passed through the tail of the magnetosphere at the moon's orbit.

At the end of November 1966, Ness told the American Institute of Aeronautics and Astronautics in Boston that magnetic-field data from Explorer 33, launched on July 1 from Cape Kennedy, showed that the magnetosphere tail extends beyond the orbit of the moon. He was struck by the analogy of the earth's magnetic tail and the tail of a comet. There were general structural similarities. Both were products of the interaction of the solar wind with the parent bodies. The earth's magnetosphere and radiation belts could be compared with the coma, the luminous cloud of gas and solid particles around the comet nucleus, and the earth itself could be compared to the nucleus.[17]

How far does the tail extend? Explorer 33 found evidence of the tail 316,000 miles from earth. As the solar wind blew, there was a curious effect: the long, magnetic tail of the earth appeared to wag.

Chapter 4

Project Mercury

In the third week of January 1962, the largest number of newspaper, radio, and television correspondents ever to gather for a single event assembled at Cocoa Beach, Florida, to observe America's first attempt to launch a man into orbit around the earth. A single-seater Mercury spacecraft, shaped like a 9-foot bell, had been mated to an Atlas launch vehicle at Complex 14 on Cape Canaveral. The combined rocket and spacecraft towered 95 feet, 4 inches above the launch platform.

Initially, the launching had been planned for the fourth quarter of 1961, but technical problems in spacecraft and missile delayed any attempt until after the new year. The National Aeronautics and Space Administration then announced it would launch the orbital flight "no earlier" than 6:30 a.m. Tuesday, January 23, 1962, with Lieutenant Colonel John Herschel Glenn, Jr., United States Marine Corps, as pilot. The phrase "no earlier" betrayed the atmosphere of uncertainty which Project Mercury had inherited from the earliest days of rocket testing at the Cape, and which characterized the early attempts by the United States to rocket a man into the interplanetary medium.

The flight which NASA referred to as Mercury-Atlas 6, or simply MA-6, attracted 1150 correspondents from the major news media in the United States, Canada, Western Europe, Africa, Australia, and Japan. At that time, the principal civilian port of entry to the Cape was the municipal airport of Orlando in central Florida. From there, it was a 55-mile drive on Highways 50 and 520 across the eastern third of the peninsula to the Atlantic Ocean beaches.

Beyond the suburban chaos outside Orlando, Highway 50 narrowed from four to two lanes. It passed through scrublands bearing the scars of subdivisions which had never matured beyond the clearing stage. Convict gangs in nineteenth-century striped trousers and jackets mowed the roadsides, under the bored supervision of straw-hatted guards, cradling shotguns. The route to the Cape veered to the southeast on 520, which passed through a wilderness of piny scrub and cypress trees, bearded with Spanish moss. The highway became the main street of Cocoa, a town of 12,000 on the main line of the Florida East Coast Railroad; it crossed the wide, shallow Indian River; traversed Merritt Island, where clumps of redwood bungalows for missilemen were growing, and spanned a mile-wide tidal estuary called the Banana River. The road ended at the ocean and Cocoa Beach, a community of motels, trailer courts, and bungalow subdivisions south of the Cape, displaying the composite disarray and luminescence of a Florida resort and the settled atmosphere of an industrial bedroom community.

Cocoa Beach was a narrow sand spit between the Cape and Patrick Air Force Base, headquarters of the Atlantic Missile Range. In the decade of rocket testing on the Cape, it had evolved into a community with a split personality. It had begun as a resort into which residential subdivisions had taken root and sprouted. These offered bungalows with two bedrooms or four, one bath or two, a carport or two-car garage, plasterboard walls, asphalt tile floors, all-electric kitchens, and 30-year mortgages. The side streets, mostly unpaved, were alive with tots and tricycles. Every house displayed the flag on national holidays and a wreath on Christmas.

For the most part, the dwelling enclaves housed young aerospace engineers and company executives and their young families. All was new: the houses, the supermarkets, the churches, the schools, the library, the police station, and the post office. The resident population moved about on wheels.

There were few sidewalks except in the area of four square blocks called "downtown." Pedestrians were mainly service employees: chambermaids, waiters, bellhops, and handymen at the motels. A bus brought most of them from Cocoa on the mainland every morning and took them back at night. Except for two taxicabs, it was the only public transportation plying between Cocoa and Cocoa Beach. Even on the beaches, bathers drove their cars and parked, now and then getting stuck in the soft sand. Few walked by the sea, except lovers, visiting correspondents, and meditative German émigrés on missions from Huntsville, who found the seashore scene, with rocket gantries distantly visible on the Cape, reminiscent of Peenemünde.

FRONTIERSVILLE

The resort personality of Cocoa Beach was expressed mainly on the ocean side of Highway A1A, which bisected the community and served as its main street. To distinguish itself from the subtropical suburbia, the resort complex referred to its array of attractions as the "Platinum Coast." It provided swimming pools, bars, and night-club entertainment to itinerant aerospace company executives, engineers, and press agents. The influx of news-media correspondents, photographers, and camera crews for MA-6 severely strained its resources.

After dark, the Platinum Coast fluoresced with carnival. There were Caribbean limbo dancers. There were jugglers, hot combos, and an Afro-Tahitian belly-dancer. In the bar across the hall from NASA press headquarters in the Starlite Motel, a Texas Guinan-style chanteuse sang bawdy ballads. The effect suggested a nineteenth-century Western town transposed in time and space to the New Frontier. Instead of the cavalry post down the road, there was the Air Force Base. And instead of mounted cowhands thundering into town, there were newsmen roaring up and down A1A in their rental cars.

In contrast to the atmosphere of carnival on the Platinum Coast, the milieu in which the astronaut prepared for his ordeal in orbit was Spartan. Glenn remained in monastic seclusion on the Cape during the weeks of preparation for the flight. The Cape had been sealed off from the beaches to the south, since its early development as a missile-test center, by fences, gates, and guards. A security system had been installed by Pan American Airways, which undertook housekeeping at the Test Center under Air Force contract.

Glenn and his backup pilot, Malcolm Scott Carpenter, 36, a lieutenant commander in the Navy, were as remote from the press as though on the moon. Twice a day, a communiqué on flight progress and the weather outlook, a critical factor, was issued by the NASA press officer, Lieutenant Colonel John A. (Shorty) Powers, the "voice" of Project Mercury. Occasionally, NASA released a comment from Glenn himself. The official verbiage made him sound like a YMCA secretary reassuring a mother that the pool was clean.

By Saturday night, January 20, 1962, 1100 news-media representatives had arrived at Cocoa Beach and registered at press headquarters. They received badges that would admit them on launch day to the press site on the Cape, to which they would be escorted in bus loads by Air Force officers. In the meantime, the launching of MA-6, which had been announced for no earlier than January 23, had been postponed until no earlier than January 24. The launch vehicle and spacecraft crews needed more time to prepare.

T MINUS EIGHT YEARS: FRICTION *

The fact was, however, that preparations in America for manned space flight had been going on for a long time. The origins of Project Mercury can be found in a seminar held by the National Advisory Committee for Aeronautics (NACA) in the spring of 1944, at which a jet-propelled aircraft that could travel faster than the speed of sound was proposed, as a response to the first German jets that were appearing over Britain. In 1945, Congress appropriated $500,000 for preliminary studies of a rocket airplane, and in October 1947, the first rocket airplane, the XS-1, broke the sound barrier by

* Since the launch of Explorer 2, the term "X" in the countdown had been supplanted by "T" for "time of launch."

flying 700 miles an hour, or Mach 1.06, at 43,000 feet altitude over Muroc Dry Lake, California.†

Studies by the RAND Corporation and the Navy on orbital vehicles directed the attention of both civilian and military aeronautics experts to the prospect of manned satellites. At the outset, these studies were clearly tied to the development of rockets, rather than to aircraft, but the possibility of an airplane that could take off from a conventional airport, fly in orbit, and land in the conventional way was considered.

Of first importance was the need to know whether a human being or any other highly developed animal could withstand flight at extreme altitudes, involving (1) massive acceleration, (2) a period of weightlessness in free fall, and (3) massive deceleration on returning to the ground. In June 1948, the Army launched a small monkey named Albert on a ballistic flight in a capsule aboard a V-2 at White Sands. Albert suffocated when his oxygen supply failed. A second monkey was lofted to a high altitude the next year but was killed on impact when the capsule landed. Two more monkeys were rocketed aloft. Telemetry signals giving their respiration and heartbeat indicated they remained alive during the flight and the descent but they suffocated before the landing capsule could be found. The first successful recovery of experimental animals rocketed to high altitudes in this country was achieved in January 1951, when one monkey and 11 mice survived an Aerobee flight to 236,000 feet.

Early in 1952, NACA proposed that a manned-vehicle flight to very high altitudes should be attempted. A working group was set up in the agency to analyze what information had been developed about such a project. H. Julian Allen, an engineer at NACA's Ames Aeronautical Laboratory, Moffett Field, California, proposed a blunt-nosed shape for a vehicle reentering the heavier atmosphere. This shape would encounter drag, or air resistance, which would help slow it down for a landing.

A manned satellite poses a problem which an unmanned one usually does not: recovery. One way of recovering a satellite from orbit is to give it braking rockets as powerful as those which launched it, so that the descent could be pow-

† Mach 1, named for the Austrian physicist Ernst Mach, is the speed of sound in the atmosphere and varies with altitude.

ered, just as the ascent was, like a movie running backward. However, the weight of the fuel alone would be prohibitive.

Consequently, theorists had to consider making use of the earth's atmosphere as a brake for a returning satellite. Once the satellite had been broken out of its orbit by small retro-rockets, atmospheric resistance would do the rest. It would dissipate most of the kinetic energy imparted to the satellite by the rocket that boosted it off the ground. But if atmospheric friction was the brake, the kinetic energy of the satellite would be transformed into heat energy. It is estimated that a satellite moving in orbit at 5 miles a second has stored energy equal to 4000 kilowatt hours of heat for every pound of mass.[1] Consequently, the kinetic energy of a satellite's motion is enough to vaporize any substance known, including a diamond, when transformed into heat. In order to bring a satellite down without incinerating it in the atmosphere like a meteorite, it was necessary to devise a means of getting rid of the heat. Studies of meteorites which had managed to plunge through the atmosphere and strike the ground without burning up showed how this could be done. Part of the surface had been ablated, or burned away, carrying the heat with it and protecting the inner portion. If a satellite could be shielded with a material that would boil off when heated, and thus carry the heat away, the vehicle might reenter the atmosphere without being destroyed.

The principle of ablation was tested on the Jupiter C nose cone which proved the ability of a missile warhead to reenter the atmosphere in 1957. A phenolic resin, stiffened with fiber glass, was found to be a good, ablative material. As it boiled off, it cooled the boundary layer of air next to the vehicle. Once the problem of a heat shield was solved, through the development of missile nose cones, it became reasonable for engineers to take manned satellites seriously.

T MINUS SIX YEARS: THE SPACECRAFT IS BORN

In 1956, the Air Force began the development of a manned space vehicle in Project 7969. A spacecraft considerably heavier than the 3000-pound Mercury was designed. It required an Atlas ICBM and a second stage, such as the "Hustler" (Agena) which was then being developed. In flight, the craft would travel nose forward, but the pilot seat would swivel so that the passenger could look about in all directions. As a demonstration of eagerness to get in on the ground floor, 11 aerospace firms submitted unsolicited proposals to build the vehicle.

Meanwhile, the Navy had begun experimenting with a human being's ability to withstand *g* forces, or multiples of normal earth gravity. It set up an Aviation Medical Acceleration Laboratory at Johnsville, Pennsylvania, with a centrifuge large enough to simulate 8 *g*s. Not to be outdone in this new field, the Air Force School of Aviation Medicine at Randolph Field, Texas, set up a space-cabin simulator.

On October 14, 1957, ten days after the advent of Sputnik I, the American Rocket Society presented a program for space exploration, including manned space flight, to President Eisenhower. The society had grown from prewar obscurity to an influential association of aerospace engineers. It advised Eisenhower that if America was to get into space, an astronautical research and development agency similar to NACA or the Atomic Energy Commission must be created, independent of the military services. The society forecast that a space budget might even run as high as $100,000,000 a year.

Later that fall, Maxime A. Faget, NACA's talented design engineer at the Langley Aeronautical Laboratory, proposed that existing ballistic missiles could be used as launch vehicles to place manned satellites in orbit. Moreover, small solid-fueled retrorockets would be feasible to reduce the satellite's orbital velocity and initiate its reentry and landing.

From a welter of industry and government proposals, a manned space-vehicle design began to take coherent form in 1958. Faget presented a detailed design of a manned orbital vehicle at the end of January. It was a ballistic capsule with a heat shield that would reenter the atmosphere by reverse thrust. Faget was convinced that the deceleration load on the pilot would not exceed 8 *g*s and that the vehicle could be landed by parachutes. At its Johnsville centrifuge, the Navy had found that men could withstand 8 *g*s quite well and perform movements required to operate an aircraft.

The Martin Company of Baltimore proposed its Titan ICBM as a manned satellite launcher. The McDonnell Aircraft Company of St. Louis presented a spacecraft akin to Faget's concept, to be launched by an Atlas with a Polaris second stage. Lockheed Aircraft proposed a conical vehicle with the pilot facing the rear, to be launched by an Atlas Agena. Convair proposed a manned space station. Republic Aviation came up with a two-ton, triangular-winged craft that would make a gliding reentry at 3600 miles an hour. A drawback of this scheme was that it required a more powerful rocket than either Atlas or Titan, and America did not have one. AVCO Corporation proposed a 1500-pound sphere, which would be lofted into orbit by a Titan and stabilized by a stainless-steel parachute. The chute's diameter would be regulated by a bellows and presumably was to operate at altitudes where there was some atmosphere, since it was to be a means of controlling the altitude or position of the vehicle. When the pilot wanted to descend, he would open the parachute wide. It would not only brake his vehicle but guide it through reentry. Bell Aircraft described a spherical satellite covered with an ablative material like a missile nose cone. It would be launched by an Atlas or a Titan using a Vanguard upper stage.

Out of this array of ideas, the dominant design theme was Faget's conception of a blunt, wingless capsule, which would execute a heat-shielded reentry into the atmosphere at near-orbital velocity.

In the meantime, a civilian space program free of the paralyzing rivalry of the armed services began to take shape. The Senate created a special committee on space and astronautics, and the committee's first task was to draft legislation for a national program of space exploration. Then, in the spring of 1958, NACA published a report by Faget and others called "Preliminary Studies of Manned Satellites—Wingless Configuration, Non-Lifting." It was the blueprint for the design of the Mercury spacecraft.

T MINUS FOUR YEARS: NASA IS BORN

During the summer of 1958, the Army proposed Project Adam, in which a Redstone would propel a man on a suborbital ballistic flight. The principal exponent of the idea was Wernher von Braun. He believed it would have psychological and political value at a time when there seemed to be no immediate prospect of catching up with the Russian vault into space. The plan gained approval by the Department of Defense but was rejected by the White House on the advice of a number of scientists. The most articulate was the late Hugh L. Dryden, director of NACA, who likened the proposal to a circus stunt of shooting a young lady out of a cannon.

Congress passed the National Aeronautics and Space Act, which President Eisenhower signed July 29, 1958. The civilian space agency came rapidly into being around the nucleus of NACA and its aeronautical laboratories at Langley, Virginia; Cleveland; and Moffett Field, California. Eisenhower named as administrator T. Keith Glennan, president of the Case Institute of Technology, Cleveland. Dryden came into the new agency as deputy administrator. Robert R. Gilruth, director of the Langley Laboratory, was appointed chief of a Space Task Group, with the specific mission of developing the manned satellite program. Abe Silverstein, director of the Lewis Research Center, NACA's propulsion laboratory at Cleveland, became head of space-vehicle development.

Gilruth's Space Task Group lost no time in drafting specifications for the first manned satellite. Under the direction of Faget and Charles W. Mathews, a young aeronautical engineer, NASA then called for industry proposals and received twelve. The Mercury development contract was awarded to McDonnell Aircraft Corporation at St. Louis, whose engineers had developed independently a spacecraft design similar to Faget's.

One of the most critical problems in the design of the spacecraft was the type of breathing atmosphere it would have. The atmosphere would be contained within the walls of the pressure vessel constituting the cabin. In case of a leak, the astronaut would be protected by his flight suit. It would provide him with a breathable atmosphere even if the cabin was fully decompressed.

During 1959, the composition of the cabin atmosphere was debated by physicians, physicists, and engineers. Should it be a mixture of oxygen and nitrogen at sea-level pressure (14.7 pounds a square inch) or pure oxygen, on which military aviation had learned to depend for life support at high altitudes?

The NASA Space Task Group and McDonnell Aircraft engineers agreed that pure oxygen at a pressure of 5 pounds per square inch would be preferable to a two-gas system of oxygen and nitrogen at sea-level pressure. It would save weight and make leakage easier to detect and control. Moreover, it would avoid decompression problems of a sea-level atmosphere in the event the astronaut had to leave the spacecraft while it was in orbit to perform an extravehicular activity. Faget explained the NASA decision: "The most important consideration in choice of a single-gas atmosphere is reliability of operation. If a mixed-gas atmosphere were used, a major increase in complexity in the atmospheric control system and in monitoring and display instrumentation would have resulted." [2] Pure oxygen at 5 psi was adopted as the breathing gas.

At the beginning of 1959, Glennan predicted that Project Mercury would cost $200,000,000.[3] A call for astronauts was issued to the military services. Physical and mental requirements were defined by a special committee on life sciences, headed by Dr. Randolph Lovelace II, who operated a clinic at Albuquerque, New Mexico. After initial interviews in Washington, prospective candidates who met general qualifications were given medical examinations at the Lovelace clinic and psychiatric evaluations by Drs. George E. Ruff and Edwin Z. Levy, Air Force psychiatrists. To qualify, a candidate had to be less than 40 years old, less than 5 feet 11 inches tall, and in excellent physical condition, have a bachelor's degree or its equivalent, be a graduate of a test-pilot school, and be a qualified jet pilot with at least 1500 hours of flight time. In addition to John Glenn and Scott Carpenter, applicants selected for the original team of seven astronauts were Leroy Gordon Cooper, Jr., 32, Air Force captain; Virgil I. (Gus) Grissom, 33, Air Force captain; Walter M. Schirra, Jr., 36, Navy lieutenant commander; Alan B. Shepard, Jr., 35, Navy lieutenant commander; and Donald K. Slayton, 35, Air Force captain.

In 1960, the German émigré rocket team headed by von Braun was transferred from the Army to NASA, along with 4000 acres of Alabama red mud outside Huntsville. This acreage "seceded" from the Redstone Arsenal to become the George C. Marshall Space Flight Center. At the time of the transfer, the von Braun team was well into the development of the Saturn family of deep-space cruisers—the first large nonmilitary rockets to be developed in this country.

Saturn was no longer a missile. It was designed for missions to the moon and the planets. It was a true space engine.

T MINUS TWO YEARS: MERCURY FLIES

The Mercury spacecraft took shape in the McDonnell plant off By-Pass U.S. 66 at the edge of the St. Louis Municipal Airport. It looked like a bell, 9 feet, 7 inches high from the base of the heat shield to the top of its conical recovery cylinder, and 6 feet, 2 inches wide at the heat shield. The spacecraft consisted of three sections: the cabin, a truncated cone; a cylinder up forward, about the size of a 50-gallon barrel, containing parachutes; and a smaller cylinder containing the radio antenna and horizon-sensing elements of the automatic-control system. The smaller cylinder also contained a drogue parachute, designed to come out before the main chute to stabilize the descending vehicle in the high atmosphere. At the opposite end, on the base of the saucer-shaped heat shield, were six solid-propellant rockets. Three of them (posigrade rockets) would fire with a thrust of 400 pounds each for one second to separate the spacecraft from its launch vehicle as orbital speed was reached. The other three (retrograde rockets) were designed to brake the spacecraft out of orbit, to begin its descent. Each had a thrust of 1000 pounds and burned for 10 seconds. Before firing the retrorockets, the astronaut would turn the spacecraft so that the heat shield and retrorockets pointed forward. The retrorockets would slow the vehicle by 322 feet a second, or from about 25,730 to 25,408 feet per second. With this loss of kinetic energy, the capsule would begin to descend at the rate of one-half foot a second until it encountered the denser (or sensible) atmosphere, where friction would speed up the descent.

On the forward end, a steeple of steel tubing held an emergency rocket, designed to pull the capsule away from the launch vehicle in the event of a catastrophic failure, such as the explosion of the booster on the pad or the failure of its engine system during powered flight up through the atmosphere. The solid-fuel escape rocket developed 52,000 pounds of thrust for one second. A smaller "jettison" rocket, was provided to cast off the escape tower after the Mercury spacecraft was well on its way.

While it could not maneuver in the atmosphere, the Mercury spacecraft was highly maneuverable in orbit. The pilot could yaw it front to rear, pitch it up or down, and roll it. These maneuvers were provided by 18 thrusters using hydrogen peroxide gas. The thrust in roll, pitch, and yaw could be throttled from 1 to 24 pounds. This array constituted the craft's attitude-control system.

The Mercury attitude-control system was, perhaps, the most ingenious ever devised for a flying vehicle up to that time. It consisted basically of two independent fuel, plumbing, and thruster systems and each one provided the astronauts with two methods of controlling the position of the spacecraft in flight relative to the ground.

In system A, the pilot could allow the automatic stabilization and control system (ASCS) to fly the spacecraft for him through an orbit or an entire mission. Or he could take control of the stick himself and fly the spacecraft as he would an airplane. This control mode was called "fly-by-wire." In it, the pilot's movements of the stick caused thrusters to fire by operating their solenoid control valves electrically. In the fly-by-wire mode, the pilot substituted his hand and brain for the electromechanical sensors of the automatic system.

In system B, the astronaut could choose two other control modes. By using one called the manual proportional system, he could regulate the flow of fuel and thus the amount of thrust to the thrusters he wanted to use. A second mode, the rate-stabilization control system, engaged the computing element of the autopilot to monitor or modify the pilot's stick positions. The two systems and four modes of controls were designed to meet a wide range of imaginable contingencies. Basically, if it turned out that the pilot could handle the ship in orbital flight, he could fly it himself. If it turned out he could not handle it, the ship could fly itself and be brought down by ground control.

To the uninitiated, the Mercury spacecraft was singularly unimpressive. It looked like a conical coke oven, and was about the same color. Behind its cylindrical recovery section, where the parachutes were stored, its bell-bottomed afterbody was covered with dull-black alloy shingles, made of an alloy called René Metal, which protected the cabin from expansion and heating. During the exit flight up through the atmosphere, the shingles were exposed to heating of 1300 degrees Fahrenheit—more heat than on reentry, when they were protected by the heat shield. During flight in orbit, variations in temperature on the vehicle ranged from 200 degrees Fahrenheit on the day side to 50 degrees below zero on the night side.[4] The exterior shingles were attached to the cabin so that they could expand and contract with drastic temperature changes without affecting the cabin structure. In what was commonly thought of as the "cold" of space, the real problem was not the cold but the heat—heat generated by the batteries, by the sun on the metal, and by the astronaut himself.

Another problem worried the NASA Space Task Group: pilot safety. No one knew how hazardous space flight might be, but it was generally assumed that it was the riskiest of all mankind's modes of transportation. In February 1959, engineers from the Space Task Group, the Air Force Ballistic Missile Division, Space Technology Laboratories, and Convair-Astronautics devised a fail-safe apparatus to prevent a malfunctioning rocket from destroying spacecraft and astronaut. The apparatus was called ASIS, for Abort Sensing Implementation System. As soon as the rocket engines began to thunder on the pad, ASIS monitored all of the launch vehicle's vital systems. And failures would instantly trigger ASIS to action. It would immediately shut down the booster engines, release the spacecraft from the launch vehicle, and ignite the spacecraft's escape rocket system—all in less than one second.

In the spring of 1959, the first of the animal-passenger experiments was launched aboard a Jupiter C to test the effects of a long ballistic flight on highly evolved species. Two rhesus monkeys were to be the passengers. They had been trained to make a series of responses to stimuli during the suborbital ride as an indication of how much voluntary muscular control a man might have on such a flight. However, the public-information apparatus of NASA made one of its rare mistakes of telling too much. It described the monkeys as "Indian-born" in a press release. The Indian embassy in Washington quickly expressed its concern, advising the State Department that to millions of Brahmans the Indian monkey is a sacred animal and its use in space tests would be offensive to the sect. A White House aide telephoned the experimenters who were preparing to launch the little beasts at Cape Canaveral and advised them to find other animals. American-born rhesus and spider monkeys were substituted at the last minute. There was no time to train them to do anything. The animals, Able and Baker, were launched 1700 miles down the Atlantic Missile Range by a Jupiter C. They did nothing, but they survived beautifully.

Then, in September 1959, two years of flight testing began on the Mercury spacecraft. The shell of one was boosted on a suborbital flight by an Atlas to test the heat shield against temperatures of reentering the atmosphere. Telemetry signals indicated that the shield had withstood heat of more than 3000 degrees Fahrenheit. The air-frame and motor system were tested on a ballistic flight from Wallops Island, Virginia, using a squat, powerful test rocket called "Little Joe" as a booster. Another rhesus monkey—American-born—named "Miss Sam" was hurled aloft from Wallops Island in a test of the Mercury escape tower. She survived the test with no ill effects. Early in 1960, NASA's space troops consisted of a squad of six chimpanzees "rated as trained and ready to support Mercury Redstone or Mercury Atlas missions," and seven astronauts.

T MINUS 12 MONTHS: THE HAM ACT

The first attempt to launch a Mercury spacecraft aboard the Redstone rocket on a ballistic flight was made November 21, 1960. The results shook up both military and NASA observers. As the Redstone lifted one inch from its steel launch ring, its engine suddenly cut off. This activated the ASIS and the spacecraft went sailing high in the sky to land on parachutes in the Atlantic. It was an unexpected but visually elegant demonstration of the escape system.

Another attempt on December 19, to launch Mercury on a Redstone, called Mercury-Redstone 1A, did succeed. The spacecraft reached an altitude of 130.68 miles and landed 243.8 miles downrange, where it was recovered by the Navy in 15 minutes. Its speed reached 4909.1 miles per hour, less than one-third of the velocity required to put it into orbit.

Mercury-Redstone 2 was launched down the Atlantic Missile Range from Cape Canaveral January 31, 1961, with a passenger, a 37-pound chimpanzee named "Ham." The fact that he was qualified to "support Mercury launches" made no impression on Ham. Once he was freed of his space harness, he skittered away and would never again go near a Mercury spacecraft, or anything even faintly resembling that dreadful contraption.

The first attempt to boost a Mercury on an Atlas failed July 29, 1960, when the Atlas fell apart just 59 seconds after launch. Mercury-Atlas 2 roared off Pad 14 at Cape Canaveral February 21, 1961. This time, the flight was successful. The spacecraft was boosted 114 miles high and sailed 1431.6 miles downrange on a test of the ability of the alloy shingles to withstand high temperatures of atmospheric friction. This flight proved that the Mercury spacecraft could reenter the atmosphere safely. The next flight, Mercury-Redstone 3, would carry a man, the space agency announced. He would be Lieutenant Commander Alan B. Shepard, Jr., United States Navy.

While NASA was preparing for this event, Major Yuri A. Gagarin of the Soviet Union flew one orbit around the earth on April 12, 1961, in the 5-ton Vostok (East) I. The length of the flight, 108 minutes, told American space experts that the Russians had evolved a manned space vehicle and recovery system similar to Mercury but considerably larger and heavier. Once more, the Russians had taken a giant step. If there had been any previous suborbital flights in the Russian program, no indication of them had been made public on either side of the Iron Curtain. Rumors that several Russian cosmonauts had died in pre-Vostok flights which failed remained only rumors. Braving the barbs and gibes of its critics here and in Europe, NASA went ahead methodically with its suborbital-flight test program.

The third Mercury-Atlas flight test was launched April 25, 1961. A "mechanical astronaut" consisting of a primitive robot simulated the pilot. Shortly after the Atlas flamed upward, observers saw that the rocket failed to roll toward its proper heading. The guidance was not operating. ASIS fired the powerful escape rocket at T plus 40 seconds and the spacecraft was pulled up and away from the Atlas on a pillar of white smoke. The Atlantic Missile Range safety officer then sent a radio signal which detonated an explosive package on the Atlas and destroyed it. Mercury coasted up to 24,000 feet, released its parachutes, and splashed into the surf, 2000 yards north of the launch pad. The spacecraft showed only superficial damage and was shipped to the McDonnell plant at St. Louis for refitting.

May 5, the day of Mercury-Redstone 3, the first American manned flight into space, dawned hot and clear at Cape Canaveral. Shepard was launched on a 15-minute ballistic flight in the spacecraft he named "Freedom 7." The figure 7 represented the seven astronauts. Despite leaks in the thrusters the spacecraft's attitude-control system worked satisfactorily when Shepard used the fly-by-wire mode. Freedom 7 reached a top speed of 5180 miles an hour and a peak altitude of 116.5 miles. It descended on its ringsail parachute 302 miles downrange, where both craft and pilot were picked up promptly by helicopters and lifted to the aircraft carrier *Lake Champlain.*

Captain Gus Grissom of the Air Force made the second manned suborbital flight in Mercury-Redstone 4 on July 21, 1961. He reached an altitude of 118 miles and landed 303 miles downrange.

Like Shepard, Grissom reported no severe discomfort at experiencing 6 *g*s on the way up and 10.2 *g*s after retrofire. He felt the boost of the rocket at liftoff and heard the noise of the rockets. Out his window he could see a deep blue sky, which suddenly changed to black. He said he saw a star—Capella, he believed. He heard a bang when the Mercury spacecraft forward-thrusting (posigrade) rockets boosted it away from the Redstone. On the horizon, Grissom saw bands of light and dark blue ringing the horizon, about 800 miles away, he estimated. He could see Cape Canaveral behind him as he turned the craft around and flew backward—heat shield forward and ready for reentry—toward the apex of his ballistic arc. The Cape seemed much nearer than he thought it should have appeared at an altitude of 100 miles.

Grissom's spacecraft, called "Liberty Bell 7," hit the water with "a pretty good bump." Within a few minutes, he was advised by radio that a helicopter was over him. "I told the helicopter I was ready to come out and to give me a call [on the radio] and I would power down and blow the hatch," he related. "I decided to get a head start. I took off the detonator cap [from a mechanism which caused the hatch to be blown outward by explosive] and pulled the safety pin [on the detonator]. All set. I lay back on my couch [to wait for the helicopter to hook onto and stabilize the bobbing capsule]. Then, pow! I saw blue sky and water coming in. I went right out. It was the biggest shock of the day."

What caused the hatch to blow out before the spacecraft could be secured by the line from the helicopter has never been determined. The powder charge which would cause the shield-shaped hatch to shear away from the bolts that held it in place had to be set off by a plunger, which the astronaut was required to push. At first, it was assumed by observers at the Cape that Grissom had pushed the plunger too soon. The initial NASA transcript reported that he told the Marine helicopter pilot, "Taking my helmet off, power down, blowing the hatch."

Later, on the aircraft carrier *Randolph,* Grissom stated that he had not blown the hatch but had simply prepared to blow it when, to his astonishment, the explosive had gone off . . . pow! A more detailed recording of Grissom's radio transmission to the helicopter pilot was then released by the space agency. Grissom had said, "You tell me to, ah, you are ready for me to blow and I will have to take my helmet off, power down, and then blow the hatch." The pilot had answered, "One Roger and when you blow the hatch the collar [flotation collar around the spacecraft] will already be down there waiting for you." Grissom then responded, "Roger."

NASA explained that the earlier version of Grissom's conversation did not represent the full transmission. Thus, what at first had seemed to be a statement by Grissom that he was blowing the hatch became merely a statement of his intention to explode it at the proper time. The first recording of the transmission suggested that he had committed an error. The later, more detailed recording indicated that the hatch detonation mechanism was faulty. The problem nearly cost Grissom's life. And that, as Gus observed later, would have spoiled the whole day.

After scrambling out of Liberty Bell 7, Grissom found himself floating high in the soft swells of the Atlantic. He felt exhilarated and swam four or five feet to the wallowing capsule to see if he could help the Navy swimmers hook a line on it from the helicopter. When he found he was too encumbered by his flight suit to do much, he paddled away. Then he discovered that his flight suit was shipping water through a vent he had forgotten to close. He began to sink as a second helicopter lowered a "horsecollar" sling to him. He tried to swim under the hovering craft to seize the sling, but the wash from the rotor blades kept driving him back. Finally, the helicopter pilot maneuvered the sling to within reach of the struggling astronaut and hauled him up into the hovering machine.

However, Liberty Bell 7, nearly filled with sea water, was too heavy for the first helicopter to lift out of the water. The pilot reported that his engines were overheating. He was ordered to cut the line to the spacecraft. When he did, Liberty Bell 7 promptly sank to the bottom of the Florida Trench, 2000 fathoms down, to join the fabled hulks of Spanish treasure ships and become the first sunken space ship.

McDonnell Aircraft engineers were unable to explain why the hatch had blown out. Perhaps rocking motions of the craft had set off the explosive after Grissom pulled the safety pin. The final Project Mercury report stated, "A postflight investigation involving a thorough analysis and exhaustive testing was conducted but the cause of the malfunction has never been established."

On August 6, 1961, the American image again was shaken. Major Gherman S. Titov of the Soviet Union flew 18 orbits in Vostok II. Again the feat was taken as a clear challenge to American technology. NASA canceled four more manned suborbital flights on the Redstone in order to move directly to an orbital flight with the Atlas. On September 13, Mercury-Atlas 4 made an unmanned orbital flight, completing the test which had been aborted by the Atlas guidance failure on MA-3. On November 29, Enos, last of the Mercury "support" chimpanzees to fly, survived two orbits in Mercury-Atlas 5.

Next up was John Glenn in Mercury-Atlas 6, the spacecraft he named "Friendship 7." It was production model No. 13.

T MINUS 30 DAYS: GO AND NO GO

To those who had observed the ballistic flights of Shepard and Grissom, the spectacle of a man climbing into a coffin-like capsule on top of an enormous rocket carried with it the curiously persistent suggestion of human sacrifice. The impression was enhanced by the sight of the victim being assisted feet-first into the conical metal sarcophagus by white-suited attendants. The ritual character of the proceedings, abetted by the chant of the countdown over the public-address system, further elaborated the fantasy that here a man was being hurled off the planet in expiation of national feelings of chagrin and inferiority at being second in the space race.

Perhaps this image of immolation haunted the high priests themselves, the Project Mercury Directorate. As preparations for the first manned orbital mission progressed, the directors of the flight exhibited extreme caution by setting rigid weather standards for the launch.

The skies had to be clear enough so that the flight of the Atlas could be observed from the ground, both by watchers and by telescopic cameras. Winds and seas in certain areas of the Atlantic had to be calm enough to make recovery possible in any of four cases: emergency descent, or first-, second-, or third-orbit descent. This was not an easy combination to find off the coast of Florida in late January, but the Project Mercury directorate stood firm on weather requirements. Consequently, the flight began to slip. The "not before" launch date of January 23, 1962, gave way to a "not before" date of January 24, then January 25. Three veteran aeronautical research and development men were calling the shots: Robert R. Gilruth, project director; Walter C. Williams, associate director in charge of operations; and Christopher Columbus Kraft, Jr., deputy operations director. All three had come into NASA from NACA and had taken part in shaping Project Mercury. Each could not have felt more committed to the flight than if he was flying the capsule.

These men would not take a foreseeable chance. Only two-tenths cloud cover over the launch site would be the maximum for a launch. They wanted to observe the early stages of powered flight and determine the correctness of the Atlas course. In addition to acceptable wind and sea conditions in the primary- and emergency-recovery areas, all eighteen Mercury tracking stations around the world had to be operating properly. Finally, all systems in the rocket and spacecraft had to be in "go" condition—including the "man in the loop"—the astronaut himself.

On Sunday, January 21, Glenn came out of seclusion on the Cape and went to church in Cocoa Beach, trailed by a stream of reporters and photographers. He drove his tan convertible into the parking lot of the Riverside Presbyterian Church and entered the church ten minutes after services had begun, taking a seat quietly on a folding chair near the door so as not to disturb the service. The congregation became aware of his presence when he rose to join in singing, "Oh Master, Let Me Walk with Thee." After the service, Glenn was surrounded by Sunday-school children in the parking lot. He autographed their Bibles and also signed the cast on the broken arm of an usher. Back home in Arlington, Virginia, his wife Anna and their children, John David, then 16, and Caroline, 14, were attending a service at the United Presbyterian Church.

The next day, January 22, NASA announced that the flight would be postponed from Thursday, January 25, until Saturday morning, January 27. A regulator in the oxygen-supply system of the spacecraft had been found to be defective. Its function was to prevent excessive oxygen flow in the event a leak developed in the cabin. It had to be replaced and the oxygen system rechecked.

Glenn, meanwhile, took the first part of his preflight physical examination at Patrick Air Force Base. The tests included an electroencephalogram to be compared with a postflight, brain-wave record, and an electrocardiogram to provide a baseline for a post-flight heart recording. For several days, Glenn had been on a "low-residue" diet so that the need of a bowel movement would not arise during the flight. There was no provision for that in the Mercury spacecraft. Urination would be accomplished on board using a "motorman's friend," a plastic bag. After use, the bag could be sealed and stowed away. The sealing was important, for if urine spilled into the cabin during weightless flight it constituted a danger as well as an inconvenience. Globules floating about in free fall could find their way into electrical connections and short circuit them.

While international attention was focused on the Glenn flight, launch traffic on the Atlantic Missile Range continued as usual. The Navy launched a Polaris A-2 missile January

23, on a 1750-mile flight, testing a fiber-glass casing on the second stage. The casing performed satisfactorily, but the second-stage rocket motor failed.

The following day, NASA launched a Thor Able-Star rocket carrying five satellites. Four of the five contained instruments to measure solar radiation, radio-wave propagation in space, and changes in the density of particles in the Van Allen belts during displays of the aurora borealis. In the fifth satellite, the Army Corps of Engineers tucked away instruments that would verify the accuracy of ground-tracking equipment. The satellite called Injun 2 was instrumented by Van Allen at Iowa to measure the radiation belts. The Thor thundered off its pad beautifully just before dawn and climbed toward the pale stars on a brilliant diamond of flame. But the second-stage Able-Star engine failed before it reached sufficient velocity to inject the five satellites into orbit. Down they came, into the sea. It was a multimillion-dollar debacle.

Failure was a persistent theme in those days. Technicians rejoiced when a valuable payload achieved orbit. But there were just as many days when they returned silently to the parking lot, climbed into their cars, and drove wearily through the gate, scarcely aware of the guards checking their badges. When some expensive "bird" had died before dawn because of an imperfect launch or an engine malfunction, sunrise was particularly glaring and the day was filled with anxiety.

TWO DAYS TO LAUNCH: RANGER 3

On Thursday, January 25, the Air Force and Weather Bureau meteorologists forecast good weather for the Saturday launching of MA-6, but warned that the weather would deteriorate after that. Liftoff time was set between 7 a.m. and 12:30 p.m. Eastern Standard Time. Walter C. Williams, the operations director, said that 12:30 p.m. would have to be the cut-off time for the launch attempt because insufficient daylight would remain for recovery operations after a 4½ hour, 3-orbit flight if Glenn lifted off any later.

Excitement among the growing crowd of observers and

sightseers mounted Thursday afternoon when a second weather advisory confirmed the earlier forecast and when a report from the Cape said that all systems in both spacecraft and rocket were "go." A Navy spokesman reported that 47 ships, 115 aircraft, and 15,000 men on station around the world were ready for recovery duty. Glenn had his final chest X ray, blood and urine tests, and a series of inner ear and balance tests. He was "go," too, the flight surgeons said. Glenn said he felt fine. Lt. Col. John A (Shorty) Powers, the NASA spokesman, said the few days of postponement had simply "honed John Glenn down to a fine edge." It was a gold-plated "go" for Saturday morning, January 27—only two days off.

But another disappointment was to occur at the Cape before Glenn was on his way. At 3:30 p.m. Friday, the 727-pound Ranger 3 was launched by an Atlas-Agena rocket as a camera probe to the moon. After the Atlas lifted the Agena and its lunar cargo up through the atmosphere, the Agena second-stage engire fired and put itself and the Ranger spacecraft in a 100-mile orbit. Agena with Ranger attached coasted across the Atlantic Ocean. Then the Agena engine ignited over Africa, boosting Ranger's velocity from 17,500 to 24,500 miles an hour, the speed required to escape the earth and go to the moon. After the Agena engine cut off, Ranger separated from the rocket and took up the journey alone. An automatic timer unfolded its solar panels so that they spread 15 feet, like flashing wings, to catch the sunshine and translate its electromagnetic energy into 200 watts of electricity to power the craft and recharge its batteries. Sensors on Ranger sought the sun, causing attitude-control jets, powered by nitrogen gas, to fire and roll Ranger until the sensors found the sun.

In this position, the main radio antenna on the spacecraft was aligned with the earth, so that the Jet Propulsion Laboratory could monitor and control the flight. Its solar panels outstretched like gleaming dragonfly wings, Ranger 3 sailed moonward. In the early hours, the mission looked good, both from the Cape and from JPL in Pasadena. Two earlier attempts to test Ranger in deep space had gone astray. Each time, the second-stage Agena engine had failed to start in space to propel the spacecraft on to the moon from parking orbit. On the third try, Ranger 3 was well on its way—too well.

Five hours after the launch, JPL reported that Ranger had been over-accelerated and would miss the moon by 20,000 miles. The miss distance was far too great to be corrected by the dragonfly's propulsion system, which had been designed to make only minor course corrections. Ranger would cross a point in the moon's orbit well before the moon got there, and then vanish into interplanetary space to become a very minor vassal of the sun.

All during that week of January, there had been little but failure on the Cape. First, Polaris had failed. Then the multiple satellites launched as Composite I had tumbled into the sea. Now Ranger had gone astray. The world waited for John Glenn.

The astronaut rehearsed his flight plan again Friday and went to bed at 7 p.m. He slept better than he had thought he would. At 2 a.m. Saturday, he was aroused by a NASA flight surgeon, Dr. William K. Douglas. After breakfast of orange juice, poached eggs, a small filet mignon, Postum, toast, and jelly, he was given a final physical examination. Small plastic pods, about the size of a circular Band-Aid, were glued to his chest and throat. They would sense his temperature, respiration rate, heart rate, muscle tension, and blood pressure during the flight. The data would be radioed directly to the ground in "real time" (i.e., as it occurred) when the spacecraft was in radio range of a ground station, or stored on magnetic tape for later transmission when Friendship 7 was out of range.

When the sensors were in place, Glenn struggled into his 20-pound B.F. Goodrich pressure suit. The $4000 garment with its silvery coating of aluminum was designed to protect him in the cabin against loss of air pressure. In a practical sense, the suit was an inner lining of the cabin itself. In the event the cabin sprang a leak or was perforated, the suit would protect the astronaut long enough to enable him to land safely.

T MINUS FOUR HOURS AND COUNTING

When he was "suited up," Glenn marched stiffly down a flight of stairs from his quarters in Hangar S carrying a portable air-conditioner. The conditioner kept him cool until he could plug the suit connections into the spacecraft's environmental control system. Despite the suit's suggestion of clumsiness, Glenn was able to walk briskly in it. He entered the rear of a small van and reclined on a contour couch as the van moved off to Complex 14, about three miles from the hangar. Glenn's backup pilot, Scott Carpenter, had been monitoring preparations on the capsule since midnight.

When the van reached Complex 14, which was brilliantly illuminated by floodlights, Glenn got out and climbed a short flight of steel stairs to the elevator in the service tower, which enfolded Atlas-Mercury in a steel embrace. At the elevator, he was greeted by Air Force Captain Donald K. Slayton, the astronaut who was to be grounded as far as space fiying was concerned because of a heart rhythm irregularity. With Slayton and Dr. Douglas, the silver-suited figure rode up 100 feet in the elevator to the open hatch of the Mercury spacecraft. Glenn spoke briefly with technicians working on the capsule. He bent awkwardly to remove plastic covers from his flight boots. The covers prevented the tracking of dirt into the capsule—particles which could float randomly about in the cabin in free fall and foul the intakes of the environment-control system or the electrical system. The astronaut worked himself feet-first into the capsule and reclined in his contour couch. Since the nose of the capsule was pointing skyward, Glenn lay on his back with his knees up. Through the spacecraft window, he could see the night sky, clusters of stars now and then blotted out by scudding clouds.

T MINUS THREE HOURS AND COUNTING

During the night, thousands of men, women, and children from the nearby towns of Cocoa, Rockledge, Titusville, Eau Gallie, Melbourne, Indiatlantic, Orlando, Daytona Beach, and even Jacksonville had gathered on the beaches south of the Cape to watch the liftoff. From Cocoa Beach, Complex 14 glowed under the lights like a Hollywood premiere. Now and then a beam would lance up to the low clouds as someone adjusted a floodlamp.

The multitudes on the beaches set up tents, wrapped themselves in blankets on the sand, or sat in their automobiles, listening to the progress of the countdown on portable or car radios. Some of the older children clustered around driftwood fires where they roasted marshmallows and hot dogs. There was a surprising degree of good humor and order among this crowd, which seemed to express a community rapport with the event. A police squad car patrolled the scene, moving slowly up and down five miles of crowded beach on the hard, wet sand near the surf. It was a cool night, in the low 60s, and a wet breeze came in from the ocean.

At the press site on the Cape, newspaper and radio correspondents began arriving shortly after midnight to report "color." A circus-style tent had been set up as an auditorium for press conferences. Inside, a half-dozen reporters pecked away on their portable typewriters, preparing copy for the early-morning editions of afternoon newspapers.

The press site was 8000 feet from the launch pad, but those with binoculars could observe tiny figures moving around near the top level of the service tower, which concealed the rocket and spacecraft. A battery of telephones had been set up on wooden platforms, and a wooden grandstand, about the size one might find on a junior-high-school baseball diamond, had been erected for the correspondents. The press site had been designed by the Air Force as though it had been arranging spectator facilities for Little League playoffs.

The radio-television network broadcasters disdained these primitive facilities. They brought in huge house trailers and parked them to one side of the press site. All night long, a coffee-and-snack wagon, which had acquired the name "Roach Coach," did a steady business. Beyond the floodlit area stood a row of portable steel outhouses, developed by the Department of Defense as remote site latrines. Electric cables snaked everywhere. One Medusa-like cluster emanated from the Western Union trailer, where a half-dozen women operators played the rhythms of several different languages on their teletype keyboards.

Across the dark ocean stretched a line of ships from Bermuda to the Canary Islands, comprising the Atlantic recovery force. Among them were the aircraft carriers *Wasp, Randolph,* and the nuclear-powered *Enterprise.* In the Atlantic and Pacific oceans, more than 15,000 men stood by under the projected flight path of Friendship 7 to recover Glenn wherever he came down.

The countdown continued through the early morning darkness as the astronaut settled himself in the tiny cabin. At dawn the skies were clear, but as the sun rose, great black clouds materialized over the ocean and began moving toward the Cape. At 7 a.m., the countdown was halted for 45 minutes for a final cleanup of unfinished minor checks and tasks. During this "hold," the clouds rolled in from the ocean. The hold was extended again "to see what the weather would do." Friendship 7 would not be launched through clouds.

By 8:48 a.m., the low clouds had cleared and the count was resumed: "T minus 26 minutes and counting," Colonel Powers announced. The crowd of observers on the press site became energized. Newspaper and radio reporters moved toward their outdoor telephones and opened lines to their offices through the Cape operator. Spectators stared at the fuming Atlas, from which the service tower had been moved back. The black spacecraft was clearly outlined in pearly morning light, its escape tower pointing toward the roiling clouds. The Atlas was wreathed in the white smoke of oxygen vapor, boiling off through a valve in its liquid-oxygen tank.

As the tension mounted, another hold in the countdown was called. The electric power had failed at 8:56 a.m. At 9 a.m., the power was restored and the countdown was resumed. "T minus 23 minutes and counting," Colonel Powers said. The clouds were moving rapidly inland to the southwest, but the skies overhead remained overcast. At Mercury Control, the weathermen presented a revised estimate of cloud conditions to the directorate: there would be no break in the cloud cover for another two days. Walter C. Williams advised Glenn of the forecast and said, "We're going to scrub, John." "Roger," said Glenn. He had been in the spacecraft 4 hours and 15 minutes when Colonel Powers announced over the public-address system without preamble, "At T minus 13 minutes, the mission has been scrubbed."

Someone remembered that the factory number of Friendship 7 was 13. Someone else observed that the launching had been postponed at 13 minutes before time of engine ignition. A jinx had appeared in the program.

Colonel Powers quoted Glenn as saying, "Well, there will be another day."

"What else did he say, Shorty?" Powers was asked at a press conference.

"The astronauts are not an emotional bunch," Powers replied. "John didn't cry."

At 6:25 p.m. Sunday, January 28, Ranger 3 flew past the moon at a distance of 22,816 miles. Controllers at JPL turned on the camera to photograph the leading edge of the moon, hoping that some photos would show a portion of the side that is always turned away from the earth. The photo transmission was so garbled, however, that no coherent picture could be assembled from the data. Ranger 3 soared off into the interplanetary medium to join the Pioneer satellites in orbit around the sun. NASA announced later Sunday night that another attempt would be made to launch John Glenn on Thursday, February 1.

Late Monday night, technicians working on the Atlas, whose liquid-oxygen tank had to be drained and checked, discovered that a plastic insulation bulkhead between the lox and kerosene tanks was soaked with kerosene. When they looked further, they found that the kerosene tank had sprung a leak. Project Mercury directors, Air Force launch officers, and engineers representing the manufacturer, the Convair Division of General Dynamics agreed that the missile would have to be dismantled in order to make a repair. No one was prepared to say how long that would take, but the engineers gave an outside estimate of about two weeks. The launch of MA-6 was then postponed until February 13.

As the news-media people trooped homeward, echoes of a generalized feeling of frustration appeared in a number of commentaries. What had gone wrong with the vaunted American industrial establishment? Why had Russian technology succeeded where ours had failed? What was the sense of going into space, anyway?

It was a time to assess the purposes and objectives of Project Mercury. Was it really the beginning of a march into a new frontier, a symbol of American technological and scientific expansion, or was it simply a stunt in the international-competition game? What was the purpose of racing the Russians in space? In that context, the postponement of MA-6 had appeared as a defeat.

Although Project Mercury did reflect the pressure of Russian technological competition, it could not be characterized as a mere stunt if its objectives were understood. The immediate ones had been spelled out by NASA: to test a space-vehicle design which had been developed by this country and to find out how well a trained man could function in orbital free fall. Beyond that, a larger objective had been defined by President Kennedy after the Shepard flight in a summons to the nation to gird itself for a landing on the moon.

In the immediate historical context, Project Mercury was clearly a response to international competition, from which the space age seemingly evolved. There was another way of looking at it, however. In the context of human development, it marked the beginning of man's expansion from earth's biosphere into the solar system. From this point of view, the flight MA-6 was a critical test of American technological readiness to penetrate a new environment.

The flights of Gagarin and Titov had proved it was possible for a trained man to endure the "zero gravity" effect of orbital flight up to 24 hours. The physiological data released by the Russians did not tell us what the effects were and how well the pilot could function in free fall. Titov had become ill, especially when he moved his head. That suggested that the weightless state deranged the vestibular functions of the inner ear which govern the "sense" of balance. If that was so, the ability of man to fly in space might be severely constrained, and long flights to the moon or the planets might be out of the question. The Russian flights had not answered a basic question of human adaptation to a nongravitational environment. Nor had they answered another key question: would a man be able to pilot a space vehicle in the weightless condition, or would his perceptions, his muscular control, and his psyche become so disoriented that he could only ride as a passenger, like the chimps Ham and Enos?

While Glenn was the third human being to attempt to fly in orbit, he was the first to address himself to the task of controlling a vehicle as a test pilot in orbital flight. The question of how well he would perform was a critical one at the beginning of 1962, when this was an unknown in the world.

There had been a number of speculations about the role of man as a pilot in space. At the Redstone Arsenal, Siegfried J. Gerathewohl, chief of bioastronautics research in the Army's Research and Development Command, had referred to man as the weakest link in a space-vehicle system.

If man was "plugged into the loop" as a control unit in a spacecraft, in lieu of a computer and electronic sensors, *he* was more likely to fail than the electromechanical systems that could be created to take his place, Gerathewohl held.[5] Was man, as the engineers claimed, merely "a noisy transmitter, an expensive nuisance" in a space machine?

In the year before Project Mercury flew the first astronauts, there had been a good deal of speculation on this point. A personnel psychologist (human-factors engineer) at Douglas Aircraft has predicted that man probably would find useful roles in flight management, scientific observation, maintenance, and repair, but he challenged the notion that manning a space vehicle would be free of major psychic problems. Evidence from studies of human responses to isolation and sensory deprivation was "hardly optimistic about human reactions to space-flight conditions," the psychologist said.[6]

Of great psychological importance was the difficulty of maintaining adequate communications with earth, according to the Douglas human-factors man. He and other psychologists warned that feelings of boredom, fatigue, suspiciousness, hostility, irritablity, withdrawal, and anxiety could be expected. The pilot might develop a sense of personal inadequacy leading to depression and suicidal impulses, other types of neurotic behavior, and paranoid feelings.

A New York psychologist flatly predicted that Glenn would be subjected to severe psychological shock which could induce psychosis.

A Florida country preacher told his Sunday radio audience, "Any fool venturing into the realm of the Lord will undoubtedly lose his mind. Let us pray for him."

Chapter 5

Fireball

Global machinery for the MA-6 mission began turning again on Monday, February 12, 1962, as the big carriers once more took up their recovery positions in the Atlantic Ocean and destroyers steamed out of Jacksonville to support them. The Project Mercury tracking network was alerted around the world, from Bermuda to the Canary Islands, across Africa from Kano, Nigeria, to Zanzibar; aboard an instrumented ship in the Indian Ocean; at the Australian cities of Muchea and Woomera; on Canton Island in the central Pacific Ocean just below the equator; on Kauai Island, Hawaii; Point Arguello, California; Guaymas, Mexico; White Sands, New Mexico; Corpus Christi, Texas; and Elgin and Cape Canaveral, Florida. From Spain to the Fiji Islands, special United States Air Force units around the world were alerted to fly emergency rescue missions. Only a world power could construct such a planetary apparatus. Yet the Russians had flown their men without one.

On the Cape, Glenn resumed his low-residue diet. Liftoff was scheduled for 7:30 a.m. Wednesday morning, February 14, in lieu of the 13th, since although the preliminary weather forecast promised fair skies in the launch area, out to sea, a storm was whipping up 10-foot waves east of Bermuda. The countdown was begun early Tuesday. The Atlas tank had been repaired, and the missile was pronounced fit to fly.

Tuesday night, the pattern of delay reappeared. Deteriorating weather persuaded the Mercury directorate to postpone the launch until Thursday, the 15th, until a procession of storm centers moving across the ocean toward Florida passed by. Early Thursday morning, the launch was postponed again, until 7:30 a.m. Friday, the 16th. "When we get the weather, we will go," Glenn was quoted as saying with patient resignation. He remained in good spirits. In the mornings, he jogged two miles along the beach, weaving in and out of colonies of jellyfish washed up by the storms, and in the afternoons he reviewed his flight plan, which he had known by heart for weeks. On Friday, a "spell" of really bad weather was forecast for the next five days and the flight was postponed until Tuesday, February 20. The Atlas launch crew drained the kerosene out of the Atlas and, leaving a standby crew of "baby-sitters" at Pad 14, took the week end off.

The week end itself came up hot and filled with sunshine, to mock the gloomy forecast. The beaches were jammed with families. Cars were parked for miles on the sand while lines of traffic moved up and down the harder sand near the surf, to the peril of sunning bathers. Glenn continued running up and down his private beach on the Cape. After ten postponements, the flight of MA-6 seemed remote. No one really believed Glenn would fly on Tuesday except Glenn and the flight directors. This time the Mercury meteorology team was positive the weather would be fair and the skies clear. Glenn went to church. Sunday night, under the bright lights, the Atlas was loaded once again with 36.5 tons of kerosene.

The first segment of a 12-hour countdown began that night. Technicians methodically plodded through a list of checks in a 200-page manual until 5 a.m. Monday. They went to bed and returned to Complex 14 Monday night to resume their work. The second part of the countdown—seven hours of it—began at midnight Monday. When it ended, Glenn would lift off in MA-6.

Glenn, who had retired early Monday evening, was awakened at 2:20 a.m. by Dr. Douglas. After a shave and shower, the astronaut sat down to a breakfast of orange juice, a small filet mignon, and scrambled eggs. Alan Shepard and Deke

Slayton had coffee with him, but Glenn drank none—only Postum. He then underwent a series of tests to record his respiration, blood pressure, heart rate, and ability to perform psychomotor tests. There was a good deal of speculation whether his eye-hand coordination would be as precise in orbital free fall as on the ground. That morning of February 20, Glenn was the living definition of a normal 40-year-old male of "old" American stock. When the sensors were attached to his chest for heart-lung telemetry, he was asked if he had the feeling he had done all this before. He grinned at the memory of the January 27 scrub. Meanwhile, up in the Complex 14 service tower, Scott Carpenter, Glenn's backup pilot, was again at his post, inspecting the spacecraft under the floodlights.

Glenn emerged from Hangar S to meet the bright glare of television lights. His helmet visor was open and he carried his portable suit air-conditioner under one arm, like a shield. He rode to Pad 14 in the van with Dr. Douglas and immediately went up in the elevator to the capsule. At the spacecraft level, Glenn took the plastic covers off his boots and slid efficiently into Friendship 7. Presently, the white-suited technicians closed the hatch. On the black shingles of spacecraft No. 13 gleamed lettering in a brilliant white paint: UNITED STATES FRIENDSHIP 7.

It was shortly after 5 a.m. A hazy moon floated full in the west, dimming and brightening as clouds scudded by, the ultimate goal of the program now beginning. From the dark sea, a fresh wind blew. At the press site, the news people were swarming, and three thousand spectators had assembled on the beaches overnight.

Trouble appeared at the pad. A 45-minute hold in the countdown was called to enable the launch crew to replace a defective transponder, a device which reflects radar waves, in the Atlas. Once this was accomplished, the countdown was interrupted for another ten minutes to replace a defective microphone in Glenn's helmet.

Dawn came reluctantly, revealing a sky well overcast by dark, gray clouds. At 6:50 a.m., a one-hour hold in the countdown to await a turn of the weather was called. During the hold, a broken bolt was found in the spacecraft hatch. The hatch had to be removed so that the bold could be replaced. Then the hatch was once more secured. At 8:05 a.m., patches of blue sky appeared. The countdown was resumed and the press people cheered. Cooler air moved in over the Cape as the sun rose and the clouds flowed off to the south and west. It was becoming a beautiful day at Cape Canaveral, and when the sun broke through at 8:25 a.m., it was reported as a major news event.

The service tower was rolled back and away from the Atlas-Mercury, showing the rocket, the spacecraft, and the steeple of the escape tower boldly silhouetted against the sky. At 9 a.m., the countdown was halted because of a stuck valve in the line feeding liquid oxygen to top-off (keep filled) the Atlas lox tank. It took 23 minutes to shut down the lox feed, replace the valve, and resume the flow. The count was then resumed. It stood at T (for engine ignition time) minus 20 minutes. At T minus 10 minutes, the Mercury spacecraft was switched to its own internal-battery power. From the capsule, Glenn, almost forgotten in the shift of attention to the countdown holds, reported that all systems were "go." At T minus 6½ minutes, the countdown was halted while flight controllers asked for details of a report that a power failure had shut down the tracking computers at Bermuda. These were required to analyze radar data for the determination of the MA-6 orbit once the spacecraft was launched. One minute later, Bermuda reported the power was on. The count was resumed. T minus 5 minutes. The Atlas-Mercury stood stark against the sky, wreathed in white oxygen vapor which the wind intermittently whisked away. Air Force jets screamed high overhead, climbing to photograph the Atlas as it lifted. T minus 3 minutes and counting. The Atlas was switched to internal power of its own batteries. Powers's voice came over the public-address speakers: "This is Mercury Control. The MA-6 countdown is at T minus one minute and counting. All systems are reported in a go condition. John Glenn is ready. The Mercury spacecraft umbilical [a line connecting the spacecraft to the ground] is out. We are at T minus nineteen seconds. T minus ten seconds, eight, seven, six, five, four, three, two, one, zero. Ignition."

Orange flame erupted from the base of the Atlas, and clouds of dense white smoke rolled out to the west. For 4 seconds, clamps held the roaring rocket fixed to the pad as its two booster engines, the single sustainer engine, and two small vernier engines reached full thrust, 360,000 pounds. The clamps released and the rocket lifted with majestic assurance. Catching the sun, it suddenly gleamed a brilliant silver. It rose swiftly on its pillar of fire and smoke, and the thunder of the engines rolled over the Cape like a tidal wave. On top of that pillar of fire a man was going into orbit. It was 9:47 a.m. Eastern Standard Time.

"Pilot John Glenn is reporting that all systems are go," said Powers. "He is reporting his cabin pressure is holding at 6.1 pounds per square inch [normal]."

Sonic booms of the chase planes sounded loud and sharp. The Atlas accelerated into the infinity of blue, until only the candle flame of its engines was visible. Then a long, white vapor trail appeared below the vanishing rocket, and it was gone from sight.

At the press site, the spectators spontaneously broke into applause.

T PLUS 30 SECONDS: THE SKY IS BLACK

In the capsule, Glenn was talking to Shepard in Mercury Control on the Cape: "The clock [the spacecraft timing device] is operating. We are under way."

SHEPARD: Roger. Read you loud and clear.

GLENN: Roger. We're programing in roll okay. [The Atlas guidance system had been preset to roll to an azimuth or compass direction of 70 degrees.] Little bumpy along about here.

Glenn had reached an altitude of about 75,000 feet, where the Atlas was encountering maximum dynamic pressure in the atmosphere. The region he was passing through is called "Max Q."

GLENN: Have some vibration area coming up here now. Coming into high Q a little bit, and a little contrail went by the window, or something there.

SHEPARD: Roger.

GLENN: We're smoothing out some now, getting out of the vibration area.

SHEPARD: Roger. You're through Max Q. Your flight path . . .

GLENN: Roger. Feels good, through Max Q and smoothing out real fine. Flight very smooth now. Sky looking very dark outside. Cabin pressure is holding at 6.1 okay. Coming up on two minutes. The *g*s are building up to six.

SHEPARD: Roger. Reading you loud and clear. Flight path looked good. Pitch 25 degrees.

The Atlas was programed to pitch over from the vertical so that at orbital speed the flight path would be parallel to the earth. In this way, once released from the Atlas, the Mercury spacecraft would continue to fall around the earth until Glenn checked its velocity by firing his retrorockets, or until residual atmosphere at his orbital altitude slowed the spacecraft down sufficiently to cause it to reenter the denser atmosphere of its own accord. As a safety measure, Friendship 7's orbit was planned at an altitude where it would decay in 18 hours (12 revolutions) naturally in the event the retrorockets failed. Glenn had oxygen for 28 hours. He had used about four hours of the supply awaiting liftoff.

At 2 minutes and 12 seconds after liftoff, Glenn reported "BECO," the acronym meaning booster engine cutoff. The Atlas-Mercury continued accelerating with the sustainer and vernier engines, but Glenn reported the g load had dropped to one and one-quarter. At this time, also, the escape tower zoomed away from him as its jettison rocket fired.

"Could not see the tower go. I saw the smoke go by the window," said Glenn.

Shepard said telemetry from the Atlas confirmed that booster engines had been jettisoned. The relief of the weight of the two booster engines and the escape tower, plus the rapid diminution in the fuel supply, caused the rocket to accelerate more rapidly. Glenn reported the *g* load had risen slightly to one and one-half.

Glenn methodically reported fuel, oxygen, electrical amperage, and cabin pressure with test-pilot precision, but he could not keep the excitement out of his voice. Neither could Shepard. He and the flight directors in the room knew that Glenn was on his way. America was in space at last.

SHEPARD: Reading you loud and clear. Seven—Cape is go. We're standing by for you. All indications here at the Cape are good. How about you?

GLENN: Roger. Cape is go and I am go. Capsule is in good shape. Cabin pressure holding steady at 5.8 [pounds per square inch]. All systems are go.

SHEPARD: Roger. Twenty seconds to SECO [sustainer engine cutoff].

As the fuel load dwindled and acceleration rose, Glenn reported he once more was feeling 6*g*s. "SECO," he reported. Then the three solid-propellant posigrade rockets at the rear of the spacecraft fired. Their one-second thrust totaled 1200 pounds. It pushed Friendship well clear of the Atlas.

SHEPARD: Roger. Stand . . .

GLENN: Roger. Zero *g* and I feel fine. Capsule is turning around. Oh, that view is tremendous!

SHEPARD: Roger. Turnaround has started.

For safety reasons, Glenn had been instructed to commence the flight with the heat shield forward, so that he would be in position to reenter the atmosphere in case of an attitude-control system failure or other emergency.

GLENN: The capsule is turning around and I can see the booster during turnaround just a couple of hundred yards behind me. It was beautiful.

SHEPARD: Roger. You have a go, at least seven orbits.

GLENN: Understand. Go for at least seven orbits.

The spacecraft's velocity and direction immediately after it separated from the Atlas, and a radar track of its arc between Florida and Bermuda, were fed to a computer at the Goddard Space Flight Center at Greenbelt, Maryland. The computer determined that the apparent orbit of Friendship 7 would not decay substantially until the vehicle had completed seven revolutions of the earth. Since the flight plan called for only three, the orbit was a good one. Initial measurements indicated that the orbit was about 100 miles at perigee, just east of the Cape, and 158.7 miles high at apogee, 180 degrees around the earth from perigee, over a point in the western Pacific Ocean.

Six minutes, 25 seconds after liftoff, John Glenn passed over Bermuda. "Roger, Friendship 7. Orbit checklist." The capsule communicator (Cap Com) was calling for the pilot's report of the condition of the spacecraft's systems. Glenn reported that all emergency landing switches were in the off position—as they were supposed to be now that he was in orbit. "This is a very comfortable zero *g*," he added. "I have nothing but a very fine feeling. It just feels very normal and very good." Glenn switched from ultra-high-frequency radio, over which reception had been clear, to high frequency to test its effectiveness.

Bermuda was not reading the spacecraft. Glenn switched his radio back to ultra-high frequency as Friendship 7 approached the Canary Islands at 11 minutes, 16 seconds after launch. He advised the Canary Cap Com that he was flying in automatic control and that the system was holding the spacecraft steady, heat shield forward, giving the pilot facing rearward a fine view of the earth as it seemed to be turning slowly under his feet.

GLENN: I have the booster in sight out of the window. [The Atlas had followed him across the Atlantic, but did not go much farther. It reentered, broke up, and large pieces of the rocket were found on a farm in South Africa]. Everything is still go. Capsule is in fine shape. Holding pressure at 5.8. Over.

Glenn also reported that the spacecraft clock was set to fire the retrorockets at 4 hours, 32 minutes, and 28 seconds after liftoff, when Friendship 7 would be flying over Hawaii on its third revolution. Automatic retrofire had been built into the automatic sequence and control system of the spacecraft in the event the astronaut blacked out and control of the vehicle could not be taken by a ground station. The clock setting was based on the assumption that Glenn would fly a full three-orbit flight, but Mercury Control could interrupt the mission and bring Glenn down at the first sign of serious trouble. Glenn was thus flying from station to station, from orbit to orbit, hoping that all systems would continue to "go" and expressing his satisfaction that they did each time he made his report.

Looking through his periscope, which enabled him to see forward when he was riding backward, Glenn remarked, "The horizon is a brilliant, a brilliant blue. There, I have the mainland in sight at the present time, coming up on the scope, and have Canaries in sight out throught the window and picked them up on the scope just before I saw them out of the window."

T PLUS 21 MINUTES: THE WATER'S FINE

The Canary Cap Com then gave Glenn readings the station had received via telemetry from the spacecraft, showing that electrical apparatus and fuel-line temperatures were normal —as though Glenn's blood pressure and spacecraft system telemetry were all part of the over-all data picture—which, of course, they were. Glenn was "the man in the loop," the human element in an integrated organic-inorganic vehicular system.

GLENN: I can see dust storms down there blowing across the [Sahara] desert. A lot of dust. It's difficult to see the ground in some areas.

CANARY: Seven, you are fading . . .

Two seconds later, at 21 minutes, 33 seconds after liftoff, another voice came into the astronaut's helmet: "Friendship 7, this is Kano [Nigeria] Cap Com. I read you loud and clear. How do you read me?"

Kano had been listening to Glenn's discourse with the Canaries.

GLENN: Roger, Kano, Loud and clear. How me?

KANO: Roger. Loud and clear. What is your status?

GLENN: This is Friendship 7. My status is excellent. I feel fine.

On the pass over Nigeria, Glenn took control of the spacecraft and yawed it 60 degrees to the right. He reported he had no difficulty doing so. In fact, he advised, it was "very easy." Then he returned the spacecraft to orbit attitude—heat shield forward.

Glenn continued with his natural-history observation, which he felt was as important to report as the readings on his instrument board: "Out the window can see some fires down on the ground, long smoke trails right on edge of the desert. Over."

KANO: Roger. We've had dusty weather here, and far as we can see, a lot of this part of Africa is covered with dust.

GLENN: That's just exactly the way it looks from up here, too.

Friendship 7 lost radio contact with Kano at 28 minutes, 40 seconds into the flight, and picked up Zanzibar 45 seconds later as the little spacecraft approached the Indian Ocean.

Glenn kept his microphone open as he spoke for his tape recorder: "This is Friendship 7 in the blind for recording. Much of eastern Africa is covered by clouds, sort of wispy, high, cirrus-looking clouds. Cannot see too much down there except the cloud decks themselves. Catch a sight of the ground underneath once in a while."

ZANZIBAR: Friendship 7, this is Zanzibar Cap Com, reading you loud and clear.

GLENN: Roger, Zanzibar.

ZANZIBAR: Message from IOS [tracking ship in the Indian Ocean] Cap Com that he will not release balloon flare this orbit. Will fire parachute flares instead. Did you copy? Over.

GLENN: Roger, Zanzibar.

These activities on the Indian Ocean Ship were designed to determine whether a terrestrial vehicle could signal a spacecraft on the night side of the earth.

At Zanzibar, the flight surgeon advised Glenn he had a good blood pressure. Glenn exercised briefly by pulling on an elastic cord to give the surgeon electrocardiographic and blood-pressure readings before, during, and after the exertion.

ZANZIBAR: Friendship 7, this is Zanzibar surgeon. Blood pressure 136 systolic after exercise, recording well and coming down now to just under 90 for diastolic. Both traces are of excellent quality. Your electrocardiogram is excellent also. Everything on the dials indicates excellent aeromedical status.

Glenn then advised he would perform a series of head movements to see if this would produce nausea, dizziness or vertigo, blurring of vision, or any other disturbance of well-being and equilibrium. Manned-space-flight experts were particularly concerned about this because several of these symptoms had been reported by Titov.

GLENN: This is Friendship 7. The head movements caused no sensations whatsoever. Feel fine. Reach test—I can hit directly to any spot that I want to hit. I have no problem reaching for knobs and have adjusted to zero *g* very easily, much easier than I really thought I would. Have excellent vision of the charts, no astigmatism or any malfunctions at all.

A few minutes later, Glenn reported he could see the dark side of earth coming up in his periscope. Zanzibar said: "Stand by to pick up Indian Ocean Ship in three minutes."

T PLUS 36 MINUTES: SUNSET

At 38 minutes, 26 seconds after liftoff, a voice came into Glenn's receiver: "Hello, Friendship 7, Friendship 7. This is Indian Com Tech [communications technician]."

GLENN: Hello, IOS Com Tech. Friendship 7. Go ahead.

INDIAN: Would you give me your status and consumable readings, please? Over.

GLENN: Roger. This is Friendship 7. My status is very good. I feel fine. Fuel is 90 and 100 per cent [actual and reserve]; oxygen is 75 and 100 per cent. Amps 21 at the present time. Cabin pressure holding at 5.5. Had a beautiful sunset and can see the light way out, almost up to the northern horizon. At this mark, at this present time, I still have some clouds visible below me. The sunset was beautiful. It went down very rapidly. I still have a brilliant blue band clear across the horizon almost covering my whole window. The redness of the sunset I can still see through some of the clouds way over to the left of my course. The sky above is absolutely black, completely black. I can see stars though, up above. I do not have any of the constellations identified as yet. Over.

The Indian Ocean ship launched its flare, but Glenn could not see it.

INDIAN: Friendship 7, this is Indian Cap Com. Do you have any feelings from weightlessness?

GLENN: Negative. I feel fine so far.

Friendship 7 sailed out of range of the voices below, and for nearly seven minutes, Glenn was alone, moving in the continuum beyond the earth he had, as a child in New Concord, Ohio, thought of as heaven.

THE BANDED HORIZON

Glenn spoke to his tape recorder: "This is Friendship 7, broadcasting in the blind. Wait a minute. Friendship 7, broadcasting in the blind, making observations on night outside. There seems to be a high layer up above the horizon, much higher than anything I saw on the daylight side. The stars seem to go through it and then go on down toward the real horizon. It would appear to be possibly some 7 or 8 degrees wide. I can see the clouds down below it; then a dark band, then a lighter band that the stars shine right through as they come down toward the horizon.*

* NASA later attributed the banding to reflection in the spacecraft window.

Glenn said he could identify the constellations of Aries and Triangulum in the southern hemisphere sky.

At 49 minutes, 49 seconds after his journey from Pad 14, Glenn picked up a voice from Down Under. Muchea, Australia, had been listening to his recital.

There was something familiar to Glenn about that Oklahoma drawl down below. It was astronaut Leroy Gordon Cooper.

GLENN: Roger! How are you doing, Gordo? We're doing real fine up here. Everything is going very well. Over.

COOPER: John, you sound good.

GLENN: Roger. Control fuel is 90-100 per cent; oxygen is 75-100 per cent; amps are 22. All systems are still go. Having no problems at all.

COOPER: Do you have any star or weather or landmark observations as yet?

GLENN: I was just making some to the recorder. The only unusual thing I have noticed is a rather high, what would appear to be a haze layer up some 7 or 8 degrees above the horizon on the night side. The stars I can see through it as they go down toward the real horizon, but it is a very visible singled band or layer pretty well up above the normal horizon. Over.

COOPER: Roger. Very interesting.

GLENN: I had a lot of cloud coming off of Africa. It has thinned out considerably now, and although I can't definitely see the ocean there is a lot of moonlight here that does reflect off what clouds there are.

COOPER: Roger. Understand you have Pleiades in sight. Have you sighted Orion yet? Over.

GLENN: Negative. Do not have Orion in sight.

COOPER: Within a few seconds you should have Orion and Canopus and Sirius probably in sight very shortly thereafter. The moon will be off.

Glenn sent another blood-pressure reading to the ground. Then he reported: I do have the lights in sight on the ground. Over.

COOPER: Roger. Is it just off to your right there?

GLENN: That's affirmative. Just to my right I can see a big pattern of lights apparently right on the coast. I can see the outline of a town and a very bright light just to the south of it. On down.

COOPER: Perth and Rockingham you're seeing there.

GLENN: Roger. The lights show up very well, and thank everybody for turning them on, will you?

COOPER: We sure will, John.

It was 1 a.m. down below and Glenn had been in flight just 55 minutes.

Woomera in south Australia picked up Friendship 7 telemetry at 26 seconds past 55 minutes into the flight, and notified Glenn his blood-pressure reading was 126 over 90. Woomera asked for any unusual symptoms.

GLENN: This is Friendship 7. I have had no ill effects at all as yet from any zero *g*. It's very pleasant, in fact. Visual acuity is still excellent. No astigmatic effects. Head movements caused no nausea or discomfort whatsoever. Over.

Glenn then made his second detailed 30-minute report on the dial readings of his instrument panel. Woomera Airport turned on its lights for him, but he was unable to see them.

As Friendship 7 sailed on a northeasterly track out over the Pacific Ocean, Glenn switched his radio to high frequency and tried potluck at raising any of the Mercury stations around the world. None of the stations read his transmission.

At one hour and nine minutes into the flight, Friendship 7 was "acquired" by the tracking station on Canton Island, 2 degrees south of the equator. Glenn reported he was still in the automatic-control mode. He reported he had seen a big storm off the south. There were flashes of lightning on top of the clouds. He then opened his visor to eat some applesauce, which he squeezed out of the tube like toothpaste. He reported to Canton Cap Com that he had no problem eating in zero gravity. Then he saw his first sunrise in space ahead of him through the periscope.

T PLUS 1:15: SUNRISE AND FIREFLIES

GLENN: On the periscope, I can see the brilliant blue horizon coming up behind me. [He was riding backward, in orbital attitude with the heat shield forward.]

CANTON: Roger, Friendship 7. You are very lucky.

GLENN: You're right. Man, this is beautiful. Oh, the sun is coming up behind me in the periscope, a brilliant, brilliant red. It's blinding through the scope on clear. I'm going to the dark filter [on the periscope] to watch it come up.

CANTON: Roger.

Glenn was silent for a few moments. Then, keeping his voice deliberately even, he reported the most baffling event of the flight: "This is Friendship 7. I'll try to describe what I'm in here. I am in a big mass of some very small particles, that are brilliantly lit up like they're luminescent.* I never saw anything like it. They're round a little. They're coming by the capsule and they look like little stars. A whole shower of them coming by. They swirl around the capsule and go in front of the window and they're all brilliantly lighted. They probably average maybe seven or eight feet apart, but I can see them all down below me, also."

CANTON: Roger, Friendship 7. Can you hear any impact with the capsule? Over.

GLENN: Negative, negative. They're very slow. They're not going away from me more than maybe three or four miles an hour. They're going at the same speed I am approximately. They're only very slightly under my speed. Over.

Glenn waited a moment for an acknowledgment, but none came from the ground. He continued to describe the little particles as though trying to depict a scene in an undiscovered country.

GLENN: They do . . . have a different motion, though, from me, because they swirl around the capsule and then depart back the way I am looking. Are you receiving? Over.

Silence.

GLENN: There are literally thousands of them.

Silence.

GLENN: This is Friendship 7. Am I in contact with anyone? Over.

Silence.

GLENN: This has been going on since about 1 plus 15 [one hour and 15 minutes into the flight]. Just after I remarked about the sunset [he meant sunrise]. I looked back up and looked out of the window and all the little swirl of particles was going by. Over.

Silence.

* Glenn later described these particles as "the color of a very bright firefly, a light yellowish-green color."

GLENN: This is Friendship 7, this is Friendship 7, broadcasting in the blind again on HF. Sunrise has come up behind in the periscope. It was brilliant in the scope, a brilliant red as it approached the horizon and came up, and just as the . . . as I looked back up out the window I had literally thousands of small, luminous particles swirling around the capsule and going away from me at maybe three to five miles an hour. Now that I am out in the bright sun they seem to have disappeared. It was just as the sun was coming up. I can still see just a few of them now, even though the sun is up some 20 degrees above the horizon.

Friendship 7 came into radio range of Guaymas, Mexico, at 1 hour, 20 minutes in flight, and Glenn reported his status was excellent.

Crossing the coast of Baja, California, Glenn discovered that the ship was yawing to the right as much as 20 degrees on automatic control. He corrected its alignment by using a manual control system called "fly-by-wire."

THE SECOND REVOLUTION

In the fly-by-wire mode, Glenn controlled the attitude of the little ship by moving his control stick, which operated the thrusters electrically. He substituted his own perceptions and brain for the electronic sensors, gyroscopes, and computer in the automatic system. After a while, at the suggestion of the ground, he switched back to the automatic mode to check it. He reported that the craft continued to drift 20 degrees to the right. Then, in correcting the drift, the automatic system would overreact, causing a thruster to fire too energetically. This would yaw the craft over to the left.

Glenn was preoccupied by this problem as Friendship 7 crossed the United States at the end of its first revolution around the earth. Shepard called up, "Would you give us the difficulty you have been having in yaw in the ASCS?" Glenn did so. Shepard then called up to say that President Kennedy wanted to talk to the pilot of Friendship 7.

"Ah," said Glenn, surprised. "The President."

"Go ahead, Mr. President," said Shepard.

There was a silence, and Glenn said: "This is Friendship 7 standing by."

"Roger, Seven," said Shepard. "Having a little difficulty. Start off with your 30-minute report."

Glenn continued with his analysis of the yaw problem. He said he believed the small 1-pound thruster in left yaw had failed. Consequently when the ship drifted too far to the right, the 24-pound thruster would cut in and overcorrect by swinging the spacecraft too far to the left. The pilot could correct this by taking control of the ship in the fly-by-wire mode, and the ground advised him to continue doing so. The yawing had no effect on the flight path, which was determined by velocity and gravity. It could affect the orientation of the astronaut, however, and interfere with the precise alignment of the vehicle, which was required for retrofire. Had a chimp been in the craft, the controllers would have brought it down for an emergency landing at that point, and the flight would have failed to achieve its objective of three orbits. But a trained test pilot was at the controls and he was able to manipulate the highly flexible control system to correct for the low-thruster failure.

For Glenn, there was no real problem. He could easily control the attitude of the capsule. On the ground, there was concern that the other thrusters might fail. This was in the minds of Walter C. Williams and Christopher C. Kraft as Glenn told Shepard, "Only really unusual thing so far beside ASCS trouble were the little particles, luminous particles around the capsule, just thousands of them right at sunrise over the Pacific. Over."

Shepard replied, "Roger, Seven. We have all that. Looks like you're in good shape. Remain on fly-by-wire for the moment. Go to Bermuda now." There was no further attempt at communication with the White House.

Glenn began his second revolution of the earth with assurances that he felt fine. One hour and 39 minutes after liftoff, he was again talking to Bermuda, and a few minutes later, to the Canary Islands capsule communicator. He attempted to tell the ground that he had seen the wake of a ship as he passed over the primary recovery area where he was to land at the end of his third revolution around the earth, but neither Bermuda nor Canary heard him. It was one of the most important observations of the flight, for it established for the first time a man's ability to see small surface details from orbit.

Canary asked Glenn to see if the motions of the particles the astronaut had been describing could be correlated with the thruster firings. Glenn's persistent description of the particles had aroused a good deal of curiosity throughout the Mercury net, and it was considered possible that he had discovered a new phenomenon at the top of the atmosphere. However, the pragmatic engineers at Mercury Control believed that the particles were somehow connected with the space craft and might be coming from the thrusters. But Glenn insisted that they had no connection with his use of the control jets.

As he coasted over the Atlantic, he felt the warmth of the afternoon sun even through his suit as its light came through his window. He tested the control jets by turning the ship completely around. He reported he liked flying frontward "so you could see where you're going." As he passed over Africa, he turned the ship around to "orbital attitude."

Then another thruster problem appeared—the reverse of the one he had encountered crossing America. Mysteriously, control reappeared in his 1-pound left thruster, but at the same time the right one failed. At 2 hours and 5 minutes into the flight, he told Zanzibar in his 30-minute report, "Head movements cause no nausea or no bad feelings at all. I am surprised that I can look as close to the sun as I can. The sun is shining directly on my face at the present time and all I have to do is to shade my eyes with my eyebrow." Three minutes later, he reported his second sunset: "There's a brilliant blue out on each side of the sun, horizon to horizon almost. I can see a thunderstorm down below me somewhere, and lightning."

THE LANDING-BAG DILEMMA

At 2 hours, 19 minutes, Glenn received a curious communiqué from the Indian Ocean Ship: "We have a message from MCC [Mercury Control Center] for you to keep your landing-bag switch in the off position. Landing-bag switch in the off position." Glenn advised that it was off. Seven minutes later, the same instruction was repeated as Friendship 7 came into radio range of Muchea. Once more, Glenn assured the ground that the switch was off. "You haven't had any banging noises or anything of this type?" Cooper asked. Glenn replied that he had not.

What did concern Glenn was his discovery that the yaw-right one-pound thruster was not working in either the automatic or the fly-by-wire mode of control. Also, the attitude of the spacecraft as shown by the gyroscope indicators did not agree with what Glenn could see looking out the window. When the indicators showed Friendship 7 to be in level flight, he could tell from looking at the horizon that the ship was canted in a slight roll. Glenn shifted to a third control mode called manual proportional, in which linkages transmitted movements of the control stick to proportional control valves regulating the flow of fuel to the thrusters. In this mode, he drew on his second fuel tank, which was 90 per cent full. The automatic and fly-by-wire control modes had exhausted 40 per cent of the fuel in the first tank.

Meanwhile, back at Cape Canaveral, concern was mounting over an indication that the MA-6 landing bag might have been deployed. One of two switches on the base of the capsule had sent a signal to the ground indicating that the locks holding the heat shield in place had opened. Glenn did not have an indicator in the cabin that would tell him the status of the heat-shield locks. He was not aware of what Mercury Control was reading on the ground.

The landing bag was a rubberized canvas cover between the heat shield and the spacecraft's rear wall. During the capsule's descent on parachute, the bag could be dropped 4 feet by releasing mechanical clamps which held the heat shield tightly against the spacecraft body. Air would fill the bag during descent in the atmosphere, and thus a 4-foot cushion of air would be provided against splashdown impact.

Consequently, any signal showing that the bag was deployed meant that the heat shield had dropped away from the spacecraft, thus exposing the base of the spacecraft to incinerating heat on reentry. The possibility that Glenn's doom had been decided by such an accident haunted the project directors, Williams, Kraft, and Gilruth. While they were dubious about the signal, they could not be sure. Glenn's insistence that the landing-bag switch was in the off position reassured them somewhat, but nevertheless, as the flight continued with the landing-bag-deployed light glowing steadily on the console at Cape Canaveral, the tension in the control room grew.

Glenn remained unaware of the anxiety that the unexplained landing-bag-deployment signal was causing at Cape Canaveral. He advised Woomera that his visual acuity was as good as ever—that there had been no deterioration in his sensory acuteness, so far as he knew, as a result of weightlessness. Over Canton Island again, he reported his second sunrise and the reappearance of "those little particles coming around the capsule."

Canton told him telemetry down below indicated that "everything is okay."

Then Canton added: "We also have no indication that your landing bag might be deployed."

GLENN: Did someone report landing bag could be down? Over.

CANTON: Negative. We had a request to monitor this and to ask you if you heard any flapping when you had high capsule rates [of movement in yaw or pitch].

GLENN: Negative! Well, I think they probably thought these particles I saw might have come from that, but these are . . . there are thousands of these things and they go out, it looks like miles, in each direction from me and they move by here very slowly. I saw them at the same spot on the first orbit.

Friendship 7, moving on a more northerly track across the Pacific than on the first revolution, came within radio range of the Hawaii station on Kauai for the first time. Glenn checked indications of attitude on his instrument panel with those on the ground. He reported that, compared to his view of the horizon, the instruments were off about 20 degrees in roll. In addition, the uneven performance of the automatic attitude-control systems continued to allow the craft to drift from one position to another until he made corrections by firing the thrusters manually.

HAWAII: Do you still consider yourself go for the next orbit?

GLENN: That is affirmative! I am go for the next orbit.

HAWAII: I understand. At present time, ground concurs.

On the ground, meanwhile, flight directors at Cape Canaveral were conferring with the McDonnell Aircraft engineers about what to do in case the heat shield had come loose. A third orbit would give them another 90 minutes to study the problem and keep John Glenn alive that much longer in the event the heat shield had dropped. Still he was not advised of the worrisome signal at the Cape and he kept sailing on, feeling fine and steering the little spacecraft easily.

GO FOR A THIRD ORBIT

At 2 hours and 57 minutes into the flight, Friendship 7 approached the coast of California south of San Diego, and Glenn said hello to his teammate, Astronaut Walter Marty Schirra, Jr., at Point Arguello, California. He reported that the erratic gyros had cleared up somewhat on the flight across the eastern Pacific after he caged (halted) them and then uncaged (released) them again.

SCHIRRA: John, the aeromeds are real happy with you. You look real good up there.

GLENN: All right, fine, glad everything is working out. I feel real good, Wally. No problems at all.

At the Cape, meanwhile, the landing-bag-deployment conference in Mercury Control was in full swing. It had been tentatively determined to order Glenn to keep the retrorocket package on, even after the rockets had fired. The package was nestled against the heat shield and held to the spacecraft frame by straps, which could be cast off by the astronaut. Controllers believed that by retaining the retropackage, the straps holding it would also keep the heat shield in place if it actually was loose.

As he flew over Texas at 3 hours and 4 minutes into the flight, Glenn reported he could see El Paso through some clouds. Three minutes later, Friendship 7 came within radio range of the Cape. Nothing was said about the landing bag-heat shield problem. Friendship 7 headed out over the Atlantic on its third revolution around the earth.

Glenn advised Canary Cap Com that the little particles had vanished. He told the Atlantic Ocean tracking ship to relay to the Cape that his attitude-control indicators were "way off" compared with his view of the horizon.

The pilot then reported he was heading into the sunset once more, and light illuminated a great deal of dirt on his window. It looked as though bugs had been smashed on the window on the way up, he said. "Looks like blood on the outside of the window, maybe. It makes it real . . . very difficult to observe anything on the sun side."

As he passed out of range of Zanzibar Station, Glenn saw a thunderstorm below. The lightning flashes, he thought, were like firecrackers going off.

SEGMENT 51

The Indian Ocean Ship flight surgeon asked Glenn whether he had been using any of his secondary oxygen supply. The quantity registering on the ship's console was 90 per cent. Glenn said he could not understand that, since he had not resorted to the secondary tank at all. His console, too, showed only 90 per cent. The tank presumably was leaking. Moreover, he told the ship, the automatic control system was slipping badly and wasting fuel. And the one-pound right thruster was out on both the automatic and fly-by-wire modes of control.

Ten minutes before retrofire, Hawaii passed up to Glenn the dreaded problem the Cape had wrestled with for nearly three hours: "Friendship 7, we have been reading an indication on the ground of segment 51, which is landing-bag deploy. We suspect this is an erroneous signal. However, Cape would like you to check this by putting the landing-bag switch in automatic position and see if you get a light. Do you concur with this? Over."

For the first time, Glenn realized what had been going on down on the ground, and he became instantly aware that he might be under a death sentence. He said, "Okay, if that's what they recommend, we'll go ahead and try it. Are you ready for it now?"

HAWAII: Yes, when you're ready.

GLENN: Roger. Negative in automatic position. Did not get a light and I'm back in off position now. Over.

HAWAII: Roger! That's fine! In this case, we'll go ahead and the reentry sequence will be normal.

Hawaii asked him to adjust the spacecraft clock, which was running one second ahead of the Cape clock, to Cape time. Retrofire would be at 4 hours, 32 minutes, and 37 seconds.

At 4 hours, 29 minutes, and 56 seconds, Glenn was in radio contact with Schirra at Point Arguello. Glenn asked for a time check with the Cape, thinking he might be several seconds off.

RETROFIRE

SCHIRRA: Will give you the countdown for retrosequence time, John. You're looking good.

GLENN: Roger. We have 50 seconds to retrograde. Over.

SCHIRRA: John, I'll give you a mark. Forty-five. Mark.

GLENN: Roger. I'm on ASCS [automatic] and backing it up manual. My fuel is 39 [per cent].

SCHIRRA: Thirty seconds, John.

GLENN: Roger. Retro warning is on.

SCHIRRA: Good. John, leave your retropack on through your pass over Texas. Do you read?

GLENN: Roger.

SCHIRRA: Fifteen seconds to sequence.

GLENN: Roger.

SCHIRRA: Ten. Five, four, three, two, one. Mark.

GLENN: Roger. Retrosequence is green.

SCHIRRA: You have a green. You look good on attitude. [This was critical, for unless the retrorockets fired in precisely the correct alignment along the flight path, their braking action would be reduced and the splashdown would be miles from the impact area where recovery ships were waiting.]

GLENN: Retro attitude is green.

SCHIRRA: Just past 20.

GLENN: Say again?

SCHIRRA: Seconds.

GLENN: Roger.

SCHIRRA: . . . five, four, three, two, one, fire.

GLENN: Roger, retros are firing.

SCHIRRA: Sure they be!

GLENN: Are they ever! It feels like I'm going back toward Hawaii.

SCHIRRA: Don't do that. You want to go to the East Coast.

GLENN: Fire retro light is green.

SCHIRRA: All three here.

GLENN: Roger. Retros have stopped. A hundred . . .

SCHIRRA: Keep your retropack on until you pass Texas.

GLENN: That's affirmative.

SCHIRRA: Check. Pretty good-looking flight from what all we've seen.

GLENN: Roger. Everything went pretty good except for all this ASCS problem.

SCHIRRA: It looked like your attitude held pretty well. Did you have to back it up at all?

GLENN: Oh, yes, quite a bit. Yeah, I had a lot of trouble with it.

SCHIRRA: Good enough for government work from down here.

GLENN: Yes sir, it looks good, Wally. We'll see you back East. Do you have a time for going to jettison retro?

SCHIRRA: Texas will give you that message. Over.

GLENN: Hello, Texas. Friendship 7, over.

It was four minutes after retrofire that Glenn saw the California coast. He reported he could see El Centro and the Imperial Valley as his spacecraft raced down toward the top of the atmosphere.

TEXAS: This is Texas Cap Com, Friendship 7. We are recommending that you leave the retropackage on through the entire reentry. This means that you will have to override that 0.05 gravity switch [the switch to release the retrorocket pack when the braking action of the atmosphere produced one five-hundredth of a gravity] which is expected to occur at 4 hours, 43 minutes, and 53 seconds. This also means that you will have to manually retract the scope. Do you read?

GLENN: This is Friendship 7. What is the reason for this? Do you have any reason? Over.

TEXAS: Not at this time. This is the judgment of Cape Flight.

Glenn asked the Texas Cap Com to repeat the instructions. The Cap Com did. Glenn said he understood he would have to make a manually controlled reentry with the retropack on. Texas affirmed and said Cape would give him the reason.

At 4 hours, 40 minutes, and 21 seconds, three minutes before reentry, Shepard's voice came in: "Friendship 7, this is Cape. Over."

GLENN: Go ahead, Cape. Friendship 7.

SHEPARD: Recommend you go to reentry attitude and retract the scope manually at this time.

GLENN: Roger. Retracting scope manually.

SHEPARD: While you're doing that, we are not sure whether or not your landing bag has deployed. We feel it is possible to reenter with the retropackage on. We see no difficulty at this time in that type of reentry. Over.

GLENN: Roger. Understand. Going to fly-by-wire. I'm down to 15 per cent [fuel] on manual.

REENTRY

SHEPARD: Roger. You're going to use fly-by-wire for reentry and we recommend that you do the best you can to keep a zero angle during reentry. Over. The weather in the recovery area is excellent, 3-foot waves, one-tenth cloud coverage, 10 miles visibility.

GLENN: Roger.

SHEPARD: Seven, this is Cape. Over.

GLENN: Go ahead, Cape, you are going out.

SHEPARD: We recommend that you . . .

GLENN: This is Friendship 7. I think the pack just let go. A real fireball outside. Hello, Cape. Friendship 7. Over. Hello, Cape. Friendship 7, over. Hello, Cape, Friendship 7. Do you receive? Over. Hello, Cape, do you receive? Over.

A sheath of ionized air which surrounded Friendship 7 as it plunged into the atmosphere blacked out radio communications for 4 minutes and 25 seconds. During that time, the deceleration built up from 0.05 gravities, where reentry began, to 7.7 gravities, and then fell back to 1 gravity over a period of 3 minutes and 30 seconds. In Mercury Control, there was strained silence. Dr. Douglas bowed his head and prayed.

At 4 hours, 47 minutes, and 16 seconds, Glenn heard Shepard.

SHEPARD: How do you read? Over.

GLENN: Loud and clear. How me? Mercury Control came back to life.

SHEPARD: Roger, reading you loud and clear! How are you doing?

GLENN: Oh, pretty good.

SHEPARD: Your impact point is within one mile of the uprange destroyer.

GLENN: Roger. We're through the peak now.

SHEPARD: Seven, this is Cape. What's your general condition? Are you feeling pretty well?

GLENN: My condition is good, but that was a real fireball, boy. I had great chunks of that retropack breaking off all the way through.

SHEPARD: Very good. It did break off, is that correct?

GLENN: Roger. Altimeter off the Peg, indicating 80,000 [feet].

SHEPARD: Recovery weather is very good. Over.

GLENN: Roger. understand. 55,000 stand by mark.

SHEPARD: Say again your altitude, please.

GLENN: 45,000. Rocking quite a bit. I may still have some of that pack on. I can't damp it, either. Going to drogue [deploy the drogue-stabilizing parachute] early. Rocking fairly . . . drogue came out. Drogue is out. Drogue came out at 30,000 at about a 90-degree yaw.

SHEPARD: Roger. Is the drogue holding all right?

GLENN: Roger. The drogue looks good. Standing by for main chute at 10 [thousand feet].

SHEPARD: Roger.

GLENN: Main chute is on green. Chute is out, in reef condition at 10,800 feet and beautiful chute. Chute looks good. On emergency oxygen and chute looks very good. Rate of descent has gone to about 42 feet per second. The chute looks good.

At 4 hours, 55 minutes, and 10 seconds after liftoff from Cape Canaveral, John Glenn reported he was nearing the surface of the ocean.

"Here we go," he said as the now-deployed landing bag hit the water. "Friendship 7. Impact." It was 2:43 p.m. Eastern Standard Time.

Glenn was instructed by "Steelhead," a voice from the recovery ship, to remain inside the capsule. Twenty minutes later, Friendship 7 was lifted aboard the destroyer U.S.S. *Noa*.

For several minutes, the Navy crew struggled to unfasten the top hatch. Glenn worked on it from the inside, but it seemed to be welded to the vehicle. Glenn decided to blow out the emergency side hatch. He pulled the pin, pushed the plunger, and the metal cover popped out with the bang of explosives and fell with a clang on the deck.

"It sure was hot in there," said John Glenn as he climbed out on deck and gazed over the calm and lovely surface of the sea.

Chapter 6

Rendezvous

The spectacular success of John Glenn's ride exhilarated the nation as no individual exploit had done since the trans-Atlantic flight of Charles A. Lindbergh in 1927. Glenn's three circuits of the earth did not match the 17-orbit flight of the Russian cosmonaut Gherman S. Titov the previous August, but the mission of Mercury-Atlas 6 stirred the world because, under the American policy of disclosure of civilian space events, the world could watch it. Sharing the national mood of elation, President Kennedy flew to Cape Canaveral to welcome the Marine pilot home. The good-humored encounter between the youthful Chief Executive and America's newest hero and his family delighted thousands of Cape workers, with whom Glenn insisted on sharing his triumph, and millions of television viewers. They saw a quick-witted, well-spoken astronaut, who combined daring with maturity. The first American in orbit was a devout, reflective man, and the public liked what it saw.

Early Monday morning, February 26, Glenn and his family flew with Kennedy from West Palm Beach to Andrews Air Force Base outside of Washington for a reception in the capital. Thousands of persons lined Pennsylvania Avenue in a cold rain to cheer and applaud the astronaut with unreserved enthusiasm. With his knack of saying what people like to hear, Glenn struck just the right balance of humility and vision in a 17-minute speech before a joint session of Congress when he suggested that his mission was merely the prologue to the great American adventure in space. "I feel that we are on the brink of an area of expansion of knowledge about ourselves and our surroundings that is beyond description or comprehension at this time," he said.

In New York March 1, four million persons were estimated[1] to have turned out to cheer Glenn, his fellow astronauts, and their families in a triumphal motorcade on Broadway, drenched in a blizzard of ticker tape. Finally, on Saturday, March 3, Glenn was welcomed by his home town of New Concord, Ohio, where more than 50,000 persons jammed the small community of 2127 to see him. After a motorcade through the town, Glenn spoke at a civic luncheon at Muskingum College, from which he had dropped out in his junior year to become a Naval Air Cadet. Glenn was so moved by his home-town reception that he seemed close to breaking down emotionally for the first time in his protracted ordeal as a hero, and when he rose to speak, to thunderous applause, he paused to recover his poise by kissing his wife Anna.

With the flight of Mercury-Atlas 6, the United States regained much of the technological "face" it had lost to earlier Russian successes in space. But of even greater significance was the lift the feat gave to Americans, who like much of the world, tended generally to regard the competition in space with Russia as a test of their own capabilities.

The flight dispelled any lingering fears that orbital free fall would disturb the central nervous system of the pilot, or cause perceptual disorientation or emotional stress that would make him unable to control the vehicle. It proved that a trained man could pilot a space vehicle in orbit for at least 4½ hours without discernible ill effects That fact alone confirmed the validity of manned space flight for Americans.

While Gagarin and Titov had emerged from their flights in good condition, Titov had complained of motion sickness after the fourth orbit. Academician A. A. Blagonravov of the Soviet Union characterized the complaint as a disorder of the vestibular apparatus of the inner ear. He said, "These data gave grounds to believe that under weightlessness disorders in interaction of spatial analyzers and changes of sensitivity thresholds of the vestibular apparatus may occur."[2] Glenn had not experienced the slightest discomfort on three revolutions of the earth. Was the vestibular disorder to be expected on longer flights, or was it peculiar to Titov? There are differences in men and differences in machines. The Vostok life-support system provided a normal, two-gas atmosphere, slightly enriched with oxygen, at a pressure of 11 pounds per square inch, or about seven-tenths of sea-level atmospheric pressure. In Friendship 7, Glenn breathed pure oxygen at 5-pounds-per-square-inch pressure, or a little more than one-third of sea-level atmospheric pressure. Each country subjected its space travelers to a different breathing and pressure environment. It was therefore not feasible from the Russian experience to predict how an American pilot would fare in orbital flight in an American vehicle.

Because of the truster problems which plagued Friendship 7, it was apparent that without Glenn's compensating skill the flight could not have completed three orbits. Walter C. Williams and Christopher C. Kraft, the flight directors, said they would have brought it down by ground radio control on the second revolution if it had been unmanned. The thruster trouble threatened loss of attitude control, which is essential to align the vehicle properly for retrofire and reentry into the atmosphere.

Beyond all this, the political effects of the Glenn flight were highly pleasing to the Kennedy administration. It had been an astonishingly good show for the White House. Moreover, the public's response seemed to be a sign of popular approval—visible for the first time—of one of the President's most critical decisions: to land men on the moon by the end of the decade and bring them back safely. Certainly, this decision, made nine months before Glenn flew, seemed considerably more feasible after the flight than before it.

Consequently, the Mercury-Atlas 6 event had the additional impact of making the lunar landing more acceptable and credible as a goal than before. The public's euphoric reaction to John Glenn's ride whetted an appetite for more space adventure. Why not land on the moon? Nothing less could surpass the triumph of February 20, 1962.

FROM THE EARTH TO THE MOON

With the advent of Mercury-Atlas 6, the space program took on the glow of a national asset in the international struggle for techno-scientific advantage, with all of its military and commercial implications. That possibility had not been well visualized during the Eisenhower administration, which had persisted in its low-key approach to space-flight development and had rationalized its attitude by minimizing the Russian space advantage.

Under Eisenhower, the program had remained an ambiguous effort, with the limited objectives of flying a man in orbit and of conducting an economical program of scientific exploration with small, instrumented satellites. The satellite program, which was paying handsome dividends in scientific knowledge, was supported so modestly that only slow development was possible. As a means of achieving national security and international prestige, the civilian space program had been conspicuously underestimated by Eisenhower and his advisers. The President insisted that despite Russian leadership in space, the United States was not engaged in a contest for space supremacy and that Soviet space prowess signified no threat whatsoever to national defense.

With the election of John Fitzgerald Kennedy, the effort in space began to acquire a new aura of urgency and mission. Kennedy found it both useful and fascinating. It was, besides, an engaging symbol of his "new frontier" program to enable America to lift itself by its bootstraps out of the doldrums of technical and economic stasis.

Kennedy and his aides had been quick to see that the American image had been seriously impaired by Sputnik and later by the Vostok flights, and that American prestige, as a result, had declined in Western Europe. The implication of technological second place did nothing to enhance the American military posture in the world or the demand for American hard goods and tools in the international market.

During the 1960 election campaign, Kennedy had raised the issue of the "missile gap" which he charged had resulted from the apathy and lack of technological foresight of the Eisenhower-Nixon regime. In vain did Nixon argue that Eisenhower had closed the gap, which Nixon charged had been inherited from Harry Truman. It was true that there had been a missile gap in 1957 when the Russians launched Sputnik. It was also true that it was closed by 1960 when the United States had more missiles than Russia had.[3] But the figures which might have proved the case for Nixon were secret. What was obvious to the American electorate in the autumn of 1960 was that Russia had bigger rockets than the United States, because Russia could orbit much larger payloads.

While the missile-gap argument involved military safety primarily, Kennedy buttressed it by alluding to America failures to match Soviet achievements in space. In this, Kennedy was supported by a special report of the United States Information Agency on "World Reaction to the United States and Soviet Space Programs."[4] It said that public opinion in most parts of the free world believed the Soviet Union was ahead of the free world in space achievements and that the USSR would still lead in space ten years hence. "The enduring, prevalent image of a substantial Soviet advantage in space appears to rest on a strong belief that the USSR has got there 'fustest with the mostest' and 'most' is apparently based on a more complex calculation than simple enumeration. Number has not captured the public imagination as range and primacy seem to have done."

The USIA survey added that in Great Britain, France, West Germany, Italy, and Norway space activities tended to be consistently reported and discussed in the context of a space race between the United States and the USSR. "Within this rivalry," the agency said, "space achievements are viewed as particularly significant because of the strong tendency for the popular mind to view space achievements as an index of the scientific and technological aspects of the rival systems and to link space capabilities with military, especially missile, capabilities."

In spite of the lack of a directive, a scheme to develop extended manned space flight beyond Project Mercury was hatched in NASA in the last year of the Eisenhower administration. It seemed inconceivable to the NASA directorate that manned-space-flight development should stop at low earth orbit. With more powerful rockets and more sophisticated spacecraft than were planned for Mercury, it would be possible to fly men around the moon and even land them on its surface, in a project called Apollo. With the aerospace industry abetting this conviction, NASA formally initiated lunar-landing studies in the fall of 1960. Envisioned was a ten-year program which would accomplish a manned lunar landing after 1970—possibly as early as 1971. The NASA task force called in a group of industrial contractors and asked them to prepare a feasibility study for a three-man space vehicle for extended flight in earth orbit and around the moon. Early in January 1961, however, NASA administrator Glennan warned his planning staff that the agency could do nothing to implement Apollo plans without a decision from the White House.[5] The Bureau of the Budget was holding NASA's request for the 1962 fiscal year to the minimum that would support only already-approved projects. All flight-development requests beyond Project Mercury were cut.

Before taking office, Kennedy appointed a nine-member Ad Hoc Committee on Space with Jerome B. Wiesner of the Massachusetts Institute of Technology as chairman. The committee made a report on January 10, 1961, listing five motives for a space program. They were national prestige, security, scientific observation and experiment, nonmilitary practical uses such as weather satellites, and possibilities for international cooperation. None of these objectives had been realized significantly by the Eisenhower administration, the report stated.

In order to settle policy problems which seemed to go beyond the purview of the space agency itself, the committee endorsed a Kennedy-Johnson plan to reactivate the National Aeronautics and Space Council, which had fallen into limbo in the last years of the Eisenhower administration. The council would coordinate civilian and military space activities, referee conflicts between NASA and the Department of Defense, and serve as liaison between NASA and other federal agencies.

The Wiesner report, however, failed to suggest what NASA was to do about manned space flight after Project Mercury. A high-level decision was essential if the plan to develop a lunar-landing capability was to be funded. In addition to the three-man Apollo spacecraft which NASA had asked contractors to study, the von Braun team at Huntsville was making steady progress in developing a truly big rocket called the Saturn C-1 for deep-space missions. Saturn consisted of eight Redstone rockets clustered around the center pole of a Jupiter. The nine tanks fed kerosene and liquid oxygen to eight Redstone H-1 engines. Since each engine developed 188,000 pounds of thrust, all eight put out a total thrust of 1,504,000 pounds—enough to place ten tons in orbit.

As was mentioned in Chapter 4, the Saturn project had been inaugurated at the Redstone Arsenal on August 15, 1958, under the Army's Advanced Research Project Agency. NASA had inherited it along with the von Braun team and the facilities at Redstone, which had become the Marshall Space Flight Center.

Here was the booster that would bring the United States up to parity with or beyond Russian rockets. In developing the Saturn, the von Braun team had enlarged upon its Jupiter C idea of clustered rockets. More powerful configurations of the Saturn cluster design were on the drawing board, using a new and more powerful engine called the F-1, which Rocketdyne was developing under a NASA contract. A single, kerosene-liquid-oxygen F-1 engine would put out 1,500,000 pounds of thrust, as much as all eight H-1 engines in the Saturn C-1. By clustering the super-powerful F-1 engines in the booster and adding high-energy upper stages, it was possible to build a rocket powerful enough to propel a 75-ton Apollo space vessel directly to the moon. At Huntsville the engineers called this super-Saturn rocket the "Nova." Except for the Saturn C1, however, NASA had no authorization to develop big rockets.

Meanwhile, Glennan submitted his resignation to the President-Elect on January 20, 1961, and Dryden followed suit. Kennedy accepted Glennan's resignation but asked Dryden to stay on. The President had a new top man in mind for NASA, someone altogether different for the scholarly administrator, who returned to the presidency of the Case Institute of Technology in Cleveland. What a new-look space program needed, Kennedy believed, was a promoter with administrative experience and keen political perception. The President selected James E. Webb, North Carolina-born lawyer, Marine Corps veteran of World War II, and former Bureau of the Budget director, whom friends and critics alike referred to as—and not without admiration—"the fastest mouth in the South."

Webb came well recommended by Vice-President Lyndon

B. Johnson and by Senator Robert S. Kerr, who had succeeded Johnson as chairman of the Senate Committee on Aeronautical and Space Sciences. Between 1936 and 1943, Webb had acquired industrial experience with the Sperry-Gyroscope Company. Following his military service, from 1946 to 1949, he was President Truman's budget director, and from 1949 to 1952 he was Undersecretary of State. At the time he was nominated to be NASA administrator, he was chairman of the Municipal Manpower Commission of the Ford Foundation. He was a director and officer of Kerr-McGee Oil Industries and a director of McDonnell Aircraft, which had the Mercury capsule contact. Webb's nomination was endorsed by the Senate space committee, whose chairman, Kerr, was a principal officer in Kerr-McGee. After divesting himself of business interests that might conflict with his new post, Webb was sworn in as administrator of NASA on February 14, 1962. He immediately went to work to sell NASA to the United States, an effort which was to become inordinately successful.

Kennedy instructed Webb to examine and if necessary increase the budget which Eisenhower had granted the agency for the upcoming 1962 fiscal year, starting July 1, 1961. The proposed appropriation of $1,109,630,000 represented a minimal funding of Mercury and scientific- and weather-satellite projects. Webb asked for an increase of $308,191,000 to support the development of Project Apollo. In the absence of White House intervention, however, the Bureau of the Budget slashed the request, allowing only an additional $125,670,000 to be submitted to Congress. This would support new rocket development, the construction of new launch pads, and a communications-satellite project. The bureau cut $42,600,000 which Webb sought to advance Project Apollo by funding development of the spacecraft. The bureau had reasoned that even if the moon project was to be undertaken, the development of the Saturn would be the pacing item for all future flights beyond Mercury. Thus, it was not necessary to begin funding the Apollo spacecraft in the 1962 fiscal year; it could be funded the following year. While this reasoning may have appealed to accountants, it was based on the mistaken belief that a spacecraft for going to the moon was easier to build than a rocket. The reverse turned out to be true.

The Budget Bureau reflected the emphasis of the Wiesner report on applications satellites,* and its trimming of Webb's budget request was evidence of the continuing vacuum in policy about post-Mercury manned flight, which had been carried over into the new administration. However, the partial increase stimulated optimism about post-Mercury planning in NASA, since it indicated an administration trend. The space agency secretly advanced its 1967 target date for a manned Apollo flight in earth orbit to 1965, and the date for a flight around the moon from 1969 to 1967. The agency then set its lunar-landing target date in the time period of 1969-70.[6]

Two days after the one-orbit flight of Yuri Gagarin in Vostok I on April 12, 1961, the House Committee on Science and Astronautics, in search of a scapegoat, zeroed in on the Budget Bureau's space stinginess. Under questioning, Robert C. Seamans, Jr., NASA associate administrator, testified that the agency had asked the Eisenhower administration for $72,100,000 for Apollo but had received only $29,500,000 with which to develop the Saturn booster.[7] The $42,600,000 which had been denied was to have financed a series of space tests to determine how much heat shielding would be required on a spacecraft reentering the atmosphere on a return from the moon at 36,000 feet per second. Seamans explained that such tests were necessary before an Apollo spacecraft could be built to fly around the moon. While the loss of funds would not slow up the projected post-Mercury schedule, he said, it would have to be made up in the budget for the 1963 fiscal year.

Representative George P. Miller asked, "No matter how you add it up, Doctor, it means that the program is going to be delayed because you can't do basic planning or you can't make basic tests that must be made before you can actually begin to design and put your vehicle in operation?"

Seamans, however, did not agree. He declined to say that the lack of funding would delay the moon shot. He simply didn't know.

Representative Overton Brooks, then committee chairman, broke in, "The Saturn program, you say, can't be pushed harder even with additional money?"

Seamans replied noncommittally, "I was saying that the Sat-

* Weather and communications satellites.

urn program as we have defined it, using liquid propellants, is a well-defined program."

In the spring of 1961, it appeared that the House space committee wanted to move faster than NASA did.

Kennedy, meanwhile, reactivated the space council and asked Vice-President Johnson to take charge of it. Under the Space Act of 1958, the council was seated as an advisory body to the President and included, in addition to the President, the Secretary of State, the Secretary of Defense, the NASA administrator, and the Atomic Energy Commission chairman. Four other members were to be appointed by the President. Eisenhower had convened the council only eight times and recommended that it be abolished in 1960. The House of Representatives passed a bill to abolish the council, but Johnson succeeded in blocking action in the Senate.

After the change in administration, Johnson, carrying out Kennedy's wishes, was able to amend the Space Act to provide that the Vice-President rather than the President head the council. Under Johnson's direction, the council became a key factor in space policy decisions, both as a force for developing policy and as a means of implementing it through NASA and in Congress. This arrangement provided the mechanism for activating one of the largest programs of technological development in the history of mankind, Project Apollo.

During the spring of 1961, Kennedy made it clear that he wanted Apollo to be developed, along with lunar and planetary reconnaissance and space-science programs. Webb had the role of contact man to awaken industrial and university support of the program in exchange for lucrative development contracts and research grants from the government. Johnson could handle Congress. With Kennedy's support, the kind of far-ranging space program these men believed the country wanted could be set in motion. What was needed, however, was a far-reaching goal which would impress all of mankind and which everyone at home would approve and understand.

A SUMMONS TO ARMS

President Kennedy personally provided the goal and the vision in a State-of-the-Union type of message to a joint session of Congress on May 25, 1961. He called upon the nation to commit itself "to achieving the goal before this decade is out of landing a man on the moon and returning him safely to the earth." The United States had both the resources and talent to accomplish such a mission, he said, but he warned it would be an expensive one. Kennedy pointed out that the effort he was asking would increase the space budget by $7 to $9 billion over the next five years. He asked Congress and the electorate in general to consider the project so that its implementation would represent a truly national decision. "For while we cannot guarantee that we shall one day be first," he said, "we can guarantee that any failure to make this effort will make us last."

In the message, Kennedy asked the nation to make the lunar-landing decision in response to his proposal, as though there had been or was to be a great national debate on the issue of going to the moon. There was no debate in 1961 and none materialized after the message. Instead, both Congress and the space agency interpreted the President's words as a signal to charge full tilt for a lunar landing. The decision to expand the space program at that time was not impelled by any public demand; it was a Kennedy-Johnson decision. Its initial effect was dulled by the shadow of the earlier Bay of Pigs debacle in Cuba. Critics said Kennedy was shooting for the moon to get the public's mind off the Bay of Pigs.

The early impact of the message was light, considering its historical importance. The newspapers generally subordinated the new space program to the international and domestic aspects of the budget message, and some of them treated it editorially as a bit of caprice not to be taken seriously.

Opposition to the lunar program among scientists became evident as soon as they realized what Project Apollo would cost. Many feared it would divert federal funds from nonspace research in the life and physical sciences. Philip Abelson, editor of *Science*, the weekly magazine of the American Association for the Advancement of Science, called it the "moondoggle." At the centennial meeting of the National Academy of Sciences in Washington, October 22, 1963, Linus C. Pauling, professor of chemistry at the California Institute of Technology, deplored it as "a pitiful demonstration of our system of values."

Pauling, a Nobel laureate in chemistry and winner of the

Nobel Peace Prize for 1962, attacked the moon program in an extemporaneous epilogue to his formal address on "The Architecture of Molecules." He asserted that for every discovery the $20 billion spent on Apollo would buy on the moon, a thousand discoveries in molecular biology could be made. Pauling insisted that scientists were on the brink of new findings that would ease mankind's suffering.

His remarks were challenged by Harold C. Urey, professor of chemistry at the University of California, also a Nobel laureate in chemistry. Urey had adopted the position that manned exploration of the solar system, which he felt was ultimately more effective than unmanned probes, was essential to the expansion of the human species. Urey also wanted men on the moon because he doubted that instruments could tell him what he wanted to know about it.

The peppery Urey wanted to launch a debate with Pauling then and there, and Pauling was perfectly willing, but Frederick Seitz, the Academy president, ruled against it. The space issue wasn't on the program. "The question of the space program has intruded itself," Seitz said, "but this occasion is our birthday party. We came here to celebrate and hear summaries of where we stand in the various fields of research. We don't want the controversy to become the main show—until our party is over. We want to focus on what we're celebrating: the evolution of science in the United States." Seitz offered Urey a chance to express his views on the space program after the Centennial, but Urey said he would find another time.

He did. His chance to answer Pauling came December 30, 1963, at the annual meeting of the American Association for the Advancement of Science in Cleveland, where he was on the program. "Pauling explained that the money being spent in space ought to be spent for biological investigation," Urey said, "but he knows the National Institutes of Health will give him all the money he wants. What he needs is trained manpower."

The Centennial meeting of the NAS might have provided the forum for the essential debate on the worth of Project Apollo, but after being shunted aside there, the issue never again became a focal point at a major national meeting of scientists—although it was frequently a peripheral one.

As scientific adviser to the White House, Wiesner had

backed the Apollo decision. Two years afterward, when he was asked at a hearing of the Senate space committee if he had supported it, he said, "Yes. But many of my colleagues in the scientific community judge it purely on its scientific merit. I think if I were being asked whether this much money should be spent for purely scientific reasons I would say emphatically, 'No.' I think they fail to recognize the deep military implications, the very important political significance of what we are doing, and the other important factors that influenced the President when he made his decision."

THE EXTRATERRESTRIAL FRONTIER

In response to the President's call to arms, NASA planned to make the lunar landing in the spring of 1968—a year after the circumlunar flight. The first three-man Apollo flights of long duration (10 to 14 days) would start in 1965.

Webb advised the Independent Offices Appropriations subcommittee of the Senate Appropriations Committee that the lunar landing would cost $20 billion. While "some people" had been saying it could be made as early as 1967, he said, "I think 1968 is as good a date as any." [8]

The President had not "taken off" for the moon without a strong background of information and technical advice that the journey was within the "state of the art" of American technology. The study group which Glennan had appointed late in 1960 under George M. Low, director of Manned Space Flight at the time, reported in February 1961 that the journey was feasible. The success of the May 5, 1961, suborbital flight of Commander Alan B. Shepard, Jr., proved the soundness of the design of the Mercury spacecraft and of recovery at sea. On May 2, 1961, NASA named an *ad hoc* task force under William Fleming, a NASA headquarters executive, to determine what had to be done to accomplish the lunar landing in 6½ to 8½ years. The study was actually a basis for program and budget planning.

In a confidential report on June 16, 1961, the task group concluded that the lunar landing could be accomplished with the President's "by the end of the decade" time frame. The committee said the "pacing" items (the hardware which would take the longest to develop) were the booster stage of the moon rocket and test and launch facilities. None of these existed. All there was in mid-1961 was the President's "call to arms" and drawerfuls of reports and drawings of rockets and spacecraft.

The Fleming report did not provide any detailed ideas about the size and shape of the launch vehicle. It assumed, however, that the mode of travel would be direct ascent to the moon. The launch vehicle which that implied was the biggest one that had been considered—the Nova.

On May 25, 1961, the day of Kennedy's moon message, NASA established a second task force to study the launch vehicle. It was co-chaired by Nicholas Golovin of NASA and Laurence Kavanau, an engineer physicist with the Department of Defense. The committee launched a study of the Nova, which was designed to develop 12 million pounds of thrust in the first stage from a cluster of eight F-1 engines. That was eight times the thrust of the Saturn 1. The committee concluded that there were so many unknowns to be faced in the development of such a large rocket, including its aerodynamic behavior in powered flight up through the atmosphere, that the Nova could not be developed within the Kennedy time frame. Consequently, serious attention had to be given to other modes of traveling to the moon than simply firing a payload directly from the earth to the target, because only a smaller and less powerful launch vehicle could be developed in time to meet the deadline.

In this way, the President's "end of the decade" time reference, which was intended to inject enough urgency into the lunar decision to get the project moving, became established as a rigid deadline. In its final report in December 1961, the Golovin Committee recommended scrapping the whole Nova idea and, with it, the direct-ascent mode of lunar flight. Instead, in order to reduce launch weight on the pad, the committee recommended that NASA develop an advanced Saturn as the moon rocket, and a mode of travel by which the launch vehicle for the moon shot could be refueled while in earth orbit. This travel mode was called Earth Orbital Rendezvous (EOR). We will hear more about it later.

The historical legacy of the Golovin Committee report was the revelation that the seemingly simplest method of reaching the moon, by direct ascent, had to be "scrubbed" because of an arbitrary deadline which had little more than rhetorical significance at the time it was defined but which quickly began to mold the entire lunar project. NASA, it became clear, was bent on putting men on the moon in 1968, if all went well, and the entire flight program, as we will see presently, was laid out to meet that objective. The lunar landing was scheduled to come while Kennedy was still in office, if he had lived to win a second term. That would have conferred upon him a unique role in history as the chief executive who presided over the conquest of the moon. Fate, however, ruled that the lunar landing, whenever it is made, will be one of many memorials to the man whose bright vision of the nation's future called for an extraterrestrial frontier.

The effect of the impulsively initiated and compulsively executed timetable was to fit the lunar-landing program into the Procrustean bed, in which the quickest way, not necessarily the best, was the only one that fit. Project Apollo and the entire manned-space flight program for at least a generation was shaped by Kennedy's single reference to time and NASA's fierce efforts to obey it.

ODYSSEY IN OKLAHOMA

On May 26, 1961, the day after Kennedy's call to arms, a well-prepared briefing on the new space program and its lunar-landing goal opened at the Oklahoma State Fairgrounds coliseum at Tulsa. It was billed as the First National Conference on the Peaceful Uses of Space. At this two-day meeting, NASA attempted to communicate its program in some detail to the American people so that they would be able to assess the lunar-landing goal.

Webb and Senator Kerr were the two chief dignitaries at the meeting, and each of 600 persons who attended a banquet where both appeared received an embossed document certifying he was an "A-OK Astronaut of Oklahoma."

In his dinner speech, Webb said, "It was the decision of the President that the key to retrieving our position lay in determining that we could no longer proceed with the Mercury one-man spaceship as if it were to the end of our program, but that we must, even in a tight-budget situation, present to Congress the urgent necessity for committing ourselves to the giant boosters required to power the larger craft needed to accommodate crews of several men on long voyages of deep-space, lunar, and planetary exploration."

The chunky, red-faced, sandy-haired Webb, with his electric-blue eyes and volubility, had the gift of making fantasy sound plausible. Whenever he spoke about the space program, he plunged through the line of incredulity and skepticism like a veteran fullback driving hard for yardage. It struck few as strange that Webb was already dispatching men to the moon and the planets with nothing more impressive in his inventory of operational rockets than the little Redstone, which had barely managed to boost Alan Shepard to the top of the atmosphere on a flight of 160 miles.

Webb spoke of the giant Nova rocket standing 360 feet high—"60 feet taller than a football field is long." The diameter of the first stage would be 50 feet and of the upper stages 25 feet. The Apollo spacecraft would carry its own propulsion system, including retro-rockets to enable it to make a soft landing on the moon, other rockets to enable it to take off from the moon and return to earth, and steering rockets. Apollo alone would weigh 75 tons, said Webb.

Robert R. Gilruth, director of NASA's Space Task Group, which was developing Project Mercury, talked about a controlled landing of the Apollo, using a stowable, flexible wing, which could be deployed in the atmosphere.* He said this device would permit the astronauts to glide down to a selected landing site.

The Air Force was represented at the briefing by Major General O. J. Ritland, commander of the Space Systems Division, Air Force Systems Command. "Our position [on space] is similar to the attitude we held toward the airplane in 1910," he said. "At that time, a few far-sighted people conceded that it might some day be practical for man to fly, but even so eminent an authority as the magazine *Scientific American* looked upon aircraft with disdain. 'To affirm that the airplane is going to revolutionize the future is to be guilty of the wildest exaggeration,' that magazine reported editorially. Such an assessment probably was valid in terms of 1910 technology, but it did not allow for the technological developments to come."

The conference resounded with determination to meet the

* Designed by Francis M. Rogallo, of Langley.

challenges of the space age. Lloyd V. Berkner, then chairman of the Space Science Board of the National Academy of Sciences, and president of the Graduate Research Center of the Southwest, remarked, "One gets a little tired these days of reading about Russian space supremacy." The mood of high optimism, in which prospects and probabilities were frequently referred to as accomplishments, evoked a slogan which seemed to justify the space program as an evolutionary inevitability. "When it's steamboat time, you steam," it went. The observation was attributed to Mark Twain. One speaker put it another way, quoting Ecclesiastes 3:1: "To everything there is a season and a time to every purpose under heaven."

There were reserved voices, talking in terms of practicalities. John R. Pierce, director of Research Communications Principles, Bell Telephone Laboratories, discussed flexibility in a communications-satellite system. "We must not settle on one sort of satellite system to the perpetual exclusion of any other," the Bell scientist warned. "Now is not the time to freeze the design of an operational satellite system." He called for experimentation with low-altitude, spin-stabilized, broad-band communications satellites. As the first step, they could add greatly to the state of the art. Experimental work, especially in controlling their attitude or position in space for maximum antenna efficiency, would enlarge the range of possibilities open to designers, Pierce explained.

Pierce disclosed that Bell Laboratories was developing a "simple, promising experiment with a realistic low-altitude satellite." It was to become known as Telstar when it was successfully launched the following year to open intercontinental communications via space.

One of the most familiar voices at the conference was that of Edward R. Murrow, then director of the United States Information Agency. "At the risk of being the one rocket on your panel that fails on the pad," he said, quenching a cigarette under his shoe, "I'm going to assume the role of Socratic interlocutor." Murrow proceeded to suggest that communications satellites were fine as long as one had something useful to say over them. It was the content of the message, not the means of transmission, that promoted international understanding.

"Space satellites will not make human communications any better," he said. "They will simply diffuse it over a wider area. Communications systems have no conscience. They will transmit both filth and inspiration with equal facility. Do we really want the world, without the context of background knowledge, to see television covering the blood-lettings and bus burnings in Alabama? It may be that the history of our day will be decided by what dreams we choose to deliver. The issue, gentlemen, is not how we deliver it but what our delivery has to say."

THE $5-BILLION BUDGET

When the House space committee met in Room 214-B of the New House Office Building in Washington on July 11, 1961, a new NASA appropriation bill awaited its approval. It called for an authorization of $1,784,300,000—$675,000,000 more than the previous administration had been willing to grant.

The biggest increase came in development funds for Apollo and for the Saturn vehicle. There was also a $60,000,000 item for a manned flight laboratory, which was to become the Manned Spacecraft Center at Houston the following year. Programs of lunar and planetary exploration were strengthened. And there was the sum of $133,000,000 for the development of the Nova launch vehicle. The new budget had evolved in two stages: an initial boost from $1.1 billion to $1.23 billion in March and a "second-stage" boost reflecting Kennedy's lunar decision to $1.78 billion in May.

As Dryden, NASA's chief scientist and deputy administrator, explained: "I emphasize again there was a major, national decision to push forward with earnestness toward this [lunar] goal. This program will require the budget level certainly to rise to the order of $5 billion in a few years."

Compared to the original NASA ten-year plan to land on the moon after 1970, Dryden said, "this [Kennedy decision] represents a moving forward of the goals and a resultant increase of the cost in this decade of $7 billion to $9 billion over what would have been spent had we been content to proceed at a slower rate."

"As I understand it," said Representative David S. King, "the present thinking is that the space vehicle will be a Nova, which will be completed sometime in 1967, 1968, 1969 . . ."

"There are hopes for earlier," Dryden said. But, he added, there are other ways than direct ascent to reach the moon, and these were being considered. There is, for example, rendezvous in earth orbit for the purpose of refueling the moon-bound rocket. The concept of rendezvous envisioned the joining of two vehicles, launched at different times, in an orbit around the earth. No one knew how difficult it would be, but it could possibly be done by ballistic interception, that is, by aiming one vehicle so that it would arrive at a point in space at the same time as its target. A more practical rendezvous method called for a vehicle with maneuvering engines powerful enough to enable it to change its orbit so as to match flight paths with and overtake its target.

In one form, he said, earth-orbit rendezvous would require the launching of seven Saturn C-2 rockets so that enough fuel could be extracted from six of them orbiting the earth to propel a seventh to the moon and home again. The Saturn C-2 was planned as a more powerful vehicle than the C-1, and was to be capable of delivering about 3 million pounds of thrust with two F-1 engines. It never was developed.

Also, Dryden remarked, there is rendezvous in lunar orbit, but he did not feel it offered any advantages. Additionally, there is rendezvous on the lunar surface. The lunar vehicle could take on fuel for the journey home from a supply vessel previously landed by radio control. The trouble with this mode, Dryden explained, is depositing the supply tanker where lunar astronauts could reach it. "A man on the moon with his supplies scattered over an area of 25 square miles is going to have a bit of a time assembling all of this to come back."

Rendezvous was a new idea, and the House committee members were intrigued by it. Abe Silverstein, the veteran engineer from NACA and then director of NASA Space Flight Programs, explained further: "Instead of going directly to the moon, you stop at an intermediate point in space, whether it be in orbit around the earth or an orbit around the moon, or on the moon, and resupply in some fashion the vehicle in which the men are being carried."

"Which is the preferable method to use, Doctor?" asked Representative Brooks, the chairman.

"Right now," said Silverstein, "I think there is some technical disagreement among the people who are studying it. I personally have a preference for a rendezvous around the earth. There are others who argue for the other point."

The dream of direct ascent to the moon aboard the monstrous Nova began to fade as it became apparent in engineering studies in Huntsville that the big rocket could not be fabricated and man-rated within Kennedy's lunar-landing time frame, a conclusion duly reported by the Golovin Committee, mentioned on p. 168. Nova, 360 feet high and 50 feet in diameter, was too big a jump from Saturn 1, only 163 feet high and 21½ feet in diameter. So was Nova's eight-fold increase in first-stage thrust.

"With Nova, we could land a locomotive on the moon if anyone wanted one there," said Eberhard Rees, von Braun's deputy.[9] "Nova vehicles give us the most direct approach to manned lunar and planetary exploration." But Nova was not to be—at least in this decade.

Nova had been designed to lift 400,000 pounds to earth orbit and 150,000 pounds to the moon. Its first stage, as we have seen, would have consisted of eight F-1 engines developing a total thrust of 12 million pounds. Its second stage was to be powered by a multiple of the projected M-1 hydrogen-oxygen engine, each designed to develop 1 million pounds of thrust. The third stage was to be propelled by a single hydrogen-oxygen J-2 engine developing 200,000 pounds of thrust.

The first two stages would insert the third stage and the Apollo spaceship in earth orbit. After a coasting period across the Atlantic Ocean, the third stage would fire to accelerate the Apollo from orbital speed of 17,500 miles an hour to escape velocity of 24,500 miles an hour. During two and one-half days of travel, Apollo would turn itself around so that the engine in its lunar braking stage would point in the direction of flight. As they neared the moon, the crew would fire the engine to slow down the vehicle so that it would fall into orbit around the moon. The stage would then be jettisoned. At a predetermined time, a second engine system in Apollo's lunar touchdown stage would be ignited to brake the spacecraft's speed further, so that it would fall out of orbit toward the moon's surface. Since the moon has no atmosphere to slow down the vehicle further, the Apollo would have to depend totally on its touchdown engine to brake its descent sufficiently for a soft landing. In this mode, the three Apollo crewmen could remain aground seven days, but on the first landing they would stay only a single day. Using the touchdown stage as the launch platform for the return trip, they would fire Apollo's Service Module engine to blast off the moon, leaving the touchdown stage behind. As they approached the earth, the astronauts would jettison the Service Module and orient the conical Command Module housing the spacecraft cabin and the parachute landing system for reentry into the atmosphere and landing at sea.

SATURN 5

Next in size to the Nova was an advanced Saturn rocket called the Saturn C-5, which was being developed in the hangar at the Marshall Space Flight Center at the end of 1961. The first stage consisted of a cluster of five F-1 engines developing a total of 7.5 million pounds of thrust. The second stage would be powered by five J-2 engines developing 1 million pounds, and the third by one J-2 engine. Instead of propelling 150,000 pounds to the moon, as Nova was designed to do, Saturn 5 could lift only 90,000 pounds to escape velocity. That meant it could not boost the Apollo spacecraft weighing 150,000 pounds with its braking and descent stages directly to the moon. For Saturn 5, a mode of travel other than direct ascent had to be found: some kind of space rendezvous.

One possibility was the Earth Orbit Rendezvous flight mode, developed at the Langley Research and Marshall Space Flight Centers. There were several ways of playing the rendezvous game in earth orbit. The most practical appeared to be the tanking of the Saturn third stage in orbit with liquid oxygen.

By leaving 95 tons of liquid oxygen out of the third stage, the launch weight of Saturn 5-Apollo could be reduced sufficiently to enable the first two stages to place the third stage and Apollo in orbit as payload. Liquid oxygen would then be pumped into the third stage after it made rendezvous with a tanker previously launched. With its tank filled with oxidizer, the third stage would be ready to burn the 18.5 tons of liquid-hydrogen fuel it carried with it and boost Apollo to the moon.

The earth-orbit rendezvous (EOR) plan called for orbiting the tanker first, while the astronauts waited in the Apollo atop a Saturn 5 on the pad. When the tanker had achieved a 150-mile-high, circular orbit, up Apollo would go, its third-stage booster carrying hydrogen fuel but no oxygen to burn it with. The first two stages of the Saturn 5 would boost the third stage and Apollo to a 300-mile circular orbit. Ground controllers would then ignite the tanker engine to accelerate it into a 300-mile orbit that roughly would match that of Apollo and the third stage. The Apollo crew would maneuver their craft and the big third-stage rocket hitched to it to rendezvous and dock with the tanker. The tanker would be jettisoned after its load of liquid oxygen was pumped into the third stage. Fully fueled, the third stage would be ready to boost Apollo out of earth orbit on a trajectory for the moon. The lunar landing by the Apollo would be carried out as in the direct-ascent method.

THE BUG MODE

A third lunar-landing mode developed at the end of 1960 at Langley was called Lunar Orbit Rendezvous (LOR). Its chief difference from EOR was that it did not provide for Apollo to land on the moon at all. Instead, two astronauts would descend to the lunar surface in a space "rowboat" while the Apollo mother ship, with the third astronaut aboard, remained in lunar orbit. The rowboat was known as the Lunar Excursion Module, or LEM. However, because of the insectile appearance the spindly landing legs conferred on it, it became better known as the "Bug."

LOR saved weight by eliminating fuel and engine components required to lower Apollo to the moon and lift it off again. These weighed about 100,000 pounds. In their place, LOR substituted the Bug, which weighed only 35,000 pounds. Shorn of landing fuel and landing stages, Apollo could be reduced to about 55,000 pounds in weight, including the command (cabin) and service (propulsion system) modules. With the Bug, Apollo represented a 90,000-pound payload, which the engineering design said Saturn 5 could boost to lunar orbit directly.*

This ingenious method of going to the moon was discussed throughout 1961. It gained favor in the eyes of Joseph F. Shea, the Apollo program manager, Gilruth, and von Braun as it became evident that direct ascent with Nova would take too much time to develop and that EOR would be more

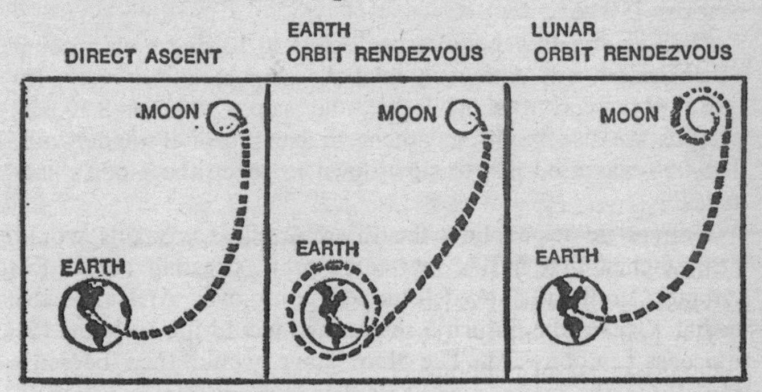

Three lunar landing modes.

costly inasmuch as it would require two Saturn 5 launchings for the lunar mission instead of one. The saving in weight was especially appealing. If Apollo did not have to sit down on the moon and get up again, its service-module propulsion system would be sufficient to push it back to earth from lunar orbit.

After repeated conferences were held in Washington, Langley, and Huntsville, it was decided by early summer of 1962 that the Bug mode was the way to go. NASA announced on July 5, 1962, that lunar-orbit rendezvous would be the mode of landing men on the moon, enabling North American Aviation, Inc., the Apollo contractor, to fix a design for the spaceship.

As D. Brainerd Holmes, director of the Office of Manned Space Flight, explained the plan to the House space committee, "We can go directly to the moon with the C-5 [Saturn] in

* Launch weight of Apollo Command and Service Modules and the Lunar Module (Bug) has since grown to 99,000 pounds. However, Saturn 5 payload capacity has been increased proportionately.

lunar-orbit rendezvous because the only thing we put down on the moon's surface is the so-called Lunar Excursion Module." [10]

Holmes had been recruited to the space agency from Radio Corporation of America, where he had been general manager of the Defense Systems Division. A 1943 graduate in electrical engineering from Cornell University, he had been appointed director of Manned Space Flight for NASA on November 1, 1961.

"I think the Bug is quite analogous to having a ship going to an island, but you don't take the ship in all the way; instead you send a small boat," he said. "Of the 150,000 pounds we use in direct ascent or earth-orbital rendezvous, 100,000 was used just to slow down to reach the moon's surface."

Holmes described how the three Apollo astronauts would be launched first into a low earth orbit, coasting across the Atlantic until, at a predetermined point over Africa or the Indian Ocean, the Saturn 5 third stage would fire to boost the spacecraft moonward. The third stage would then be jettisoned. On the two-and-one-half day journey, the crew would use the service-module propulsion system to make midcourse corrections and to slow the craft so that it would fall into lunar orbit.

In orbit around the moon, at an altitude of 60 miles, two of the three crewmen would climb into the Bug through a tunnel in the nose of the conical Apollo spacecraft, which was to be joined tightly against the hatch in the Bug. Hatches would then be sealed in the Lunar Module and the Apollo, and the two would separate. With an initial burn of the LM's descent engine, the two crewmen in the landing vehicle would maneuver it into an elliptical orbit. The apogee or high point of the ellipse would be 60 miles, the same as Apollo's circular orbit, but the low point, or perigee, could be 50,000 feet or less above the lunar surface. Since there is no atmosphere on the moon to impede orbital flight, the perigee could be dropped to an altitude just high enough to clear the lunar mountains.

From this orbit, the astronauts in the Bug could observe their landing site at close range and still be able to return to the mother ship at apogee in an emergency. They would not be committed to land until they were fully satisfied that the landing area was reasonably clear and level and that all systems aboard the Lunar Module were operating properly. Once aground on its four spindly legs, which end in large, circular pads, the Bug would be shelter for a stay of two to four days. To depart, the two men would blast off with the ascent engine, using the descent stage as a launch platform. Leaving the descent stage on the moon, the upper half of the LM would be propelled back into the 60-mile orbit, where the crew would dock with the waiting Apollo. They would reenter Apollo and cast off the LM, which would remain in lunar orbit. Then the reunited Apollo team would fire the Apollo's service-module engines to move the vehicle out of lunar orbit for the return to earth.

Objections to this LOR mode focused on the critical rendezvous around the moon, which seemed inordinately difficult and dangerous. Holmes did not try to minimize this aspect of the scheme. "If you don't rendezvous you are in real trouble up there, of course," he admitted. "So we do everything possible to have redundant equipment, both in the Bug and in the mother ship. If the Bug can't dock with the mother ship, the Apollo can come and get it. Because it is an unknown, it is quite natural for people to focus on the rendezvous part and worry about it. Nevertheless, all of our studies make us feel it will not be a difficult task to do."

Holmes said that an analysis by NASA showed that the probability of mission success was about the same for direct ascent as for lunar-orbit rendezvous. However, success probability was cut by 50 per cent in the earth-orbit mode because it required two Saturn 5 launchings. Mission success and mission safety were not the same in these modes. EOR was about as safe as LOR, since the astronauts in EOR would not be launched until their liquid-oxygen tanker was in orbit. It was the requirement of launching both the tanker and the astronauts separately, however, that reduced success probability in EOR.

The time frames of the three modes had been studied thoroughly. Development of direct-ascent capability with Nova would take the longest time, Holmes said—at least 20 months longer than lunar-orbit rendezvous. The reason was the time required to develop the Nova, which was not under contract.

EOR would take 6 to 15 months longer than LOR, he estimated. In EOR, as in direct ascent, it was necessary to develop the lunar braking and touchdown stages for the Apollo. Holmes seemed to feel these would take longer to develop than the Bug, the only new item which had to be developed in the LOR mode.

All other components of the mission in LOR were under development. North American Aviation had won the Apollo contract, although it was not the choice of the Apollo Source Evaluation Board which studied the qualifications of a number of contractors to perform the contract. The Board, consisting of 190 panels of NASA department heads and staff engineers, had selected the Martin Company to build Apollo. Its judgment was overruled, however, by Webb, Seamans, Dryden, and Gilruth on the grounds that the panels had underrated North American's technical competence. The NASA chiefs also preferred North American because its cost estimate was lower than Martin's.[11] The estimate later was to prove too low, and North American, much later, was to ingest additional technical competence from Martin by hiring some of its top management people.

In addition to the Apollo spacecraft, North American had been awarded the contract to build the powerful second stage of the Saturn 5, and the company's subsidiary, Rocketdyne, was developing the Saturn H-1, F-1, and J-2 engines. Chrysler had received the contract to develop the "small" Saturn 1 and 1B vehicles for Apollo earth-orbit flights. Boeing in Seattle had the contract for the giant booster stage of the Saturn 5, and Douglas Aircraft had been awarded the Saturn third stage—the one which, after earth orbit was reached, was to boost Apollo to the moon. So far as the Saturns were concerned, however, the basic development work was well under way at the Marshall Space Flight Center.

The relative costs of EOR and direct ascent "come out equal," Holmes said, while the estimated cost of the Bug mode was 15 per cent less. On a project costing $20 billion, the estimated saving by going Bug mode could be $2 to $3 billion—or so it seemed.

Representative James G. Fulton asked Holmes when he was scheduling his first moon shot.

"Our statements on this," Holmes replied carefully, "are

that we will get there within this decade. Even if I had a more definite answer, I really think that the question has such import nationally that the administrator rather than I should give the answer."

"Well," said Fulton, referring to Holmes's comparative time estimates for the three modes of lunar landing, "when I subtract 20 months from the end of the decade I get about 1968."

"History will tell us whether it is right," observed Representative Ken Hechler.

"I am completely satisfied we are on the proper pad," said Representative Richard L. Roudebush. "And I think it is a fine decision."

THE KENNEDY EFFECT

The "fine decision," according to Holmes, had required more than a million man-hours of study by NASA, industry personnel, and university scientists who had been asked to examine the relative advantages and disadvantages of the lunar-travel modes. As Joseph F. Shea, the Apollo project manager put it, LOR was a trade-off against the launch of two complete Saturn 5 rockets in earth-orbit rendezvous. Cost had to be considered, for it was estimaed that the cost of launching one Saturn 5 would be about $100 million. Shea added, "When we had Nova in the plan, we found that the requirements for the vehicle and the length of time it would take for a design would use one and one-half to two years in accomplishing the lunar mission." [12] Such was the impact of the President's timetable—the Kennedy effect.

There were a number of dissenters to LOR. Some said it was too hazardous. Others complained it froze the weight and size of the lunar payload, which could be increased in earth orbit simply by adding more propulsion units.

This view was elaborated to the House committee by Norman V. Petersen, technical director of the Air Force Flight Test Center at Edwards Air Force Base, California, home of the X-15 rocket plane.[13] Lunar-orbit rendezvous was inflexible, he said, because it used a single launch vehicle, and the payload was determined by what that vehicle could lift moonward. In the case of the Saturn 5, it was 90,000 pounds. On the other hand, earth-orbit rendezvous offered the opportunity of enlarging a payload for the moon, because propulsion sufficient to lift double that weight to the moon could be assembled in earth orbit.

Moreover, Petersen argued, EOR was safer because it allowed a return to earth in the event of a failure in the Apollo while it was in earth orbit and before it was accelerated to the moon.

One of the most influential opponents of LOR was Wiesner, chairman of Kennedy's Ad Hoc Committee on space, who preferred direct ascent, with earth-orbit rendezvous as his second choice. He felt that LOR was too complicated and offered too many opportunities for failure. The rendezvous modes were hotly debated in NASA, while North American impatiently waited for a decision so that it could complete its design for the Apollo. Even the final decision on July 5, 1962, did not end the argument. It erupted one hot morning on September 11, 1962, when President Kennedy, Vice-President Johnson, Webb, Harold Brown, then Air Force chief scientist, Wiesner, and the British Minister of Defence, Peter Thorneycroft, inspected the Marshall Space Flight Center on a tour of NASA installations. In the administration building, von Braun showed the party a model of the Saturn 5, which cunningly could be disassembled into the rocket's stages.

"This is the vehicle which is designed to fulfill your promise to put a man on the moon by the end of the decade," von Braun said to Kennedy. "By God, we'll do it!" [14]

Von Braun then began to recite the details of lunar-orbit rendezvous, and Wiesner began to argue, "No, that's no good." He pressed with a recital of LOR's disadvantages, as the President stood looking off into the middle distance with his arms folded and Johnson stared at the floor. Lord Thorneycroft seemed to be embarrassed. Kennedy broke the tension with a joke, and the party moved out of the building, but Wiesner continued to insist that LOR was neither the best nor safest mode of lunar voyaging.[15]

While this episode suggested that the "fine decision" was not fine all around, it seemed to be the last public controversy on the issue. Von Braun and his ex-Peenemünde team had taken part in the basic studies of EOR, which they had favored because of its flexibility. One day, they said, when the United States considered flying men to the planets the EOR capability would be a great advantage, for this would be the mode for launching a manned interplanetary flight. Nevertheless, when LOR began to look like an answer to the problem of accomplishing the lunar landing within the decade, the Huntsville engineers made a thorough study of it as well. In the autumn of 1962, after the decision was made, von Braun was asked by Representative Teague if there was disagreement on the lunar voyage mode.

"None whatsoever" replied von Braun.[16]

TEAGUE: In Houston * we were told that the astronauts were unanimous in their belief that this was quicker, cheaper, and safer.

VON BRAUN: We believe so, too. I am aware that there have been some statements to the effect that it was a bit surprising that Marshall, after having advocated earth-orbit rendezvous, came around and recommended lunar-orbit rendezvous. The fact is that at first we put a lot of work into the earth-orbit rendezvous studies and we found this mode entirely feasible. But later we put a lot of work in the lunar-orbit rendezvous mode also, and now we are convinced this is the fastest and safest way to go. The mode selection is not a question of basic feasibility. There are many ways to go to the moon. It is a question of time, cost, and so forth.

A MARTIAN ANALOGY

The lunar-orbit rendezvous mode using the Bug was similar to a plan for landing on Mars which von Braun had described in 1952 in an article titled "Das Marsprojekt" in the German magazine *Weltraumfahrt*. In it, the rocket pioneer had proposed a Mars expedition using a flotilla of ten space vessels. These would be assembled in earth orbit and would carry a crew of 70 men. Three-stage ferry rockets would deliver parts, equipment, supplies, and crews to the vessels or-

* By then the Manned Spacecraft Center had been shifted from Langley to Houston.

biting the earth at 300 miles. When fully loaded and fueled, the vessels would thrust out of earth orbit into an elliptical orbit around the sun, shaped to intersect the orbit of Mars. As they approached Mars, the ships would decelerate into an orbit around the planet. Three of the vessels would have landing boats for taking men down to the Martian surface, while the mother ships would remain in orbit about the planet.

Von Braun had assumed that the Martian atmosphere was dense enough to support a glider landing, so he equipped his landing boats with wings. These would be discarded when the landing parties blasted off the surface to rendezvous with the orbiting spaceships for the return to earth orbit—from which the explorers would descend into a spaceport in the shuttle rocket.

The main advantage of assembling interplanetary vessels in earth orbit was the reduction in weight which had to be boosted away from the planetary surface. It was another version of the principle of making big ones out of little ones—small modules lifted into orbit could be fastened together to form very large vehicles. Similarly, propulsion units could be joined to provide powerful launch vehicles. The von Braun plan was based on assumptions that the technology and technique of construction in free fall would not prove beyond man's capacity, and that the orbital mechanics of rendezvous could be mastered, enabling crews to assemble in space prefabricated units of space stations and interplanetary ships, launched by any number of rockets.[17]

From this conception of the true space vessel, plying the interplanetary sea but never touching shore and carrying its own landing boats, the Bug was born. It was the lunar version of the Martian landing boat.

Von Braun has prefaced his plan with the comment, "It is time to explode once and for all the theory of the solitary space rocket and its little band of bold, interplanetary adventurers. No such lonesome, extraorbital thermos bottle will ever escape earth's gravity and drift toward Mars."

As the next step after the LOR decision, Holmes called for bids on the Lunar Module, or Bug, and in November 1962 announced that the $350-million contract had been awarded to the Grumman Aircraft Engineering Corporation at Bethpage, New York, which had no other major space work. While the design of the Bug was dictated by its function—it did not require an aerodynamic shape since it never entered air—there was one mystery no one had the answer to at the beginning. That was the nature of the surface on which the Bug was to land. Whether it was deep dust in which a vehicle might be drowned, or treacherous rubble on which it might overturn, or simply coarse, grainy material, like a beach or a plowed field, strewn here and there with big boulders—no one knew.

No astronomer on the surface of the earth could see the fine detail of the lunar surface. The atmosphere limited resolution of surface detail on the moon by even the biggest telescopes to little less than half a mile. Consequently, the landing gear of the Lunar Module had to be designed tentatively on a best-guess basis. And the entire contract was based on the assumption that it was possible for the 15-ton vehicle to land safely on its four spindly legs, each terminating in a circular pod, like a pie plate, of crushable aluminum honeycomb. The honeycomb would help absorb landing shock.

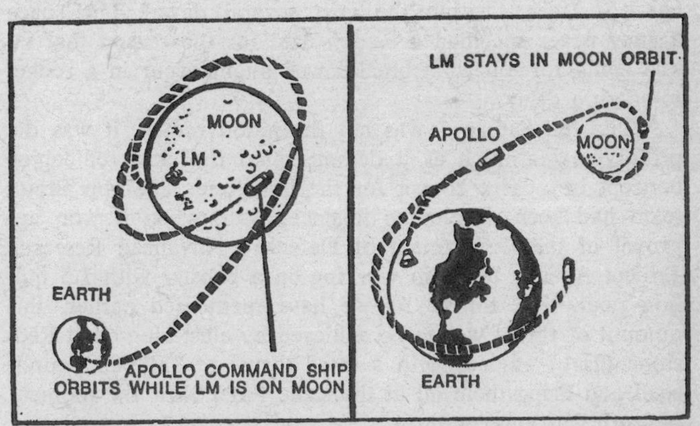

Diagram of landing by lunar orbit rendezvous and return from the moon.

In view of the controversy over whether the surface of the moon was covered with deep dust, rock, or rubble, NASA and Grumman faced a problem analogous to one of designing a landing gear for an airplane without knowing whether it was to land on terra firma or water. Nevertheless, the space agency and the contractor forged ahead, planning a landing gear that would support such a vehicle on any terrestrial land surface except quicksand or bog. Eventually, scientists would know what the lunar surface was like when NASA succeeded in landing reconnaissance vehicles on the moon, as it planned to do in advance of Apollo. And they might know in time to make any necessary changes in the Bug before the landing deadline. Or they might not. It was space-age gambling on an incredible scale. The entire $20-billion lunar adventure amounted to a bet that it was possible for something like the Bug to land on the moon. By the time this could be determined scientifically, the investment in Project Apollo would exceed $10 billion and it would have become an important facet of the economy.

AN EAR PLUG FOR SATURN

The first flight in the lunar-vehicle program—the maiden voyage of the Saturn 1—was announced by NASA on October 15, 1961, "within the next several days." The space agency never specified a launch date for the reason that no one could foretell what breakdowns might occur in a rocket to cause a delay.

While the Saturn 1 was not the moon rocket, it was the prototype, inasmuch as it demonstrated the clustered-engine concept in a large rocket for the first time. The von Braun team had been developing it since 1958, when it won approval of the Department of Defense's Advanced Research Project Agency to begin working on a booster with 1.5 million pounds of thrust. As we have mentioned earlier, this amount of thrust was to be achieved by clustering eight Redstone (H-1) engines with a rated thrust of 188,000 pounds each and firing them all at the same time. Now the moment of truth was approaching.

The Marshall engineers had proved that the cluster concept was quite dependable on the upper stages of the Jupiter C as early as 1956. In 1958, Explorer 1 rode into orbit on the upper-stage Sergeant rocket clusters of Jupiter C. In the fall of 1961, the whole moon-landing program was riding on the workability of clustered engines in the booster stage. If the eight H-1 engines worked properly in the Saturn 1, there was no reason to expect that the five F-1 engines would cause unforeseen problems when fired in unison in the Saturn 5 moon launcher.

Fabricated in the shops at the Marshall Space Flight Center outside Huntsville, the first stage of the Saturn was 21½ feet in diameter and 80 feet long. Early in August 1961, it was hauled to the Tennessee River by truck and loaded aboard the barge *Promise*, which had been equipped with a semi-cylindrical canopy to protect the rocket as it was fitted into a cradle lengthwise.

The water-borne covered wagon, pushed by a towboat, carried the rocket to the Mississippi River, down the river into the Gulf of Mexico, around Florida and up its east coast through the intracoastal waterway into the Banana River estuary west of Cape Canaveral. The stage was unloaded and erected by a 60-ton bridge crane on a brand-new launch pad at Launch Complex 34, which had been completed in June by the Jacksonville District of the United States Army Corps of Engineers.

When its dummy second and third stages were mated to the booster, Saturn 1 stood 162 feet tall. But it was dwarfed by its 310-foot-tall service tower in which it was cradled during months of checkout. The huge, red-painted steel tower, then the largest movable structure in the United States, and perhaps the world, could be eased back 600 feet from the launch pedestal to allow the Saturn to lift off. The tower moved on steel rails.

This test was the first of ten research and development flights of the Saturn 1. In the first four launches, the H-1 engines were rated at only 165,000 pounds of thrust each, producing a total of 1.3 million pounds. Later, starting with the fifth launch, they would be brought up to 188,000 pounds to produce the design thrust of 1.5 million pounds. Even at 1.3 million pounds, however, the Saturn was believed to be the most powerful booster anyone had attempted to fly. It had more than three times the thrust of the United States Atlas and Titan ICBMs and double that of the known Russian ICBMs.

From a distance of a mile and a half, the Saturn 1 looked like a gigantic rifle bullet, cleanly shaped and tapered. The first stage was painted white with black stripes running up and down, and UNITED STATES was painted vertically, in red block letters on a white portion of the booster. The upper stages gleamed plain white. Fully fueled and ready to go, Saturn weighed 925,000 pounds. The first stage was loaded with 600,000 pounds of propellant, consisting of kerosene fuel and liquid oxygen to burn it. The second stage, called the S-4, was inert—no fuel or oxidizer. It weighed only 25,000 pounds. To give the vehicle the weight it was designed to carry and to stabilize it during powered flight up through the atmosphere, the S-4 was ballasted with 90,000 pounds (11,000 gallons) of water. A dummy third stage, called the S-5, weighing 3000 pounds, was ballasted with 100,000 pounds of water (12,000 gallons).

About three hundred news-media correspondents and photographers gathered at Cocoa Beach, the largest assembly of press at the Cape since the suborbital flight of Virgil Grissom the previous July. The launch date was finally set for October 21, but, characteristically, it slipped to October 27. The test was to be an 8-minute flight to 90-miles altitude and a peak velocity of 3700 miles an hour. The rocket was to splash 225 miles downrange in the Atlantic east of Grand Bahama island. Of prime importance in the test were the performance of the engines, particularly the inner four, which could be gimbaled to provide steerage, and the aerodynamic behavior of the vehicle. Observers believed they were witnessing an historic first, for nothing this heavy had ever left the earth before. Saturn 1 weighed as much as two B-52 bombers.

The Saturn team was essentially the same one which had developed Jupiter C and the Jupiter IRBM. With von Braun and his deputy, Eberhard Rees, were Oswald H. Lange, who headed the Saturn systems office; Ernst D. Geissler, the aeroballistics expert; Helmut Hoelzer, computations chief; Werner Kuers, in charge of fabrication and assembly; Heinz H. Koelle, director of future projects; Walter Haeussermann, guidance and control; Hans Hueter, light and medium vehicles; Hans H. Maus, lunar-program planning; Ernst Stuhlinger, research projects; William A. Mrazek, structures and mechanics; Dieter Grau, quality control; Karl L. Heimberg, chief of testing; and Kurt H. Debus, director of launch operations.

Debus ruled the fortress-like control room at Pad 34 with its concrete walls 12 feet thick and its 2-foot steel door. His assistants were Hans F. Gruene, deputy launch director, Albert Zeiler, mechanical office chief, and Carl Sandler, chief of tracking. The test conductor was Robert Moser, who had presided at the launch of Explorer 1. U.S. Army Major Rocco Petrone observed as chief of the Saturn Project Office.

Kerosene was loaded into the booster tanks Thursday night, October 26, and the countdown proceeded smoothly throughout the night, under the daylight glare of floodlights. Complex 34 could be seen for miles as a giant glow against the dark sky. Before dawn, correspondents and official observers began bouncing onto the Cape aboard blue Air Force buses. As the sun rose out of the ocean, the rifle-bullet shape of the first Saturn appeared silhouetted blackly against the lightening clouds. The skyscraper service tower had been rolled away from it and loomed massively to one side. The upper stages of Saturn took on a pearly gleam as the sun rose higher, and liquid-oxygen vapor curled ethereally around the top of the first stage.

The information officers at Patrick Air Force Base, Lieutenant Colonel Kenneth Grine and Major John Whiteside, passed out plastic ear plugs to the reporters and photographers at the press site. No one could predict how shattering the roar of those eight engines would be. On the test stand at Marshall, the static firings of the Saturn were conducted only when atmospheric conditions would minimize vibration damage in Huntsville.

The launch had been set for 9 a.m., but there was an hour's delay as the launch directorate in the blockhouse waited for low clouds to drift on by. There were no mechanical delays in the countdown. Shortly after sunrise, military vehicles equipped with sirens raced around the fringes of the launch pedestal and a flock of ungainly, stork-like birds scampered away. They were 21 roseate spoonbills which had been nesting so close to the pad that the Brevard County Audubon Society feared they would be roasted by Saturn's exhaust. The Society had prevailed upon the Air Force to chase the birds away before the launch.

In the blockhouse, Moser, alert and taciturn, called out, "Twenty-five minutes and counting. Verify ground-measuring voltage." There were 62 persons in the spacious control room. Moser sat at a control desk with John Twigg, his assistant, and John McGough, director of pad safety. Von Braun and Debus sat opposite them, with two test-support officials,

Major Ray Clark of the Army and Jack Abercrombie. Gruene sat at one of the desks watching the guidance-and-control console. Zeiler stood near consoles which would report the performance of the H-I engines. Sandler sat before an array of tracking displays which would show the progress of the rocket and its flight attitude after it was launched.

At T minus 6 minutes and 14 seconds, Moser ordered the operator at the firing console to push the firing-command button, and from then on the final countdown was conducted by an automatic sequencer. It checked fuel-tank pressure and hydraulic pumps. It opened a valve which purged the base of the rocket with nitrogen to get rid of contaminants. It closed the liquid-oxygen pressure valve, shutting off the halo of vapor. It switched the Saturn from external power to its own batteries and it ejected the umbilical power cable which linked the vehicle electrically to the pad. At zero, the sequencer commanded the engines to fire. They ignited in pairs in a sequence of one-tenth of a second apart. The effect was like that of ten thousand automobile engines starting up simultaneously. A blinding wave of yellow-orange flame issued from the rocket and seemed to surround it as the eight engines built up to 1.3 million pounds of thrust in one second and a quarter. The ground began to shake. Still, the Saturn remained on the pad, secured by huge clamps until the automatic sequencer confirmed that all engines were firing properly. At any point during this 3.6-second hold-down, trouble in one of the eight engines would have triggered the automatic shutdown of all of them. Then the big clamps holding down the rocket flipped back and the Saturn began to rise like an elevator. It was 10:06 a.m. Eastern Standard Time. The vehicle ascended smoothly on a tower of sun-bright flame. It looked magnificent and flawless in the pale-blue sky, titanic in contrast to a Navy photographic jet aircraft which flashed by it. For the first several seconds, it climbed with the sound of distant thunder, then the full-throated roar washed back over the spectators. It was not the crushing sound most of them expected but there was awesome power in it.

As the big bird dwindled into the sky on the inverted candle-flame of its exhaust, the observers at the press site burst into the kind of prolonged applause one hears after a concerto has been beautifully performed. Executives of Rocketdyne stood gazing raptly at the vacant launch pad for a few moments and then began pounding each other's back with wild glee. No one had ever seen anything like this performance of a new machine. Saturn had worked perfectly the first time. The United States became the possessor of the most powerful launch vehicle in the world. The booster gap at last had been closed.

At 10 seconds after the liftoff, the four inboard engines swivelled to tilt the vehicle toward the east-southeast. Saturn gradually inclined 43 degrees from the vertical, as programed. The inboard steering engines cut off at 111 seconds, and the four outboard ones 6 seconds later. Saturn then entered a ballistic coasting arc.

At 6 minutes, 50 seconds after liftoff, Moser looked up from the consoles and announced, "This concludes the flight test. Stand by all stations for results."

A happy assortment of space-agency officials greeted the press with unusual warmth. "It has been a very happy occasion," said Seamans. Von Braun stood beside him, beaming. Accustomed to explaining lucidly why rockets don't work, the Marshall Center director found himself in the unique postlaunch position of having nothing to explain. "So far as we know, the Soviets have boosted vehicles weighing 750,000 to 800,000 pounds," Seamans said. "Our vehicle weighed 925,000 pounds. Their booster would not have lifted the C-1 you saw today."

PORKBARREL FLATS

As the lunar project shaped up, it became apparent that it was to become the largest program of technical development in American history. It would funnel vast amounts of federal money into regions where the apparatus of the $20-billion lunar mission was being fabricated, tested, and launched. Much of the spending was concentrated in California, where North American was developing the Apollo spacecraft and the second (S-2) stage of the Saturn 5, where Douglas was building the third (S-4) stage, and where many of North American's subcontractors were located. At a renovated government aircraft plant at Michoud, near New Orleans, Boeing was to build the booster stage of the Saturn 5 and

Chrysler was to fabricate an advanced model of the Saturn 1 called the Saturn 1B, which would fly after the ten research-and-development flights of the Saturn 1 were completed. In the Middle West, McDonnell Aircraft in St. Louis was the prime contractor for the Mercury spacecraft and was slated to build the next generation of manned vehicles, the two-seater Gemini.

The South was getting its share of space work and testing. In addition to Michoud in Louisiana and the Marshall Center in northern Alabama, NASA planned to construct a large test center in Mississippi and to build a spaceport, the John F. Kennedy Space Center, on 80,000 acres of scrub land and citrus groves on Merritt Island, Florida, across the Banana River from Cape Canaveral. This four-state complex costing more than $1.5 billion was to attract thousands of engineers, technicians, and skilled machinists to the South.

As the program advanced, more than a quarter of a million people became employed in it, all working toward the goal defined by John F. Kennedy of landing a man on the moon by the end of the decade.

One section of the country had been left out of this bonanza, which scientist Philip Abelson referred to as "technological leaf raking."[18] The busy chambers of commerce in Texas and Oklahoma complained of not getting their share. A site on the Gulf Coast near Brownsville, Texas, had been considered for a spaceport by a NASA-Department of Defense survey team. So had Christmas Island in the Pacific Ocean, the island of Hawaii (at South Point), Mayaguana Island in the Bahamas, Cumberland Island, Georgia, and the White Sands Missile Range in New Mexico. But all of these had been dismissed because of specific problems of overflight hazard, of the booster stages falling on land near cities, and of accessibility and cost. The Cape Canaveral area was judged the best launch site.

No one in the government was particularly surprised, consequently, when Webb announced on September 19, 1961, that a site southeast of Houston had been selected for the Manned Spacecraft Center. Astronauts would be trained there for the lunar flight, and their flight would be monitored and controlled from the Center. The selection of Houston as the training headquarters (in lieu of Langley Research Center, Virginia) and as the command center (in lieu of the

Cape) was one of the most conspicuous examples of pure porkbarrel in the space program. There was no reason connected with space flight to transfer the control center a thousand miles from the launch site. As long as NASA was spending $60 million to acquire 80,000 acres for the spaceport on Merritt Island, its failure to use some of this land, which was far in excess of spaceport requirements, for a manned spacecraft center seemed inexplicable, except in terms of regional porkbarrel. In a program as big as this one, political expediency required the space agency to follow the old Chicago city hall dictum on public works: a little something for everybody. Houston got the Manned Spacecraft Center, which initially was budgeted for $120 million.

The men most influential in NASA's decision to locate on the Gulf plain were Lyndon Johnson, the Vice-President and chairman of the Space Council, and Representative Albert Thomas, a Houston Democrat, who presided over the House appropriations subcommittee which held the purse strings on space-agency funds. Ostensibly, the selection was made on the basis of a survey of 20 cities, including Boston, by a committee of NASA payrollers, so that an administrative justification was available in the event anyone wanted to study it.

The plan for the Center as designed by Brown & Root, Inc. of Houston, under a $1,500,000 contract, called for an array of 15 buildings, including an auditorium, on 1600 acres at the western edge of Clear Lake 22 miles from downtown Houston. "Part of this land was donated by Rice University in Houston," a NASA announcement said.[19]

The space agency created the impression that the land was essentially a gift from a university, but this was not entirely the case. Initially, when the Houston area was being considered for the Center, NASA stated its land requirement at 500 acres. Rice—then Rice Institute of Technology—agreed to provide it. However, Gilruth, the Center director, wanted more land than that. The requirement was raised to 1000 acres. Rice agreed to provide it free of charge.

The Institute, it turned out, didn't own the land. It was receiving the acreage as a donation from the Humble Oil Company, the marketing and refining subsidiary of Standard Oil of New Jersey. The Institute was performing the role of a broker. In handing over Humble land to NASA, Rice lent academic prestige to the deal and carried out a civic function which the City of Houston, the County of Harris, or the State of Texas might have performed.

The original tender of land from Humble was actually 1020 acres. It reserved a 20-acre plot at the center of the tract for Humble's use in drilling for oil.[20] Thus it was foreseeable that the complex of Manned Spacecraft Center edifices might one day be clustered around a forest of oil derricks, providing earthly contrast to the enterprise of voyaging to the moon.

At first, a lease arrangement was proposed, but the government attorneys rejected it. Rice then agreed to convey the 1020 acres to NASA in fee simple, subject to mineral rights, pipeline easements, and a reverter clause. The reverter provided that the land would revert to the Institute if NASA stopped using it.

This arrangement, however, was upset when NASA planners found they would need another 600 acres, or a total of 1600. Humble balked. Rice notified the agency that the additional acreage could not be provided free of charge. After considerable negotiation, a price of $1.4 million was agreed to by all parties.[21]

This sum was to be paid to Rice Institute, which would swap 33 acres of industrial property it owned for 600 acres on the porkbarrel flats southeast of Houston which Humble was prepared to part with, for a price. Rice valued its 33 acres at $1.50 to $1.75 a foot, or $2,000,000 to $2,500,000 for the tract.[22] From the Institute's viewpoint, both Humble and NASA were getting a bargain. However, one Texas congressman, Representative Olin Teague, wasn't so sure.

At a hearing of the Manned Spaceflight Subcommittee April 4, 1962, he stated that three appraisals of the 600 acres had been made. They ranged from $635,000 to $751,000 before NASA announced the Center's location in September 1961 and from $720,000 to $1,246,175 afterward. The agreed price of $1,400,000 was 12.4 per cent above the highest appraisal, he observed.

The general counsel for NASA, John Johnson, replied: "I would like to say that in negotiating the price of purchased real estate, no government agency is bound by an appraisal." He went on to say that the 600 acres could not be separated, so far as price was concerned, from the total of 1600 NASA was receiving. "We are obtaining here 1600 acres for the total price of $1,400,000, which is considerably less than the appraised market value of the total of 1600 acres even before the NASA announcement."

As the counsel explained it, NASA was getting a bargain if one considered it was buying the entire tract instead of receiving five-eighths of it as a gift. Even so, the space agency was paying an average of $875 an acre for the vacant, undeveloped land in Texas, compared with an average of $750 an acre it was paying in Forida for the John F. Kennedy Space Center.

By the autumn of 1962, according to the Houston *Chronicle*, the biggest land boom in the history of Harris County was in full swing between the Gulf Freeway, the expressway to Galveston, and the NASA site.[23] Land that had been selling for $300 to $500 an acre was selling for $2500 an acre, the newspaper reported. Land fronting on FM (Farm-to-Market Road) 528, the space center's frontage road, was being priced at $10,000 an acre, and on State Road 146 bordering Clear Lake City it was priced from $5000 to $12,000 an acre. Housing and shopping-center developments were springing up all over the area. The largest was Clear Lake City, a $750,000,000 project inspired by the NASA location. It was being developed into an all-around community by the Friendswood Development Company, which was jointly owned by the Del E. Webb Corporation and Humble Oil. Clear Lake City was to become a metropolis of 250,000 people, according to the promoters. It was to be built on 15,000 acres, including 3500 acres adjoining NASA.

The good fairy of land speculation waved a fabulous wand. The wife of a high-school janitor sold 41 acres she had inherited from her mother for $207,000, or about $5000 an acre, the *Chronicle* reported. The newspaper also reported that a three-acre nursery on the FM road was sold for $40,000 and then resold a month later for $60,000.

NASA began to move personnel out of Washington and Langley Field in May 1962. By July 1, 1152 employees were settled in temporary buildings in downtown Houston or at Ellington Air Force Base, not far from Clear Lake. One happy group found itself officed in a brand-new luxury apartment building, with a swimming pool.

On July 4, 1962, the Houston Chamber of Commerce, which had proclaimed NASA's choice "the most significant single event" in Houston's economic history, sponsored a parade of welcome and a barbecue for NASA. The parade began on a hot morning with a motor caravan starting off from the Sam Houston Coliseum, led by police cars with sirens wailing. The astronauts and their families and Representative Thomas, who waved his hat, rode through the downtown streets. Several hundred space-agency payrollers brought up the rear, along with a Mercury spacecraft mounted on a trailer. After circling the downtown area, the caravan returned to the coliseum, where 5000 persons jammed inside to hear speeches and partake of a mass barbecue to the blare of high-school bands. "It was the most meaningful Fourth of July parade Houston has had in recent times," observed the Houston *Post*.

On another hot morning—September 12, 1962—President Kennedy spoke in Houston on the same tour of space facilities that had taken him to Huntsville, where Wiesner had criticized lunar-orbit rendezvous. The President did not have to tell the Texans who massed in the Rice stadium that the Manned Spacecraft Center would bring them nothing but good. They already knew. It would provide a payroll of $60,000,000 a year and invest $200,000,000 in plant and laboratory facilities. More than $1-billion worth of contracts would be administered from the new Center, he said.

The President's address reverberated in the oval arena with epic cadences, which sounded at times like those of Longfellow's *The Song of Hiawatha*. The Chief Executive neared the end of his speech, perspiring and virtually chanting:

> If we are, my fellow citizens, to send to the moon, 240,000 miles away, from the control station here in Houston, a giant rocket more than 300 feet high, made of new alloys that have not yet been developed, capable of standing heats and stresses several times that ever before experienced, fitted together with a precision many times finer than the finest watch, carrying all the equipment needed for propulsion, guidance, control, communications, food, and survival, on an untried mission to an unknown celestial body, and then returning safely to the earth, reentering the atmosphere at speeds up to 25,000 miles an hour,

causing heat about half the temperature of the sun—almost as hot as it is here today—and do all this and do it right and do it first before this decade is out, then we must be bold and daring. And unflinching.

There were 50,000 in the Rice stadium that morning. They clapped their hands, stomped their feet, and shouted their approval.

Chapter 7

Men in Orbit

OVERSHOOT

At 1:15 a.m. on May 24, 1962, Malcolm Scott Carpenter, 37 years old, a lieutenant commander in the Navy, father of four, and pilot of the Mercury spacecraft Aurora 7, was awakened in the crew quarters on Cape Canaveral by the firm hand of Dr. Howard A. Minners, NASA flight surgeon. The countdown for the second United States manned orbital space flight was going well under the bright lights of Complex 14. There was a pungent wind blowing from the northwest, bringing smoke from brushfires which were burning inland over miles of Florida scrub. Otherwise, the night was clear and the stars were out.

If an American movie director had to pick one of the seven Mercury astronauts to play the role of space hero, his choice probably would have been Scott Carpenter. Five feet, 10 inches tall, weighing 160 pounds, with brown hair, green eyes, and clean-cut, regular features, Carpenter fulfilled the breakfast-food image of an American astronaut. Perhaps a European movie director might not have selected him for the role under the mistaken impression that his conventional handsomeness indicated a lack of sensitivity. In a number of respects, the pilot of Aurora 7 seemed to those who worked with him to be the most sensitive, introspective, and self-critical of the original Mercury seven. Now and then his inner feelings about himself and the space program came through the overlay of pilot banter, with its code of understatement and impersonal view of life, death, and difficulties. He once said that he volunteered to become an astronaut because it was a chance for immortality which few men ever get.

Carpenter's chance came when Air Force Major Donald K. (Deke) Slayton, who had been selected initially for the flight, was grounded because of a chronic irregularity in heart rhythm (idiopathic atrial fibrillation). While this condition had been known to the medical staff and had never interfered with Slayton's training, some staff physicians felt it might affect his performance under flight stresses, which had not been assessed fully. They deemed it safer to leave him on the ground.

The second manned orbital flight was planned as a duplicate of the first, with a goal of three orbits. Still, it was enough of a novelty to call forth the image of the hero. Carpenter played the role with the correct mixture of modesty and forthrightness, and with a touch of irony.

Carpenter grew up in Boulder, Colorado, at the foot of the front range of the Rocky Mountains. His parents separated when he was a child and he lived with his mother, Florence, and her parents. His grandfather, Victor Noxon, editor of the *Boulder County Miner and Farmer*, represented a stabilizing influence in Scott's life. But this wasn't enough. Scott viewed himself as a "drifter." He told *Life* magazine,[1] "I didn't study hard and I had to quit high-school football because I couldn't devote myself to learning the plays. I stole things from stores and I was just drifting through, sort of a no-good."

Scott's mother, suffering from tuberculosis, was sent to a sanatorium, and while she was away, he was often alone. He recalls solitary, rainy afternoons in his mother's house, "in the great, big, dirty cellar with little light and spiders . . ." where he played. He had a horse, "Lady," and a rebuilt Ford coupe, which he loved to drive. He was inspired to enlist as a Navy air cadet in 1943 at the age of 18 by the movie *Wake Island*. He entered the Navy's V-5 aviation program while a freshman at the University of Colorado, but the war ended before he won his wings. He then entered engineering school at the university. During the summers, he worked as a hod carrier and as a lumberjack in northwest Colorado, and one summer he ran a jack-hammer in a construction crew. In 1948, he married Rene (pronounced "Reen") Price of Boulder. He continued at the university, working toward a degree in aeronautical engineering, but he failed to win the degree because he flunked a course in heat transfer. He left school then, but it was not the last he was to hear about that heat-transfer course.

In 1949, he rejoined the Navy. During the Korean war, he flew with a Navy patrol squadron on antisubmarine-patrol, ship-surveillance, and aerial-mining missions in the Yellow Sea, the South China Sea, and the Straits of Formosa. After the war, he became a Navy test pilot and was selected as a Mercury astronaut in April 1959.

Now, the greatest day of his life had come. He dressed quickly and went into the dining room for breakfast with Dr. Minners and John Glenn. It was the big meal for the condemned man: orange juice, soft-boiled eggs, filet mignon, buttered toast, and coffee. Minners checked Carpenter and pronounced him fit to fly. The medical sensors that would record his heartbeat, blood pressure, and breathing rate were attached to his chest. Their data would be radioed to ground stations as he flew within radio range. Shortly before 3 a.m., Carpenter was driven in the little transfer van with Dr. Minners and Glenn to Pad 14, where his backup pilot, Navy Commander Walter M. Schirra, Jr., told him how the countdown was progressing. After last-minute adjustments to his flight suit and helmet, Carpenter rode up the pad elevator to the 100-foot level, where he eased himself into the tiny Mercury cabin. His wife, Rene, had come to the Cape to observe the launch, but she remained secluded from the newsmen.

ROUGH THROUGH MAX Q

Mist shrouded the base of the pad service tower as the sun rose out of the dark sea. A pall of black smoke hung heavily over the palmetto scrublands to the west but did not move toward the Cape, because the wind had shifted around to the east. In the capsule, Carpenter, lying on his back in his contour couch, could look into the brightening sky through the Mercury window. Below him, the mist coiled up from the concrete pad, around the base of the silvery Atlas booster. The day brightened and the birds fell silent. Then the public-address system brayed: "T minus 50 minutes and counting." Eight hundred miles to the southeast, the 1941-vintage aircraft carrier, U.S.S. *Intrepid,* and two destroyers, the *Pierce* and the *Robinson,* hovered on station in the prime recovery area, 180 miles northwest of Puerto Rico.

Mercury-Atlas 7 lifted off at 7:45 a.m. on gouts of orange flame and white smoke. As it thundered up above the mist, Carpenter's voice came back, flat and businesslike: "I feel the liftoff. The clock [in the spacecraft] has started." "Roger," acknowledged Virgil I. Grissom, the capsule communicator at Mercury Control Center.

Carpenter reported: "Clear, blue sky . . . 32 seconds . . . 9000 [feet], fuel and oxygen steady . . . A little rough through Max Q . . . Mark one minute."

As he gained altitude rapidly, Carpenter saw the sky turn black. After the Atlas's powerful booster engines cut off, and acceleration slowed as the rocket proceeded under the lower thrust of its sustainer engine, he heard the emergency escape tower jettison, and then saw it receding from him toward the far horizon. When the sustainer engine cut off, Carpenter heard two distinct bangs. The first was the clamp ring which held the spacecraft to the rocket being opened by explosive bolts to free the capsule. The second was the firing of small solid-fuel thrusters at the base of the spacecraft to push it away from the dead Atlas. With the end of powered flight came weightlessness and silence.

Now Carpenter's first piloting task was to turn the spacecraft around, so that it would fly with heat shield forward, in "orbital attitude." As in the Glenn flight this was a precaution. If the attitude-control system failed, the vehicle was in position so that it could reenter the atmosphere, heat shield first, after its retrorockets fired to slow it down.

Carpenter swung the spacecraft around with the manual controls, using only 1.6 pounds of hydrogen-peroxide fuel, compared to the average of 4 pounds burned when the turnaround was made by the ship's autopilot.[2]

"Following turnaround, I watched the expended launch vehicle through the window as it fell behind me, tumbling slowly," he related.[3] "It was bright and easily visible. After the initial sensation of weightlessness, it was exactly what I had expected from my brief experience with it in training. It was very pleasant, a great freedom, and I adapted to it quickly. . . . My only cues to motion were the instruments and the view through the window and periscope. At times, during the flight, the spacecraft angular rates [roll] were greater than 6 degrees a second, but, aside from vision, I had no sense of movement. I was never disoriented. At times, when the gyros were caged [not operating] and nothing was visible out the window, I had no idea where the earth was in relation to the spacecraft. However, it did not seem important to me."

Carpenter experimented with drifting flight during his second and third revolutions, allowing the spacecraft to turn or roll "whichever way it wanted." He reported that the motions of the spacecraft were slow and deliberate and created no discomfort. There had been fear that drifting flight, with the thruster-control system off, would produce nausea or vertigo in the pilot. Carpenter experienced neither symptom. He released a small balloon, which was tethered to the spacecraft, as a means of measuring air drag at his orbital altitude of 144.96 nautical miles apogee and 86.87 nautical miles perigee (166.7 and 99.9 statute miles), but the balloon did not inflate properly and little could be determined from its random motions.

A VIEW FROM ABOVE

Experimenting with his attitude controls, Carpenter found that he could align the spacecraft in pitch (vertically) quite easily by sighting on the horizon without using the gyroscopes to give him position information. A reticle or scale was etched in the window, and the horizontal lines were marked off in degrees above and below the center horizontal line. When this line was on the horizon, the pilot knew his ship was flying parallel with the surface of the earth. The horizon also provided a good roll reference and was visible on the night side of the planet as well as on the day side. Yaw (left or right) orientation was more difficult to determine visually. Carpenter believed he could determine it by pointing the ship's nose down 50 to 70 degrees and then watching the horizon through the window to see if the vehicle was aligned along the flight path. Both the vertical (pitch) and horizontal (yaw) alignments were critical for retrofire, as Carpenter, and the world which was watching the flight, were soon to discover. For if the retrorockets did not fire precisely along the flight path, they would not change the trajectory in a predictable way and the spacecraft would not come down in the planned landing area. It would, instead, tend to overshoot the impact zone where the recovery ships were waiting, leaving a lone astronaut and a tiny spacecraft to the uncertainties of search aircraft on the very big ocean.

So far as appearance and color were concerned, Carpenter observed, there was little difference in the view of land, water, and clouds from orbital altitude of 500,000 feet and the view at an aircraft altitude of 30,000 feet. There was simply more to see from orbit.

"The South Atlantic was 90 per cent covered with clouds, but all of eastern Africa was clear," he reported. "I had a beautiful view of Lake Chad. Other parts of Africa were green and it was easy to tell that these areas were jungle. . . . Over the United States on the second revolution, I noticed a good amount of cloudiness, but after retrofire, I could see the area around El Centro, California, quite clearly. I saw a dirt road and had the impression that had there been a truck on it, I could have picked it out."

This latter portion of Carpenter's flight report did not attract much attention at the time, but in retrospect it is one of the most significant observations he made. Glenn, too, had indicated that from altitudes of 100 miles or more he could see objects on the ground with surprising clarity in unclouded areas. These reports suggested that visual acuity from orbit was considerably enhanced over that on the ground. The reconnaissance implications of this suggestion were interesting, and were to be brought out more vividly later in Project Mercury.

Like Glenn, Carpenter also saw the space "fireflies," those bits and motes of greenish, yellowish, or whitish stuff that floated and danced around the capsule at sunrise. "They appeared to be like snowflakes," Carpenter said. "I believe that they reflected sunlight and were not truly luminous." Shortly before he prepared for retrofire on the third revolution, his gauntlet inadvertently hit the spacecraft hatch when he reached for an instrument and suddenly a cloud of particles flew by his window. Always quick at putting two and two together, he knocked on the hatch and other parts of the ship several times to see what would happen. Each time, he reported, a cloud of particles swarmed past the window. They were gray or white, varying in size, one a half inch long, and some had curlicue shapes, like wood shavings from a lathe. Later analysis of Carpenter's observations, and of photographs he took, indicated that Glenn's space fireflies came from the spacecraft itself and probably were nothing more than bits of ice or frozen fuel.

Carpenter was so busy observing and photographing the fireflies that he fell behind the flight-plan schedule in his preparations for firing the retrorockets. Aurora 7 was crossing the Pacific Ocean on its third revolution, and the Mercury station at Hawaii was urging him to hurry with his retrofire checklist and to start stowing equipment. Carpenter busied himself with aligning the ship in the correct attitude for retrofire, which called for elevating the heat shield 34 degrees above the horizon so that the retrorockets would be pointing squarely along the flight path when they fired.

At 6 minutes before retrofire, Carpenter engaged the autopilot to hold the ship steady. He then reported it was not working properly. The vertical alignment of the ship as indicated by the automatic horizon scanner did not match his window reference with the horizon. "I've got to evaluate this ASCS [Automatic Stabilization Control System] problem before we go any further," he told Hawaii. A few minutes later, Carpenter heard the California tracking station calling him. Alan Shepard came on the air and asked if Aurora 7 was ready for retrofire.

"Yes, but I don't have agreement with ASCS in the window, Al," said Carpenter. "I think I'm going to have to go to fly-by-wire and use the window and the scope. ASCS is bad. I'm on fly-by-wire and manual."

"Roger," said Shepard. "We concur. About 30 seconds to go."

Carpenter aligned the ship in retrofire attitude. At Shepard's suggestion, he then engaged the autopilot to check its orientation system once more. There was a 24-degree discrepancy between his horizon window reference and the electronic horizon scanner. When the minus 34-degree mark on his window rested on the horizon, the electronic scanner registered a pitch of only minus 10 degrees. Carpenter elected to rely on his own perception. It was well that he did, for a malfunction was later found in the automatic horizon-scanning circuit.

In shifting his gaze back and forth between the horizon reference and the cockpit indicator, Carpenter found it difficult to readjust his vision from the brighter horizon to the dim light in the cockpit. The seconds ticked by, and while the pilot was making these checks, the autopilot allowed the spacecraft to pitch down excessively. Carpenter knew he was running out of time until retrofire. He quickly switched from automatic to the fly-by-wire mode, in which he could control the thrusters electrically by moving the stick, to reposition the spacecraft in retrofire attitude, using his horizon window reference. In switching to the fly-by-wire mode, Carpenter did not shut off the manual proportional mode of controlling the thrusters, so that in realigning the ship, he used both manual control systems and about twice as much fuel as was necessary. During this period, the gyroscopes indicated "a significant excursion in yaw." But Carpenter no longer had confidence in the automatic attitude indicators. He relied entirely on what he could see out the window.

"Then seconds to retrofire," Shepard called. "Six, five, four, three, two, one . . ."

Nothing happened.

"If your gyros are off, you will have to use attitude bypass," Shepard advised.

He meant that Carpenter would have to override the autopilot in order to fire the retrorockets, which would not go off in the automatic mode without gyroscopic confirmation that the ship was in proper retrofire attitude. This safety device had prevented the retrorocket system from operating.

"The gyros are off," Carpenter said.

"But you'll have to use attitude bypass and manual override," Shepard repeated.

"Roger," Carpenter acknowledged.

"Four, three, two, one, zero," Shepard counted down.

"Okay," said the pilot. "Fire one, fire two, and fire three. I had to punch off manually. Have a little bit of smoke in the capsule." Instead of the "big boot" he expected when the rockets fired, he felt only "a gentle nudge."

"Attitudes hold, Scott?"

"I think they held well, Al. I think they were good. I can't tell you what was wrong about them because the gyros were not quite right."

Radar data showed that the spacecraft's pitch during retrofire was correct but that it was yawed 27 degrees to the right, so that the rockets did not fire directly along the flight path. Perhaps this was the reason Carpenter felt only the gentle nudge instead of the big boot he had expected. Glenn had said that when his retrorockets fired, he felt that he was being boosted back to Hawaii.

OUT OF FUEL

It was immediately apparent to Mercury Control at the Cape that because of this misalignment and also because retrofire came three to four seconds late that Aurora 7 would land some distance beyond the planned impact zone, called Area H, in the Atlantic Ocean near Grand Turk Island. The delay in retrofire alone would cause a 15- to 20-mile overshoot. What the effect of the yaw excursion would be had to be determined by radar.

As Aurora 7 soared over the California coast, Point Arguello radar tracked its new flight path. The descending arc, when extended, showed that Carpenter would land 250 nautical miles (287.5 statute miles) beyond Area H.

The pilot reported he was out of fuel in the manual proportional system—the penalty for using too much in repositióning the craft for retrofire—but had some left in the fly-by-wire system to stop some of the oscillations which were beginning to build up. As Aurora 7 plunged ever deeper into the atmosphere, he asked Shepard for an estimate of the time it would take the capsule to reach atmospheric resistance equivalent to one-twentieth of a gravity. At this point, an accelerometer would activate the automatic control system which would realign the craft to point the heat shield down slightly for reentry. It would also damp out excessive swinging, prevent tumbling, and impart a slight roll to the ship, unless Carpenter chose to override it. Shepard estimated that Carpenter had seven minutes until the 0.05-gravity switch closed. "Take your time on fly-by-wire to get into reentry attitude," he advised.

As Aurora 7 crossed the Southern states and neared the

Cape, Grissom came on the air, asking Carpenter if his face plate was closed. "Negative," responded the pilot. "It is now. Thank you."

Grissom said that the weather in the recovery area was overcast, there were three-foot-high waves, and visibility was 10 miles. He was describing Area H—not where Aurora 7 was actually to splash down. Aurora 7 arced across Florida and hit the denser atmosphere over the Atlantic. At 0.05 gravity, Carpenter noted he still had 15 per cent fuel in the automatic system. But the manual control tanks were empty. He began to hear the hissing outside that Glenn had described. The spacecraft was fairly well aligned in the reentry position so that the heat shield now pointed along the descending flight line in order to protect the cabin from the 3000-degree-Fahrenheit heat of atmospheric friction.

Carpenter felt that the sturdy little vessel would reenter properly without any effort to control its attitude. The gradual increase of aerodynamic forces acting on the bell shape of the vehicle seemed to be sufficient to align it properly, he believed. Shortly after experiencing 0.05 gravity, Carpenter began to notice some pitching and yawing. These were about the same magnitude as those he had experienced in the Mercury trainer. He consequently was reassured that his reentry attitude was good enough to take the heat. He allowed the automatic system to damp out the oscillations. As heat began to accumulate, Carpenter reported, "I've got an orange glow. I assume we're in blackout, now. Gus, give me a try. There goes something tearing away." Aurora 7's plunge had built up a sheath of ionized particles of atmosphere around it, effectively blocking radio transmission. Carpenter continued talking, but only to his tape recorder. To the Cape, he had vanished—not only on the radio, but also on the radar screens. Aurora 7's extended flight path had carried it over the Cape radar horizon.

"My fuel, I hope, holds out. There is one *g*. Getting a few streamers of smoke out behind. There's some green flashes out there. We're at one and one-half *g*s now. There was a large flaming piece coming off. Almost looked like it came off the tower [the cylindrical nose housing the parachute]. Oh, I hope not."

As denser air slowed the spacecraft's descent, Carpenter reported that the accelerometer had reached a peak of 6.8

gravities—nearly seven times the normal feeling of weight. Then the pressure eased.

"Back to 5 *gs* and I'm standing by for altimeter off the peg. Cape, do you read yet? Altimeter is off the peg, 100,000 feet. Rate of descent is coming down. Cabin pressure is . . . cabin pressure is holding okay. Smoke pouring out behind. Getting ready for the drogue at 45 [thousand feet]. Getting some pretty good oscillations now, and we're out of fuel."

While the spacecraft appeared to be able to stabilize itself at supersonic speed in the high atmosphere, it also seemed to lose aerodynamic stability when its descent slowed to subsonic speed. Carpenter noticed an increase in its oscillations as the craft reached 40,000 feet. Aurora 7 began to swing so wildly that the sun appeared to be whirling around the sky and at one point the pilot feared the craft would flip over.

"I think I better take a try on the drogue," he told his tape recorder. "Drogue out manually at 25 [thousand feet]. It's holding and it was just in time."

He switched on the main parachute-deployment fuse at 15,000 feet and waited for it to release the chute. When the deployment switch failed to close by 9500 feet, he closed it by hand. The parachute came out reefed, became taut, and then unreefed beautifully.

DOES ANYBODY READ?

"I see a perfect chute, visor open," Carpenter reported to his tape recorder. "Cabin temperature is only 110 degrees at this point. Helmet hose is off. Does anybody read? Does anybody read Aurora 7?"

There was no answer. When Aurora 7 emerged from the ionization blackout, it was so far downrange that the Cape radio could no longer receive it. Carpenter tried to reach a recovery ship.

"Hello, any Mercury recovery force. Does anyone read Aurora 7? Over."

There was no reply to this transmission, but in a few moments he heard the Cape calling him. "Aurora 7, Aurora 7, Cape Cap Com, over." He replied: "Roger, I'm reading you. I'm on the main chute at 5000 [feet]. Status is good. I am not in contact with any recovery force. Do you have any information on recovery time?"

But the Cape could not hear him. Gus Grissom kept calling, "Aurora 7, Cape Cap Com. Over."

"Roger, loud and clear," Carpenter replied. "Aurora 7 reading the Cape loud and clear. How me, Gus?"

No answer.

"Gus, how do you read?"

No response. However, Grissom kept broadcasting in the blind: "Aurora 7, your landing point is 200 miles long. We will jump the air-rescue people to you."

"Understand," said Carpenter. "I'm reading."

"Aurora 7, Aurora 7," Grissom repeated. "Cape Cap Com. Be advised your landing point is long. We will jump air-rescue people to you in about one hour."

"Understand, one hour," said Carpenter. But the Cape did not hear. The spacecraft radio signal was not strong enough to reach the Cape. Aurora 7 drifted down to an empty ocean.

Meanwhile, at the Cape, six hundred news correspondents, cameramen, and military and industry observers waited for the finale. But Mercury Control remained silent. The public-address system cleared its throat and Powers came on. "This is Mercury Control. We are still attempting to reestablish contact with the Aurora 7 spacecraft. We expect to establish contact momentarily. This is Mercury Control standing by."

The silence this time was longer. Powers came on again: "We are beginning to receive indications that perhaps he is triggering his transmitter. We expect to reestablish contact with the Aurora 7 spacecraft momentarily."

The minutes ticked by. Newspaper correspondents were still writing rapidly on their portable typewriters on picnic tables set up in a big revival-style tent which the Air Force had erected for a postflight press conference. It was extremely hot and windy in the tent, and the newsmen were desperately trying to catch up with the story, which had been unfolding faster than they could write it, up to the ionization blackout. No one, however, could write a lead paragraph saying that Carpenter had landed safely. The radio reporters were stalling, talking about the delay in communications, and trying to explain what an ionization communications blackout was. Reporters on afternoon newspapers with imminent deadlines could not tell their editors what had happened to Carpenter.

Was he up? Down? Alive? Dead? Who knew? So the telephone exchanges went between the men on the scene and the news editors in the office, who were tending to blame the man on the scene for the dilemma of not knowing what headline to write.

"This is Mercury Control," Shorty Powers resumed. "Our data at this time indicates that it is distinctly possible that the Aurora 7 spacecraft may land considerably longer downrange than was planned. Our present estimate of his landing point may go as far as 200 miles downrange. For that reason, we feel that the ionization period is over now, that he has exceeded the range of our transmitting equipment here at Cape Canaveral. We are diverting aircraft into the area both for the purpose of reestablishing communications and effecting rescue."

The blackout of communications with the spacecraft had begun at 12:28 p.m. For 53 minutes the world press remained in the dark about the astronaut's fate. As the minutes oozed by in the hot May afternoon, the correspondents stopped writing. They had caught up with the story. Now they needed the ending. A British newspaperman said, "Isn't that a tragedy? Is he lost? I can't believe it." A tall, gangling boy, who was picking up the paper coffee cups and sandwich debris in the tent, asked, "Will they find him? Did he burn up?" On the grassy mound outside, where a bleachers had been erected, two women were standing in front of a world map on which Carpenter's orbital paths were shown in red tape. One said, "Oh, think of Rene. She had to watch!" The other began to weep.

In the oppressive heat of the tent, fear grew that Carpenter had perished. Some of the reporters began to prepare their editors for a tragic climax. Then, a few seconds after 1:21 p.m., Shorty Powers returned to his microphone. "This is Mercury Control. We have just received a report through our Recovery Operations branch that an aircraft in the landing area has sighted the spacecraft and has sighted a life raft with a gentleman by the name of Carpenter riding in it."

NEPTUNE TO THE RESCUE

Aurora 7 had splashed down at 12:41, but at the time only Carpenter was aware of it. Only a few bits of telemetry recording the astronaut's heartbeat had been received by the Cape following the ionization blackout. Then there was nothing from Aurora 7.

Immediately after the impact on the sea, which was milder than Carpenter expected, the astronaut was alarmed to see that water had splashed on the face of the tape recorder. He suspected that Aurora 7 was leaking, and that seemed to be confirmed by its failure to right itself in the water. The craft was listing halfway between pitch down and yaw left. He waited for several seconds for this angle to change, but the craft continued to list. Since the temperature inside was 105 degrees Fahrenheit and he knew he would have to wait at least an hour before recovery craft arrived, Carpenter decided he would be safer and more comfortable outside. He opened the hatch and climbed out with a rubber life raft attached to his suit. He carefully deposited his camera, with which he had taken a number of ground pictures, on top of the spacecraft's cylindrical recovery section, where the parachutes were stored, so that he could get it quickly in case the spacecraft sank. Then he slid into the sea and, holding onto the spacecraft with one hand, inflated the raft. He climbed into it, but it didn't feel quite right. After a moment of "assessing the situation," he decided that the raft was upside down. He climbed back onto the spacecraft, flipped the rubber raft over, and got back in, making sure it was moored to Aurora 7. He started a small, powerful radio transmitter called a "Sarah" beacon. As the raft rocked on the gentle Atlantic swells, Carpenter settled down to think. It had been a full day.

The intermittent radio signal from the Sarah beacon was picked up six minutes after Carpenter landed by a Navy Neptune P2V aircraft pilot, at 12:47 p.m. But this information was not immediately made public by the Cape. At 1:21 p.m., 40 minutes after splashdown, the Neptune pilot sighted Carpenter in the raft. As soon as Carpenter heard and then saw the aircraft, he signaled with a mirror. The Neptune began to circle him.

Carpenter then heard someone calling him from the water. Turning, he saw a swimmer stroking toward him, eyes obscured by goggles but mouth split by a wide grin. Then a second swimmer hailed him. The Air Force para-rescue team had arrived in a C-54 airplane and had plunged efficiently into the water. Carpenter was glad to have company. The para-swimmers were Airman John Heitsch of Madison, Wisconsin, and Sergeant Ray McClure of Mount Sterling, Kentucky. Carpenter offered them some condensed rations, but the swimmers said they had just finished lunch. It was a polite little parley out there in the gray Atlantic. Carpenter helped the swimmers put a flotation collar around the capsule to keep it afloat until it could be picked up by a destroyer, the U.S.S. *Pierce,* which was steaming toward the scene. At 3:30 p.m., a twin-turbine HSS-2 helicopter from the aircraft carrier U.S.S. *Intrepid* arrived to pick up the nation's second orbital spaceman.

On May 29, 1962, Malcolm Scott Carpenter was welcomed back to his home town of Boulder in a manner befitting its most illustrious son. A special ceremony had been prepared for the astronaut in the University of Colorado football stadium, which was filled to standing room only. Officiating was the university's president, Quigg Newton. He explained that Carpenter had failed to get his degree in aeronautical engineering 13 years before because he had missed the final examination in one subject: heat transfer. Now, as a result of Astronaut Carpenter's recent experiences, no one could doubt that he had become an authority on that subject. Therefore, said President Newton, the University of Colorado was pleased to award Commander Carpenter his degree of bachelor of science in aeronautical engineering.

THE TEXTBOOK FLIGHT

In contrast to the cliff-hanging suspense of the first two orbital flights, the third, Mercury-Atlas 8, had its troubles at the start. When these were solved, it continued nearly trouble-free to become Project Mercury's "textbook" mission. It followed the flight plan and there were no breakdowns. MA-8 was flown to a new American space record of six revolutions of the earth by Navy Commander Walter M. Schirra, Jr., 39 years old, on October 3, 1962. Nine hours and 14 minutes after he was boosted in the capsule, Sigma 7, off Cape Canaveral, Schirra brought his spacecraft down in the Pacific Ocean 330 miles northeast of Midway Island and four nautical miles (4.6 statute miles) from the aircraft carrier *Kearsarge.*

"Wally," called Gus Grissom, capsule communicator at Hawaii, as Sigma 7 was descending on its 63-foot ringsail main chute, "you landed 9000 yards from the carrier. How about that!"

"That's pretty close, isn't it," said Schirra. "Boy, this is a sweet, little bird."

He waited inside the spacecraft, rocking gently in the three-foot Pacific swells, as it was hoisted aboard the *Kearsarge.* Following Navy protocol, he radioed the carrier's commander, Captain Charles Rankin, for permission to come aboard. "Permission granted," replied Rankin. He and Schirra had been next-door neighbors when the astronaut was stationed at Patuxent, Maryland.

The high performance of the spacecraft and Schirra's flawless piloting sent a surge of optimism through the manned-flight program. The United States program still lagged considerably behind the 64-orbit flight of Adrian Nikolayev in Vostok III and the 48-orbit flight of Pavel Popovich in Vostok IV the previous August. But it was clear now that Project Mercury was hitting its stride.

Of French Huguenot descent, Schirra was born in Hackensack, New Jersey, March 12, 1923. His father, Walter Marty Schirra, Sr., an engineer, had flown aircraft on bombing and reconnaissance missions over Germany in World War I. After the war, he barnstormed in a biplane at New Jersey county fairs. Young Wally developed a similar passion for flying. He spent much of his time building model airplanes and hanging around the Teterboro Airport, talking to pilots and mechanics. He also took trumpet lessons and built a kayak, which he floated in a reservoir, but his first love was flying. He was given his first flying lessons at age 13 by Wally senior and soloed at 16. In 1942, he was accepted by the United States Naval Academy for its three-year, accelerated course and was graduated in June 1946 with a bachelor of science degree and an ensign's commission. He was 215th out of a class of 1045. Schirra completed pilot's training at the Pensacola Naval Air Station and was assigned to a Navy fighter squadron. During the Korean war, he flew 90 combat missions and was credited with shooting down one MIG fighter and with a second "possible." He was awarded the Distinguished Flying Cross and two air medals.

After Korea, Schirra became a test pilot at the Naval Ordnance Training Station, China Lake, California. Once, a Sidewinder air-to-air missile he launched on a test doubled back to chase his jet, and it was all he could do to evade it. He completed test-pilot training at Patuxent River and was assigned to development work on the F4H jet fighter there. When he first heard about the prospect of test pilots flying in space, his reaction was, "What a crazy thing to do." But the more he thought about it the less crazy the prospect became. Eventually, he saw it as a "sensible" extrapolation of the speeds and the altitudes he was accustomed to. He was one of the seven Mercury astronauts whose selection was announced by NASA April 9, 1959.

Schirra was married to Josephine Cook Fraser of Seattle in 1946. They have two children, Walter M. III and Suzanne Karen. Stalwart in appearance, Schirra is 5 feet, 10 inches tall, weighs 185 pounds, and has brown eyes and hair. His hair is graying now above the ears. He has the capacity of becoming completely absorbed in a problem until he solves it —or decides it cannot be solved. This characteristic saved the flight of Sigma 7 from being aborted on the first orbit.

The final countdown was flawless. There was an aura of confidence at Pad 14 and at the press site some 8000 feet away at sunrise of flight day that liftoff would be on schedule —at 7 a.m. Eastern Standard Time. Dark rainclouds which loomed up on the sea horizon at first light drifted off as the sun climbed out of the Atlantic. The morning sky brightened to the color of freshly laundered blue jeans, and the wind was light.

Then, at 6:15 a.m., the count was halted because of a breakdown in the Canary Islands radar, which was essential to determine the shape of the orbit after Sigma 7 was launched. At 6:31, word was received at the Cape that the radar had been repaired and the count was resumed. Tropical Storm Daisy, 285 miles north of San Juan, Puerto Rico, menaced the second-orbit landing zone but in the third-orbit zone, the aircraft carrier *Lake Champlain* reported moderate seas and calm weather.

The mission of Sigma 7 was to test the Mercury systems and the performance of the astronaut on a six-orbit flight. If both functioned as hoped, the next Mercury flight would be programed for at least 24 hours, or 17 orbits. Since the main objective of the flight was an engineering evaluation of the Mercury systems for extended flight, Schirra, with fine logic, had named his craft "Sigma," which he said symbolized the idea of "summation." The figure 7, of course represented the seven astronauts.

A HOT SUIT

Liftoff came at 7:15 a.m. A few seconds after the Atlas cleared the pad, it executed a brief clockwise roll, which momentarily alarmed controllers until it stopped. It was later determined to have been caused by a misalignment of the booster engines. Also, the rocket overaccelerated the spacecraft by about 15 feet a second, causing the orbit of Sigma 7 to be 8.6 nautical miles higher at apogee than planned. These deviations, however, were considered relatively minor.

"Looks real fine from down here," Deke Slayton, the capsule communicator at Mercury Control, advised Schirra as the Atlas thundered upstairs. "Ah, she's riding beautifully," replied Schirra. He watched the emergency escape tower sail away and described it: "Hey, that tower really did a sayonara."

SLAYTON: You have a go from Control Center.

SCHIRRA: Roger. You have a go from me, which is real fast.

SLAYTON: This looks real good. Are you a turtle today?

The query was a bit of cocktail-lounge repartee. It called for a specific response, and the penalty for the inappropriate one was the obligation to pick up the tab for drinks. Schirra's response has eluded history. A television tape of the launch was broadcast to Europe over Telstar, the Bell System's experimental communications satellite, which had been placed in orbit the previous July.

Early in the first revolution, it became apparent to Mercury Control that the temperature in Schirra's flight suit was getting too high. It increased from 74 degrees Fahrenheit at liftoff to 90 degrees at Muchea, Australia, 45 minutes later. Schirra worked with the coolant control valve steadily, but to no avail. No matter how high he turned it up, the suit temperature inexorably increased. At Mercury Control, Dr. Berry was concerned lest the astronaut become dehydrated. Concern increased when the body-temperature monitors, reporting the astronaut's temperature to the ground, failed, so that the flight surgeons could not determine this parameter of the pilot's condition. The combination of difficulties led Williams and Kraft to consider calling Schirra down at the end of the first revolution. They reasoned that if the suit air-conditioner was not working properly at the end of the orbit, it never would, and they did not want to risk an emergency landing with an astronaut suffering from heat exhaustion. The source of most of the heat was the batteries, and the astronaut himself contributed to the over-all temperature, too, along with sunshine on the dark-hued metal plates of the cabin.

In the absence of telemetered temperature data, the word of the astronaut was controlling. Schirra kept assuring the ground as he worked with the coolant valve that while he felt a bit warm he was not suffering any serious discomfort. Between Muchea and Canton Island, the suit temperature stabilized. The possibility that the cooling system was taking hold and would soon begin to function properly persuaded Kraft, who was directing the flight, to allow Schirra to fly a second revolution. "It was the best decision we made all day." Kraft said later.

GLENN'S FRIENDS

The suit temperature began to drop as Sigma 7 crossed the Pacific Ocean. Schirra advised Scott Carpenter at the Guaymas, Mexico, tracking station that he was feeling fine. He had turned up the coolant valve to double its normal reading and had succeeded in reducing the suit temperature by two degrees. "Scott, I feel we're in very good shape for one more orbit at least, and we'll see how we can hack this suit circuit here," he said. "Understand, Wally," replied Carpenter. "We have a go."

Schirra crossed the United States in what he called the "chimp configuration," that is, on autopilot control. The ship, he said, was "flying beautifully." He reported the appearance of some of Glenn's mysterious fireflies. "I saw some of John's friends up there, but I'm afraid to say, although I knocked them off the way you did," he told Carpenter. "Ha, ha."

"Roger," said Carpenter. "Interested in your report."

"I imagine," said Schirra. "John listening to some of that, too?" Glenn, who was capsule communicator at Point Arguello, had kept insisting that the luminous particles he had seen were not connected with the spacecraft, in spite of Carpenter's report that it was possible to produce them by knocking on the spacecraft hull. "Basically, what I saw," said Schirra, "was the firefly color that John saw, which I could create at other times as white color. I'm definitely convinced it's capsule—a capsule derivative and once in a while, even now, I see one go by."

When he came within radio range of Cape Canaveral, Schirra reported he was making slow but steady progress in reducing the temperature in his suit. "Drink some water," advised Slayton. Schirra, however, said he was reluctant to open his helmet visor to take a drink because it would let into the suit some of the heat in the cabin, which was over 100 degrees Fahrenheit. When he reached Australia on the second revolution, he advised the Woomera station that the suit temperature was down to 72 degrees, and all was well. It was found later that the coolant control valve was partly blocked by hardened lubricant. That delayed the system's ability to reduce the suit temperature to a comfortable level. The bit of solidified lubricant came within a hair's breadth of terminating the flight.

The pilot was delighted with the responsiveness of the spacecraft. He said he fell in love with the little ship. Only a light touch and a few pulses of the thrusters were necessary to change its attitude. He experimented with the manual controls in the fly-by-wire mode after Sigma 7 was first injected into orbit by the Atlas. Watching the spent rocket tumble away from him, Schirra shrewdly calculated the problem of attempting to make a rendezvous with it. Rendezvous followed by docking was the critical technique of linking two vehicles in orbit which would have to be learned for the lunar journey. While Schirra found that he could turn the spacecraft any way he wanted with ease, he concluded that the relative-motion problem between two vehicles in orbit would be complex. He did not believe "I would be able to steam along and join up with it," he reported, referring to the

Atlas. In fact, he felt that the problem of rendezvous would be quite difficult to solve by one's own visual analysis, and would require time and instrumentation to aid the pilot.

In controlling the craft manually, he found also that a tailing-off in the thrust from the attitude-control jets caused the vehicle to overshoot the desired position. "You have to counteract and recounteract," he told the ground.

So engrossed was he with the spacecraft's performance that he did not see much of Africa on the first revolution, except to note the brown color of the Sahara desert. Even the magnificent sunsets did not enthrall him as they had his predecessors. He noted that lightning strokes looked like "big blobs" of light or like "antiaircraft shots" down below as he flew over a night thunderstorm in Australia. He advised Muchea that he was using the moon as a reference to orient his spacecraft. When Schirra came around to California on the second revolution, this time within radio range of Point Arguello, Glenn passed the word from the Cape to continue with the normal flight plan.

"Yeah, I feel real happy with everything," Schirra told Glenn. "I stopped everything to get a hold of that suit circuit and that seemed to fix it up."

GLENN: Roger. That had everybody concerned for a while. But it looks like it is in good shape now.

SCHIRRA: I was sure everybody was jumping up and down on that one.

GLENN: You were right.

SCHIRRA: It's kind of hard to describe all this, isn't it, John?

GLENN: Yeah, it sure is, Wally. You can't describe it.

A month earlier, an appearance of strain had been reported between the two astronauts because Schirra had remarked in a television interview that the social and civic affairs John Glenn was attending "had just about wiped him out of the program." [4] This was construed as criticism of Glenn and of the space agency for encouraging him to overplay the space-hero role, which each astronaut was required to play for the sake of the program's public image as soon as he completed a flight.

Schirra had not intended personal criticism of Glenn, but of the way the space agency was using him. At this moment, the two were linked in a communion of mutual experience as orbital-flight pioneers. It was a bond only a few men of their time would know.

The California coast was obscured by fog, but beyond it, the air was clear. Schirra said he could see distinctly the Salton Sea. On his third revolution, he reported the feeling that Sigma 7 was "off in pitch." He surmised this was an illusion created by "that damn horizon airglow line—it makes you think it [the horizon] is higher than it is." As photographed by sounding-rocket cameras, the airglow is a band of misty light visible at night above the horizon. It appears to begin 54 to 71 miles above the earth, and to be about 14½ miles thick, although these numbers may vary from one observation to another. Carpenter had given a clear description of it, and Glenn had reported it also. Schirra observed it while he was allowing the spacecraft to drift. He said he believed at first he was looking at clouds until he saw stars shining underneath the glowing layer. Carpenter had described the layer as comparable in brightness with that of clouds illuminated by a quarter moon. Physicists surmised that the glow is a product of the excitation of gas molecules in the high atmosphere by particles from the sun, so that the phenomenon represents an aspect of solar-terrestrial relationships.

As he passed over Cape Canaveral on his fourth revolution, Schirra heard from Slayton: "For your information, we are going to start calling you Venus." Slayton explained that the tracking ship, coastal sentry *Quebec,* in the Indian Ocean had reported that Sigma 7 was visible for five minutes as it sped across the night sky. The ship's officers said the spacecraft looked as bright as the planet Venus.

"How about that," said Schirra, amazed.

"Did you have your steak?" asked Slayton.

"Yeah," responded Schirra. "Did you?"

"Yeah, it was okay. Did you eat it?"

Schirra did not respond. Instead, he asked for the Greenwich Mean Time. The exchange referred to a steak sandwich which Schirra had smuggled aboard in his ditty bag. Later, he said he never ate the sandwich because it had become stuck to the bag lining and he could not get it out without pulling it apart. Because of the tendency of crumbs to float in the zero gravity, and possibly clog apparatus, the astronauts' food was specially prepared as a paste, which would be squeezed into the mouth, like toothpaste, or in bite-sized chunks of concentrated meat, vegetables, and fruits. Dehydrated food had to be moistened with water in its cellophane bags.

Flight director Kraft was pleased with the way things were going. "Been a real good show up there," he said. "I think we are proving our point, old buddy."

"I hope so, Chris," replied Schirra. "I'm enjoying it."

DRIFTING AND DREAMING

During much of his fourth revolution, Schirra allowed Sigma 7 to drift. As he came up on California he told John Glenn how it felt: "I've got a real weird attitude at this point. I'll clue you. Ha, ha, I'm looking down at the earth. I'm sort of coming toward you, head first, inverted . . . which is an unusual way for any of us to approach California, I'll admit."

GLENN: Roger, Wally. You got anything to say to everyone watching you across the country on this thing? We're going out live on this. [The transmission was being carried by the radio and television networks from Mercury Control Center.]

SCHIRRA: That sounds like great sport. I can see why you and Scott like it. I'm having a trick now. I'm looking at the United States and starting to pitch up slightly with this drifting rate. And I see the moon, which I am sure no one else in the United States can see as well as I right now.

GLENN: I think you are probably right.

SCHIRRA: Ha, ha. I suppose the old song "Drifting and Dreaming" would be apropos at this point. But, at this point, I don't have a chance to dream. I'm enjoying it too much.

GLENN: Things are looking real good from here, Wally.

SCHIRRA: Thank you, John. I guess what I'm doing now is a couple of Immelmanns across the U.S.

GLENN: Roger, Wally. Have you had any chance to observe a haze layer?

SCHIRRA: Yes, I have. It's quite fascinating. In fact, it's misleading in the evening. Gives you the feeling that you are pitched down quite far. Have you noticed that?

GLENN: Roger.

As he came within range of the Cape radio to start his fifth revolution, Schirra reported that when the spacecraft drifted to a 90-degree yaw so that it was flying sideways, the view was "just like looking out a train window . . . that's all there is to it." He described the sensation of yawing around as "just walking right around the horizon." He said he had eaten some of the processed peaches and meat cubes, but wasn't very hungry.

"How are you feeling in general?" asked Slayton.

"Very fine, Deke. It's the first time I've had a chance to relax since last December."

The sixth revolution carried Sigma 7 over the lower bulge of South America. Schirra asked the ground station at Quito, Ecuador, to advise the Cape that he could hear Mercury Control's radio transmissions "loud and clear" even though MCC couldn't hear him.

"Yes," responded the Quito communicator. "You are coming through fine. You don't have any word to pass on? Can you say anything in Spanish to the fellows down here?"

SCHIRRA: I'm afraid I can't. Except I would like to come down and visit you. I'm enjoying a beautiful sight of the country.

QUITO: Certainly nice to hear that, but could you say just a few words of greeting to them? They would appreciate it so much. They want to put you on their radio down here. Would you say, "buenos días," or something like that to them?

SCHIRRA: Right. All I can do on that now is say, "Buenos días, you all."

DEATH OF AN AUTOPILOT

Next up was Mercury-Atlas 9—"Faith 7"—with Air Force Major Leroy Gordon Cooper, Jr., age 36, at the controls.

It was 1:30 p.m. on the afternoon of May 16, 1963, that the first word of trouble in Faith 7 reached press observers at Cape Canaveral. Shorty Powers, the Voice of Project Mercury, reported that the 0.05-gravity-indicator light was shining green on the console of the spacecraft. It should not have been, because Faith 7 was flying in orbital free fall.

The flight of Mercury-Atlas 9 had been running like a good watch for 18 revolutions. The fourth manned flight of Project Mercury was programed to go 22 revolutions. Cooper had spent nearly 29 hours in orbit, moderately uncomfortable but delighted with the ride. His suit overheated, as had Schirra's, but he had managed to reduce the temperature. However, a malfunction in the suit's dehumidifier caused moisture to collect in the suit and made the astronaut feel clammy. Before long, the tips of his fingers became wrinkled, like a bather's who has been in the water too long.

At the 28th hour and 59th minute of the flight, Faith 7 had just passed over Houston on its 18th revolution and Cooper had reported that he had the area in sight "loud and clear." He could also see all of Baja California. The flight had been going well. The Cape radio had sent this message: "Please pass to Major Cooper in flight from Air Force Secretary Zuckert and Chief of Staff General Le May: 'It is with great pride and enthusiasm that the entire United States Air Force is following the progress of your historic flight—a dramatic contribution to aerospace exploration. Good luck and Godspeed.' "

As he passed over the Cape to begin his 19th revolution, Cooper reported he had a magnificent view of Miami Beach. He turned on a slow-scan television camera which had been mounted in the front of the cockpit to transmit pictures of him to the ground, but the Cape reported it was not receiving a picture, possibly because the light in the cabin was too dim, and advised him to turn it off. Cooper did so and then reported his bad news: "At 28:59, my 0.05 *g* telelight came on after I turned my warning lights off and back on to dim."

There was no response. Faith 7 had presumably passed out of radio range. Cooper hoped vainly that the light would go away, but it glared persistently at him. The light signaled the return of $^{1}/_{20}$ of a gravity, an event that should occur only when the spacecraft would reenter the atmosphere to land. He described the problem to Scott Carpenter when he passed over Hawaii and asked Scott to relay it to the Cape for advice. California reported that "our panel looks good . . . telemetry does not indicate 0.05 *g*."

"Roger," said Cooper. "It must be . . . I just threw a glitch * into the light when I was turning my warning lights off and on, then, probably." But he knew there was more to it than that, and the ground knew it, too.

The light signaled that the autopilot had shifted into the

* A "glitch" is an unexplained electrical phenomenon.

reentry mode of operation. That meant it was no longer available to maintain the spacecraft in orbital-flight attitude or to adjust it for retrofire attitude. It would function only to damp out oscillations of the spacecraft as it was falling through the atmosphere. If Cooper wanted to change the position of the ship in flight, he would have to use the manual-control modes. He would also have to align the ship manually for retrofire.

Somehow, the light indicated, the autopilot had disengaged itself from the gyroscopes and the horizon sensor and therefore could no longer function in either the orbital- or retro-fire-flight modes. The "glitch" was later found to be a short circuit caused by moisture accumulation in the amplifier calibrator, a device where electronic signals from the automatic-sensing systems were converted into commands which fired the jet thrusters.

On advice from the ground, Cooper made a series of tests to see if the autopilot would work in the orbital-flight mode. "I determined that indeed I had lost the autopilot and that our logic system had latched into the position after retrofire," he said. "As far as it was concerned, I was ready for reentry." Cooper found himself the first astronaut to become the victim of a system designed for a chimpanzee. It was clear then that he would have to align the spacecraft manually for retrofire, as Carpenter had done. Recalling Carpenter's misalignment in yaw, Cooper pulled out his star charts. He would begin retrofire on the night side and take his yaw alignment from the stars. He could position Faith 7 in pitch by looking at the horizon, which was quite distinguishable at night.

During his 20th revolution, Cooper conferred with John Glenn, who was stationed on the coastal sentry *Quebec* in the Pacific Ocean south of Japan; with Gus Grissom at Guaymas, Mexico, and with Alan Shepard at the Cape. Details of how he would align the ship manually for retrofire so that he would descend in the impact zone in the Pacific, where the *Kearsarge* was awaiting him, were worked out. Everyone agreed that after retrofire, Cooper would allow the autopilot to take over for reentry, since it was engaged in that mode. At the advice of Dr. Berry, Cooper took a dextro-amphetamine sulfate tablet to pep him up. This was no job for a tired astronaut, although the rugged pilot insisted he felt perfectly fine. He had even slept on his second revolution—thereby proving it was possible for a man to sleep in zero gravity. His only worry had been that his arms had been outstretched and he had feared his fingers might accidentally trigger a switch.

In weightlessness, his arms felt as relaxed outstretched as at his sides. In order to keep his hands from touching anything, he anchored his arms to his sides by hooking his thumbs in the restraint harness.

"Gordo," called Grissom, as Faith 7 passed over Mexico, "this is your last pass over us."

"Roger," said Cooper. "I'll see you in a couple of days."

GRISSOM: You're doing an outstanding job. I'm proud of you.

COOPER: Roger. Thank you, Gus.

GRISSOM: Your friends in Mexico say adios.

COOPER: Roger. Muchas gracias. That's French for thank you.

On his 21st revolution, Cooper reviewed his retrofire procedure with Glenn aboard the coastal sentry. Glenn warned him that in addition to firing the retrorockets manually, he would have to jettison them manually, too, afterward. Glenn said he and Shepard would count down with Cooper for retrofire on the next pass.

On his 22nd revolution, Cooper had another dose of bad news, which he gave to Zanzibar: "I have an item for you. My ASCS a-c inverter has failed, so I will be making a manual reentry."

"Have you tried the standby inverter?" asked the voice from the east coast of Africa.

"Roger," said Cooper. "The standby inverter will not start."

When this intelligence was relayed to the Cape by the Mercury network, controllers were deeply concerned. The automatic Stabilization Control System inverter changed direct current from the batteries to alternating current to power the autopilot. When both it and its backup failed, the entire autopilot system was dead. Cooper would have to stabilize the ship in reentry, too. It would be the first time a pilot would have attempted to damp out the swinging of the ship as it descended in the atmosphere by using the manual controls. In addition, the ship had to be rolled continuously during the descent to prevent yawing and tumbling. Profiting from Carpenter's experience with severe oscillations below 40,000 feet before the drogue parachute opened, Mercury engineers had set an atmospheric-pressure switch (barostat) on Faith 7 to deploy the drogue at 42,000 feet. Until then, it was up to Cooper's piloting skill to hold the ship reasonably steady during reentry.

Swinging northeast across the Indian Ocean, Southeast Asia, and China, the disabled craft sailed out over the East China Sea and was hailed by John Glenn aboard the Coastal Sentry "How's the window attitude?" he asked. "Does it check okay?"

"Right on the old gazoo," responded Cooper confidently.

GLENN: That's the way, boy. Okay, our procedure, Gordo. I'll give you the one-minute hack before retrofire and then there will just be a countdown to a 30-second point and then a 10-second count to retrofire. At the 5-second point I'll tell you to arm squib [prepare to fire].

COOPER: Roger. That's fine.

GLENN: How's your carbon dioxide doing? [The level had been rising and the Cape had warned Cooper about it.]

COOPER: Oh, it's coming on up. And my ASCS inverter has failed, and a few other little odds and ends. I'll shoot the retros on manual and I'll reenter on fly-by-wire.

GLENN: Roger. Okay.

COOPER: I'm looking for a lot of experience on this flight.

GLENN: You're going to get it. Okay, one minute to go, on my mark. Stand by. Mark.

The flight time was 33 hours, 59 minutes, and 24 seconds.

COOPER: Roger. I got it.

GLENN: Roger. I'll give you a 10-second count here down to the 30-second point. Ten, nine, eight, seven, six, five, four, three, two, one. Thirty seconds. The next 10-second count will be a countdown to your manual retro. Ten, nine, eight, seven, six, squib arm, four, three, two, one. Fire.

At the Cape observers heard Shepard counting down with Glenn. It sounded like a chant.

GLENN: Roger! A green one here. [The rockets had fired.]

COOPER: Roger. I think I got all three [retrorockets].

GLENN: Roger. How did your attitudes hold, Gordo?

COOPER: Well, pretty fine.

GLENN: Good show, boy, real fine. Looks like they came off right on the money on time. On the next mark at 60 seconds from that retro you should jettison retros and you'll do that one manually. Right?

COOPER: Roger.

GLENN: You can go ahead and jettison retros.

COOPER: Roger. Jettisoning retros. And off they came.

GLENN: Okay. Dealer's choice on reentry here, fly-by-wire or manual. I think you said you're coming back in on fly-by-wire?

COOPER: I think I'll come back in fly-by-wire.

GLENN: Roger, okay. You can hold retroattitude now for a while here. Your 0.05 *g* time is 34 hours, 9 minutes, 19 seconds [since liftoff]. Just before you get to that, you can come up to your zero reentry attitude. And you can establish a roll at that time, too. It's been a real fine flight, Gordo. Real beautiful all the way. Have a cool reentry, will you?

COOPER: Roger, John. Thank you.

A CHIMNEY IN TIBET

Gordon Cooper had nine minutes before he struck the atmosphere, which appeared quite blue in color below him, as blue as it does from the ground. For a few moments, he could think.

The youngest of the seven astronauts, Cooper was born March 6, 1927, in Shawnee, Oklahoma. His father, Leroy Gordon Cooper, Sr., was a pilot in the Army Air Corps. Gordo could remember the time his father first let him handle the stick of an airplane at the age of six. Cooper senior knew the greats in aviation of his day—Wiley Post and Amelia Earhart among them—and Gordo loved to listen to flying talk. The senior Cooper died of cancer as an Air Force colonel in 1960, but he lived to see his son become an astronaut. As a student at Shawnee High School, Gordo worked after school and week ends at the local airport to pay for flying lessons. Like Schirra, he soloed when he was 16. After graduation from high school in 1945, Cooper entered the Marine Corps for a short hitch and then enrolled in the University of Hawaii, which he attended for three years. He married Trudy Olson of Seattle, taught her to fly, and the two hopped through the Hawaiian Islands in a Piper Cub. They have two daughters, Camala and Janita.

From the university, Cooper entered the Army and promptly transferred to the Air Force. After training, he was assigned to the 36th fighter-bomber group in Munich, Germany. When he was reassigned to the United States, the Air Force sent him for technical training at Wright-Patterson Air Force Base, Dayton, Ohio, and he received a bachelor's degree in aeronautical engineering from the Technical Institute there. He was then assigned to the Air Force experimental flight test school at Edwards Air Force Base in California, where he was designing and testing experimental aircraft when he was selected for astronaut training in 1959.

Cooper's most remarkable contribution to the body of knowledge concerning a man's adaptation to space was his insistence that he could see fine detail on the ground from altitudes of more than 100 miles when the weather below him was clear and dry and the light was good. He reported: "I could detect individual houses and streets in the low humidity and cloudless areas such as the Himalaya mountain area, the Tibetan plain, and the Southwestern desert area of the U.S. I saw several individual houses with smoke coming from the chimneys in the high country around the Himalayas. The wind apparently was quite brisk and out of the south. I could see fields, roads, streams, and lakes. I saw what I took to be a vehicle along a road in the Himalaya area. I could first see the dust blowing off the road, and then could see the road clearly, and when the light was right an object that was probably a vehicle. I saw a steam locomotive by seeing the smoke first, then I noted the object moving along what apparently was a track. This was in northern India. I also saw the wake of a boat in a large river in the Burma-India area." [5]

Cooper's observations over the Himalaya-Tibet region were made about 7:30 a.m. local time at an altitude of 101 statute miles. He said that as he passed over the south edge of Houston, Texas, he could see the lakes near the Manned Spacecraft Center there. He added, "I couldn't see my own home [nearby] because we left too many trees up around there." [6]

Cooper's claims were doubted by several experts in optics, who insisted that what he said he saw was beyond the resolution of the human eye. One specialist, Dr. W. Ross Adey, director of the Space Biology Laboratory at the Brain Research Institute, University of California at Los Angeles, suggested that Cooper might have been having hallucinations as a result of neuromental disturbance from extended weightlessness.[7] Cooper remained unperturbed by these opinions. He knew what he had seen. His report confirmed experimental work at the National Bureau of Standards' Central Radio Propagation Laboratory in Colorado, that optical resolution is better when looking down through the atmosphere than when looking up through it. One explanation is that the viewer in orbit is farther away from image-distorting turbulence in the lower atmosphere than is the viewer on the ground. The closer the viewer is to turbulence the more it blurs the image beyond it. Conversely, the blurring effect is reduced with distance. The implications of this "good seeing" from space for orbital reconnaissance were interesting, not simply for military but also for scientific purposes and mineral prospecting. For significant geological patterns that were not visible from the ground or even from aircraft altitudes were apparent from orbit.

As reentry time approached, Cooper held the spacecraft with the heat shield canted down. When the vehicle hit the atmosphere, he had the impression that it was stabilizing itself in reentry attitude. The controls became "mushy." Cooper then began to roll the ship as it plunged into thicker atmosphere. Burning particles came past his window, including one of the straps which somehow had been left behind from the retrorocket package he had jettisoned. He noted the fireball John Glenn had first described. Cooper felt no particular discomfort at experiencing weight again after 34 hours of weightless flight.

Then Scott Carpenter came on the radio from Hawaii: "What is your status?"

Cooper replied, "Doing fine. Reentering."

At 42,000 feet, Cooper manually deployed the drogue parachute and he heard it come out with "a big rattle, a roar, and a thump." Then the big main parachute deployed at 11,000 feet. He reported that his main chute was good.

CARPENTER: Good show! Everything looks good. Preparation for impact. Urine-transfer shutoff valve closed. Transfer hose, disconnect. Blood-pressure hose, disconnect. Aeromed connector, disconnect. Helmet outlet hose, disconnect. Are you staying with me, Gordon?

COOPER: Roger. I've got my list right here, Scott.

A new voice entered the astronaut's earphones: "Hello, Astro. This is One Indian Gal. Over."

"Roger, this is Astro. Go ahead."

"Roger. We are circling you at 500 feet. You're coming down very nicely. Sea state is about five- to eight-foot waves, a few whitecaps. How do you feel? Over."

"Roger. I'm in fine shape. Excellent."

"Thank you, Astro. This is Indian Gal. We are still circling you, very nicely. You are steadying up quite nicely, about 400 feet. You are passing my starboard side. Have three helos right around you. Got the swimmers with me. They'll be out just about the time you're setting down on the water. The carrier is only about three miles away. Couldn't be a nicer shot."

"Roger. I'd like to come aboard the carrier if they will grant permission to an Air Force troop."

"Roger. Begonia, this is Indian Gal. Gordon Cooper desires to come aboard the carrier if they will let an Air Force officer aboard. Over."

From the *Kearsarge:* "Roger. Permission granted. Estimate 45 minutes to have him on deck. Over."

So ended Project Mercury, four years and eight months after it began in October 1958. At a cost of $384,100,000 and with the combined efforts of 2,020,528 workers, technicians, engineers, and scientists it showed that "man can function ably as a pilot-engineer-experimenter without undesirable reactions or deteriorations of normal body functions for periods up to 34 hours of weightless flight." [8] The United States had passed its first milestone on the new frontier.

Chapter 8

Mare Nubium

"With regard to the side of the moon facing us, let it be said first that one part of it is noticeably brighter, the other darker . . . so that we could perceive that the surface of the moon is neither smooth nor uniform, nor very accurately spherical, as is assumed by a great many philosophers about the moon and other celestial bodies, but that it is uneven, rough, replete with cavities and packed with protruding eminences, in no other wise than the earth, which is also characterized by mountains and valleys." So wrote Galileo Galilei in 1610.[1]

Three and one-half centuries later, men who were contemplating a manned landing on the moon knew little more about the nature of its surface than Galileo did. They did not know whether the 15-ton lunar-landing vehicle would be supported in an upright position on a reasonably firm, smooth field, whether the craft would sink deep in a sea of dust, or be toppled over by sliding rubble, or break through a thin, brittle crust to crash into a hidden crevasse. All of these possibilities had been predicted.

The lunar scientists of the twentieth century, who peered at the moon through giant telescopes and groped at its surface with fingers of radar, were still as ignorant of its true nature as were the lunar philosophers of the seventeenth century.

The state of ignorance concerning this matter, however, did not deter President Kennedy from issuing a call for a manned lunar landing nor did it inhibit NASA from responding with a well-shaped, well-conceived, $20-billion program. All that it lacked was evidence that it was possible to land on the moon and to blast off it again once the landing was made.

Because of the blurring effect of earth's atmosphere, no one on earth could see surface detail on the moon much smaller than an area about the size of Chicago's Loop or of Central Park in New York City. Even the most powerful terrestrial telescopes could resolve the lunar detail down to only six-tenths of a mile. On rare occasions, when a clear, calm atmosphere made the seeing exceptionally good, details down to two tenths of a mile could be ascertained.* This was not good enough, however, to show whether the surface was dust, sand, rock, or a combination of them. Radar reflections from the moon were interpreted by some observers as indicating a rough, rocky surface, and by others as indicating a layer of dust. It was theorized that the infall of micrometeoroids and the impact of meteorites over several billions of years had produced a great layer of finely-divided particles on the lunar surface.

The space agency directorate was fully aware of the urgency of developing automatic spacecraft which could take photographs close up and analyze the structure, composition, and bearing strengths of typical surfaces on the moon well in advance of Apollo. The Kennedy time frame required that the development of the manned-landing hardware and of lunar reconnaissance vehicles be done concurrently. Consequently, the lunar-landing vehicle had to be designed on an assumption that the moon's surface would support loads of 4 pounds per square inch.† In the light of prevailing knowledge, or ignorance, about the surface, this was the most expensive technological gamble in the nation's history.

If NASA was to see what the surface of the moon was like, it had to send a camera there. This was the mission of Project Ranger, the first of three American space enterprises designed to identify the nature of the lunar surface. Ranger essentially was a photo probe, designed to take pictures for

* By the Lick Observatory in California in 1963.

† Earth weight, or 0.66 pounds moon weight.

about 15 minutes and transmit them to earth as it hurtled at the moon and crashed. Later, a more sophisticated spacecraft called Surveyor would execute a soft landing. In addition to photographing the lunar ground from distances as close as four feet, it would be equipped with instruments to determine the bearing strength and chemical composition of the ground —provided it was not buried in a lake of dust. Finally, by placing a camera into a low orbit around the moon, NASA hoped to develop a photo map from which to select a suitable landing site for Apollo astronauts. This enterprise was budgeted as Project Lunar Orbiter.

Ranger was conceived in 1959, when the fledgling NASA was evolving plans for a manned fly-by of the moon. On December 21, Abe Silverstein, then director of the Office of Space Flight Programs, asked Pickering at the Jet Propulsion Laboratory to study spacecraft design for a mission that would acquire and transmit a number of images of the lunar surface. He also requested that Pickering evaluate the probability of obtaining useful data from a survivable package incorporating a lunar seismometer of the type NASA was then developing. The seismometer would be dropped on the moon to measure moonquakes and the vibrations caused by meteorite hits.

JPL began on the project in 1960. Initially planned as a five-flight program, it was expanded to 15 flights after the manned lunar landing became a national goal. However, unforeseen problems, costs, and a sequence of disastrous failures in the program resulted in a cutback to nine flights.

In retrospect, it becomes clear that Ranger, rather than Apollo, was the beginning of interplanetary exploration by the United States. The Ranger called for a new level of technological sophistication and engineering excellence. The sun-powered vehicle had to be capable of aligning itself automatically in two axes so that its solar-energy panels faced the sun and its directional antenna pointed toward earth. This required the development of a delicate, photoelectric eye-sensing system which could command the gas-fed, attitude-control jets when it saw the sun in one axis and the earth in the other. Ranger was more than an automatic vehicle. It was also a robot photographer equipped with the most sensitive cameras ever made—cameras capable of taking pictures far exceeding the quality of the display on the finest television set. Ranger required an electronic brain, the central computer and sequencer, that could tell it what to do, and a powerful, precise motor that could correct its course after it had been launched by an Atlas-Agena rocket on a lunar trajectory.

Ranger, as a reconnaissance system, incorporated new features which had not been tested. These were an attitude-control system, referenced to the sun, to the earth, or to a star; a solar-power system, which was tied to attitude control; an on-board, automatic computing and sequencing system which controlled the vessel with only a minimum of commands from earth; a midcourse correction system; the concept of a space-frame and subsystems design which could serve both lunar and planetary investigations; and a two-step launching technique of placing Ranger first in a parking orbit around the earth and then boosting it moonward from a more favorable geographical point than Cape Canaveral.

In designing Ranger, Pickering and his associates at Pasadena were aware that they were commencing the development of a new generation of vehicles that would reconnoiter not only the moon but the planets beyond. The aerospace industry was aware of it too, and particularly conscious of the fact that the pioneering program was being developed in-house by a university-operated government laboratory.

Ranger looked as though its design had been inspired by a channel-marker buoy, with a pyramidal or conical tower rising out of a floating base. The spacecraft had a hexagonal base 5 feet in diameter. From it a telescope-like cone extended about 8 feet, containing the television cameras and an omnidirectional antenna. Hinged to the base were two solar panels which unfolded on opposite sides in flight to a wingspan of 17 feet. Their 9800 solar cells were designed to convert sunlight into 200 watts of electricity, which provided power for two radios, the cameras, the attitude-control and sensing systems, and the central computer. A silver-zinc storage battery powered the vehicle when the panels were not facing the sun or were folded up against the cone.

Precise control of the spacecraft from the ground and picture transmission depended on the directional antenna, a 4-foot dish which unfolded itself from the base hexagon during flight and pointed earthward when Ranger was properly aligned. JPL controlled the spacecraft by signaling its central computer from Deep Space Instrumentation System radio stations at Goldstone, California, Johannesburg, South Africa, or Woomera, Australia. The radio commands could be stored in the spacecraft computer until a preset time or until a JPL signal ordering the computer to act was received. Telemetry radio would keep earth advised whether the commands had been executed. Telemetry told the JPL control room at Pasadena a complete story about Ranger as it flew: when its solar panels unfolded, how its high-gain, directional antenna was positioned, what the craft's attitude was, how the power supply was operating, what the temperature was at a number of points on the "bus," or body of the vehicle, and the status of the cameras.

To prevent electronic parts from freezing when facing away from the sun, or melting when in its glare, Ranger was given a "passive" temperature control system. It distributed solar heat evenly throughout the vehicle at nearly room temperature. The system employed gold plating, polished aluminum, and white paint on the vehicle surfaces to balance the amount of heat received from the sun-facing side with the amount lost on the shadowed side.

THE GOLD-DUST DISPUTE

In the 350 years between the time Galileo looked at the moon through a 30-power telescope and the development of a machine to fly there, the principal theories about the lunar surface held that it was hard rock, deep dust, rubble, or vesicular (bubbled) lava. On other features of the moon, however, there was better agreement. The moon's diameter had been calculated at about 2163 miles, nearly a quarter of earth's, and lunar gravity at one-sixth of one earth gravity. A 180-pound astronaut would weigh only 30 pounds on the moon. A 15-ton lunar-excursion module would weigh only 2.5 tons when it landed on the lunar surface. The average distance of the moon from earth had been measured at 238,857 miles

Because it takes about the same time to rotate on its axis as it does to revolve around the earth, 27⅓ days, the moon always presents the same face to earth. However, the moon wobbles a bit on its axis, an effect of the gravitational pull of the sun and other planets called "libration," and thus terrestrial observers can see about 59 per cent of the lunar surface. Only 41 per cent is hidden—or was until the Soviet Luna III photographed it October 7, 1959.

The lack of a sensible atmosphere on the moon is generally agreed. Temperatures are extreme, from 230 degrees Fahrenheit at midday in the tropics to 300 degrees below zero Fahrenheit at night.

Since Galileo, observers have seen more than 300,000 craters on the moon. The largest, Clavius, is more than 146 miles in diameter, with walls more than a mile high. However an astronaut standing in the center of Clavius would not see the walls—they would be over the near lunar horizon. The smallest craters that could be seen appeared to be about 1000 feet in diameter.

With the naked eye, anyone can see that the moon has two kinds of surfaces, as Galileo said, the light and dark. The light surfaces show up in the telescope as mountainous, rugged highlands, while the dark ones appear to be lower and relatively smooth. Astronomers of the Renaissance referred to the dark areas as "maria," or seas, and to the lighter, lunar highlands as the "terrae," or land. This nomenclature has persisted even though it has been realized for many centuries that there are no "seas" on the moon in the terrestrial sense.

The origin and evolution of the moon have been problems with high priority in solar-system investigation for a long time. There is general agreement among astronomers and physicists that the moon's structure may be an important clue to the origin and evolution of the solar system. Was the moon formed in the same way as the earth and then captured by the earth? Was it formed of condensing dust clouds already in orbit around the earth after the earth was formed? Was it torn out of the earth, leaving a depression on the order of the Pacific Ocean basin? Disagreement persisted also about the origin of the craters on the moon, and whether the dark material in them was lava. Most moon experts have in the last 25 years accepted the view that craters on the moon are principally the products of meteorite impacts, but a strong minority argues that while some may have been formed in this way, many are volcanic in origin. In the impact theory, the lava would have been produced by the enormous heat of impact energy, melting the rocks and causing them to flow into the vast depression of the crater. On the other hand, the lava might be a product of extensive lunar volcanism in the past.

For many years, it was assumed that the maria generally lay below the level of the continental terrae, in the manner of ocean basins, so that lava would have pooled in these low-lying regions. However, recent contour measurements may have discredited this idea. The higher regions on the moon include a large part of the maria, while the continental areas surrounding the mountains called the Caucasus, Alps, and Jura are actually lowlands.

There are two kinds of maria, the circular "seas," like Imbrium, Crisium, and Serenitatus, and the irregular ones, like Tranquillitatus, Fecunditatus, Procellarum, and Nubium.

More specific features of the surface are crevasses, rilles (long valleys), ridges, and faults, where the surface has cracked. Among the most curious are the rayed craters, radiating streaks of whitish material for hundreds of miles. The most conspicuous rayed craters are ones called Copernicus, Kepler, Aristarchus, and Tycho. Lunar mountains frequently appear around the circular maria, with peaks rising 10,000 to 20,000 feet above the floor. Near the south pole, the Leibniz Mountains appear to be of Himalayan dimensions.

By the end of the 1950s, the amount of scientific attention the moon was getting began to increase at virtually an exponential rate as it became an obvious target of manned-space-flight planning. By 1963, conferences on the nature of the lunar surface evoked technical papers in a dozen fields, from cartography and civil engineering to geology and chemistry. It was generally agreed that the maria offered the smoother landing sites. But that was as far as agreement on the surface seemed to go.

In 1955, the British astronomer Thomas Gold, then of the Royal Greenwich Observatory, had suggested that the maria were covered with a layer of dust that might be several miles deep. Space travelers, he said, would simply sink into the dust and become buried in it with all their gear. Gold theorized that the particles were so small that electrical and thermal effects would cause the surface to heave and billow like a liquid. Over the eons, dust falling on the moon and fine particles produced by meteorite impacts would tend to accumulate in the maria "basins," which were then assumed to be low. In this view, the maria were indeed seas—seas of dust.

The dust hypothesis was derived mainly from the manner in which the moon absorbed and radiated solar energy, but analysis of reflected light in the bright regions also indicated "a rough, loose surface of cracked and shattered stone and smaller particles down to and including a fine dust." [2]

In the opinion of the chemist-cosmologist Harold C. Urey of the University of California, "probably the surface is more nearly like fine sand rather than dust." [3] A third view by Gerard P. Kuiper, astronomer of the University of Arizona, held that the maria surface was lava, probably produced by the enormous heat of meteorite impacts. Various other hypotheses were advanced from time to time that the surface was covered with boulders, with large rock fragments from crashed meteorites, or with dust compacted like cement several inches below the surface.

For many years the general appearance of the moon was visualized mainly in the terms of the work of the artist Chesley Bonestell, as a steep, angular landscape, of escarpments abrupt and sharp, knife-edge cliffs and pinnacles, all boldly lighted or shadowed. On a body where no air existed to diffuse the light, or where no erosion, it was supposed, had ever rounded the edges of the rocks or softened a precipitous landscape, all would appear to be bare and faceted, like a diamond, and with a diamond's hardness. Most artists' conceptions of the lunar landscape followed Bonestell's original and dramatic visualization. Whether accurate or not, it was unearthly, and, therefore, lunar.

Observations of radio-wave reflections from the moon in the wavelength range of 0.1 to 3 meters drew another picture of the surface. They indicated that instead of being steep and sharp it was smooth and undulating, with slopes averaging only six degrees, and all but ten per cent of the areas observed were covered with some kind of small objects, possibly rocks, which were well below the limit of telescopic resolution from earth. "Heat absorbent and reflecting characteristics of the surface may be interpreted as representative of finely divided, granular material in vacuum, or, at least, a thin layer of this material, a few millimeters thick, overlaying a better conductor, such as pumice or loose gravel." [4]

Whether the lunar dust was thin or thick, it appeared from those data that the surface was not characterized by knife-edge crags and sharp clefts. It was more like a gently undulating desert.

THE WAYWARD RANGERS

It was the expectation of both NASA and Congress that Ranger would provide details of the lunar surface on which to verify the design of the Lunar Excursion Module landing gear in plenty of time. An official NASA release emphasized the importance of the Ranger reconnaissance to the whole Apollo effort: "Peering through the thick layer of atmosphere around the earth, astronomers would not be able to detect an aircraft on the lunar surface. . . . With such limited knowledge about the moon's surface, it is virtually impossible to design a manned spacecraft to land on the moon. Ranger is designed to provide information as to the nature of the lunar surface." [5]

The first two Rangers, 1 and 2, were designed to be used only on engineering test flights. They were not to fly to the moon. They were to be launched to test the parking-orbit mode of launching to the moon or to a planet, and to determine how well the solar cells and the automatic controls functioned.

Ranger 1 was launched August 26, 1961, by an Atlas-Agena rocket. After the Atlas engines cut off at 280 seconds, a shroud protecting the Ranger from aerodynamic buffeting was jettisoned by spring-loaded bolts. The spacecraft, with Agena, then separated from the Atlas and pitched its nose down from an angle of 9 degrees above the horizon to a flight path almost level with it. The Agena engine then lit for its first burn of 2½ minutes. The firing accelerated the rocket and spacecraft to earth-orbital speed of 17,500 miles an hour. Agena then shut down, and rocket and payload coasted down the Atlantic Missile Range. Over Ascension Island in the South Atlantic, Agena-Ranger reached a point from which it could proceed directly to the moon with another burst of acceleration. Agena was programed to provide this burst by starting up its engine and firing to increase velocity to nearly 25,000 miles an hour—or about the speed required to escape from earth. But the Agena engine failed to restart when the time came for Ranger to be lifted out of earth orbit. Ranger 2, launched October 18, 1961, met the same fate. Both vehicles remained in low earth orbit until they fell back into the atmosphere and disintegrated. However, JPL reported the tests were not a total loss. The engineers had been able to get some telemetry information on the performance of the spacecraft systems, even though the systems could not operate fully in earth orbit.

The first Rangers were written off as engineering test models, in Block 1. Ranger 3 represented a fully instrumented vehicle in a series called Block 2, which was designed to perform a lunar reconnaissance in three modes. It would take television photos of the moon as it approached the surface, measure gamma radiation from the rocks, sound the surface with pulses from a small radar set, and drop a seismometer encased in balsa wood to report on moonquakes. The gamma radiation would indicate whether the lunar rocks were terrestrial or meteoritic in nature and the radar signals would provide some information on surface roughness. The seismometer in its balsa-wood housing was carried in a 56.7-pound subcapsule. It was to be separated from the moonward-plunging Ranger automatically when the radar reported an altitude of 70,000 feet. The descent of the subcapsule was to be checked from 6000 miles an hour to zero velocity at 1100 feet by a retrorocket developing 5080 pounds of thrust. The subcapsule would then fall the rest of the way. It was hoped that the seismometer would report any moonquakes or large meteorite impacts for 30 days. The Ranger 3 camera system was able to transmit one photograph every 13 seconds. This spacecraft was by far the most complex the United States had ever attempted to launch.

Ranger 3 was boosted off the Cape at 3:30 p.m. Eastern Standard Time on January 26, 1962, before an international audience of newsmedia correspondents which had assembled to watch the flight of John Glenn. This time, the Agena second-stage rocket engine ignited for the second burn after an eight-minute coast down the Atlantic Missile Range. Ranger was at last hurled on its 66-hour flight to the moon.

As Ranger 3 began its lunar flight, its central computer and sequencer issued commands which had been stored in its electronic brain before the launch. One command increased the power. which had been kept low during powered flight up through the atmosphere to prevent arcing. This event was telemetered to earth and noted at the Johannesburg station of the Deep Space Instrumentation Facility. A second command unfolded and extended the solar panels. A third ordered the spacecraft to align itself with the sun, so that the solar panels would receive maximum sunlight. The sun was perceived by light-sensitive diodes which command the nitrogen-gas jets to turn the spacecraft until the diodes see the sun. When the power system began receiving sufficient energy from the solar panels, a switch automatically disconnected the battery from the power supply.

In the meantime, the directional antenna, a 4-foot dish which had been tucked under the hexagonal base, was extended outward on command of the automatic controller. Three photomultiplier tubes serving the directional antenna then began to look for the earth, the brightest object in the heavens after the sun. Their search caused the nitrogen-gas jets to roll the spacecraft around its long axis without disturbing its lock on the sun. When the tubes saw the earth, they shut off the jets. Ranger 3 was then stabilized so that the solar panels faced the sun and the directional antenna faced earthward. The first American photo-reconnaissance probe was in its cruise mode, en route to the moon.

Five hours after the launch, NASA announced that Ranger would miss the moon by as much as 30,000 miles. It had been over-accelerated by the Atlas as a result of an error in the rocket's guidance system. The miss distance was too great to be corrected fully by Ranger's midcourse-correction motor, capable of developing 50 pounds of thrust for 68 seconds. Nevertheless, JPL controllers decided to make the maximum correction possible. They radioed a four-word command which was stored in the central computer and sequencer. One word spelled out the direction of the new course, another described the amount of roll required to align the spacecraft on it, the third represented the change in velocity, which was determined by the firing time of the hydrazine motor and the fourth word, "go," told the autopilot to act.

The maneuver was executed 16 hours after liftoff, on the morning of January 27, 1962. Ranger 3 then realigned itself with the earth and sun and continued on its new flight path. Analysis of the new path showed that the miss distance had been held to 23,000 miles. Ranger would cross the orbit of the moon at a point in space and time ahead of the moon. However, Project directors decided to attempt to get pictures of the moon's leading edge, and also of its hidden side, which had been photographed by the Russian Luna III in 1959.

As the vehicle flew past the moon, JPL controllers transmitted commands from the Goldstone station to turn the spacecraft so that its camera pointed squarely at the moon. The gas jets fired properly and the camera-lens opened on schedule. The television system came on and began transmitting what the camera saw. But the spacecraft failed to perform one critical task. It failed to reposition its high-gain antenna so that it pointed toward earth. Picture-transmission signals, consequently, were so weak that they could not be separated from the background radio noise the big Goldstone antenna also picked up. Ranger 3 flew within 22,862 miles of the moon at 6:23 p.m. January 28, 1962, at a velocity of 4188 miles an hour. It sped on to join Pioneers 4 and 5 in solar orbit.

Overshadowed by the drama of the impending Glenn flight, the fate of Ranger 3 seemed to be of secondary consequence. The JPL team felt that fate had dealt them a cruel blow. The Block 2 Rangers were the most complex automata ever built, capable of acting, reacting, and, to a limited extent, making decisions. Now the first of these fantastic machines had been rendered useless first by an error in the relatively simple-minded guidance system of its launch vehicle and then an antenna-pointing failure. Of some consolation was the fact that the other automatic controls had worked beautifully during the flight. Ranger 3 went down in the records as a failure, but it demonstrated that the United States had achieved a functional machine that could reconnoiter the moon and the planets.

The second Block 2 moon probe, Ranger 4, was launched April 23, 1962. It was identical to Ranger 3. Shortly after it was injected into a good lunar flight path, the control-system sequencer failed because of a short circuit. Inert, Ranger 4 crashed into the far side of the moon—the first United States vehicle to hit the lunar target.

On October 19, 1962, NASA launched Ranger 5. Shortly after the Agena B injected it into a lunar path, the spacecraft's power system failed and another dead machine hurtled into oblivion. It missed the moon by 450 miles.

THE PROBERS PROBED

The failure of Ranger 5 precipitated a Congressional investigation of the program's spectacular bad luck, which had not been equaled since Vanguard. The investigation brought into the open a simmering dispute between NASA headquarters and the Jet Propulsion Laboratory over management.

Because of its history, which antedated the formation of NASA, and its operation by the highly prestigious California Institute of Technology, JPL occupied an anomalous position as a NASA laboratory. Its own management was generally unresponsive and contemptuous of NASA's bureaucratic red tape, which grew exponentially, like the agency itself. The laboratory operated with a university viewpoint, which emphasized the interdisciplinary approach to space research. It catered to the intellectual independence and initiative of the scientist-engineer, and refused to subject him to buraucratic controls. This atmosphere attracted an unusually gifted group of young men and women to the laboratory. It was from this group that Pickering, who nurtured talent when he saw it, built the team that created a new class of interplanetary reconnaissance vehicles, starting with Ranger.

However, the Ranger failures not only made this genteel system vulnerable to criticism from the parent agency but also exposed it to attack from industry forces which opposed in-house development by the government. In developing Ranger itself, JPL set up a yardsick on cost for interplanetary vehicles of this type, and industry cost estimates could be compared with it. "Contractors were restive with the glaring differences in cost and effectiveness between contractor and JPL work," according to H. L. Nieburg, associate professor of political science at the University of Wisconsin.[6] These were differences, he added, "which could not be explained away merely by the tax-exempt and nonprofit status of the university-operated facility."

As was customary, the space agency set up an official

board of inquiry to examine the Ranger failures. The board, headed by Albert J. Kelley of NASA headquarters, made a 30-day study. Its report criticized JPL's work, charging that the laboratory had not applied the high standards of technical "design and fabrication which are necessary prerequisites to achieving a high probability of successful performance." It also found deficiencies in spacecraft design, construction, systems testing, and checkout. It criticized the managerial relations between NASA and JPL, and JPL's managerial technique. In 1959, JPL had established a "matrix" system of management organization, consisting of seven divisions, each concerned with a particular engineering or scientific discipline. All divisions worked together in developing Ranger. In this way, as Pickering told the House of Representatives Committee on Science and Astronautics, the talent in each division could be applied to a number of similar projects simultaneously—such as Ranger and its interplanetary counterpart, Mariner. NASA headquarters, however, would have preferred a "projectized" organization in the development of Ranger, with the concentration of a single-purpose team of scientists and engineers on that project only. "We have been pushing on the Laboratory for a stronger, projectized type of approach to things since early in the program, since late 1960 or 1961," complained Homer Newell, the director of the NASA Office of Space Science and Applications, when he appeared at a meeting of the House Subcommittee on NASA Oversight.[7]

The Kelley report recommended that an industrial contractor should be brought into the program to take over systems management and to redesign a series of Rangers called Block 4, which included spacecraft 10 to 14. NASA negotiated a contract with the Northrop Corporation on March 11, 1963, to fabricate and test the Block 4 Rangers, under JPL supervision. This move would have transferred the Ranger program to the aerospace industry if the Block 4 spacecraft had ever been built. They never were, however, because of a tightening down in the NASA budget required by rising costs of Project Apollo and a decision to move quickly into the Surveyor craft which would make a soft landing on the moon.

In addition, the Kelley report called for postponement of the Ranger 6 launching, which had been scheduled for January 1963, so that the spacecraft's design and JPL's testing procedures could be reviewed. Ranger 6 was thus delayed for one year while it was revamped. All scientific experiments except the camera system were eliminated. An array of six cameras was installed to take thousands of photographs with wide- and narrow-angle lenses as Ranger plunged into the moon. Ranger was put through 13 simulated 66-hour lunar missions in a simulator at JPL. Backup electronic sub-systems were installed to prevent a minor failure from wrecking the entire mission.

In the meantime, JPL broke its string of failures with the amazing flight of Mariner 2 to Venus, a feat unequaled in the space age up to the end of 1962. The spacecraft, a first cousin of Ranger, passed Venus at 21,648 miles on December 14, 1962, all systems working. Nevertheless, JPL still faced the prospect of losing the Ranger program to industry, and several persons at the Laboratory reported that Northrop had invited them to join its staff.

Ranger 6, first of the Block 3 vehicles equipped with only a battery of television cameras, was launched January 30, 1964. It was injected into an accurate lunar flight path and functioned perfectly, performing a precise midcourse correction on the morning of January 31. The new course would take Ranger 6 into Mare Tranquillitatis (the Sea of Tranquility). Excitement at Pasadena began to mount as Ranger 6 approached the moon and controllers prepared to turn on the two full-scan and four partial-scan cameras at an altitude of 1000 miles. The first photographs would have a resolution comparable to that of the best earth telescopes. During the 10-minute photographic sequence, the cameras could take at least 3090 photographs which would be radioed back to Goldstone before the vehicle crashed on the moon.

Shortly after midnight Sunday, February 2, 1964, JPL employees began to take seats in the Von Kármán auditorium at JPL to await the grand finale of five years of work. Now that the moment of triumph was near, all of the disappointments and frustrations of the past seemed to be erased. A few minutes after 3 a.m., one of the spacecraft television channels was switched to the warm-up mode by the spacecraft sequencer. The other was turned to warm-up by a signal from Goldstone. Then both cameras were commanded on. Scientists, engineers, technicians, secretaries, and clerks in the auditorium waited silently for the announcement that the pictures were coming—the first close-up pictures of the moon. But no picture signals came. Goldstone operators sent a series of emergency commands from the big antenna to trigger the balky phototransmission system into operation. No signals came, and Ranger 6 smashed into the moon a few miles from a crater named Julius Caesar. Even though Ranger 6 had been checked and rechecked as thoroughly as Americans knew how, it too failed, and it appeared that the Ranger mission was beyond the capability of the space program.

An investigation of the camera failure by JPL and the Radio Corporation of America, the television-system contractor, showed that the TV system had burned out after being turned on briefly during powered flight through the atmosphere, possibly by vibration at the time the Atlas booster engines were jettisoned. This could have caused a relay to close or produced a momentary electrical discharge which would have triggered a control circuit. JPL had known earlier that if the TV system was turned on during powered flight, electrical arcing was likely to occur at altitudes of 180,000 or 250,000 feet. A coronal-type electrical discharge in the components of the high-voltage system would destroy the television channels in both the partial- and full-scan camera system.

Again, NASA investigated JPL. A four-man board, headed by Earl D. Hilburn, NASA Deputy Associate Administrator for Industry Affairs, reported that JPL's testing and fabrication procedures still were "an area of concern." The Hilburn report said that rigid and complete standards "have yet to be established" and "there is some question as to whether this deficiency may not be attributable to the diverse delegation of responsibilities under the type of organization employed by JPL." The report also opined that NASA headquarters might exercise specific direction of JPL in setting up testing and quality-assurance standards. In testimony before the House committee, Oran Nicks, director of lunar and planetary programs in the Office of Space Science & Applications at NASA headquarters, admitted that "we don't tell them what nuts and bolts to use, or anything detailed of that sort."

"What you are saying, really, is that the design and manufacture and the integration of the systems is almost wholly a JPL responsibility?" asked Representative Joseph E. Karth.

"Yes, sir, I think that is a fair statement," said Nicks.

In its own report on Project Ranger, the Subcommittee on Oversight, headed by Representative Olin E. Teague, observed that "if there was a single thread which consistently ran through the testimony of NASA officials regarding the management of Project Ranger it was JPL's resistance to NASA supervision." NASA Administrator Webb had stated in a letter to Teague that "NASA has felt that administrative management of the laboratory was slighted in favor of the campus method of operations." From this point of view it was the "campus method" and JPL's intransigence at conforming to NASA bureaucracy that lay at the bottom of the Ranger failures. These were the roots of a debacle which had cost the taxpayers $260,000,000 and which, in Webb's view, was likely to delay the lunar landing. Webb had no need to define "campus management." Members of the House committee knew what he meant. The committee was sympathetic to Webb's tribulations. Its report stated, "This extraordinary show of independence on the part of a contractor operating a government facility and doing approximately a quarter billion dollars' worth of business per year for NASA is considered the most significant fact developed during the subcommittee's investigation of Project Ranger."

Coupled with the failures, the committee criticism laid the foundation for tighter control of JPL by NASA headquarters. The space agency was able to bring the Laboratory to heel in a new contract. The three-year agreement under which the California Institute of Technology had operated the Laboratory expired in December 1964. In the new contract, provision was made whereby Webb could get closer supervision over JPL and thus make sure it would be responsive to NASA headquarters direction. The position of general manager was set up in the Laboratory under Pickering, who appointed Air Force Major General Alvin R. Luedecke, Retired, to the post. General Luedecke had been general manager for the Atomic Energy Commission and had had experience in handling relations between government agencies and contractors.

With the subcommittee's approval, NASA moved the Surveyor project outside of JPL, awarding the development contract for the soft-lander to the Hughes Aircraft Company.

Lunar Orbiter, designed to photograph the moon from altitudes as low as 28 miles in a lunar orbit, was placed under the direction of Langley, with the Boeing Company as spacecraft developer.

MARE COGNITUM

With the impending flight of Ranger 7, second of the Block 3 spacecraft, JPL approached a new crisis. Another failure might well mean the end of the Laboratory, in spite of its Mariner 2 success. It was generally understood that if NASA took over full direction of JPL and placed its personnel under civil service, about half of the engineers and scientists would leave. The flight of Ranger 7 would decide the Laboratory's fate.

In order to avoid unintended switch-on of the TV cameras, a circuit was added to inhibit the camera system from operating until Ranger 7 had separated from the Agena and was flying on its own. The practice of sterilizing Rangers so that they would not contaminate the moon's surface with terrestrial microorganisms was abandoned. It was suspected that the disinfecting process, which required heating, could degrade sensitive electrical components. NASA headquarters had been persuaded to sterilize the Rangers so that if biology experimenters found microorganisms on the moon later on they could be reasonably sure they had discovered lunar life rather than immigrants from earth. But now, when the chips were down, sterilization went out the window. Ranger 7 had to make it.

At 12:50 p.m. EST on July 28, 1964, the 806-pound Ranger 7 was launched by an Atlas-Agena from Cape Kennedy on a true course. When its solar panels were outspread in space and its high-gain antenna extended, Ranger resembled a 10-foot dragonfly with a wingspread of 15 feet. Its six cameras were designed to take more than 4000 photographs in the final 13 minutes and 40 seconds of flight before the spacecraft crashed into the moon at 5800 miles an hour.

So accurately was the gold-plated vehicle injected on its lunar flight path by the Atlas-Agena that an early calculation showed it would hit the moon without a midcourse correction —but on the far side. The midcourse motor could easily adjust the flight path so that Ranger 7 would fall into the earth-facing side of the moon, in a region called Mare Nubium, the Sea of Clouds. After the midcourse maneuver was accomplished, JPL announced that the spacecraft was flying in an optimum attitude, with the cameras pointed exactly right, and that no terminal maneuver would be necessary. Impact was expected at 6:29 a.m. Pacific Daylight Time on July 31, in the northwest portion of the Mare Nubium.

About 4 a.m. July 31, newsmen and television-camera crews began to assemble in the JPL auditorium, which was filled before dawn by JPL personnel, talking quietly and exuding vast hope. The audio component of Ranger's telemetry was being piped into the auditorium from Goldstone and over the public-address system. It sounded like a metallic moaning.

Since no terminal maneuver was necessary, the flight controllers resorted to an interesting deception in order to make sure that the cameras would turn on. The turn-on command was tied directly to the terminal-maneuver sequence in the spacecraft computer. The cameras could be turned on also by direct radio command from Goldstone. However, in order to have the automatic command in the terminal-maneuver sequence available as a backup, in case direct communication should fail, the controllers sent a fake terminal-maneuver sequence which was stored in the spacecraft computer. It consisted of minimal changes in pitch and yaw, but before Ranger could act on them, the commands would be nullified by a second signal telling the central computer to ignore them. The countermanding signal, No. 8, would not interfere with the automatic camera turn-on command, however. If countermanding command No. 8 did not reach the spacecraft or if it failed to cancel the earlier sequence, the amount of maneuvering the spacecraft would do would not impair the camera view of the surface.

At 4:15 a.m. JPL controllers began sending the first "phony" pitch command. After it was acknowledged by a return signal from the spacecraft, a second order to execute a brief roll of one second was transmitted and acknowledged. Then a second pitch command was sent. At 5:25 a.m., Inhibiting Command No. 8 was transmitted. Shortly thereafter, the computer aboard the vehicle issued the first pitch command, and telemetry showed that the control system ignored it.

Fine. "Inhibition working as desired," announced the JPL commentator. "Stand by for roll command. It is issued. It is ignored."

Anyone entering the auditorium at this point would have gained a clear impression that Ranger 7 had failed. At 5:52 a.m., the spacecraft computer issued the second pitch command and the JPL control room commentator announced, "Attitude-control system not responding as desired." It was a strange, inverse kind of remote controlling, but it played into the logic of this strange space vehicle. JPL was covering all bets.

An infusion of observers from NASA's Lewis Research Center in Cleveland arrived at the auditorium, and Pickering, standing near the front, made them publicly welcome. It was fitting that the Lewis team, which had perfected the Agena to propel Ranger to the moon, should be on hand at the finale, he said. The commentator began to speak and the noisy auditorium instantly became dead silent.

"The spacecraft continues its cruise with solar panels locked on the sun and high-gain antenna locked on the earth. Goldstone has turned on the magnetic tape and kinescope recorders in readiness for video signals."

The voice broke off, and there was only the sound of breathing in the auditorium, spiked by nervous coughing. The red eyes of the television cameras glared at rows of taut, white faces, as though observing a cult that had assembled to await the end of the world. Ranger 6 had come this far on the morning of February 6, only to fail completely in the final minutes. Another failure would crush this Laboratory.

At 6:07 a.m., the commentator spoke: "The backup clock [on the spacecraft] has turned on Channel F [the full-scan cameras]. Channel F is in the warm-up phase. Goldstone is recording!"

A surge of excitement rippled through the crowd. The room was filled with the sound of chimes, indicating that the recorders at Goldstone, 100 miles away across the mountains, were on, waiting to receive the first picture signals from the morning sky.

At 6:08 a.m., the commentator announced that the Channel F cameras were shifted to the operating mode. All recorders were on. The cameras were operating. The sniffling and weeping of some of the women in the audience blended with the unearthly moaning of the spacecraft and the chimes of the recorders at Goldstone. A new kind of ritual, with its own hymns, had come into being. The chiming grew louder.

The commentator said, "Video signals are being received." The crowd broke into a roar of cheers and applause. "They're strong and clear." Cheers. "We are now receiving signals from the P channel [the four partial-scan, close-up cameras]." Applause. "We are receiving pictures on both channels!" That brought down the house. The commentator's voice was lost in the uproar as Ranger 7 plunged on toward the lunar surface at 5850 miles an hour.

"Twelve minutes from impact," the commentator was saying as the pandemonium abated. "Goldstone magnetic tape and kinescope recorders are all go. Strong and clear video signals are being received. We are receiving pictures from both the F and P cameras. There are no interruptions. It appears that all six cameras are functioning properly. Eight minutes to impact. Channel F and P cameras in go condition. Entire mission at this point is go. Seven minutes to impact. All camera systems continue to transmit excellent video signals. All recorders at Goldstone are operating."

"Six minutes . . . there have been no interruptions. Five minutes to impact . . . signals strong, no interruptions. We have maximum coverage. Four minutes. Video signal still excellent. No abnormalities indicated. Three minutes. No interruptions. Excellent video transmission. Ranger is go. Two minutes. All six cameras go. One minute to impact. We are receiving pictures all the way!"

A new burst of cheering and applause filled the auditorium. Pickering appeared on the television monitors, smiling.

"Twenty seconds to impact, ten seconds, five . . ."

The moaning of Ranger's telemetry stopped at 6:25:49 a.m. Pacific Daylight Time. Ranger 7 was down on the moon, aground in the northwestern quadrant of the Sea of Clouds, between the Riphaean Mountains on the west and the Craters Gueriche and Parry-Bonpland on the east. It came to rest eight miles from its aiming point, and the place forever after would be known as the Mare Cognitum, the known sea. For Goldstone had recorded 4316 photographs, showing the surface of the moon as a man might see it, first from a spacecraft, then from an airplane. The greatest photographic achievement in history was accomplished.

Outside, the sun was rising in the mist. It was going to be a hot, smoggy day in Pasadena. But it was great day for science, for the Jet Propulsion Laboratory, and for the United States. Later, at a press conference, Pickering was asked what he thought now about the future of JPL. "I think it has improved," he said.

AT SEA ON THE MOON

The first Ranger pictures to be printed showed that man was now able to see the moon's surface a thousand times better than before. As Dr. Kuiper, director of the Lunar and Planetary Laboratory of the University of Arizona, explained it, Ranger had brought the moon from a distance of 500 miles, as seen by terrestrial telescopes, to one-half mile. Craters as small as 3 feet in diameter could be seen clearly. The final photograph, taken with a P camera a fraction of a second before impact from 1600 feet, revealed features as small as 10 inches. A striking revelation was that cratering continued on down in size below the resolution of the terrestrial telescopes to craters of small dimensions. There were craters nested within craters, like a nesting of pots. Most of the craters were rounded, rather than sharp, as though worn by erosion. What kind of erosion could there be on the moon, without wind or water? Kuiper, who headed the Ranger scientific team, and Eugene M. Shoemaker, of the United States Geological Survey, the team geologist, suggested that erosion was caused by meteorite impacts. The rain of space dust (micrometeoroids) might also account for the rounded appearance of some of the craters.

Ranger 7 had come down in one of the fan-shaped rays of the huge crater Tycho, where there were clusters of secondary craters punched by the splash of rocks that had been hurled out of Tycho as far as 600 miles to the south when the crater was formed. These rays or outsplashings of material from the great impacts had been believed to consist of finely pulverized material, such as rock flour, but the Ranger 7 pictures showed that the rays contained rocks of all sizes. The rayed regions must be avoided as Apollo landing sites—that much was immediately clear from the preliminary inspection of the Ranger pictures.

Ironically, however, the peerless performance of Ranger 7

shed little light on the question of whether the lunar surface could support the weight of a 15-ton spacecraft. Instead of resolving the argument of whether the surface was dust, lava, or rocky debris, the Ranger photos intensified the debate. As another member of the Ranger science team, Harold C. Urey, remarked, each partisan found in the Ranger pictures confirmation for the views he had always held.

Kuiper, who believed that the dark maria represented extensive lava flows, said that the photographs confirmed this view. He noted that the small craters, one a foot deep, had the same shape as the big craters. This indicated a hard, lava-like surface that would preserve the shapes of craters of all sizes. Shoemaker concurred that the surface appeared principally to be hard and dense. However, there might be an upper layer of porous or pulverized material, he speculated. But he doubted that it could be much deeper than 18 inches, since the Ranger photos showed inverted cone-shaped craters which were three feet in diameter and hardly more than 18 inches deep, and these had retained their contour. A grainy, porous overlay might account for the waffling or graham-cracker appearance of much of the surface in the photos, other observers suggested. In any case, Shoemaker surmised that the maria would be a favorable region in which to land the Lunar Excursion Module as long as the Bug stayed clear of boulders and holes in the rayed regions. Generally, the scientific team agreed that the surface in the Ranger 7 landing zone was much smoother and more level than expected. That observation enhanced the prospect—at least in their view—that the ground would be suitable for a landing.

The British astronomer Tom Gold, who had emigrated to the Center for Radiophysics and Space Research at Cornell University after he proposed his lunar deep-dust theory in 1955, was not so sure. He found support for his theory in the Ranger photos. The rounded rims of the older craters, compared to the sharper ones of craters superimposed on them, suggested dust accumulation to him. While erosion rates would be low on the moon, Gold agreed, the motion of tiny particles from high to low ground under lunar gravity at the rate of 1 micron (40 millionths of an inch) a year could lead to dust deposits 2.4 miles thick throughout geologic time, according to Gold.[8] However, he conceded, the dust might be in various stages of cementation. He suggested that one cementing process might be vacuum welding, by which particles pressed together without intervening air for long periods of time would fuse. Even with the assumption that dust layers had indeed hardened, "such an origin raises the question of bearing strength of the surface for exploration purposes, while an igneous [lava] origin would have implied an adequate strength."

Urey tended to agree with the idea that loose material on the surface would tend to consolidate underneath, but he did not accept Gold's hypothesis of deep dust. Nor did Urey accept Kuiper's theory of lava. Urey said he found no evidence of volcanism in the Ranger 7 photos, as had Kuiper.[9] "I am reasonably sure that we can land on the moon," he said, after he had studied the Ranger pictures. "The surface, as I say, is finely divided material, but it will compact and bear weight. I would say a man can walk there. It would seem crunchy."[10]

Urey suggested that a missile might be propelled out ahead of a descending Ranger which would photograph the missile's impact. That would provide some idea of the bearing strength of the lunar soil. Urey also proposed that Ranger 8, the next photo probe, be aimed at the Mare Tranquillitatis in the northeastern quadrant of the moon. If there was a lava flow of any size on the moon, he said, it would have been produced there by the heat of the impact that created the vast depression.

At Bethpage, Long Island, New York, where Grumman Aircraft Corporation was developing the lunar-landing module for Project Apollo, no one appeared discouraged by the dispute over deep dust versus lava. Joseph Gavin, the Grumman vice president in charge of program development, said his engineering team had created a landing gear that would enable the astronauts to set the Lunar Module down on virtually any combination of rocks, holes, fissures, and spurs that the most imaginative pessimist could suggest. "Of course, if the vehicle is buried by dust, you've had it," Gavin admitted, "but we do have a safety factor here. We have designed a craft that can land on chocolate pudding and pull out again before it sinks in. It will hover and maneuver over the airless surface of the moon like a helicopter in the atmosphere." Thus, if the pilots found that the surface they were landing on proved to be unstable, they could throttle up and lift off again to seek a safer landing site.[11]

RANGER 8

Ranger 8 lifted off Cape Kennedy February 17, 1965, and landed 66 hours later within 15 miles of the aiming point in the western side of Mare Tranquillitatis. Again, the vehicle was phenomenonally successful. It returned 7137 photographs. Some of them showed rilles and "dimple" craters that indicated that subsurface activity was taking place or had taken place. The rilles, which appeared as long streaks or cracks up to two or three miles wide, were hundreds of miles long and estimated to be 1400 feet deep. Examining the Ranger 8 photographs at his home in Grand Rapids, Michigan, the industrialist-astronomer Ralph Baldwin noted that the rilles paralleled the edges of the maria.[12] This suggested isostatic adjustments on the lunar surface caused by the transfer of masses of material. On earth the process is carried on in part by streams carrying sediments from the mountains and the highlands down into the ocean basins over many millions of years. In this way, billions of tons of material are shifted from one part of the planet to another. On the moon, a transfer of mass might be caused by lava flows.

The shift in weight would exert stresses that might produce cracks such as these as the surface adjusted to transfer of load. Thus, the indication that the rilles ran parallel to the edges of the maria, where lava flows seemed to be evident, supported this idea. The waffling or graham-cracker appearance of the surface suggested to Baldwin a porous, foamy material. He compared it to a "fairy castle" structure, with interconnecting voids. The dimple craters were another clue to a supposed bubbly surface structure. Close-up photos indicated that soft material was draining out of the bottom of the dimple craters, like sand from an hourglass, into caverns below, and lines of dimple craters suggested that deep canyons might underlie them, covered only by a thin, brittle crust, like glacial crevasses.

Kuiper was intrigued by the dimple craters, which suggested to him terrestrial "karst" formations.[13] Named for the Karst Plateau in Yugoslavia, these are limestone potholes, dissolved out by water, but covered by a weak roof. At some places, where the roof has caved in, dirt and debris drain down into the space below, leaving a conical dimple that looks like the dimple craters in the Ranger photos. Perhaps the openings into which the supposed dimple craters might be draining were created by large bubbles in the lava flows. This hypothesis of a graham-cracker surface on the moon added no confidence at all to the Apollo mission.

THE LAST RANGER

Ranger 9, end of the line of vehicles which had failed so dismally at first and succeeded so brilliantly at last, was boosted off Cape Kennedy March 21, 1965, and landed March 24 on the floor of the crater Alphonsus in the lunar highlands east of Mare Nubium.

The landing site was of interest for several reasons. It would enable scientists to get a close look at the lunar terrae, the light-colored highland region, and also it was the area where a reddish gaseous emission had been reported by the Russian astronomer N. A. Kozyrev. The emission had appeared to come from the vicinity of the central peak of Alphonsus. This and similar reports in the past suggested that the moon was not a cold, dead body at all, but had a hot interior, like the earth.

Theories of a "hot" versus a "cold" moon were closely allied to the controversy over the origin of the craters and the structure of the maria. Proponents of the "hot" moon insisted that many of the craters were formed by volcanic eruptions, and that the dark maria were "seas" of lava. The sighting of gaseous emissions was nothing new. Observers in Europe had been reporting them for nearly four hundred years and the reports lent substance to the theory of lunar volcanism as the principal agency in shaping the lunar landscape. Opponents of volcanism accounted for all of the craters as primary or secondary products of meteorite impacts. Urey had been intrigued by reports that spectographic analyses of emissions observed by Kozyrev and others showed a carbon molecule called C-2. "If it is indeed C-2," said Urey, "it is a very curious situation because this molecule with two atoms of carbon in it does not escape from any known volcano that has been recognized on earth. It is a very different thing and it is very difficult to understand how one gets two carbon atoms together from a single molecule under these circumstances. Most unusual indeed." [14]

Shoemaker suggested that volcanic features did indeed appear in the Ranger 9 photos, specifically "halo" craters, which were surrounded by matter ejected from them, and chains of craters which looked from afar like rilles.[15] The halo craters resembled a type of volcano found in the Eifel region of Germany, Shoemaker said. Urey agreed that the halo craters were a result of "some sort of plutonic activity beneath the surface of the moon," but they did not resemble any terrestrial volcano, in his opinion.

The persistent question—can a spacecraft land on the moon—was still not settled by the 17,225 photos of Rangers 7, 8 and 9. Kuiper said it was not possible to conclude with certainty from the pictures that the lunar surface would support a spacecraft and a man, but one could make guesses. "I would be willing to guess that the hardness would be sufficient to support a landing vehicle, that is, of the order of maybe a ton or maybe a few tons per square foot," he said, "but, honestly, this is not a measure; this is obviously just a guess." [16] It proved to be a good one.

FINAL RESOLUTION

The answer was supplied by the Surveyor spacecraft, which was designed to land on the moon gently, take photographs, test the hardness of the soil, and eventually determine its chemical composition. Surveyor was the prototype of robot planetary landers which one day would fly reconnaissance missions to Mars and possibly to Venus and the larger moons of Jupiter.

Surveyor was built by the Hughes Aircraft Company of Culver City, California, under the general supervision of JPL. Like Ranger, it was plagued with developmental problems. Its full evolution took twice as long and cost ten times as much as the three years and $72,500,000 which NASA had estimated. Begun in 1960, Surveyor was scheduled to land on the moon in August 1963, but cuts in funding the program and problems in developing both the spacecraft and the Centaur rocket which was to propel it to the moon caused a delay of 35 months. The primary responsibility for the delay was shared by management at Hughes, JPL, and NASA headquarters, according to the House Committee on Science and Astronautics.[17] "Poor management" had become a catch-all in Washington to account for a variety of faults in space work, including engineering incompetence and shoddy work in the factory. It betrayed Congressional impatience with the slowness of the industrial system in adapting itself to the rigors of designing and building vehicles which represented a new technology.

The delay in Surveyor had political impact. It raised a question about American technological capability, vis-à-vis Russia, which was also building a lunar soft lander in its Luna series. Also, the schedule for Surveyor development was tied to the Apollo landing schedule in the Kennedy time frame. By the end of 1965, after 28 months of delay in Surveyor, both Congress and NASA headquarters were becoming extremely anxious. But the Russians were also having problems with their landers which were crashing, one after the other, on the moon as they came in for a soft landing.

Both the American and Russian soft landers depended on radar altimeters to control the firing of their descent engines, which braked the fall to the surface. It was theorized that loose, sandy rubble was masking the surface and that the radar beam was going through it and being reflected by rock or a compacted mass at some distance below, thus "fooling" the radar guidance control into accepting a subsurface formation as the surface.

Surveyor was developed as a triangular structure of tubular aluminum weighing 2200 pounds at launch. Like Ranger, its course was corrected by small engines and gas jets. However, instead of the earth-sun orientation system used by Ranger, Surveyor could take its bearings from the sun and the star Canopus, visible from earth's southern hemisphere. After the 66-hour flight to the moon, Surveyor's main descent engine, a 1444-pound solid-fuel rocket which developed 10,000 pounds of thrust, would be ignited by a signal from the radar altimeter at an altitude of 52 miles. It was designed to slow the fall from 5900 miles an hour to 250 miles an hour. After burnout, the engine was to be jettisoned at 37,000 feet and three small engines would continue to decelerate the vehicle to zero velocity at 13 feet. The engines would then shut off to avoid disturbing the surface and causing the vehicle to bounce, and the Surveyor would fall the final 13 feet, landing on its shock-absorbent, tripod landing gear. Each of the three pods was a foot in diameter and the vehicle's weight was distributed over them at 2 pounds per square inch. In the first Surveyor, the only experiment would be a television camera to make close-up surface photos.

By the end of 1965, it was clear that Surveyor could not be flown until the spring of 1966 at the earliest. The Russians launched Luna IX, sixth in their series of soft landers, from the Tyuratam spaceport on January 31, 1966. The 225-pound automated probe settled down successfully in Oceanus Procellarum (Ocean of Storms) west of the crater Reiner at 7.1 degrees north latitude and 64.4 west longitude. Battery-powered, the craft was spherical, protected by panels which unfolded like petals to uncover a camera and an antenna system. To avoid rocket-blast bounce, Luna IX was hurled to one side as it detached itself from its descent engine. In contrast to the tripodal Surveyor, which was 5 feet tall, the Luna IX spacecraft stood only 2 feet above the ground.

The Russians reported in Moscow that Luna IX came down on hard, porous soil, which might have been formed of lava, but which had apparently crumbled, possibly under millions of years of meteoritic and radiation bombardment. The impression of ancient lava flows was given in several photos by lines suggesting surface boundaries of a viscous material which had cooled. No evidence of dust appeared in the photos, nor in the dynamics of the touchdown, for the spacecraft did not sink into the surface.

NASA officials interpreted the Luna IX data as signifying the existence of a safe landing site. "We've got a rational surface," Joseph Shea, the Apollo program manager, concluded.[18] "There are no fundamental problems standing between us and our landing on the moon."

Other experts were not so sure. L. D. Jaffe of JPL and R. F. Scott of the California Institute of Technology's Division of Engineering and Applied Sciences expressed an opinion that "the landing statics and dynamics are not inconsistent with the properties expected under conditions for highly porous fairy-castle structures which had been surmised in the Ranger 9 pictures." [19]

Gold and an associate, B. W. Hapke, at Cornell University, said that the dust hypothesis had not been demolished.[20] They said that the surface shown by Luna IX was "almost indistinguishable" from a pulverized-rock surface subjected to bombardment by meteorites and radiation.

While the Russians achieved the first soft landing of an automatic machine on the moon, the feat did not settle the ancient argument about the nature of the surface. In the final analysis, it required the landing of Surveyor 1 to accomplish that.

Surveyor 1 lifted off Pad 36A at Cape Kennedy at 9:41 a.m. Eastern Standard Time May 30, 1966. It was boosted on a direct ascent to the moon by the Atlas-Centaur launch vehicle. The main retro and vernier engines fired precisely as programed and Surveyor fell the final 12 feet in one second, hitting the surface at 1:17 a.m. June 2. The three footpads landed almost simultaneously within 19 milliseconds of each other. They sank into the surface to a depth of three to eight inches, and the entire vehicle rebounded two and one half inches above the surface once before coming to rest.

Surveyor stood in a fairly smooth, nearly level, and almost bare area, in an unnamed crater about 60 miles in diameter, filled with maria material. It was 2:41 degrees south of the lunar equator and 43:34 degrees west longitude, nine miles from its aiming point. The nearest large landmark was the Flamsteed crater. The first of the 11,000 photos taken by Surveyor showed a gently rolling surface of granular material, littered with fragments of rock ranging in size from 0.04 inches to more than three feet. The crests of low mountains on the horizon were visible to the spacecraft camera, which stood four feet above the ground. Comparing the positions of Sirius, Canopus, and Jupiter, and points on the lunar horizon in the photos, the JPL photo-interpretation team concluded that the general slope of the ground around the landing site was less than 3 degrees. The imprints of the footpads and the appearance of small craters nearby showed that the surface material "has appreciable cohesion." The impact of the footpads before the bounce produced clods. "The mechanical behavior of the lunar surface appears qualitatively similar to that of a damp, fine-grained, terrestrial soil," JPL reported.[21]

At Pasadena, Pickering compared the texture of the surface where Surveyor stood to that of a freshly plowed field.

Robert J. Parks, the Surveyor project manager for JPL, said, "I think you can state that in this particular area where Surveyor landed, the way it landed, you certainly would expect the Lunar Module to be able to land." [22]

THE LITTLE GOLD BOX

The riddle of the lunar landing was solved. The question of the nature of the material in the maria was not, however. It remained for a chemical analyzer aboard Surveyor 5 the following year to discover that the dark, lava-like material in the lunar seas was probably basalt, with a chemical composition the same as terrestrial basalt.

Although it narrowly escaped a crash because of a leak in a pressure regulator valve, Surveyor 5 was maneuvered by JPL to a soft landing in a small crater in Mare Tranquillitatis at 8:45 a.m. Eastern Daylight Time September 10, 1967. By lunar nightfall, 14 days later, Surveyor's cameras had taken 18,006 pictures, more than the total taken by Surveyors 1 and 3. (The third Surveyor had been launched April 17, 1967. Surveyors 2 and 4 had failed.)

The chemical analysis of the moon soil was a limited one made by an alpha-particle scattering method proposed 50 years ago by Lord Rutherford. It was able to report the relative abundance of elements between hydrogen and silicon in the periodic table. Elements heavier than silicon could not be identified singly and the instrument could not detect hydrogen.

The findings of basalt in Mare Tranquillitatis was extrapolated to indicate the composition of virtually all of the maria lowlands. It suggested at once that the moon was not a primitive and undifferentiated body as many investigators had supposed, but had undergone at least a partial melting. Basalt is formed by chemical fractionation of an ultrabasic type of rock. Thus, in the lowlands, the moon appeared to be differentiated, like the earth, with a crust and a more primitive layer underneath which might be considered analogous to the mantle.

The analyzer itself was a gold-plated box 6 inches on a side, with a glass mirror highly resistant to temperature changes on the top to reflect solar energy during the long lunar day. After Surveyor landed, the gold box was lowered automatically on a cable to the surface, where it rested on a white-painted base.

This instrument was designed and built at the University of Chicago under the direction of Anthony Turkevich, professor of chemistry in the Enrico Fermi Institute of Nuclear Studies, and James H. Patterson, research chemist at Argonne National Laboratory, the Atomic Energy Commission research institution 25 miles southwest of Chicago. Turkevich was the principal investigator on lunar-soil analysis, and Patterson and Ernest Franzgrote of the Jet Propulsion Laboratory were co-experimenters. Franzgrote at Pasadena developed a method to analyze the telemetry data from the little gold box.

The instrument emitted alpha particles through a 4-inch aperture in the bottom from a radioactive source, curium-242, supplied by Argonne. Alpha radiation consists of positively charged particles with two protons and two neutrons—the structure of the nucleus of the helium atom. Most of the particles simply buried themselves in the ground. A small fraction, however, bounced back to detectors on the instrument, which registered their energies and fed the results to the telemetry radio system for transmission to earth.

Each chemical element in the soil reflects the alpha particles back to the sensors in a different way. From the numbers and energies of the alpha particles scattered back from the target, it was possible to deduce the identity of some of the elements. Also, the instrument registered the energies of protons given off by some of the elements as the product of alpha-particle bombardment. The energy distribution of the protons was characteristic of their source. That provided a check on the alpha-particle reflection data.

The gold box began sending a torrent of data as soon as Surveyor lowered it to the surface. While it readily revealed some of the elements present, it could not provide any enlightenment on their chemical state. The experimenters could not tell whether carbon was in the form of diamond, graphite, organic matter, or calcium carbonate. They all looked alike in the alpha-particle backscatter technique.

The data showed that the most abundant element on the moon at the site of the gold box was oxygen, comprising 58 per cent of all the atoms reported. Oxygen is also the most abundant element on earth. The next most abundant element seen by the analyzer was silicon. Its occurrence in the maria rock was 19 per cent, about the same as its percentage in some earth rocks. Aluminum made up 6.5 per cent of the atoms the gold box saw; carbon appeared to be less than 3 per cent; sodium, less than 2 per cent; magnesium up to 6 per cent, with the probability of 3 per cent; iron, cobalt, and nickel, 3 per cent, and heavier elements, one-half of one per cent. Turkevich commented, "The most abundant chemical elements on the moon appear to be the same as the most abundant on earth and their relative abundance looks very much like that of a silicate rock, a rock similar in composition to many that are available on earth." [23]

The basaltic floor of Mare Tranquillitatis was akin to the rock of the Palisades of the Hudson River, the Columbia River Plateau, the Hawaiian Islands, and Iceland—an affinity which could be interpreted as lending new credence to the theory of George Darwin, the English astronomer and second son of Charles, that the moon was torn by tidal action from the earth early in its history. Even if the moon were not a piece of the auld sod, it would not seem so strange a world to the first men to land on it as had been imagined. They would see familiar rock types, implying a basic unity among the terrestrial planets of the solar system in origin and evolution.

The moon was as solid a place to stand on as the earth.

Chapter 9

No Hiding Place

THE TROUBLE WITH CENTAUR

On the afternoon of May 8, 1962, a Centaur rocket on its maiden test flight from Cape Canaveral blew up 52 seconds after it was launched atop an Atlas booster. Rocket explosions were nothing new at the Cape, but the Centaur explosion shook the space program. It signaled a new crisis in Project Apollo and in the programs to reconnoiter the moon, Venus, and Mars.

Centaur was the first of a new generation of rockets using liquid hydrogen, the highest-energy chemical fuel known. It provided 40 per cent more thrust per pound than conventional hydrocarbons. As we have seen in Chapter 2, its potential as a rocket fuel had been proposed as early as 1895 by Ziolkovsky, but the development of a technology for using it in a rocket did not begin until Centaur. Consequently, the 30-foot Centaur, with two RL-10 hydrogen engines delivering 15,000 pounds of thrust each, represented the most important advance in propulsion since the V-2. The upper stages of the Saturn moon rocket were based on hydrogen technology. Centaur itself, as a second stage of the Atlas, had the mission of lofting Surveyor to the moon and the Mariner probes to the planets.

Now Centaur was in trouble. The test flight had been delayed 17 months by technical problems and its disastrous climax showed that these had not been solved. In four and one-half years of effort, the nation had failed to master hydrogen-rocket technology. Until it was mastered, the lunar-landing and planetary-exploration expectations of this country were stalled.

One of the problems with Centaur was the fabrication of a tank that would hold liquid hydrogen without leaking. At 423 degrees below zero Fahrenheit—126 degrees colder than liquid oxygen—the fuel was extremely difficult to obtain. The hydrogen tank had to be well insulated from the "warmer" oxygen tank in order to maintain low temperature. In addition, hydrogen would leak through microscopic, hairline cracks as through a sieve, and these cracks would frequently escape detection even under a microscope. A principal unknown to be solved was the behavior of liquid hydrogen in zero gravity. The Centaur engine had to be re-startable in space after being shut down for a while as the rocket coasted. The conventional way of starting the fuel flowing again into the engine was to apply acceleration to the rocket by small "ullage" engines, so that the liquid fuel would reach the main engine pump inlet by inertia. Would liquid hydrogen flow as expected under these conditions? No one knew.

The Navy had considered a hydrogen-powered rocket to place a satellite in orbit as early as 1946, but development had not gone very far and these problems had not been encountered. It was not until the end of 1956 that General Dynamics-Astronautics submitted a proposal to the Air Force to build a hydrogen rocket as an upper stage to the Atlas ICBM. The Centaur program was approved by Advanced Research Projects Agency in August 1958. General Dynamics was awarded a $36,000,000 contract to begin developing Centaur in November, and Pratt & Whitney Division of United Aircraft Corporation received a $23,000,000 contract to develop the hydrogen-oxygen engine. Six development flights were then scheduled to begin in January 1961.

As the Centaur program got under way at General Dynamics in San Diego and at Pratt & Whitney at West Palm Beach, ARPA assigned a mission to it: Centaur was to place a large military communications satellite called Advent into a synchronous orbit over the equator. A satellite in orbit at about 22,300 miles over the equator appears to hang motionless over one point on the earth's surface because its orbital speed is synchronous with the earth's rotation. A communications satellite in that position would thus perform the function of a microwave relay tower 22,300 miles high and would relay radio or television signals from one point to any other over one-third of the planet.

After NASA was set up, the Centaur program was transferred to the civilian agency where it was tagged as the upper-stage booster for Surveyor and Mariner. The application of hydrogen technology was then expanded for the Saturn program. A more powerful engine than the Centaur RL-10 was designed by Rocketdyne for the S-IVB second stage of the Saturn 1B. The engine, The J-2, was to develop 200,000 pounds of thrust. Five J-2 engines developing a total of 1,000,000 pounds of thrust would be developed for the second stage of the Saturn 5 moon rocket on which the S-IVB would be the third stage. The Saturn upper stages—and Project Apollo—thus were based on the new technology evolving with Centaur.

The first delay in the much-delayed Centaur program came during the winter of 1959–1960, when Pratt & Whitney failed to get the additional funds it had requested for engine development. NASA did not provide the money because at the time Centaur did not have a high priority. Accordingly, the initial test flight was rescheduled from January to June 1961. The second delay resulted from an explosion during a duel-engine test November 7, 1960, at West Palm Beach. One of the two engines failed to ignite and exploded when simultaneous ignition was attempted on both. Believing that a mistake had been made in the firing sequence, Pratt & Whitney repeated the dual-engine test on January 12. Again, only one engine ignited and the second one blew up. It was not until the end of January that the engineers found that insufficient oxygen to ignite one of the engines was the cause of its failure to fire when the engines were mounted vertically. This engine would then explode when an accumulation of gases was touched off by the engine which did ignite. The problem of providing more oxygen to the spark igniter was solved, but it forced another flight-test postponement until December 1961. An additional five months of delay accumulated from test-stand problems and last-minute adjustments in the rocket.

At the time of the John Glenn flight, a Centaur rocket had been standing on the pad for a year, and Cape technicians were calling it "The Public Servant," explaining "it won't work and you can't fire it."

The detonation on May 8, 1962, when Centaur was launched was caused by a rupture of the rocket's plastic weather shield, which punctured the thin, stainless-steel hydrogen tank. Escaping hydrogen was lit by the flaming exhaust of the Atlas as it was boosting Centaur up through the atmosphere. The flame of the explosion was brighter than the sun, but it dimmed hopes that Centaur would fly its lunar or planetary missions on the schedule NASA had programed. ARPA scrubbed Centaur for the Advent program and redesigned the satellite so that it could be boosted by Atlas-Agena.

The rupture of the weather shield, a plastic collar protecting Centaur from atmospheric buffeting on the way up through the atmosphere, had been the result of a design error which miscalculated the aerodynamic loads on the shield.[1] Bad luck now dogged the program. A second Centaur, scheduled for flight test in October 1962, was damaged on the test stand by the explosion of a nearby Atlas. The first test flight of Centaur thus slipped into 1963.

Meanwhile, General Dynamics engineers and NASA wrestled with the problems of fuel leakage and heating, which were causing the liquid hydrogen to evaporate so rapidly that the rocket could not perform the missions it was designed for. Krafft Ehricke, who had directed the early phases of the Centaur program at General Dynamics, told the House Space Sciences subcommittee that the company had gambled on a thin bulkhead between the hydrogen and oxygen tanks to save weight—but it was now obvious that the heat transfer from the liquid oxygen to the liquid hydrogen was greater than expected. Moreover, the leakage through hairline cracks which developed after the liquid hydrogen was loaded in the tank had not been anticipated. Ehricke explained that General Dynamics was using for Centaur the manufacturing techniques it had developed for Atlas, but these were found to be not adequate for hydrogen technology. He explained, "The Atlas manufacturing technique and quality control permits the detection of leaks . . . down to the order of 1/10,000th of an inch. No such leaks were detected. For oxygen, this is completely satisfactory. We found also that a bulkhead which is tight under normal temperatures can open up very minute holes, less than 1/10,000th of an inch under extreme, cryogenic [low] temperature. It was this precise thing that happened."[2]

The House subcommittee criticized management of the Centaur program, which had been shared by Air Force and the Marshall Space Flight Center, as "weak and ineffective." An Air Force management team which had been administering the industrial program on the West Coast under nominal NASA direction was dissolved and the entire Centaur program was transferred to Abe Silverstein at NASA's Lewis Research Center, Cleveland. Silverstein was ordered to get Centaur off the ground. In time, he did.

The trouble with Centaur added at least $24,000,000 to the cost of the Mariner and Surveyor programs. About $4,000,000 of this total was spent on redesigning and reducing the weight of the Mariner spacecraft so that it could be boosted on planetary reconnaissance missions by the less powerful Atlas-Agena. Approximately $20,000,000 was spent reducing the weight of the Surveyor spacecraft from 2500 to 2100 pounds so that the redesigned Centaur could lift it to the moon.

When Atlas-Centaur roared off Pad 36-A on November 27, 1963, without blowing up, the cost of the nation's first hydrogen rocket had escalated from $59,000,000, as estimated by ARPA five years before, to $350,000,000. Centaur did not become operational until May 30, 1966, when it lifted Surveyor 1 to the moon—seven and one-half years after development began.

MARINER R

The trouble with Centaur threatened an indefinite postponement of NASA's program to fly instrumented Mariner probes by Venus in 1962 and Mars in 1964. The time periods of these flights were dictated by solar-system geometry as well as by technological readiness. Venus came within shooting distance of earth every 19 months, when it approached to within 25,702,000 miles. Mars came within shooting distance of about 50,000,000 miles every 25 months. In terms of rocket development of the early 1960s, "shooting distance" implied a minimum-energy flight path. The spacecraft would be hurled into a solar orbit on which it would pass near the planet. On a minimum-energy trajectory, a Mariner spacecraft weighing 1250 pounds would require 109 days to reach the vicinity of Venus and 228 days to reach the vicinity of Mars after being launched by Atlas-Centaur.

The Soviet Union had taken advantage of the decade's first launch opportunity to send the 1419-pound Venus I spacecraft out to the clouded planet on a minimum-energy trajectory February 12, 1961. Venus I was injected into solar orbit from a parking orbit around the earth by the seven-ton Sputnik VIII. Radio contact with the probe was lost at a distance of 4,700,000 miles. The Russian attempt left no doubt that the Soviet Union was bent on being first to make a reconnaissance of another planet. This demonstration of intention exerted pressure on NASA to adhere to its planetary-exploration schedule not only for the sake of the national image but to protect itself from Congressional and public criticism. When it became apparent in the summer of 1961 that Centaur would not be available for the Mariner program, NASA headquarters asked JPL to cut the weight of the spacecraft so that it could be sent to Venus by the trusty Atlas-Agena.

JPL redesigned Mariner A (for Venus) and Mariner B (for Mars) by incorporating some of the light-weight features of Ranger and succeeded in cutting the weight by almost two-thirds to 447 pounds—including 40 pounds of instruments. The revamped Venus probe was a hexagonal structure of magnesium and aluminum tubing 5 feet in diameter at the base and 9 feet, 11 inches high. Hinged to this frame, its solar panels had a wingspan of 16½ feet when extended. Ten gas jets provided attitude control in flight, and course correction was made by a liquid-fuel engine, which was commanded by a central computer and sequencer. Designated Mariner R (Revision), the probe would be launched from a parking orbit by the Agena, as Ranger had been, when it reached the most favorable position to begin the interplanetary journey.

JPL completed two Mariner R vehicles in 11 months under a crash program. The team which accomplished it consisted of Robert J. Parks, JPL planetary program director; Jack N. James, project manager; Dan Schneiderman, spacecraft systems manager; Eberhardt Rechtin, space tracking chief; and Marshall S. Johnson, director of Mariner spaceflight operations. Nicholas Renzetti supervised the Deep Space Instrument Facility which would track, direct, and monitor the space vehicle and receive the data from Venus. The overall program was under Pickering, director of JPL.

In the initial Mariner program, it was planned to make measurements of the solar wind, cosmic rays, and space dust en route to Venus, and to scan the surface of the planet with a radiometer to read temperature, determine the constituents of the atmosphere with an ultraviolet spectrometer, and measure the planet's magnetic field with a magnetometer.[3]

In spite of its reduced size, Mariner R carried an array of instruments to detect water vapor and to measure temperature at the surface and in the clouds, the magnetic field strength in interplanetary space and at Venus, cosmic radiation in space and trapped radiation near the planet, the flux of space dust, and the intensity of the solar wind en route. The vehicle did not, however, have instrumentation to determine the composition of the atmosphere.

VENUS, VENERIS

In the early religion of Rome, Venus was the goddess of vegetation. She did not acquire the characteristics of the Greek Aphrodite and become identified with ideals of love and beauty until the third century B.C. After observing the planet Venus with his telescope, Galileo wrote 350 years ago that "the mother of loves imitates the forms of Cynthia," a cryptic announcement of his discovery that Venus underwent change in phase like the moon (Cynthia). This observation implied that Venus revolved around the sun and not, as dogma of the day asserted, around the earth. When it is between the earth and the sun, Venus presents only a thin, glowing crescent to observers on earth because its illuminated side faces the sun. On the opposite side of the sun from earth, its full disc is illuminated, but at that point in its orbit, the planet is so far away (162,000,000 miles) that it appears much smaller than it does in the nearer, crescent phase.

The first United States probe of Venus in 1962 was one of the nation's most ambitious undertakings of the space program, but it failed to capture the public's imagination. Perhaps it was over-shadowed by Project Mercury.

With a diameter of 7700 miles, compared with 7926 for earth, Venus is nearly the twin of earth in size and mass. It is the nearest planet to earth and the easiest to reach. But, perpetually veiled by clouds, the surface of Venus was a mystery. What lay below the clouds? In science fiction, Venus had been most commonly imagined as a planet of warm seas and steaming jungles, with a fauna and flora analogous to Mesozoic earth: giant reptiles crashed through the tall ferns, plowed through the seas, or glided through the thick atmosphere. This image became less tenable in the 1960s when a series of radar soundings by JPL and other institutions indicated that Venus was turning on its axis once every 225 days, its period of revolution around the sun, so that it always presented the same face to the sun—as the moon does to earth. Further, analysis of microwave radio emissions indicated that the surface temperature might be hotter than the boiling point of water—much too hot for life as we know it. Russian radar measurement, however, indicated that the planet's rotational period was only nine to eleven days. Spectroscopic analysis showed only minute quantities of free oxygen in the atmosphere and abundances of nitrogen and carbon dioxide. But the data were inconclusive, and only indicative. It was essential for the scientist to take a much closer look at Venus, and that was the role of Mariner R. By scanning Venus at close range with a microwave detector (radiometer), it should be possible to verify the temperature indications reported by terrestrial instruments.

Mariner's essential mission was to check measurements already made but not fully accepted and to analyze the Venusian cloud structure. If the magnetometer on board reported a magnetic field similar to earth's, scientists could assume that the planet's rotation and internal structure were similar to earth's. In that case, the terrestrial radar probes were wrong. From earth, it could not be determined whether the microwave indications of high temperatures emanated from the surface or from an ionized region in the Venusian atmosphere. This distinction could be made, however, by Mariner. The spacecraft also could detect water vapor. This could not be done by earthbound spectroscopes because the earth's atmosphere contains water vapor.

Mariner's key instrument was a microwave radiometer—a device to measure the electromagnetic energy emitted by the planet. In the microwave region of the spectrum, the brightness or intensity of the energy radiated by a planet indicates its temperature at the surface. The radiometer observed the planet with a 19-inch dish antenna designed to pick up radio waves in wavelengths of 13.5 and 19 millimeters. If there was an abundance of water vapor in the atmosphere, it would absorb the radio energy coming up from the surface in the 13.5 millimeter wavelength. Thus, if the radiometer failed to report this wavelength, observers could assume the presence of water vapor in the Venusian atmosphere. Energy radiating from the surface in the 19-millimeter wavelength, however, would pass through water vapor like light through glass and would indicate surface temperature at various points on the planet as Mariner flew by.

The antenna would be nodded in an arc of 120 degrees by a small electric motor. It had a diplexer which enabled it to receive both wavelengths at once. Also, it carried two "reference" horns which pointed away from the planet to pick up radio energy from space for contrast with the energy coming from the planet. It was expected that the temperature indication from interplanetary space would be absolute zero, so that the emanations from Venus could be calibrated in relation to it.

How would the radiometer settle the question of whether the temperature indications were coming from the surface or from an ionosphere miles above it? The experimenters reasoned that if there was an ionized layer in the high atmosphere, the radiometer would see the thinnest part of it when looking straight down and the thickest part along the curvature of "limb" of the planet. The atmosphere would appear thicker at the limb because of the angle at which the antenna was observing it. If a dense ionosphere accounted for the radiowave indications of temperature, it would appear to be hotter or brighter at the limbs of the planet than straight down. This effect was termed "limb brightening." Conversely, if there was no appreciable ionosphere, and if the surface heat was producing the microwaves which terrestrial instruments had recorded, the radiometer would see higher temperatures when looking straight down than at the limbs. This effect was called "limb darkening."

Hitched to the microwave antenna was an infrared radiometer which measured radiation in the infrared part of the spectrum at wavelengths of 8 to 9 and 10 to 10.8 microns. This radiant energy would indicate temperatures in the great cloud decks of Venus.

THE CROSSING

The space agency had instructed JPL to prepare two Mariner R vehicles for launching to Venus in order to reduce the odds on failure, which were quite high. Both had to be launched within a 56-day period or "window" in order to arrive at the vicinity of Venus at the time of inferior conjunction, when the planet was between earth and the sun. The closer Venus was to earth when Mariner arrived, the better were the chances of radio reception. The radio put out a signal of only 3 watts and it had to span distances of more than 26,000,000 miles.

During the week of May 28, 1962, three Mariner vehicles were shipped to Cape Canaveral, the two designated to fly to Venus and a third "backup" which had been put together out of spare parts. They were designated as Mariners 1, 2, and 3, and by June 11, the firing dates were established on Mariners 1 and 2. So it was that 324 days after the Mariner R program was started, two of the scaled-down probes were ready to fly.

At 4:21 a.m. EST July 22, 1962, an Atlas-Agena bearing Mariner 1 lifted off Complex 12 bound for Venus. The flight looked good for several minutes until the Range Safety Officer observed that the silvery ICBM was veering off course toward the northeast. The controllers radioed steering commands to the yawing Atlas, but it ignored them. When it became likely that the rocket might crash into shipping lanes or possibly fall into maritime Canada, the Safety Officer pressed a button and the first Mariner mission erupted into a brilliant flash as the "destruct package" blew up the wayward Atlas. Even as it tumbled down into the Atlantic Ocean, Mariner 1 continued to transmit data for 1 minute and 20 seconds. The cause of the guidance trouble was traced to a transcription error in which a hyphen was omitted from the coded instructions to the Atlas guidance computer. The omission allowed false information to reach the guidance system, commanding a hard-left, nose-down maneuver, which the autopilot obeyed. The error caused a loss of $18,000,000, NASA admitted.

Thirty-six days later, on August 27, 1962, Mariner 2 was boosted off the pad at 1:53 a.m. This time the Atlas flew true, and the 447-pound spacecraft began a four-month journey to the "mother of loves."

Mariner had been aimed with great care from a launch pad which itself was moving around the sun (with earth) at 66,000 miles an hour to pass near a target speeding around the sun at 78,300 miles an hour. (Venus travels 12,300 miles an hour faster than earth because it is closer to the sun.) Even though the clouded planet was less than 40,000,000 miles from earth at the time Mariner was launched, the spacecraft would travel 180,200,000 miles to get there on a course requiring three different orbits, of the earth, the sun, and Venus.

Mariner 2 was injected into its first orbit by the initial burn of the Agena engine after the Atlas engines had burned out. For 16.3 minutes, Agena-Mariner coasted in earth orbit at an altitude of 116.19 miles, moving down the Atlantic Missile Range between Brazil and Africa. At a point over the Atlantic about 360 miles northeast of St. Helena, controllers signaled Agena to fire its engine a second time for 1 minute and 34 seconds. The boost injected Mariner 2 into a Venus trajectory at a velocity of 25,420 miles an hour, more than enough to enable it to escape from the earth's gravitational field. At this point, it was 4081.3 miles from Cape Canaveral.

Heading directly away from earth, the spacecraft seemed to observers on the ground to be drifting to the west because the planet it was leaving was rotating toward the east. After a few hours, Mariner came within line of sight of the JPL antenna at Goldstone, California.

All was well with Mariner 2 as it came into radio view over South America. Its central computer and sequencer commanded the solar panels and dish antenna to unfold and the directional, radio-communications antenna to extend. Responding to sun sensors, the gas-jet attitude-control system adjusted the vehicle's position so that its long axis pointed toward the sun. In this attitude, the solar panels were illuminated fully to transmute sunlight into 195 watts of electricity. The gold-plated, aluminized, and white-painted surfaces of the vehicle balanced the absorption and reflection of solar heat, so that in the heatlessness of interplanetary space the instruments were kept at room temperature.

Climbing "uphill" against the earth's gravitational pull, Mariner's velocity slowed steadily. By the time it had left, effectively, earth's gravitational field and had become a true satellite of the sun, at a distance of a million miles from home port, Mariner 2 was moving at 6500 miles an hour relative to the earth. However, its velocity relative to the sun was what counted now, as it came under the sun's gravitational control. Back on Pad 12 at Cape Canaveral, Mariner had had a sun-relative velocity of 66,000 miles per hour—earth's approximate orbital speed. It had retained this sun-relative velocity while coasting in parking orbit around the earth. But once it was propelled out of earth orbit, its sun-relative velocity changed, because it had been aimed in the direction opposite to that in which the earth was moving around the sun—in a direction called retrograde, in relation to earth's motion around the sun. At the earth-relative speed of 6500 miles an hour, Mariner's sun-relative speed was decreased by 10 per cent as it departed from earth. This was enough loss in the orbital energy imparted to Mariner by the earth to cause it to fall behind the earth, and begin to curve inward toward the sun—and Venus. In this respect, the launching of Mariner 2 was a retrorocket maneuver which dropped the quarter-ton vehicle into a "lower" orbit around the sun than that of earth. If Mariner had been launched in the same direction as earth's motion around the sun, the boost would have increased its sun-relative velocity by 10 per cent. The additional energy would have swung it outward in a "higher" orbit than earth's—toward Mars.

Soon after the spacecraft entered the Venus trajectory, flight-path computations at JPL's Space Flight Operations Center showed it would fly by Venus at 233,000 miles. This was beyond the effective range of the scientific instruments, but well within the correction capability of the midcourse motor. Following an adjustment of the spacecraft's attitude so that its directional antenna was pointed earthward, the midcourse maneuver commands were transmitted September 4. They instructed the craft to roll 9.2 degrees, swing around 139.8 degrees, and fire the motor 27.8 seconds to produce a 69.5-mile-per-hour velocity change. This increased Mariner's sun-relative speed by 45 miles an hour, from 60,117 to 60,162, and decreased its earth-relative speed by 59 miles an hour, from 6748 to 6689 miles an hour. At the time of the midcourse correction, Mariner had dropped 1,492,500 miles behind the earth and was traveling in a solar orbit about 70,000 miles closer to the sun than earth.

THE WIND

It was during its 109-day cruise to Venus that Mariner 2 made the measurements of the structure of the solar wind discussed on p. 90, and established that this stream of plasma was continuously ejected from the sun.

The spacecraft carried a cosmic-ray detector designed by H. R. Anderson of the Jet Propulsion Laboratory and H. V. Neher of the California Institute of Technology. It registered high-energy cosmic-ray flux of about 3 particles per square centimeter per second. This measurement remained constant during the flight in the absence of solar events which screened out the cosmic-ray particles. Was the flux of cosmic radiation constant throughout solar space? Similar measurements had been clearly indicated in the opposite direction, toward Mars and Jupiter. In addition to the cosmic-ray detector, which counted protons above 10 million electron volts energy and electrons above one-half million electron volts, lower-energy particles were counted by an instrument provided by Van Allen and Louis A. Frank of the State University of Iowa. These were protons of about 500,000 electron volts and electrons of 40,000 electron volts. The detector was a cigarette-size Geiger counter which had proved effective in measuring the radiation belts in Explorer and Pioneer satellites. Its main purpose was to determine whether there were radiation belts at Venus.

While the flight proceeded without serious problems, there were moments when its success was in doubt. During the launch, the Atlas made an unprogramed roll but quickly stabilized itself before it went far enough off course to exceed range safety limits. The roll turned the Agena to within 3 degrees of a position at which its horizon sensors would not work to adjust its course. Had this position been reached, the flight would have failed.

Twelve days out, Mariner 2 lost attitude control and its solar panels drifted away from the sun. Power dropped sharply, but the gyroscopes came on automatically and corrected the attitude so that power was restored. It was not clear in the control room at JPL what had happened. Controllers guessed that Mariner had collided with a small object, disturbing its position. On September 12, 16 days out, Mariner was cruising at a distance of 2,678,960 miles from earth and its earth-relative velocity had dropped to 6497 miles an hour. However, as the spacecraft moved more deeply into the sun's gravitational field, its speed relative to earth increased and by October 24 it was moving at 10,547 miles an hour relative to earth. It was then 10,030,000 miles out and 21,266,000 miles from Venus on an elliptical course.

Controllers were alarmed October 31 when current in one of the solar panels dropped abruptly. The entire power-supply load was placed on the other panel, which was operating near its limit. To reduce power drain, controllers turned the cruise scientific instruments off. A week later, the "dead" panel came back to life. No one knew why. Radiation detectors, dust counters, and the magnetometer were turned on again. Then, on November 15, the panel went dead again and never revived. The trouble was later diagnosed as a short circuit. By the time of the second failure, however, Mariner was close enough to the sun so that a single panel could supply enough power.

Meanwhile, the Soviet Union launched the 1976-pound Mars I probe on November 3, 1962. The vehicle's mission was to photograph the Martian surface and make scientific measurements of radiation and dust en route and near the planet. The Tass news agency announced that Mars I would pass within 115,800 miles of Mars in June 1963. Before that event, however, the spacecraft's radio died and nothing more was heard from Mars I.

By November 6, Mariner was midway between earth and Venus, at a distance of 13,900,000 miles from each. The controllers at JPL began to dare to hope the mission would succeed. Repeated orbit computations confirmed the one made after the midcourse correction that the spacecraft would pass within 22,000 miles of the planet, close enough to scan it with the radiometers and measure the magnetic field. On December 7, it became apparent at Flight Operations in Pasadena that Mariner had entered the gravitational field of Venus. This effect altered its orbit for the third time, bending it into the path the tiny vehicle would follow in its orbit around the sun until the end of solar time. The long journey was almost over so far as Mariner's scientific mission was concerned.

Then, on December 9, five days from closest encounter with Venus, a fuse blew out and cut off telemetry which reported the position of the high-gain antenna, fuel pressure in the midcourse motor tank, and nitrogen-gas pressure in the attitude-control system. Also, temperature readings in the spacecraft were higher than expected. Some portions of the vehicle were getting too hot. The JPL controllers grimly awaited the worst as the heat load built up and threatened to damage the circuitry. Relative to earth at this point, Mariner was moving 35,790 miles an hour and, relative to the sun, 83,900 miles an hour. It was only 635,525 miles from Venus. For the first time, the central computer and sequencer balked, failing to respond to signals that were to start the instruments scanning on December 14.

JPL reported: "On its 109th day of travel, Mariner approached Venus in a precarious condition. Seven of the overheated temperature sensors had reached their upper telemetry limits. The earth sensor brightness reading [the sensor kept the high-gain antenna pointing toward earth to maintain communications] stood at 3 [zero was the threshold] and was dropping. Some 149 watts of power were being consumed out of 165 watts still available from the crippled solar cells." [4]

THE ENCOUNTER

Twelve hours before the encounter, or beginning of radio contact with the planet, the central computer and sequencer failed to start the radiometers scanning and to shut off the transmission of engineering data so that scientific data would be sent. A radio signal from Goldstone flashed across 34,000,000 miles of interplanetary space, commanding the encounter sequence "on." At 2:59 p.m. EST December 14, the spacecraft responded. The radiometer dish antenna began to nod as it swept up and down across the planet as Mariner raced by at a velocity relative to the sun of 88,400 miles an hour. The spindly spacecraft was approaching the planet from above and behind it, to a distance of 21,598 miles. Within 35 minutes, the length of time the instruments could observe Venus, the radiometer antenna completed three scans. The first looked at the dark side, the second at the terminator, the boundary between the dark and illuminated sides, and the third looked at the sunlit side.

On the first scan, looking at the dark side, the radiometer reported a brightness reading equivalent to temperature of 369 degrees Fahrenheit.[5] The second scan, at the terminator, showed a temperature of 566 degrees, and the daylight scan produced a brightness temperature of 261 degrees.

Because only a fraction of microwave energy produced at the surface was being received by the spacecraft antenna, these readings were interpreted as approximations. Some of the energy was not being reflected by the surface and some was absorbed in the atmosphere. For this reason, JPL researchers concluded that brightness temperature, as reported by the spacecraft, was considerably lower than the actual surface temperature. When the readings were adjusted, they showed no significant difference in temperature between the light and dark sides of the planet. Higher temperatures were reported when the antenna looked straight down than when it looked at the edges or limbs of the planet. This observation disposed of the possibility that the energy indicating surface temperature was actually coming from an electrified layer of the high atmosphere; it was clearly coming from the ground. This was confirmed by the fact that higher temperatures were recorded on the day-night boundary or terminator as Mariner swept past it than at the dark or illuminated edges.

HELL IN HEAVEN

Calculations at JPL showed that from Mariner data the surface of Venus was 800 degrees Fahrenheit, hot enough to melt lead and 200 degrees hotter than the radar measurements had indicated. There was no limb-brightening effect to suggest that the radiometer was reading a highly charged ionosphere. Instead, the edges of the planet looked cooler than the center—as they would if the energy was generated by a hot surface.

While the microwave radiometer looked at radiation coming from the surface, the infrared radiometer scanned the clouds. From the results of these readings, it appeared to the scientists * who were operating the experiment that the clouds were about 15 miles thick, starting at an altitude of 45 miles and rising to 60 miles. The cloud layer might be as warm as 200 degrees Fahrenheit at the bottom. At the top, the radiometer readings indicated variations from 30 degrees below zero in the center of the planet to 60 or 70 degrees below zero along the curvature.[6] Temperature distribution on the cloud tops appeared to be fairly evenly distributed, showing that the infrared radiation from the surface was being blocked by the dense cloud deck and that there were no holes through which the surface could be seen.

However, the infrared detector found a region at the terminator which seemed to be 20 degrees cooler than the general cloud layer. Either the clouds were thicker or there was a hidden irregularity on the surface. Perhaps it was a mountain which forced the clouds upward to cooler altitude. If so, the scientists reasoned, the mountain must be gigantic. That was the only suggestion of a surface feature.

Evidence that night- and day-side temperatures were the same suggested a massive heat exchange—violent winds—if the day side always faced the sun. A very slow period of rotation was indicated by the magnetometer's failure to detect a magnetic field. From terrestrial experience, scientists believe that a planetary magnetic field is the dynamo effect of a rotating planet with an iron core. The apparent lack of the field at Venus tended to confirm the American radar measurements showing that the Venusian day was at least as long as its 225-day year or longer.

Atmospheric pressure at the surface might be twenty times that of earth at sea level, it was speculated. If there was water vapor in the dense atmosphere, Mariner did not detect it. Without an appreciable magnetic field, Venus would not be expected to have Van Allen radiation belts, and none was detected.

During the Mariner flight, JPL bounced radar signals off Venus from October to December 1962 in order to make

* The team included L. D. Kaplan and G. Neugebauer of JPL, and Carl Sagan, then of the University of California, Berkeley.

comparisons of radar and Mariner data. The returning signals were scattered in such a way as to indicate that Venus has a surface roughness comparable to that of the moon. On the basis of these measurements, JPL scientists estimated that Venus made one complete turn on its axis every 230 earth days in a clockwise direction, which is the reverse of earth's counterclockwise rotation. In the spring of 1966, Richard M. Goldstein, chief of JPL Communications System Research, reported that further radar research showed Venus rotates clockwise at the rate of one revolution every 243 days.[7]

When Mariner passed beyond detection range of Venus, its instruments were turned off by a radio signal from Goldstone and the engineering telemetry was switched back on. Thirteen days after the encounter, January 27, Mariner 2 reached perihelion—its closest approach to the sun. Then, at 2 a.m. EST January 3, 1963, the DSIF Station at Johannesburg, South Africa, broke off contact with the spacecraft after receiving 30 minutes of engineering telemetry. And that was the last heard from Mariner 2. The Woomera and Goldstone stations tried to raise it, but it was gone.

The 109-day, 180,000,000-mile flight of Mariner 2 showed that Venus, which had seemed so alluring from afar, looked like hell close up—hell in heaven. It could be visualized from the temperature data as a dry, lifeless ball, blanketed by dense clouds, its surface as hot as a blast furnace, swept by convection hurricanes of unimaginable fury.

This Luciferan image of the "mother of loves" was confirmed five years later by an instrumented capsule which the Soviets contrived to parachute to the surface of Venus October 18, 1967, from a 2427-pound, automatic spacecraft. Launched June 12, 1967, the Russian probe, Venus 4, recorded temperatures from 104 to 536 degrees Fahrenheit and a pressure of 15 earth atmospheres during its 90-minute descent. According to Soviet reports, the Venusian air is principally carbon dioxide, with 1.5 per cent hydrogen and water vapor. The capsule ceased transmitting when it reached the surface.

The next day, October 19, America's second Venus probe, Mariner 5, flew by the planet at 2480 miles and went into solar orbit between Venus and Mercury. As the spacecraft rounded the planet, it beamed its telemetry radio signal at about 2298 megacycles through the Venusian atmosphere. At

Firing of a German V-2 rocket after World War II in Britain's Operation Backfire. *(Imperial War Museum photo)*

First missile launched at Cape Canaveral, July 24, 1950. *(U.S. Air Force photo)*

Unsuccessful test of the Vanguard launch vehicle on December 6, 1957. *(NASA photo)*

(Below, left) The Jupiter-C rocket that launched the first U.S. satellite, Explorer 1, on January 31, 1958. *(U.S. Army photo)*

(Below, right) Launching of the Vanguard 1 into orbit, March 17, 1958. *(NASA photo)*

Dr. James A. Van Allen of the State University of Iowa, who postulated the existence of the earth's intense radiation belts in 1958, based on data from Explorer satellites.

Launching of the Mercury-Redstone rocket, which carried Alan B. Shepard, Jr., into space, on May 5, 1961. *(NASA-Marshall photo)*

President Kennedy congratulates Alan B. Shepard, Jr., first U.S. man in space, for his suborbital flight in the Freedom 7 Mercury spacecraft. *(NASA photo)*

Capable of placing 10 tons into earth orbit, Saturn C-1 is seen at the Marshall Space Flight Center, Huntsville, Alabama *(NASA-Marshall photo)*

A Saturn booster being moved on the Tennessee River by a barge during its 2200-mile voyage to Cape Canaveral. *(NASA-Marshall photo)*

Drawing of the launch complexes at the Atlantic Missile Test Center. *(NASA-USAF photo)*

NASA LAUNCH COMPLEXES ATLANTIC MISSILE RANGE

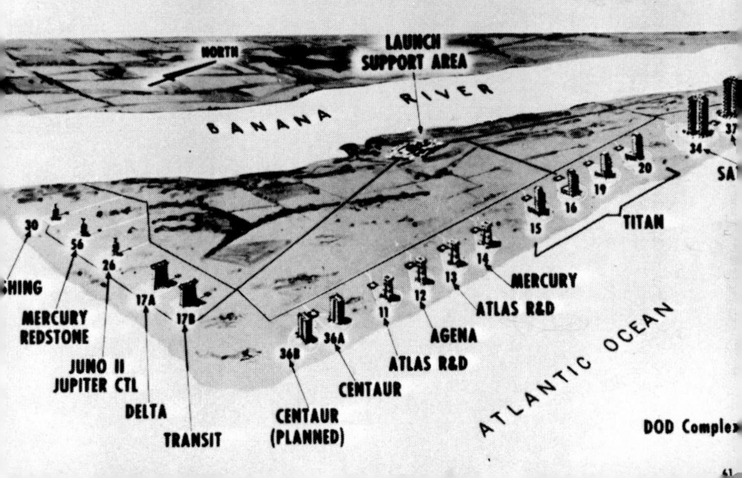

Dr. Kurt H. Debus, director of NASA launch operations at Cape Kennedy. *(NASA-USAF photo)*

Astronaut John H. Glenn, Jr., with Vice-President Johnson, riding down Broadway in New York City, in a ticker-tape parade honoring his three-orbit Mercury flight on February 20, 1962. *(NASA photo)*

Glenn explaining the astronaut space glove to President Kennedy on a tour of U.S. space installations in 1962. *(NASA photo)*

President Kennedy pinning the NASA Distinguished Service Medal on L. Gordon Cooper, Jr., at the White House. *(NASA photo)*

The Gemini-Titan launch vehicle at Cape Kennedy during the countdown for the first Gemini flight. *(Martin Company photo)*

Left to right: Dr. William H. Pickering of the Jet Propulsion Laboratory; Dr. Gerard Kuiper of the University of Arizona; the late Dr. Hugh L. Dryden, deputy administrator of NASA; and Dr. Robert C. Seamans, Jr., associate administrator of NASA, shown with a full-scale model of Ranger 7. *(NASA photo)*

Launching of a Delta rocket carrying Syncom 1 into orbit on February 14, 1963. *(NASA photo)*

View of John W. Young, left, and Virgil I. Grissom inside the Gemini-3 spacecraft, just before the hatches were closed. *(NASA photo)*

Liftoff of the Gemini-3 spacecraft on March 23, 1965, for its three-orbit mission. *(NASA photo)*

Astronaut John W. Young awaits pickup by helicopter near Grand Turk Island, while frogmen prepare the Gemini-3 spacecraft for recovery. *(NASA photo)*

First American extravehicular activity was performed by Edward H. White II, during the third orbit of the Gemini-4 flight. *(NASA photo)*

James A. McDivitt, at right, and Edward H. White II, stepping from helicopter, are greeted aboard the U.S.S. *Wasp* on June 7, 1965, after their Gemini-4 flight. *(NASA photo)*

Astronauts L. Gordon Cooper, Jr., and Charles Conrad, Jr., during preparations for their eight-day Gemini-5 mission. *(NASA photo)*

Astronauts Thomas P. Stafford and Walter M. Shirra, Jr., having their suits checked for their Gemini-6 flight. *(NASA photo)*

Photograph of the Gemini-7 spacecraft, taken at an altitude of 160 miles through the hatch window of the Gemini-6 spacecraft during rendezvous maneuvers on December 15, 1965. *(NASA photo)*

Astronauts John A. Lovell, Jr., and Frank Borman on the deck of the U.S.S. *Wasp,* after completing their 14-day Gemini-7 mission. *(NASA photo)*

Astronauts David Scott and Neil Armstrong, still seated inside their Gemini-8 spacecraft, wait to be picked up by the U.S.S. *Mason*. *(NASA photo)*

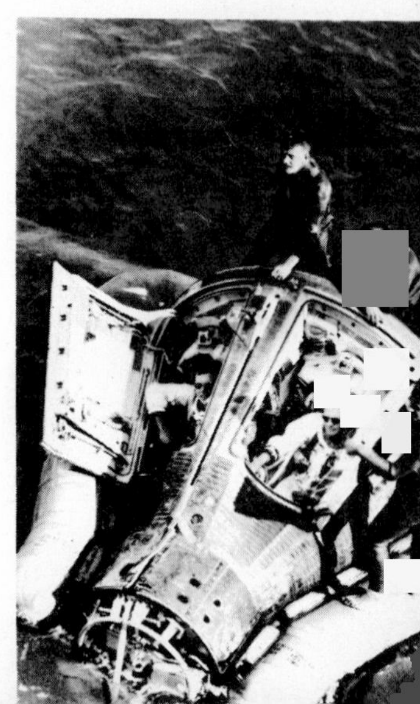

Photograph taken from the Gemini-9 spacecraft of the Augmented Target Docking Adapter, with the fiberglass cover partly open around its docking end, making it look like an "angry alligator." *(NASA photo)*

Astronauts Thomas P. Stafford and Eugene A. Cernan in their Gemini-9 spacecraft wave to the crew of the approaching U.S.S. *Wasp*, following a flawless reentry on June 6, 1966. *(NASA photo)*

Astronauts John W. Young and Michael Collins aboard the U.S.S. *Guadalcanal* on July 21, 1966, after their Gemini-10 flight during which they docked with an Agena target vehicle. *(NASA photo)*

Astronauts Richard Gordon and Charles Conrad, Jr., on board the U.S.S. *Guam,* after docking with an Agena target satellite during the first revolution of their Gemini-11 flight. *(NASA photo)*

View of the Agena target satellite at end of the tether securing it to the Gemini-12 spacecraft. *(NASA photo)*

Jubilant astronauts Edwin Aldrin and James A. Lovell, Jr., aboard the U.S.S. *Wasp,* following their 94-hour Gemini-12 flight—the last in the Gemini program. *(NASA photo)*

A 365-foot Apollo Saturn 5 being moved to the launch complex from the huge Vehicle Assembly Building at Cape Kennedy. *(NASA photo)*

View of the giant Apollo Saturn 5 and its mobile launcher, being transported by a 3000-ton crawler to Pad A of Cape Kennedy Complex 39 *(NASA photo)*

Left to right: Astronauts Virgil I. Grissom, command pilot; Roger Chaffee, pilot; and Edward H. White II, senior pilot, garbed in spacesuits for a test of the Apollo spacecraft command module. *(North American Aviation photo)*

The prime crew of the first manned Apollo space mission, Apollo 7. *(Left to right)* Astronauts Donn F. Eisele, Command Module Pilot; Walter M. Schirra, Jr., Commander; and Walter Cunningham, Lunar Module Pilot. *(NASA photo)*

Apollo 8 crew at the Apollo Mission Simulator, Kennedy Space Center. *(Left to right)* James A. Lovell, Jr., Command Module Pilot; William A. Anders, Lunar Module Pilot; and Frank Borman, Commander. *(NASA photo)*

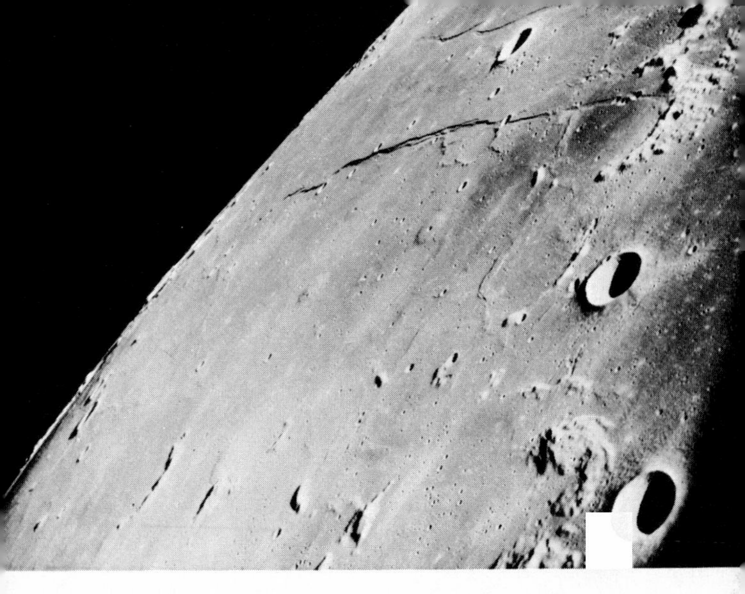

From the Apollo 8 Spacecraft, looking generally northwest into the Sea of Tranquility. The lower linear feature about in the center of the picture is Cauchy Scarp. *(NASA photo)*

The three astronauts of the Apollo 9, *(left to right)* are James A. McDivitt, Commander; David R. Scott, Command Module Pilot; and Russell L. Schweickert, Lunar Module Pilot. *(NASA photo)*

Excellent view of the docked Apollo 9 Command Service Module and Lunar Module "Spider," with earth in background, during Astronaut David B. Scott's extravehicular activity on the fourth day of the Apollo 9 earth-orbital mission. Scott, Command Module pilot, is standing on the open hatch of the Command Module "Gumdrop." Astronaut Russell L. Schweickart took the photograph. *(NASA photo)*

Lunar Module "Spider" on the fifth day of the Apollo 9 earth-orbital mission. Bubbles are vapor condensation on the window of the Command Module. *(NASA photo)*

For the Apollo 10 flight, NASA named *(left to right)* Eugene A. Cernan, Lunar Module Pilot; John W. Young, Command Module Pilot; and Thomas P. Stafford, Commander. *(NASA photo)*

Triesnecker crater, taken from the Command and Service Module, Apollo 10. Beyond the highlands, the smooth floor of the Sea of Vapors extends almost to the horizon some 375 miles away. *(NASA photo)*

The lunar farside from the Apollo 10 Lunar Module. Note difference in terrain. The identity of these craters has not yet been established. *(NASA photo)*

The Apollo 10 Command Module carrying Astronaut John Young is photographed by the Lunar Module with Astronauts Tom Stafford and Eugene Cernan aboard. Both spacecraft are in a 60 mile high lunar orbit crossing the farside of the Moon. *(NASA photo)*

Edwin E. Aldrin, Jr.,
Lunar Module Pilot.

Neil A. Armstrong,
Commander.

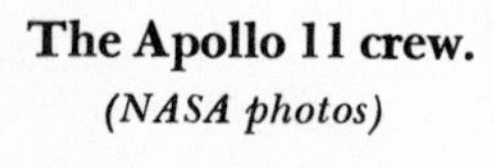
The Apollo 11 crew.
(NASA photos)

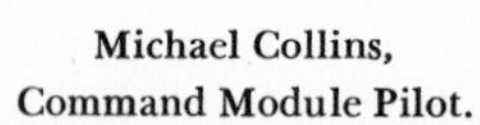
Michael Collins,
Command Module Pilot.

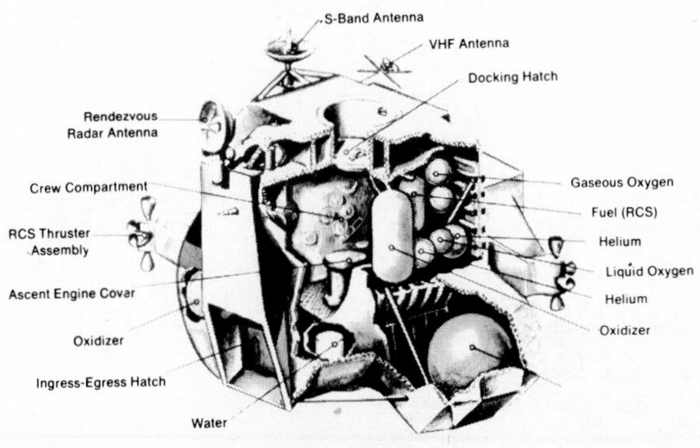

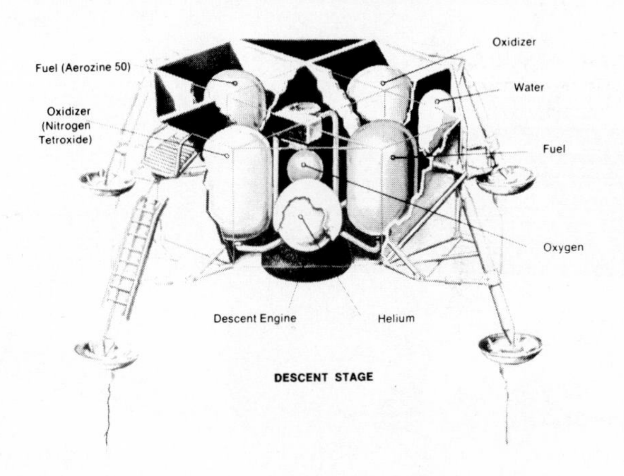

The Lunar Module

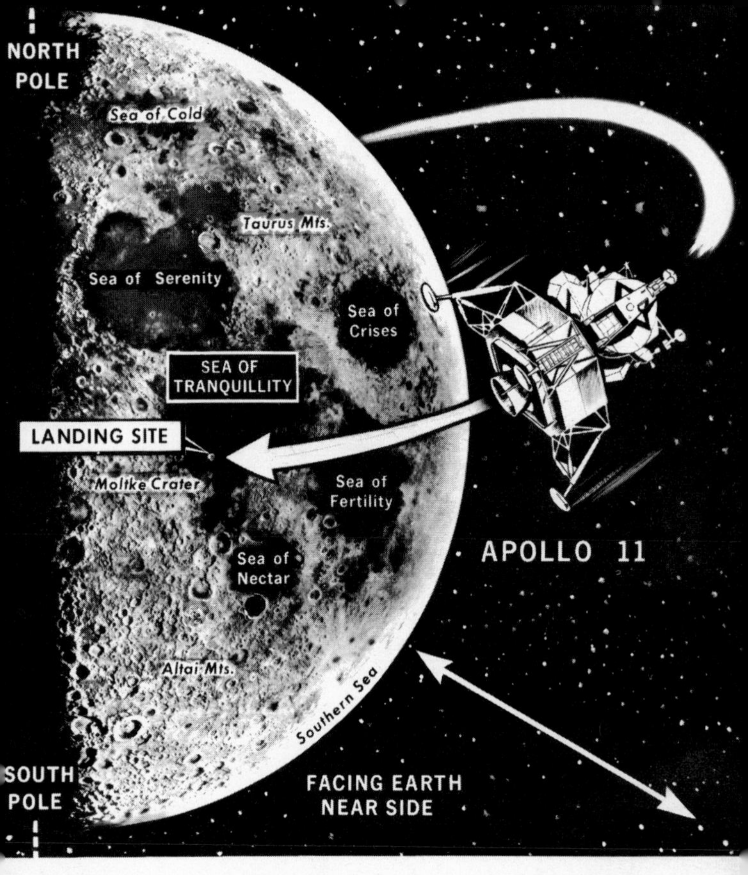

ABOVE: Artist's conception of Apollo 11 Lunar Module showing landing site near Moltke Crater in the Sea of Tranquility. *(Wide World Photos)*

LEFT: Liftoff of Apollo 11 at 9:32 a.m., July 16, carrying Astronauts Armstrong, Collins and Aldrin on their way to the moon. *(Wide World Photos)*

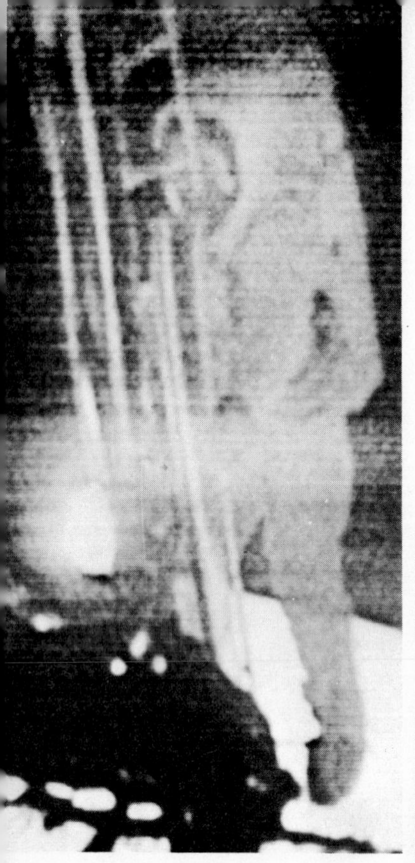

"That's one small step for Man . . . one giant leap for Mankind." Astronaut Neil Armstrong becomes the first human being ever to set foot on the moon. *(Wide World Photos)*

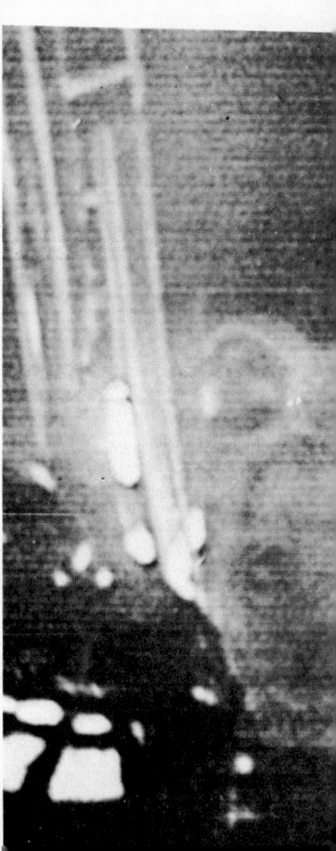

"Here men from the planet earth first set foot upon the moon, July, 1969. They came in peace for all mankind." Astronauts Edwin E. Aldrin *(right)* and Neil Armstrong read the plaque attached to the ladder of the Descent Stage which will remain on the moon for all time.

BELOW: Aldrin deploys the solar wind experiment near the Lunar Module while Armstrong practices movement in moon's low gravity. *(Wide World Photos)*

To mark the success of the United States's historic Moon landing, Astronauts Neil Armstrong and Edwin E. Aldrin plant the American flag on the surface of the moon. *(Wide World Photos)*

the same time, Stanford University transmitted two radio signals at 49.9 and 423.3 megacycles to the vehicle. These were stored on magnetic tape when Mariner 5 received them and played back after the spacecraft cleared the planet. The degree to which the signals were altered as they passed through the Venusian atmosphere provided data on the height, density, and temperature of the atmosphere and the ionosphere.

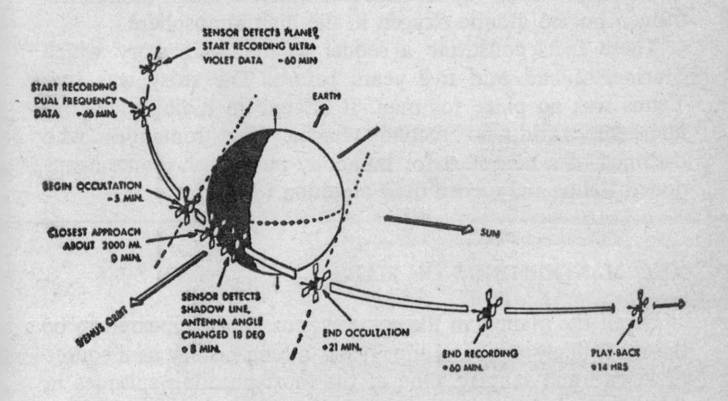

Mariner 5 fly-by of Venus.

While the Russians initially reported that the Venusian atmosphere was 90 to 95 per cent carbon dioxide, the preliminary data from Mariner's telemetry signal indicated 69 to 87 per cent carbon dioxide if nitrogen could be assumed to be the second major gas.[8] This indication was based on the refractivity of model atmospheres of nitrogen and carbon dioxide.

Like Mariner 2, the second American Venus probe found no measurable magnetic field at the planet. In the opinion of Van Allen and his colleagues at the State University of Iowa, the lack of a magnetic field was attributable to its slow rotation and "the consequent weakness of the dynamo, electromotive forces." [9]

However, the Stanford radio-probe results indicated that the solar wind does not impinge on the planet even though there is no detectable magnetic umbrella. The plasma stream from the sun is deflected by the Venusian ionosphere around the planet and hence does not collide directly with the main body of the atmosphere.

In the case of the earth, the solar wind is deflected by the geomagnetic field at a distance of more than 40,000 miles, but at Venus, the deflection by the electrified portion of the upper atmosphere appears to take place at 300 miles—the altitude at which the ionosphere is compressed by the impact of the solar wind.[10] Spacecraft photometers found atomic hydrogen but no atomic oxygen in the high atmosphere.

These facts constitute a sequel to the main story which Mariner 2 had told five years before. The story was that Venus was no place for man. It offered no hiding place, no alternate world, no second chance. The romantics who dreamed of a new start for humanity on a fresh planet abandoned Venus and turned their attention to Mars.

THE MASTERMINDS OF MARS

Of all the planets in the solar system, Mars appeared to be the most likely abode of life. It has a long history as a source of legend and fantasy. One of the most puzzling episodes in the history of science concerns the discovery of the two moons of Mars, Deimos and Phobos. Scientifically, they were first reported and named in 1877 by the American astronomer Asaph Hall, using the 26-inch Naval Observatory refracting telescope. However, these moons were described 156 years earlier by Jonathan Swift's Laputan astronomers in *Gulliver's Travels,* published in 1721. "The Laputan astronomers . . . have likewise discovered two lesser stars, or satellites, which revolve about Mars," Swift wrote. He included data on their distances from the planet, obtained "by the assistance of glasses far excelling ours in goodness." It was a remarkably accurate guess.

With a diameter of 4216 miles, slightly more than half that of the earth, and a mass only 10.8 per cent as great, Mars has a surface gravity that is only 0.38 that of the earth's. It orbits the sun at an average distance of 141,500,000 miles, or 1.4 astronomical units. Every 15 to 17 years, it comes to within 35,000,000 miles of earth. In the big telescopes, the Martian surface appears mostly dull red or orange, with splotches of bluish green covering three-eighths of the surface near the equator and most of the southern hemisphere. On the poles of Mars, which is inclined on its axis like earth, there are white caps, suggesting polar ice and, hence, the presence of water. However, it is speculated also that these caps may be frozen carbon dioxide.

Like earth, Mars has seasons. In the spring, the north polar cap seems to vanish and portions of the ochre surface become dark as though vegetation was growing. While the Martian year is 687 days long, the Martian day is curiously similar to earth's. It is 24 hours and 37 minutes long.

In the same year that Hall found Deimos and Phobos, 1877, an Italian astronomer, Giovanni Schiaparelli of the Brera Observatory at Milan, announced that he had discovered numerous fine, dark, straight lines on Mars' surface. He referred to them as "canali," meaning channels, but in the translation and popularization of his reports, the word came to mean canals. Thus, Schiaparelli's discovery acquired the implication, which he did not intend, that there were canals of Mars, structures obviously created by intelligent beings.

Fascinated by this idea, Percival Lowell, who founded the Lowell Observatory at Flagstaff, Arizona, devoted many years to studies of Mars. He advanced a theory that the canals were indeed the work of intelligent Martians who had created a vast irrigation network to supply water from the polar ice caps for agriculture on a planet that was slowly losing its water. Lowell's theory provided an explanation for the seasonal changes in coloration and the disappearance of the polar caps in spring, all of which could be observed: Since the resolution of earth telescopes for Mars was no less than 50 miles, Lowell suggested, the lines that were visible were not necessarily canals themselves but great bands of vegetation growing along them.

Lowell's books, *Mars and Its Canals* (1906) and *Mars as the Abode of Life* (1908), published after he had become nonresident professor of astronomy at the Massachusetts Institute of Technology, provided an authoritative background for generations of science fantasy and fiction. Together with H. G. Wells' unforgettable *War of the Worlds*, appearing in 1898, or 20 years after the "canali" were reported by Schiaparelli, Lowell's work created a widespread popular belief that Mars was a habitable planet. And this belief influenced the orientation of NASA's program of planetary exploration, which gave Mars priority.

Academically, the question of life on Mars stirred profound interest. In the 1920s, the Russian biologist Aleksander Ivanovich Oparin proposed that life was a natural consequence of the physical evolution of a planet and had risen on earth from a primitive reducing atmosphere.[11] If that was so, living systems like those on earth would have evolved on other terrestrial planets, like Mars. Venus seemed to be ruled out by Mariner 2.

In 1956, evidence indicating plant life on Mars from spectroscopic observation was described by W. M. Sinton of the Harvard Observatory. In the spectrum of infrared radiation reflected by the planet, Sinton found absorption bands suggesting the presence of hydrocarbons and aldehydes, complex molecules associated with living organisms.[12] Sinton hypothesized that in order to survive in a low-pressure atmosphere, such as Mars was assumed to possess, under cold, arid conditions, vegetation might be shaped like a barrel cactus. This structure would enable the plant to absorb moisture in its outer cells during the day, when the temperature might reach 80 degrees Fahrenheit, and to retract the moisture inward where the outer cells would insulate it against freezing during the night, when temperatures fell to 150 degrees below zero Fahrenheit.

In 1964, the Space Science Board of the National Academy of Sciences and the National Research Council recommended to NASA that an unmanned exploration of Mars, emphasizing a search for extraterrestrial life, be made the primary objective of the national space effort after the manned landing on the moon. The recommendation was based on a study of the biological justification for Mars exploration made by the Board at NASA's request. Chairmen of the study group were biologists Colin S. Pittendrigh of Princeton University and Joshua Lederburg of Stanford University. The philosophy of the study reflected the ideas of Oparin. It stated at the outset that "Implicit in the evolutionary treatment of life is the proposition that the first appearance of organisms was only a chapter in the natural history of the planet as a whole." The study group summarized the knowledge of Mars as follows: Atmospheric pressure at the surface ranged from 10 to 80 millibars (1 to 8 per cent of earth's) and the major constituents, while not identified, are believed to be nitrogen and argon. Oxygen has not been detected. Water vapor has been identified spectroscopically as a minor constituent of the atmosphere. The report added, "Our knowledge of what lies between the polar caps is limited to the distinction between the so-called dark and bright areas and their seasonal changes. The latter, usually considered deserts, are orange-ochre or buff color. The former are much less vividly colored. Biological interest . . . continues to center in the dark areas. In several respects, they exhibit the kind of seasonal change one would expect were they due to the presence of organisms absent in the bright [desert] areas. . . ."

The study concluded that "The exploration of Mars, motivated by biological questions, does indeed merit the highest scientific priority in the nation's space program over the next decades." The study group said that while the exploration should be commenced with automatic spacecraft, "the ultimate scientific exploration of Mars will require that man be present when it becomes technologically feasible to include him." The Board's recommendation provided a rational for Project Voyager, which contemplated the soft landing of an automatic spacecraft on Mars in the 1970s. The landing capsule would be equipped with a robot collector capable of identifying the existence of microorganisms.

The assumption underlying the recommendation—that life was waiting to be discovered on Mars—stirred dissent among biologists. One of the most outspoken dissenters was Barry Commoner, professor of plant physiology and chairman of the Botany Department of Washington University in St. Louis, Missouri. He criticized the Space Science Board in a paper he delivered at a symposium on the ethics of science at the Montreal meeting in December 1964 of the American Association for the Advancement of Science. On the basis of what is known about Mars, Commoner said, the existence of life there is highly improbable. The planet has no liquid water, which is essential to the most fundamental of life processes, he contended, adding, "If liquid water is lacking on Mars, there is no reasonable basis for the expectation of life and no good reason to design programs and instruments to seek it." In essence, Commoner's paper accused the Board of trying to sell NASA and Congress a biological Brooklyn

Bridge. The Board, he said, failed to balance the Mars biology picture by not presenting the probability that there is no life on the planet as well as the probability that there is. "The grave danger in this situation," said Commoner, "is that in our eagerness to convince Congress, and, perhaps, ourselves, of the importance of searching for life on Mars we may forsake an attribute of scientific discourse which endows science with its great capability for successful analysis of nature: the objective discussion of all the evidence and open confrontation of alternative interpretations. This has not been done by the Board."

The Board replied simply, "Given all the evidence presently available, we believe it entirely reasonable that Mars is inhabited with living organisms and that life independently originated there."

At the time of Commoner's dissent in Montreal, the 575-pound United States spacecraft Mariner 4 was passing through the Ursid meteorite stream, 3,900,000 miles from the earth at 6970 miles an hour, earth-relative velocity, en route to photograph the gray-green and ochre surface of Mars. If all went well, the 9-foot space vehicle, carrying four solar panels instead of two because it was traveling away from the sun, would pass within 5400 miles of the red planet July 14, 1965.

THE MARTIAN MARINER

The principal new addition in Mariner 4 over its predecessor, Mariner 2, was a television camera which the Martian Mariner carried to photograph the surface of the planet through a small reflecting telescope. It was designed to take 22 black-and-white photographs of Mars as Mariner 4 flew by, through orange-red and blue-green filters. Mariner 4 also had more than twice the solar-energy collection area on its solar panels of No. 2. Each panel was 6 feet long and 3 feet wide. They were hinged to the top or sunward side of the octagonal framework of magnesium tubing, which was divided into seven compartments and a rocket bay containing the midcourse-correction motor. The motor had a thrust of 50.7 pounds for 100 seconds. Each of the four solar panels had attached to its outer edge a 7-square-foot vane of aluminized mylar, shaped like the business end of a fly swatter. It was called a "solar pressure vane." The vanes were set in position by the attitude-control system so that radiation pressure from the sun would help maintain attitude control. The vanes were the first interplanetary application of the principle of the solar sail, using light pressure or the solar wind as a means of propulsion or stabilization.

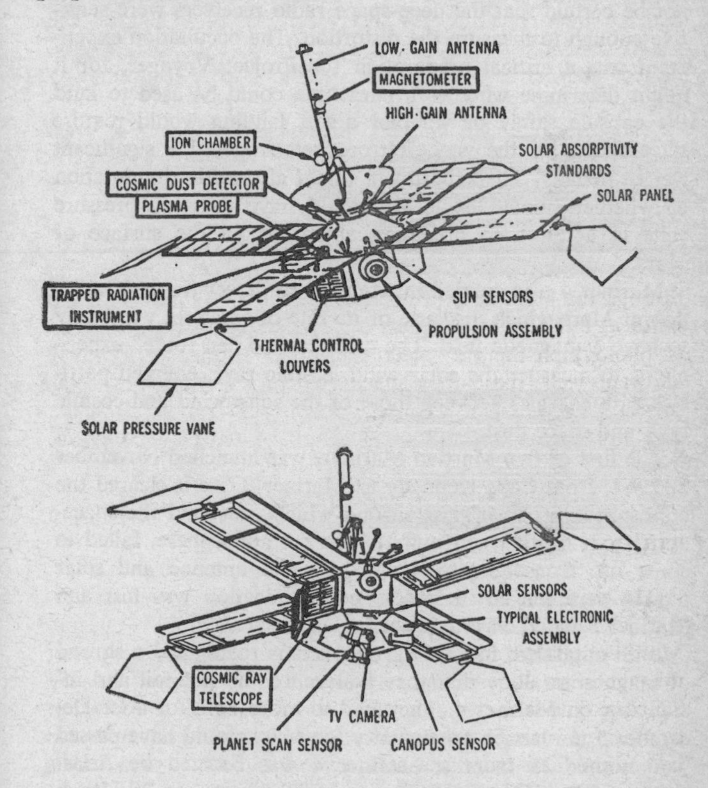

Mariner 4 spacecraft.

In contrast to the life-seeking Voyager planned for the next decade, Mariner 4 was crude and its reconnaissance goals modest. After the 22 TV pictures were taken, the spacecraft would transmit radio signals as it began to fly behind the planet so that the signals would pass through the

Martian atmosphere. The degree to which the signals were "bent" or distorted by the atmosphere would tell something about its density, pressure, and the rate at which it became attenuated with altitude. It was to be the first attempt to perform a planetary "occultation" experiment. There was no doubt that the Martian atmosphere would distort the signals, but the experimenters from JPL, Stanford, and Cornell could not be certain that the deep-space radio receivers were sensitive enough to measure the distortion. The occultation experiment was a critical preparation for Project Voyager, for it might determine whether a parachute could be used to land the capsule safely or whether a soft landing would require retrorockets all the way. Retrorockets imposed a significant weight penalty. The experiment might also settle the question of whether human explorers would have to wear pressure suits in addition to breathing apparatus on the surface of Mars.

Mariner 4 also carried instruments to detect trapped radiation at Mars which, because of its rate of rotation, was likely to have a magnetic field. The craft carried "en route" experiments to measure the solar wind, cosmic rays, charged particles with energies between those of the solar wind and cosmic rays, and space dust.

The first of two Martian Mariners was launched November 5, 1964, from Cape Kennedy as Mariner 3. As it cleared the atmosphere, its fiber-glass shroud which protected the spacecraft from heating and buffeting in the atmosphere, failed to come off. Consequently, both the radio antenna and solar panels were locked in. Radio communication was lost and Mariner 3 sailed off into space, a dead issue.

With anguished haste, JPL technicians redesigned a shroud of magnesium alloy that they believed could not fail and installed it on Mariner 4. They had to move fast, for after December 5 the launch opportunity for Mars would have passed for another 25 months. Mariner 4 was boosted by Atlas-Agena off the Cape at 9:22 a.m. EST November 28, 1964, and was injected into the Mars trajectory at 10:04 a.m. at a velocity of 25,598 miles an hour, about seven miles an hour faster than the desired velocity. This time, Mariner was boosted in the direction of earth's motion around the sun, so that as it began its flight its sun-relative velocity was the sum of 25,598 miles an hour and 66,000 miles per hour (earth's

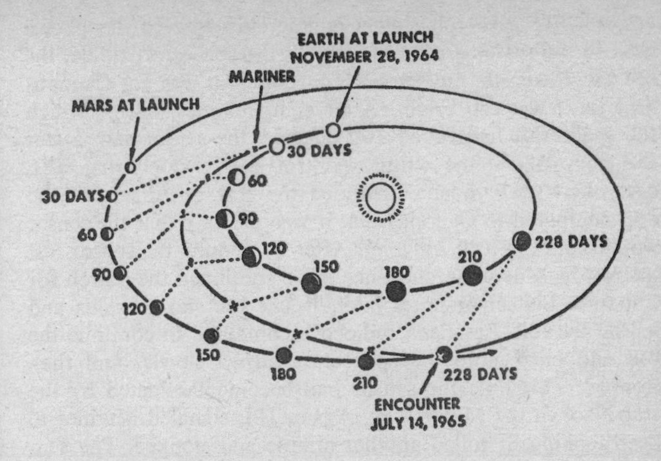

The path of Mariner 4 to Mars.

orbital velocity around the sun), or 91,598 miles an hour. With a higher orbital velocity than earth's, Mariner would assume a "higher" orbit around the sun, moving outward to intersect the orbit of Mars near the planet 228 days after launching. On its curving flight path, Mariner 4 would cover 325,000,000 miles. Its electronic and mechanical control systems would be subjected to stresses which Soviet vehicles had been unable to survive.

Like Mariner 2, the Mars probe was equipped with sun sensors which enabled the attitude-control system to keep the solar panels facing the sun. However, the system of pointing the high-gain antenna toward the earth used the star Canopus as a reference. In order to assure ground controllers that the Canopus sensor actually was seeing that particular star, the control system monitored the brightness which the sensor perceived and radioed the value to earth. This remarkable device, which was used later on Surveyor spacecraft, operated as a multimillion-mile extension of the human eye. But it was not infallible.

At 16 hours and 37 minutes after the launch, the central computer and sequencer ordered the orientation system to begin seeking Canopus. The spacecraft jets began expelling gas, causing it to roll slowly counterclockwise as the photoelectric eye scanned the heavens. As the spacecraft rolled, the eye saw the earth, but sensed immediately it was not Canopus because it was too bright. After eight minutes, during which the spacecraft had rolled 100 degrees, the sensor saw a star and commanded the attitude-control jets to stop firing. JPL controllers at Pasadena identified the star as Alderamin, so dim compared to Canopus that it was unlikely that the sensor would hold the lock for long. After five hours, the sensor lost interest in Alderamin and once more continued the search for Canopus. Sixteen minutes later, it saw the star Regulus and halted the roll. Pasadena radioed a command to continue the roll and once more the spacecraft turned slowly, and then stopped. This time the sensor had become fascinated by the star Naos in the Milky Way. Again, JPL signaled Mariner to keep looking. It rolled another minute and stopped. The sensor was fixed on a cluster of stars in the central galaxy. Keep rolling, JPL commanded once more. At 6 a.m. Pacific Standard Time on November 30, the sensor found Canopus, the friendly Yellow Giant, glowing steadily from 200 light years away in time and space. Mariner stopped rolling. It was stabilized at last in its cruise attitude at a distance of 380,000 miles from earth. At this point, earth's gravity had slowed the vehicle to 7360 miles an hour.

On December 4, JPL controllers determined it was time to make the first course correction. Mariner 4 was following a path on which it would fly past Mars, ahead of the planet, at a distance of 151,000 miles on July 16, 1965. JPL wanted the vehicle much closer than that for sharp pictures. They prepared for the course-correction maneuver early in the morning by storing commands in the spacecraft's central computer. However, before the signal could be sent to start the maneuver, Mariner lost its Canopus fix and began to roll in search of it. The stored commands were canceled until the roll could be stopped. The mindless Canopus sensor locked on seven stars in succession in its search for the friendly Yellow Giant it had been programed to love. Seven times, JPL radio commands ordered the craft to keep looking. By the evening of December 4, Mariner found Canopus again and JPL once more prepared the computer for the course adjustment with commands that instructed the vehicle how to position itself for the firing of its midcourse motor.

On the morning of December 5, the course correction was executed by a 20-second firing of the motor. It added 28 miles an hour to Mariner's earth-relative velocity, then 7019 miles an hour. While the course change was relatively small, its final effect would be large at the interplanetary distance. On the new trajectory, Mariner 4 would arrive within 5400 miles of Mars on the evening of July 14 from behind the planet, crossing the equator on the leading edge and passing over the southern hemisphere almost as far as the south pole. At the time of the midcourse maneuver, Mariner was 1,267,613 miles from earth, traveling in a solar orbit with a sun-relative speed of 74,108 miles an hour.

By December 30, Mariner had reported 7,500,000 scientific and engineering measurements to earth. It had passed through three meteorite streams, the Gemini, Leonid, and Ursid. Sunlit space dust passing close to the Canopus sensor confused it several times, and for ten days, the spacecraft had been locked by mistake on the bright star Gamma Velorum. But as always, in the end the eye of Mariner found Canopus again.

At 3 a.m. on April 29, 1965, Mariner 4 set a new long-distance-communications record of 66,000,000 miles, topping the Soviet Mars I record of 65,000,000 miles set in 1963. At 66,000,000 miles, the radio signal took six minutes to reach earth. For months, Mariner's velocity had been decreasing, at first under the major pull of earth and later under the major pull of the sun. It was literally traveling "uphill" against the sun's gravitational influence since it was moving away from the sun. By May 5, 1965, its sun-relative velocity had dropped to 52,159 miles an hour. Unless the electronics system failed, Mariner 4 would accomplish its mission and man, for the first time, would see close up the face of a planet that might be explorable if not habitable by man. Tension and excitement were beginning to build up in Pasadena as the eight-sided spacecraft, now faithful to Canopus, approached the orbit of Mars.

ANOTHER WORLD

As July 14 approached, the television-photo experimenters at JPL prepared to receive the first close-up photographs of another planet. They were Professors Robert B. Leighton, physicist, Bruce C. Murray, geologist, and Robert P. Sharp, geophysicist, of the California Institute of Technology, and Richard K. Sloan and J. Denton Allen of JPL. Not since Galileo turned his telescope on the moon to discover that it was not a smooth ball had mankind had the opportunity of seeing the landscape of another body in space. Would the photographs confirm the existence of canals? Would they reveal evidence of plant life? The excitement of being on the edge of sensational discoveries was muted by the ever-present possibility that failure could wreck the whole experiment.

On the morning of July 14, 1965, newspaper, radio, and television reporters began to assemble in the Von Kármán auditorium before the smog had cleared from the San Gabriel Mountains. At 10:10 a.m. Pacific Daylight Time, the Johannesburg Deep Space Instrumentation Facility station sent a radio signal commanding Mariner to turn its camera lens into photographic position. It took 12 minutes for the signal to reach the spacecraft, which was 77,000 miles from Mars at that time and closing at the rate of 10,000 miles an hour under the acceleration of Martian gravity. The climax of the eight-and-one-half-month flight would come about 5:20 p.m., when an electric eye in the camera would perceive the edge of the planet and start the camera. The camera was set to take the 22 photographs in 25 minutes as Mariner swept over the equator and crossed the southern hemisphere.

To make sure the camera was turned on, JPL sent a backup radio signal at 5:13 p.m., which would activate the camera when it reached the spacecraft at 5:25 in the event the camera was not on. However, at 5:18, a signal arrived from Mariner showing that the camera had turned itself on and was seeing Mars. The camera methodically took photos through its blue-green and orange-red filters as Mariner flew across the world below. At 6:01 p.m., it made its closest approach at 6118 miles, rather than the 5400 previously estimated, but the difference was not significant so far as picture quality was concerned. Only the moaning of telemetry signals from the spacecraft indicated that an historic event was being enacted. There were no immediate indications of photographs from Mars, as there were in the dramatic finale of Ranger 7. It would take 8½ hours for Mariner's 10-watt radio to transmit the 40,000 bits (binary digits) comprising a single picture, for the radio could send the data at the rate of only 8⅓ bits a second. Following the picture-taking sequence, Mariner would pass behind Mars, which would block any further communication from the spacecraft until it left the planet behind.

Shortly after the spacecraft flew behind the planet, JPL engineers reported indications of trouble with the tape on which photographs were to be stored until the spacecraft was in the clear to transmit them to earth. The tape assembly was designed to run only during the one-fifth of a second it took the camera to make a picture, then the tape reel was to shut off between pictures, and when the camera shutter lifted for the next picture, the tape was supposed to start running again. Telemetry gave an ambiguous report that the tape had not stopped once it started. If that had happened, the tape had run out long before all 22 pictures could be recorded on it, and only the early photos would be available for transmission. In that event, the best that could be hoped for was a recording of seven or eight photographs. Meanwhile, the telemetry suggested also that the tape spool would continue to run, on and on, and that might ruin what pictures were on the tape. In order to make certain the tape was not running after all the pictures were taken, JPL controllers sent a signal commanding the tape recorder to shut off. They repeated it twice to make sure it would be received. Mariner, in due time, acknowledged receipt of the first command. Its effect, in fact, was overwhelming: it shut off all the scientific experiments, including the magnetometer, radiation instruments, and the space-dust detector. When JPL discovered that, they quickly sent another signal ordering Mariner to turn on these experiments. The tape, however, was left off until the early morning of July 15, when it was started up again for picture transmission.

In the meantime, other scientific data provided a series of surprises. The magnetometer detected no magnetic field at Mars, a fact that strongly suggested that the interior of the planet was different from that of earth. The magnetic field of earth is thought to be created by its rotational energy affecting its molten-iron core. Mars had one ingredient for a magnetic field: rapid axial rotation, like the earth's. But if it displayed no magnetic field, it must lack a liquid-iron core. The lack of a magnetic field signified that no radiation belt would be found around Mars, Professor Van Allen told a press conference at noon on July 15. The State University of Iowa instruments confirmed this. Additional confirmation was provided by the University of Chicago's cosmic-ray telescope which had been observing cosmic radiation en route to Mars, but could observe also lower-energy particles trapped in a magnetic field. Professor John A. Simpson of the University of Chicago said the data suggested that Mars received a considerably heavier cosmic-ray barrage than earth. The radioactive products of the disintegrations (caused by collisions of cosmic-ray particles with atoms and molecules in the atmosphere and soil) would be different from those on earth and more intense, he added. Van Allen estimated that the thin atmosphere of Mars meant that the surface received 50 times as much radiation as does the earth's surface. Even this amount, however, should not be "frightening in any sort of radiation sense," he said.

Did the higher radiation bombardment diminish the prospect that life might exist? Pickering did not think so. "I have always felt that we will find some forms of life on Mars," the JPL director said. "I still look forward with great anticipation

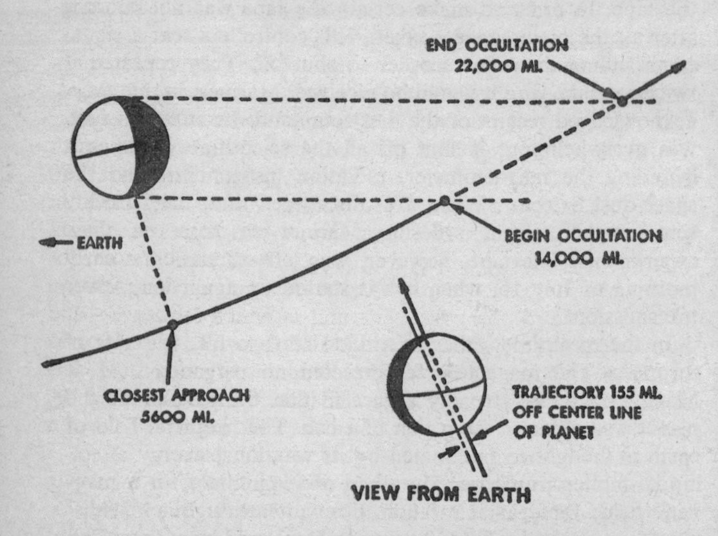

Geometry of Mariner 4 occultation of Mars.

to experiments which will involve landing capsules on the surface and searching for life."

"If there are any men there," commented Van Allen, "you can be sure they don't find their way about by using compasses."

The results of the occultation experiment provided the second major surprise of the Mariner expedition. The degree of distortion of Mariner radio signals sent through the Martian atmosphere indicated a surface atmospheric pressure of 10 to 20 millibars, instead of the expected 10 to 80. The Martian atmosphere appeared to be thinner than anticipated. Since earth's atmosphere at sea level is 1013 millibars, this finding suggested that Mars' atmosphere was only 1 to 2 per cent that of earth's. The occultation experiment had been calibrated with great care by Arvydas J. Kliore of JPL and his associates, and it was regarded as the most definitive density analysis of the atmosphere of another planet ever made. Under such a tenuous atmosphere, man could not roam about on Mars without a pressure suit, and the use of a parachute to land a Voyager life-detecting capsule did not look so promising. Mars, however, did possess an ionosphere that would reflect radio waves around the planet and make long-distance radio communication possible.

Picture data from Mariner began coming in at Johannesburg early on July 15. It was received in conventional pulse-code bits. Six bits would be decoded by the JPL computer into a number which described the position and brightness of one of 40,000 dots making up a photograph; there were 200 dots per line and 200 lines per picture. As the data came in from Johannesburg, a complete picture was represented by a column of figures three yards long. The numbers told the computer where each of the 40,000 dots belonged in relation to all the others—on which line and in which sequence—and how bright it should be on a scale of 0 to 63, where 0 was bright and 63 was jet black. The computer organized the dots so they could be transposed to a magnetic tape. Scanning the tape with a beam of light, a film converter imprinted the dot pattern on 35-millimeter film. That was the picture.

Professor Leighton showed the first picture with a projector at the Von Kármán auditorium on the night of July 15. The picture looked like the inside of a barn at midnight. But as everyone stared at the screen, the picture slowly seemed to sort itself out into a dim representation of the curved edge of a planet against the backdrop of space. It was a poor photo and it was difficult to imagine how anyone could make anything out of it. Leighton explained that it was the first picture the camera made and it showed a 200-mile arc of the limb of Mars. The region shown on the planet is a bright one called Elysium.

"You will note," he said, "also there are a few little smudges which might be something on Mars. There is a kind of a long, dark smudge with a little of a light thing next to it over there. This streaking through here is a magnetic transient due to the action of the shutter and it is not anything on Mars." He said he hoped no one would think it was a canal.

"I hope you note that it is very hard to see anything on that picture," he said. A television cameraman remarked, "You can say that again, doc." The audience broke up in laughter. "On the other hand," Leighton continued, unperturbed, "we were overjoyed to get that picture. It showed us, first of all, that Mars was there and I understand from some of the previous reports that, except for the trajectory people, that might have been somewhat in doubt."

There was a ripple of laughter. Then Leighton pulled the scientific rabbit out of the hat. By using a different translation procedure, he said, the photo development team had managed to improve the picture from the data. "Next slide," he commanded. It came on. A breathtaking view of the planet now appeared. It was still dim, still without deep contrast, but much clearer. There was instant applause from the audience. Now, JPL had something! Pickering announced that the second photograph was being transmitted and would be shown the next day.

Processing problems delayed the showing of the second and third photos until Saturday, July 17. As they were projected on the auditorium screen, a dim, dreamlike vista appeared of the other world men hoped they might one day visit and be able to use in some way. Picture No. 2 showed a region 186 miles wide from east to west and 550 miles long from north to south between Elysium and another bright area called Amazonis, believed to be a desert, like Elysium, but details could not be made out. In picture No. 3, surface detail became clearer in an area 175 by 310 miles at the western edge of Amazonis. The sun was directly overhead. It was high noon on Mars, and a feature appeared in this photo which seemed to some observers to be a ridge or a hill about 12 miles long. Other observers said it looked more like a crater or depression. A winding valley appeared rather prominently and it seemed to be similar to rilles or cracks which Rangers 8 and 9 had shown on the moon.

Together, the three photographs formed a mosaic of a vague, undulating landscape, like something in a dream. Leighton said he suspected that although the scene thus far appeared relatively featureless, compared with the moon, further pictures and improved processing would bring a lunar aspect to Mars. This was beginning to become evident in the first three, but the extent of it could not be guessed at that time. The shock was to come later.

Photo reception continued for the next ten days. The tape problem which telemetry had indicated had not actually occurred. Mariner's camera had taken 21 photos and part of the 22nd. It was a remarkable achievement—the first photographs of another planet from a space vehicle passing close by.

The National Aeronautics and Space Administration displayed the entire set of Mariner photos at the White House before President Johnson on July 29, 1965. Picture No. 4 showed little more detail than the third photo had, since the sun was still directly above the scene being photographed and contrast was poor. In the fifth picture, circular features began to appear. Unmistakably, they were craters. The detail improved in the sixth picture. In the seventh, the contrast was better still. Many craters were appearing. In the eighth, the lunar quality of the surface of Mars was revealed beyond question. Here the craters were large, some of them 20 miles in diameter, and apparently very ancient. The ninth picture enhanced the cratered aspect of the landscape. At the eleventh picture, geologist Bruce Murray commented that "this begins to almost everybody here to look very much like lunar photography." One of the craters was 75 miles across, and a small crater was only three miles across, with intermediate ones of various sizes pocking the landscape. Picture 12 substantiated the battered appearance of the landscape. The camera was looking at the southern hemisphere, where it was winter, and in subsequent pictures, which became less distinct, there appeared to be frost rimming the ragged crests of mountains

13,000 feet high. Beyond picture 13, the quality of the photos dropped off and the land became blurred and featureless. The photo team could not explain why this was so, and they were unable to improve the vague images. At picture 20, the spacecraft crossed the terminator, and 21 and 22 were indistinguishable blurs on the night side of Mars.

The revelation of the cratered face of Mars was the most astonishing event in science since the definition of the Van Allen radiation belts in 1958. Curiously enough, nothing even faintly resembling a canal showed up in the photographs. That seemed to spell the end of the canal theory, even though Mariner sampled only 1 per cent of the surface.

If the Mariner sample is representative of the entire surface, Leighton speculated, Mars must have more than 10,000 craters. The larger ones observed had rims rising several hundred feet above the surrounding surface and were several thousand feet deep. The number of craters per unit area, Leighton said, was comparable to the densely cratered lunar highlands.

No terrestrial features such as mountain chains, major valleys, ocean basins, or continental land masses appeared. No clouds could be seen in the atmosphere. It seemed doubtful that rain had ever fallen. From the apparent lack of earthlike erosion, it appeared that no atmosphere denser than the present one had existed on Mars for millions of years, nor had any free water in amounts large enough to have formed streams, lakes, or seas ever run on the planet. There was no sign of internal activity such as volcanoes, a finding which Leighton equated with the lack of a detectable magnetic field. Earth, he pointed out, is internally dynamic, giving rise to mountains, continents, and other features, while, evidently, Mars has long been inactive. The President and Congressional space-committee members listened attentively and watched the photographs in silent awe. It was clear to everyone that Mariner 4 had found a larger edition of the moon. The discovery, said Leighton, "further enhances the uniqueness of the earth within the solar system."

President Johnson spoke on this point, as "a member of the generation that Orson Welles scared out of its wits." He said, "It may be—it may just be that life as we know it with its humanity is more unique than many have thought, and we must remember this. I believe it is very clear that in this day when we are reaching out among the stars that earth's billions will not set their compass by dogmas and doctrines which reject peace and embrace force and rely upon aggression and terror for fulfillment."

Now, at last, it was clear that there was only one planet in the solar system which man could inhabit. If he destroyed it, or made it uninhabitable, there was nowhere to go. There was no hiding place.

Chapter 10

The Tall Towers

Beyond its impact on science, military strategy, and national status, the new space technology provided a means of linking mankind more intimately than ever before—communications satellites. Potentially, these instruments, which had been thought of 20 years before Sputnik, could be of enormous value, not only in promoting international commerce and industry, but also in enhancing human relationships. Satellites in orbit have created a true revolution in global communications, negating the traditional hegemony of Great Britain and France over the international telephone circuits. The orbital switchboards opened world communications to any country which built a ground station. They made possible network television on a global basis and promised to increase a thousandfold the exchange of data, news, and pictures. There was a general belief among communications experts that the economic benefits of communications satellites alone would justify the entire cost of the space program.

When the space age began in 1957, the first trans-Atlantic telephone cable was only a year old and a cable linking Australia with Canada was in the planning stage. Most intercontinental telephone communication was carried by high-frequency radio, which depended on the ionosphere, the electrified region in the high atmosphere, to reflect the radio waves around the curvature of the earth. Since the ionosphere is affected by solar activity, commercial radio telephony was always at the mercy of events on the sun and subject to being blacked out during severe disturbances of the ionosphere.

Like urban expressways, intercontinental communications pathways have tended to become overloaded as fast as they were completed. In 1927, when commercial radiotelephone service was started across the Atlantic, there were 11,000 telephone calls between the United States and Great Britain. The number reached 250,000 a year in 1957. By 1961, the volume of overseas telephone calls reached 4,300,000 a year.[1] At the beginning of the 1960s, telephone use required additional circuits, for there were delays before most calls could be put through, and the delays were growing longer. Interest grew among the television networks for intercontinental transmission. The transmission of television, however, required greater bandwidth than was available in the cables. A single television program required a band wide enough to carry 600 separate telephone conversations. In 1961, the Bell System was operating 600 telephone circuits for overseas communication. It estimated that 1200 would be needed by 1965 and 3000 by 1970. For telephony alone, the Bell System estimated that more than 10,000 circuits would be required by 1980 at expected growth rates, and an additional 2000 circuits for data transmission and for television. To meet the anticipated 1980 telephone demand alone, without data or television, at least 50 more undersea cables would be needed.

The answer to this demand was the microwave, broadband radio system, which AT&T was using to supplement its land lines and coaxial cable system. Microwaves could carry vastly more information than the longer, lower-frequency waves.

Compared with microwaves, the entire bandwidth of an ordinary high-frequency radio is small. To get voice channels in a microwave transmission, the engineer cuts the band into slices, like a loaf of bread. Each slice represents a channel. Microwave transmission had another important advantage in that it could be done with relatively low power. For example, the transmission of microwaves between relay towers about 30 miles apart requires about one watt of power, the amount in a small flashlight.

The microwave segment of the electromagnetic spectrum was ideal for commercial communications systems, but it suffered from the same limitation as the old-fashioned heliograph. The waves traveled in straight lines, like light waves. They did not follow the curvature of the earth, and were not ordinarily reflected by the ionosphere. Consequently, as the earth curved, the microwave signal went straight out over the horizon into space. AT&T developed cross-country systems of microwave relay towers, which marched across the countryside from coast to coast and border to border, every 20 to 30 miles. The function of the relay tower was to receive the microwave signal, amplify it, and send it on its way to the next tower.

To transmit microwaves across the Atlantic Ocean, however, it was necessary to establish a single relay tower at least 400 miles high, or station ships every 30 miles, all the way across the ocean. Before the space age arrived, the 400-mile-high tower seemed to be as unlikely a prospect as the spectacle of 100 communications-relay ships strung out across the ocean.

That possibility that a very high tower could be established by launching an orbital satellite was suggested first by the British science writer Arthur C. Clarke in an article, "Extra-Terrestrial Relays," in *Wireless World* in October 1945. Nine years later, John R. Pierce, who directed communications research for the Bell Telephone Laboratories, described a practical communications-satellite system to the Princeton section of the Institute for Radio Engineers.[2] As the space age opened, the concept of a communications satellite as a potentially practical instrument of microwave relay was firmly established. Indeed, technicians at Bell Laboratories were beginning to design one.

PROJECT DIANA

The idea of bouncing radio waves off a satellite actually had been tested by the Army Signal Corps in 1946 when it made radar contact with the moon in Project Diana. The Navy applied the principle in 1954 to develop a communication system it called CMR (Communication by Moon Relay). Signals were relayed between Washington and Honolulu, using the moon as a "backboard" to reflect them, and between 1959 and 1963 the project operated a regular communications link between the two stations. The first live voice transmission was achieved via the moon between Bell Telephone Laboratories at Holmdel, New Jersey, and the Jet Propulsion Laboratory at Pasadena, California, November 23, 1959.

Meanwhile, the Advanced Research Projects Agency (DoD) developed an advanced communications satellite called SCORE, an acronym for Signal Communications by Orbiting Relay Equipment. The satellite attached to its empty Atlas booster in orbit weighed 8750 pounds. It was capable of relaying voice, code, and teletype transmissions, live or delayed. On December 19, 1958, SCORE retransmitted a message of peace and good will from President Eisenhower. It had been stored on tape and broadcast by the satellite on a radioed command from the ground. SCORE operated for 12 days until its batteries became exhausted.

The next Army satellite venture was the Signal Corps' Courier 1-B. Weighing 500 pounds, this vehicle was powered by 20,000 solar cells. It was designed to demonstrate the active-repeater mode for relaying messages, by which signals were received on one frequency and transmitted after being amplified on another. Courier 1-B could also store signals on tape and transmit them at another point in its orbit when it was in sight of a station in another part of the world. This highly sophisticated experiment was launched from Cape Canaveral October 4, 1960, by a Thor-Able Star rocket. It received and relayed 118,000,000 words during the 18 days it operated before the power supply failed.

While the Army tinkered with active satellites, NASA developed a technique for putting a large, passive satellite in orbit, one which would merely reflect radio waves, like a mirror. Echo 1, the most conspicuous satellite of the space age, was the first demonstration of what a passive satellite could do. It was launched from the Cape on August 12, 1960, a silvery 100-foot balloon made of aluminized mylar. After it was injected into an orbit of 941 to 1052 miles by a Delta rocket, it inflated automatically. Because of its size and high reflectivity, it was easily visible at night as a bright star sailing serenely across the heavens.

On Echo's first pass over the United States, Bell Laboratory technicians at the utility's experimental site, Holmdel, New Jersey, received the taped voice of President Eisenhower from Califonia. They said it was as clear as a local radio broadcast. The signal was transmitted by the Jet Propulsion Laboratory's radio station at Goldstone in the Mojave Desert and reflected to the East Coast by the big balloon. A series of two-way telephone calls were exchanged on ensuing passes of the satellite, and ordinary home telephones were connected into the ground stations in California and New Jersey for cross-country conversations by satellite.

The Echo experiment demonstrated that the Bell System's microwave relay technology, used overland, could be applied to transocean communication if a microwave repeater in the sky could be provided by satellites.

TELSTAR

The Bell System turned to the active-repeater satellite as the answer to intercontinental communications. In orbit around the earth, it would serve the same function as the microwave relay towers that stretched across the country on land. It could relay a microwave signal in line of sight, not 30 miles, but 3000 miles. Distance was not the problem. The problem was straight-line transmission between two points. And the orbital satellite solved it, so far as overseas communications was concerned.

On July 6, 1960, the Bell System proposed to the Federal Communications Commission a plan for a world-wide communications system, using 30 to 50 active-repeater satellites in orbit at altitudes of 4000 to 7000 miles. These would provide up to 600 additional telephone circuits. The system would have the broad-band capability for transmitting television also. Engineers at the Bell Laboratories knew that one properly designed satellite could handle more telephone circuits than all the cables then in use. Moreover, the addition of television programing seemed to offer a new dimension in communications revenue.

The FCC authorized AT&T on January 19, 1961, to establish an experimental communications link across the Atlantic, and arrangements were initiated by AT&T to win the cooperation of Britain, France, Italy, and West Germany at the other end. Bell Laboratories had been working on a prototype satellite it called Telstar. It was spheroid, 34.5 inches in diameter, and weighed 170 pounds, incorporating in its design two Bell Laboratory inventions: the transistor and its descendant, the solar cell.

The exterior was faceted like an ornamental ball on a crystal chandelier, with 72 flat surfaces. Sixty of them were covered with 3600 solar cells, which supplied a total of 15 watts of electricity at full capacity. The remaining 12 facets were covered with special electronic equipment to measure the effects of radiation on transistors and the solar cells. Telstar, built under Pierce's direction, expressed his belief that the low- or medium-altitude satellite would provide high-quality, long-lived, and reliable communications links across the world's oceans. Pierce and the Bell System had rejected synchronous satellites which would appear to hang stationary over one point on earth at an altitude of 22,300 miles over the equator. These posed technical problems which Pierce believed would delay the establishment of a communications-satellite system.

In order to reduce damage from radiation and space dust, the solar cells on Telstar were covered with a thin layer of artificial sapphire. The faceted sphere thus gleamed like an enormous blue jewel. Two bands of rectangular openings ran along the equator of the sphere, containing 72 receiving antennas and 48 transmitting ones. These allowed the machine to transmit and receive equally well in all directions except along its poles. In addition, Telstar had its own antenna for command, tracking, and telemetry. Through it, the satellite transmitted reports on the radiation environment through which it passed and data on micrometeoroids, or space dust.

In substance, Telstar was a scientific experiment, as well as a demonstration. While its main purpose was to test a satellite design for intercontinental telephone and television transmission, it was also equipped to make two critical radiation studies. One study was designed to find out how much damage would be done to solar cells and transistors by radiation in the Van Allen belts. The other was to determine how many energetic protons and electrons the satellite encountered in a medium-altitude orbit. The data would enable satellite designers to build long-lived electronic components for the machines passing through zones of high radiation. Longevity was a criterion of satellite feasibility. While no one expected a satellite to last 20 years—the life expectancy of a submarine cable—a lifetime of two to five years seemed to be a reasonable goal. First, one had to know what the space hazards were.

Telestar's electronic circuits contained more than 1000 transistors and 1500 semiconductor diodes, plus one electron tube—the traveling-wave tube—for amplification. The frame of the satellite was magnesium, covered with aluminum panels. All electronic components were housed in an aluminum canister 20 inches in diameter. It was attached to the interior frame by nylon lacings to reduce launch vibration. After the components had been packed in the canister, it was filled with liquid polyurethane foam, which hardened to a light, rigid solid. Thus the electronic parts were held in a solid matrix which would protect them from vibration. In addition to the solar cells, the power supply contained a nickel-cadmium, rechargeable storage battery, which would provide power when Telstar was in darkness.

Telstar required an elaborate ground station with equipment to track, receive, and transmit to the satellite. A horn reflector antenna which Bell Laboratories had built at its Holmdel branch for Project Echo became the model for a huge antenna which AT&T built for Telstar near the village of Andover in western Maine. The antenna and the complex to house it arose in 1961 on a 900-foot hill in the White Mountains surrounded by ridges with an elevation of 3800 feet. In this natural bowl, the antenna was shielded by the surrounding ridges from a great deal of radio interference but at the same time it was high enough to "see" the satellite close to the horizon.

The entire antenna structure was 177 feet long and 94 feet high and weighed 380 tons. Its most striking feature was the cornucopia-shaped horn, which was 90 feet long. Pointed skyward, the open end of the cornucopia was 68 feet in diameter. At the small end was a cab, about the size of a one-car garage, containing the transmitting and receiving equipment. This entire complex rode on steel rails which allowed it to be rotated in a full circle. The horn could be swung independently on a horizontal axis from the horizon to the zenith. The structure was protected from the weather by a radome, a bubble made of dacron and synthetic rubber one-sixteenth of an inch thick, 210 feet in diameter, and 160 feet high. It was supported by air pressure of about one-tenth of a pound per square inch above normal.

Signals were beamed through the horn to the satellite on a frequency of 6390 megacycles by a traveling-wave tube with an output of 2000 watts. Signals from Telstar were received on a frequency of 4170 megacycles at a power level of one-trillionth of a watt. The transmission from the satellite was amplified by a ruby-crystal maser operating at the temperature of liquid helium (268.9 degrees below zero centigrade).

The entire Telstar experiment depended on a mosaic of technologies developed since the end of World War II, from the rocket which was to launch it to the solar cells that were to provide it with electrical power and the ruby maser that was to amplify its signals. The antenna complex at Andover cost the Bell System $10,000,000 plus $4,000,000 for a tie line.[3]

A ground station of similar design was constructed by the French National Center of Telecommunications Studies at Pleumeur-Bodou in Brittany. The British General Post Office built a satellite ground station of different design, using a parabolic dish antenna, at Goonhilly in Cornwall. The station was near historic Poldhu Point where the first radio message was transmitted across the Atlantic December 12, 1901, to Guglielmo Marconi, standing on a hilltop with a receiver outside of St. John's, Newfoundland.

In July 1962, an experiment comparable in magnitude to Marconi's was ready to begin. Earth Station No. 1 at Andover was completed. The British reported that their dish at Goonhilly was ready. The French station was not finished, but there was a chance it might be rigged in time to join the experiment. Telstar was scheduled for launch by a Delta rocket from Cape Canaveral July 10, 1962.

AT&T had invested about $50,000,000 in research and development[4] of two Telstar satellites and the Andover station. In addition, the company paid NASA $2,680,982 for services in launching Telstar.[5] This amount included $36,892 paid in reimbursement of salaries and wages to Goddard Space Flight Center personnel for management services and $6690 for reimbursement of travel expense by NASA engineers and technicians. AT&T paid the entire launch bill on

February 16, 1962, five months ahead of the launch date! The utility's arrangement with NASA was the first of its kind in space work, wherein a private utility subsidized a major technological experiment in which the federal space agency itself was interested. NASA, meanwhile, had let a contract to Radio Corporation of America to develop an active-repeater satellite called Relay, also designed to fly at medium altitude. Thus, two communications-satellite prototypes were under development at about the same time, one privately funded, the other financed by the taxpayers. This redundancy might be expected to bear many black-and-blue marks of hard competition, pitting AT&T against NASA. If this was the case, and many observers believe it was, the Bell System satellite got there first with the most.

When Telstar was launched from the Delta pad on the Cape at 4:35 a.m. EDT July 10, 1962, it inaugurated the communications revolution—the outcome of which is not yet in sight. The jeweled satellite was hurled accurately into an orbit with an apogee of 3503 miles and a perigee of 593 miles, and a period of 2 hours and 37.8 minutes.

For five revolutions of the satellite around the earth, Bell engineers tested their machine. Its orbit took it through dense radiation regions of the Van Allen belt, but those turned out to be only a part of the radiation peril to the satellite. The day before Telstar was launched, the Department of Defense rocketed a 1.5-megaton nuclear bomb from Johnston Island in the Pacific Ocean and exploded it at an altitude of 250 miles. The experiment, Project Starfish, which was mentioned earlier, in Chapter 3, was designed to determine the effects of injecting high-energy nuclear particles into the radiation zones on missile-defense systems and communications. The explosion's effects on the belts were grossly underestimated by the scientists advising the Defense Department, and the radiation zones quickly became surcharged with energetic electrons and protons which damaged three satellites and reduced the longevity of Telstar.

The first public test of Telstar began on the evening of July 10 as the satellite rose over the Atlantic from the west on its sixth revolution. Most of the AT&T officials were assembled at Andover, with an entourage of newspaper and television correspondents. Pierce and his associates observed the test at the Holmdel station, historic site of the first satellite horn, on a promontory overlooking New York Bay and Brooklyn. I arrived there late in the afternoon as a motion-picture crew was filming background scenes for a movie of the event. The Holmdel horn, looking like a toy compared to the huge one at Andover, was swinging, as though searching for invisible electromagnetic bounty to fill its cornucopia-shaped aperture.

The Bell people had contrived an elegant communications hook-up between Andover and Holmdel, so that we were able to see Eugene F. O'Neill, AT&T chief engineer, explaining at Andover what was happening. Like all such hook-ups in a highly technical environment, this one was subject to periodic shrieks and groans as the engineers tinkered with it. The television monitors showed vague, shadowy figures milling around in a room at Andover, and a commentary recited a somewhat dramatized account of what the experiment was about for those who didn't know. In Washington, Vice-President Johnson and Senator Everett Dirksen of Illinois, the Republican leader in the Senate, were standing by for the event. In contrast to the bustling activity of AT&T executives at Andover, Pierce, who had led the development of the satellite, sat calmly at one of the picnic-style tables which the Holmdel laboratory staff had rounded up for the visitors. For more than eight years, the rangy, taciturn scientist had been waiting for the day when satellite communications across the ocean would begin, and now his moment of truth had come on this humid evening. There was a good deal of pure tension at Holmdel which was not being discharged in the purposeless meandering around that was going on 400 miles to the north in Andover. One experiences a curious dilation of time while waiting for satellites to appear, and then an astonishing compression as events occur more rapidly than expected. So it was with Telstar 1. At 7:31 p.m., a photograph of the satellite itself materialized on the television monitors at Holmdel. It had been transmitted to the satellite by Andover and was being returned, and the old Echo horn at Holmdel had received it. Applause filled the room and the picture of the satellite stayed steady on the screen.

Then began a ceremonial telecast. Frederick R. Kappel, chairman of the board of AT&T, called up Johnson in Washington—via Telstar—and told him what was going on, as though the Vice-President didn't know. Johnson replied in the usual formal prose, suitable for engraving in granite: "I want to congratulate AT&T for launching this dramatic opening of a new frontier for international telephone, international television, and most of all international understanding."

In the background of these stylized exchanges, excitement was quite high at earth station No. 1 and at Holmdel, for the experiment had materialized into an historic demonstration. Telstar was working perfectly. The communications between Andover and New York and Washington, going via the satellite, were brilliant. News correspondents at Andover lined up to talk to New York and Washington over the satellite. The demonstration was contrived inasmuch as the signal went up to the satellite from Andover and then came back before being sent over conventional ground-transmission equipment in the tie-line, but one reporter remarked that the circuit he had through Telstar was the best one he had been able to get all day.

On Telstar's seventh revolution, my turn came to talk over the satellite. At Andover, the other party on the line was Ken Berry, a reporter for the Portland, Maine, *Evening Express*. His voice was transmitted from Andover to the satellite, which relayed it to the horn at Holmdel, where I heard it. The Holmdel horn could receive but not transmit. My voice was carried overland to Andover, boosted to the satellite some 3600 miles distant, and then returned to the Big Horn. Berry said he heard me without difficulty.

The only trouble was the one which Ed Murrow had predicted at the first NASA conference in Tulsa in 1961: It isn't how you transmit it—it's what you have to say that counts. Our conversation went something like this:

LEWIS: Hello, this Don White of the Boston *Globe?*

BERRY: No, this is Ken Berry from Portland.

LEWIS: Oh. Well, hello, Bob.

BERRY: I said Ken—Ken Berry. How did it go up there in New Jersey?

LEWIS: Good. How did it go down there in Maine?

BERRY: It was quite a moment here.

LEWIS: I'm getting a signal to sign off. Someone seems to be twisting my arm. Fine to talk to you.

BERRY: Fine to talk to you.

GOONHILLY ON THE LINE

As observers at Andover and Holmdel were watching TV transmission, an excited voice broke in on the audio circuit to say that pictures were being received in France. They looked as though they were being telecast locally, a spokesman for the French National television stated. The French picked up the TV signals from Telstar at 7:47 p.m. and Goonhilly terminal came on the line at 8 p.m.

The show ended when Telstar set over Spain on its seventh revolution and the engineers at Andover resumed their test program, which was not made public. I asked Pierce how he felt about the outcome of the experiment.

"Relieved," he said. "The most it could do was work. We really didn't expect to hit the jackpot with it. It shows us we're on the way."

Thursday night, July 11, the French climbed aboard Telstar with a refreshing disregard of protocol by sending a nine-minute, taped musical telecast from Paris on the satellite's 15th revolution. It looked and sounded like the Late Show at the Lido on the Champs Elysées. A formal exchange of European and American telecasts had been arranged for July 23, but France's lively offering upstaged it. It featured the French popular singer Yves Montand singing "Chansonette," among lesser known entertainers, and it was as Parisian as a travel poster. A picture of the French ground station in Brittany was shown, its high radome looking unearthly on the flat, pastoral landscape. Later that evening, the British transmitted six minutes of live television—the first over Telstar. It showed their antenna, the consoles, and the transmitting equipment. Picture quality was comparable to good local reception. There was no longer any doubt that the age of the tall communications towers in space had opened.

THE ZERO GATE

Now it remained to be seen how long the satellite would live in the nuclear-blast-enhanced radiation of the Van Allen belts. One indication of trouble showed up in a command circuit in August. There were two such circuits in the satellite —the only operation which Bell engineers had believed should be made redundant. Telemetry advised the ground station that the No. 2 command chain was not transmitting certain commands to the satellite. The chain became intermittent in its operations and then failed.

Since the ground crew could still turn the satellite on and off through the No. 1 command circuit, there was not a great deal of concern until mid-November, when No. 1 became intermittent in its operation. On November 23, 1962, it failed and Telstar was dead for communication relay, although its telemetry radio was still working.

Now the experiment entered a new phase, raising the questions: what caused the satellite to fail and how could it be fixed? A number of causes could be supposed, such as a loose connection, premature aging of the components, extreme temperature variations, or radiation damage. All of these causes were carefully examined and all but one eliminated. The answer was radiation. Several clues pointed to it. During the fall of 1961, researchers at Bell Laboratories and the Brookhaven National Laboratory of the Atomic Energy Commission had found that radiation can cause deterioration of performance in a transistor in the following way. When it penetrates a transistor's outer shell, radiation ionizes the gases inside.[6] The ions collect on the surface of the transistor and change its electrical properties. This effect was especially noticeable when the transistor was operating under reverse-bias voltage. In each of the two command-decoder circuits of Telstar there were 37 transistors operating under continuous reverse bias. Moreover, they had less metal shielding than had the semiconductors in the telemetry and the receiver circuits. The command circuit decoded ground signals which arrived as pulses that the decoder identified as zeros or ones. In the decoder circuit, sequences of zeros and ones were translated into operating instructions for the satellite.

Another clue was the unexpectedly high radiation level which Telstar was reporting—more than 100 times that predicted for the Van Allen belts. Some of the additional radiation had been injected into the magnetic field, where natural radiation was trapped, by the Starfish shot of July 9, 1962, but another surge of high-energy particles had been reported by Telstar in November. The report confirmed other evidence of Russian high-altitude testing of nuclear weapons.

In the laboratory, the engineers tested the theory that heavy dosages of radiation would cause transistors to fail, like those in the command-decoder circuits. The most sensitive part of the command-decoder circuit was a segment called the Zero Gate. It recognized the zeros in the binary code of ones and zeros used to command the satellite. The Zero Gate, which admitted zeros into the circuit, had been blocked by radiation damage. The One Gate, which admitted the pulse identified as a one, was still working. Bell engineers concluded they could regain control over Telstar if they could bypass the Zero Gate and get an instruction into one of the decoders through the still-operating One Gate.

THE NOTCHED ZERO

To accomplish this, they resorted to a ruse, an electronic trick, designed to fool the command-decoder circuit. And it became the basis of one of the monumental "fixes" in the history of technology.

The difference between a zero and a one in Telstar language was the length of a radio pulse. The pulse for zero was only one-half as long as the pulse for one.

The engineers reasoned that if they could disguise a zero to look like a one pulse they could slip it through the One Gate and thus inject into the command decoder the proper sequence of zeros and ones required to form a command. They adjusted the transmitter to send a special long pulse that would look like a one except for a dip or a "notch" in it. This pulse they called the "notched" zero. In the laboratory, it worked. The One Gate accepted it and the pulse advanced the code sequence just as a true zero pulse would.

The first attempt to fix a disabled satellite was made December 20, 1962, on Telstar's 1492nd pass. The notched zeros were radioed to Telstar from Andover. The first two attempts produced no response, but on the third, the command circuit came to life once more and Telstar was reborn. Difficult as it must have been, the Bell System kept this amazing accomplishment a secret for several weeks while the engineers experimented further. They had surmised that if they could remove all voltages from the ionized transistors that had failed in the command decoder, the ionization layer would dissipate and the transistors would get well. In order to remove all voltages, they had to disconnect Telstar's storage battery, so that when the satellite was in darkness, without power from its solar cells, the complete absence of voltage would allow the ionization to dissipate and restore the transistors to working order. However, as though anticipating what the groundings were planning to do, Telstar turned off the battery itself before the ground crew could send the special command to do so. Apparently, the decoder had misinterpreted one of the trick commands.

Now the satellite was dead, with even the telemetry off. When Telstar reappeared over the Atlantic, the engineers hailed it by radio. This time, the Zero Gate in the No. 2 command decoder was working perfectly again, and it was no longer necessary to send notched zeros. Apparently the complete electronic "rest" had cured the circuit of radiation sickness. On January 1, 1963, the controllers were able to disconnect the battery, using the normal code through the No. 2 circuit. The second rest cure restored the No. 1 circuit and both command chains were once more operating normally. Television relays were resumed on January 3 and the world was advised that Telstar was back on the air.

However amazing, the cure was not permanent. Intermittent blockages reappeared in the command-decoder circuits and the engineers had to resort more and more often to the notched-zero ruse to keep the satellite under their control. Even with periodic resting, the transistors continued to deteriorate, and by February 14, 1963, resting them failed to restore them. A week later, Telstar misinterpreted a command, disconnected its storage battery and expired. This time, it could not be revived.

RELAY

In the meantime, another active-repeater, medium-altitude communications satellite was up and taking Telstar's place: RCA's Relay 1, developed under a NASA contract. It had been launched December 1, 1962, by a Delta rocket from Cape Canaveral into an orbit of 820 to 4612 miles.

RCA's octagonal, bottle-shaped machine proved to be less spectacular but considerably longer-lived than the trail-blazing Telstar, once it recovered from an initial loss of power shortly after the launch. Relay was 35 inches long and weighed 172 pounds. It contained two of everything—receiving, transmitting, and amplifying circuits—and thus provided more redundancy than Telstar. The power system consisted of 8740 solar cells, backed up by nickel-cadmium batteries. The two receiving-amplifying-transmitting transponders each had an output of 10 watts.

Shortly after the launching, a voltage-regulator switch for one of the transponders partially failed, allowing excessive drain on the power supply. NASA controllers switched it off and turned on the second transponder, which worked properly and allowed the satellite to operate trouble-free for two years. Relay proved the necessity of having backup systems in a communications satellite—a lesson that NASA later had to learn all over again in its next experimental satellite, Syncom.

Relay was phenomenally successful. It transmitted its first television program January 9, 1963, consisting of the ceremony unveiling the *Mona Lisa* in Washington and a 10-minute segment of NBC's "Today Show." During its two-year lifetime, which was twice as long as planned, Relay provided the first space communications link between North and South America. It could receive and retransmit a single television program or 12 simultaneous, two-way telephone conversations. It was used for live TV transmission to Britain of the Washington ceremonial of April 9, 1963, when Sir Winston Churchill became an honorary citizen of the United States. Relay transmitted television coverage of Astronaut Gordon Cooper's liftoff in the Mercury spacecraft, Faith 7. It was used by American networks in coverage of the death of Pope John XXIII and the coronation of Pope Paul VI. In June 1963, RCA set up apparatus in the Palmer House in Chicago from which typesetting instructions were radioed via Relay to a computer which set a line of hot type in the composing room of the Glasgow *Scotsman*. The line read, "Hot type set by paper tape." Brain waves of a patient in Bristol, England were transmitted via Relay to the National Academy of Neurology meeting at Minneapolis, and the satellite retransmitted the first network color-television program, the Disney version of *Kidnapped*. The film was beamed up to the satellite, which beamed it back down across the Atlantic. Relay had a time switch designed to turn it off automatically at the end of 1963, but the switch failed to work and the satellite continued to operate until its transponder became unusable in February 1965.[7]

AT&T sent Telstar 2 aloft May 7, 1963, with improvements suggested by the experience of Telstar 1, in an orbit with an apogee of 6717 miles—nearly twice as high as that of Telstar 1—and a 604-mile perigee. This time, the satellite showed greater resistance to radiation damage. Its higher orbit gave it three hours of trans-Atlantic visibility a day during which signals could be exchanged with Europe, compared to about an hour and three-quarters for Telstar 1. Also improved, Relay 2 was boosted aloft January 21, 1964, and continued to be used until September 26, 1965.[8]

SYNCOM

The experience of Telstar and Relay demonstrated that a medium-altitude system of trans-ocean communications, which Bell System had visualized in 1960, could have been established in 1963. By that time, however, the question of communications satellites was no longer simply a technological one. It had passed into the arena of politics. Telstar had proved there was gold at the end of the satellite rainbow, and in the summer of 1962 the issue of who should control the communications satellites of the United States was being hotly debated in Congress. The issue became substantive in a bill creating a Communications Satellite Corporation to develop a world-wide system. In its final form, the bill provided that half of the corporation's common stock would be owned by the common carriers and half by the public at large.

The proposal angered many of the old-line liberals in both the House of Representatives and the Senate, who charged that the bill was a giveaway of technology developed at public expense. Senators Wayne Morse and Estes Kefauver led a filibuster against the bill, arguing that the fruits of rocket and satellite technology paid for by the people of the United States were being served up to private interests. It was obvious that the communications utilities would dominate the new corporation and it was deemed quite likely that disagreements among them might retard the development of a workable system. After 19 days of debate, the Senate invoked cloture for the first time in 35 years to shut off the filibuster. The Senate passed the bill on August 18, and the House on August 20. When he signed the measure, President Kennedy stated that "no single company or group will have the power to dominate the corporation."

With the birth of the satellite corporation, the question of whether medium- or synchronous-altitude satellites would become the basis of a commercial system remained up in the air. The decision would involve a number of entities: the corporation itself, the Federal Communications Commission, and Comsat's 52 foreign partners in the International Telecommunications Satellite Consortium (Intelsat), which was created to activate the global communications system.

NASA had awarded a contract to the Hughes Aircraft Company for construction of three synchronous satellites in a project called Syncom. In contrast to the bevy of 30 to 50 satellites required in a medium-altitude relay system, under the proposal of AT&T, it would take only three satellites in synchronous orbit over the equator to provide coverage of the entire globe, short of the poles. A synchronous orbit, as mentioned in Chapter 2, is one in which the satellite's orbital period is the same as the period of the earth's rotation—24 hours. If the satellite's orbit is in the plane of the equator, it would appear to hang stationary over one point on the earth's surface at synchronous altitude. This altitude was computed to be 22,300 miles, in a precise circular orbit. The computation was derived from Kepler's third law of planetary motion which stated three hundred years ago that the square of a planet's period of revolution is proportional to the cube of its mean distance from the sun. The law can be applied to satellites of the earth to show that a satellite's period of revolution increases with orbital altitude. Also, the higher the orbit the slower the satellite's velocity appears to be in relation to any point on the surface of the earth.

In Project Mercury, for example, the astronauts orbited the earth at 100 to 160 miles altitude at a velocity of 17,500 miles an hour, flying rapidly eastward. They were moving much faster than the earth was rotating. The moon, at a mean altitude of 238,857 miles, revolves around the earth in

27 days, 7 hours, and 43 minutes at an average velocity of 2287 miles an hour. Although it is orbiting the earth in the same direction as were the Mercury spacecraft, the moon appears to be moving west because its orbital motion is slower than the earth's rotation. Between Project Mercury and lunar altitudes, at 22,300 miles, a satellite would appear to move neither east nor west, but remain stationary. Although it would be traveling at a velocity of 6800 miles an hour, its orbital motion would be synchronous with the earth's rotation. By adding velocity to increase the altitude of the satellite, ground controllers could cause it to drift to the west. The increase in altitude would enlarge the orbit and thus lengthen the time it takes the satellite to encircle the earth. Conversely, by slowing down the satellite, the controllers could make it drift to the east, because the satellite would fall into a lower orbit with a shorter period of revolution around the earth.

The orbital mechanics of a synchronous satellite provided a means of moving it from one station to another. It could be placed by ground command to hover over any point on the surface in its orbital plane. This degree of flexibility was not possible with medium-altitude satellites. It made the synchronous machines (Syncoms) quite attractive, for one could be stationed over the Atlantic, a second over the Pacific, and a third over the Indian Ocean to cover the world.

On paper, the synchronous system looked feasible, but it was fraught with unknown problems and several known disadvantages. These were recited by James B. Fisk, president of Bell Laboratories, in support of AT&T's medium-altitude satellite proposal.[9] In order to place a satellite in synchronous orbit over the equator from Cape Canaveral, 28½ degrees north of the equator, it was necessary to turn the satellite to the left as it passed the equator in order to place it into an equatorial plane. Fisk contended that this maneuver, which had never been done before by the United States (as of mid-1961), was beyond the capability of existing rockets and guidance systems.

While advancing rocket technology would unquestionably make such a maneuver possible, Fisk cited a second drawback about which science and engineering could do nothing. This was a delay of about six-tenths of a second between the time one speaker said hello and the time he heard the other speaker reply in a synchronous-satellite relay. This delay was imposed by the limitation of the speed of light, at which the radio microwaves traveled. At 186,000 miles a second, the microwaves would take about three-tenths of a second to go up to the satellite from America and be relayed back down to Europe on a dog-leg path of about 55,000 miles. It would take another three-tenths of a second for the answer to come back. Fisk predicted that this delay was too long for ordinary conversation and would induce a requirement for echo suppression. In ordinary conversation over the telephone, a speaker hears his own echo, but it returns so fast that it is nearly simultaneous with his speech and doesn't bother him. Over a longer distance, the echo must be stopped because the slight delay makes it annoying. "With as much as six-tenths of a second delay, most people are quite irritated," Fisk said. "They feel they must slow down or stop to let the echo catch up."

In very long-distance calls, the Bell System eliminates the echo by suppressing the sound from the other direction as long as one person is talking. The other person has difficulty breaking in until the first speaker stops. The system works satisfactorily most of the time in coast-to-coast calls in the United States because the delay is short. But with a six-tenths-of-a-second delay, Fisk insisted, complications in conversation became too unpleasant and confusing.

The Syncom satellite built by Hughes under the direction of the Goddard Space Flight Center was shaped like a bass drum. It was 28 inches in diameter and 15½ inches high, without its antennas and motor. It weighed only 79 pounds. The problem of establishing the satellite in a 24-hour orbit was solved by attaching a small but powerful solid-fuel rocket motor to the satellite. In addition to the three-stage Delta rocket which would launch Syncom, this fourth or "kick" stage would provide the needed energy to establish the machine in a synchronous circular orbit. The satellite in orbit could be controlled by two gas-jet propulsion systems, one using nitrogen and the other hydrogen peroxide. These systems enabled the ground controllers to move the satellite from one point in longitude to another by increasing or decreasing its orbital velocity and to control its altitude so that its antenna pointed toward the earth.

Syncom's vertical surface was covered with 3,960 silicon solar cells which supplied 20 watts of power. Nickel-cadmium storage batteries provided backup power, but the satellite at synchronous altitude would be in sunlight all but a small fraction of the time and could operate entirely on its solar cells. Syncom, like Telstar and Relay, was an active repeater, receiving incoming signals at 7400 megacycles and relaying them at 1800 megacycles. There were two receivers, each with an output of 2 watts. Communications capacity in the experimental version was limited to a single two-way telephone conversation or 16 one-way teletype circuits. However, Hughes had designs for an advanced Syncom with a capacity for 1200 two-way telephone circuits and four television channels.

LOST IN SPACE

The establishment of a satellite in a synchronous orbit was the most delicate launching operation the United States had attempted. It required the utmost precision in the functioning of the launch vehicle and in the propulsion of the satellite, and called for meticulous guidance. The launch vehicle, the Delta, had proved itself by 1963 as the most reliable in the national inventory. Its first stage was a modified Thor IRBM and the second and third stages were modifications of Vanguard's upper stages. While Vanguard had not survived as an operational vehicle, the technology developed in that project was proving an invaluable asset to the nation's space goals.

Syncom 1 was launched at 12:35 a.m. EST on February 14, 1963. All three stages of the Delta performed apparently well and injected Syncom into a looping orbit over Mozambique. The orbit was an extended oval with an apogee of 22,300 miles and a perigee of 150. That was the best the Delta could do, besides imparting a spin of 150 revolutions a minute to stabilize the satellite. The task of circularizing the orbit then fell to Syncom's motor. It was designed to fire as Syncom reached apogee. With the capability of adding 4696 feet a second to Syncom's velocity, the motor would kick the spacecraft into a circular orbit by boosting perigee from 150 to 22,300 miles.

All night long, controllers listened to Syncom's telemetry. Everything was going well as the satellite coasted upward, hour after hour, to apogee, where its kick stage would cut in. At 5:42 a.m., when it was due to reach apogee, a radio signal was transmitted to the spacecraft to fire its apogee motor. The motor fired, either in response to the signal or to an automatic timer on the vehicle. Then, 20 seconds later, the telemetry stopped. Syncom 1 was never heard from again.

There was no immediate clue to what had happened. Project engineers feared that either the apogee motor had exploded or that the vibration it had induced in the spacecraft had broken a circuit. Newsmen who sought out Hughes engineers at the Cape Colony motel in Cocoa Beach found the contractor contingent had folded its tents and fled back to Culver City, California. There was a vast silence concerning the Syncom project, for the good reason that no one knew what to say beyond the fact that the first Syncom had failed.

Radio antennas searched the heavens in vain. Syncom was lost. Then, on February 20, 1963, probable sightings of the satellite were reported by Smithsonian Astrophysical Observatory telescopic cameras at Curaçao, Netherlands West Indies, and at Nani Tal, India. Subsequent optical observation and tracking data derived from them matched what Syncom's orbit would have been had its motor fired properly. It now appeared that Syncom had achieved a nearly synchronous orbit, the first one of the space age. Its motor had not exploded. The power supply presumably had failed, probably because of intense vibration during the firing of the apogee kick motor. The orbit was 21,375 miles perigee and 22,823 miles apogee, with an orbital period of 23 hours and 45 minutes. Testing at Hughes suggested that bursting of the nitrogen-gas-jet tank could have caused the blackout. No one would ever be certain.

Goddard and Hughes went ahead with preparations to launch Syncom 2. A small silver-zinc battery was added to the power supply to operate a backup beacon and telemetry system for 30 or 40 minutes in the event the main power supply failed. The power distribution system was installed in duplicate. A new motor, made by the Jet Propulsion Laboratory, was installed in Syncom 2, in place of the Thiokol TE-375 on Syncom 1. The JPL motor was designed specifically for Syncom while the Thiokol motor had been adapted from one designed for another vehicle.

The launching of Syncom 2 was set for July 23, 1963, but shortly before the countdown started, NASA officials ordered a change in the electronic system by the addition of a capacitor. This was a delicate job of soldering, especially since the satellite was already mounted atop the 90-foot Delta rocket. Hughes Aircraft flew one of its most experienced solderers, Charlie Benson, to the Cape from the Culver City plant, and he made the fix from a perch atop the gantry.

The countdown began as scheduled, but was scrubbed after unexplained oscillations showed up in the satellite's circuits. An investigation showed that the oscillations were the result of the antenna being folded for the launch and would disappear when it was extended in flight. The launch was then rescheduled for July 24, and a second countdown was started. It had to be stopped because of a gyroscope problem in the Delta. By this time, nearly everyone connected with the launch was developing the classical symptoms of the jitters. The third countdown began on the evening of July 25 for a launch on the morning of the 26th. Shortly after midnight, a malfunction in ground-test equipment caused one of the jets on the satellite to fire by mistake. The countdown was held while technicians searched for damage. They found that all was in order, and the count was resumed.

On the morning of July 26, as the countdown neared zero, a dozen observers gathered at Press Site No. 1, a rickety wooden platform about 30 feet high which had been erected on the site where hundreds had watched the Project Mercury suborbital launchings. About 1200 yards away, Delta No. 20 with Syncom 2 aboard, stood naked in the rising sun. The gantry had been rolled back and the rocket glowed like a bright trophy urn without handles, its midsection wreathed in traceries of liquid-oxygen vapor. The countdown reached zero at 9:33 a.m. and Delta's booster engines fired. It roared smoothly off the pad and vanished into the dappled morning sky.

Syncom 2 entered its initial transfer orbit—the high, looping one—on schedule after being "spun up" to 147 revolutions per minute by small solid-fuel rockets on a turntable between Delta's second and third stages. This was the old Explorer-Jupiter C technique of spin-stabilizing an upper stage so that its motor would be aimed in the proper direction.

As the satellite coasted up toward apogee for 5 hours and 15 minutes, the United States Army communications ship *Kingsport,* a converted Liberty Ship of World War II, began beaming messages to the satellite. The *Kingsport*, at anchor in the harbor of Lagos, Nigeria, was the eastern terminal for the Syncom test; the western terminal was an Army ground station at Lakehurst, New Jersey. During the climb to apogee. Syncom received messages from the ship and relayed them back perfectly.

At 5 hours and 32 minutes after liftoff, the Cape held its breath as Syncom's apogee motor fired. This time, telemetry signals continued arriving, sweet and clear. Syncom 2 was in a nearly synchronous orbit. Like the ill-fated Syncom 1, however, it had not been turned into an equatorial plane. Its orbit was inclined 33 degrees to the equator and for this reason it would appear to move in a figure 8 north and south of the equator even though it would remain stationary in longitude when its orbit was finally adjusted. But for experimental purposes, the figure-8 double loop across the equator was as good as an equatorial orbit.

Measurements throughout the day of July 26 showed that Syncom 2 was drifting eastward in longitude at the rate of 7 degrees a day. This meant it was moving too fast, relative to the earth; its orbital period was too short: instead of 24 hours, it was 23 hours and 46 minutes. The following day, July 27, the hydrogen-peroxide gas jet was turned on for 140 seconds to boost velocity and raise the apogee of the orbit enough to increase Syncom's period to 24 hours and 23 minutes.

Now the eastward drift had been reversed to a westward one, amounting to about 7 degrees a day. From the control room on the *Kingsport*, the engineers positioned the satellite by radio commands to the gas jets so that its antenna faced groundward. Communications experiments were carried on as Syncom drifted westward across the Atlantic Ocean toward its final parking place at 55 degrees west longitude—over Brazil. When the satellite reached 35 degrees west, it was mutually visible to California and East Africa. This made it possible for the experimenters to establish the longest microwave communications link in history, from a station at Paso Robles, California to the *Kingsport* in Lagos harbor—a distance of 7700 miles.

On August 11, 1963, the controllers began to slow the westward drift of Syncom from 7 to 2.7 degrees a day by pulsing the jets so as to shrink the size of the orbit slightly.

The following day, the jets were used again to slow the westward drift to 1.2 degrees a day. At 1:45 a.m. August 15, on command from Lakehurst, the nitrogen-jet system was pulsed 264 times to bring the drift down to zero. Syncom 2 was on station in synchronous orbit and the feasibility of the synchronous communications satellite was proved.

In its first month of operation, Syncom 2 accumulated 470 hours of message time, more than that of all the previous communications satellites combined. NASA meanwhile prepared to take the next step in the Syncom experiment: to launch Syncom 3 into a 24-hour equatorial orbit, requiring a left turn as Syncom reached the equator.

A new and more powerful Delta rocket was used to launch Syncom 3, the Thrust-Augmented Thor, or TAD. The vehicle's usual first-stage thrust of 170,000 pounds was increased to 330,550 by three Thiokol solid-fuel rockets strapped to the booster. In addition, the third stage was equipped with a more powerful rocket motor, the X-258, and a more powerful hydrogen-peroxide jet system was installed.

Syncom 3 was launched at sunrise August 19, 1964, into a transfer orbit which was inclined 28 degrees to the equator. In order to swing the satellite into the equatorial plane, the second and third stages of the Delta were yawed to the left. The turn reduced the inclination of the orbit to 16 degrees.

During the first revolution, the satellite's new orbit was carefully calculated. When the vehicle reached apogee on the second revolution, the space craft was turned by its hydrogen-peroxide jet system so that the thrust of the apogee motor would complete the left turn and reduce the inclination of the orbital plane to zero. The motor was fired by ground command from Salisbury, Australia, at apogee on the third revolution. The maneuver succeeded beautifully and, while it was little noted by the public, it will be long remembered in the history of technology as a significant "first" in space flight.

In nearly synchronous orbit, Syncom 3 was allowed to drift to its permanent station over the International Date Line in the Pacific Ocean. It reached its station on September 11, three weeks after the launch, and there it was halted. In October, the satellite transmitted television pictures of the Olympic Games in Japan to Point Mugu, California. From there, the pictures were relayed to Europe via Relay 1.

Four years after the launching of the Echo balloon, a new technology of communications satellites had been created, perfected, and put into commercial use. Tall microwave relay towers, 22,300 miles high, appeared over the Atlantic and the Pacific Oceans in 1965 and 1966 and commercial service began through the new Communications Satellite Corporation. Here was the first commercial dividend from the national investment in space.

Less spectacular than the communications satellites but of potentially even greater value were the meteorological satellites. The first Tiros (Television Infrared Observation Satellite) was launched April 1, 1960, by a Thor-Able rocket from Cape Canaveral into a slightly oval orbit of 436 to 461 miles. The two television cameras on the drum-shaped, 263-pound satellite, developed for NASA by the Radio Corporation of America, permitted meteorologists for the first time to see big-scale weather systems. Cyclones with spiral bands a thousand miles wide and hurricane systems became visible from above.

In addition to the television cameras, Tiros 2 carried infrared radiometers which could measure the radiation emitted and reflected by the earth. On these experimental weather stations, the picture and radiation data were transmitted to Command and Data Stations at Wallops Island, Virginia; San Nicolas Island, California; and Fairbanks, Alaska. However, Tiros 8, launched December 21, 1963, carried automatic picture-transmission equipment which made it possible for inexpensive ground stations, usually at airports, to receive local cloud-cover pictures by radio command from the satellite as it was passing overhead.

A larger and more complex machine called Nimbus was also developed experimentally for the space agency by the General Electric Company's Missile and Space Division. Cost factors, however, persuaded the Environmental Science Services Administration (ESSA) of the Department of Commerce to adopt RCA's Tiros as the weather bureau's operational satellite. The first Tiros Operational Satellite, Tos 1, was launched from Cape Kennedy, February 3, 1966.

The meteorological satellite has proved to be an essential tool in weather forecasting. But it has not revolutionized the art of forecasting, as some enthusiastic promoters hoped. Its use in long-range forecasting appears to lie in the collection of data on temperature, moisture, and other matters for the development of mathematical models which simulate large-scale atmospheric data processing in the computer. From such models, forecasts of one to two weeks are now feasible and probably are worth more than a billion dollars a year to the American economy.[10]

NASA also developed a stationary weather and communications satellite which makes a continuous photographic record of weather systems on a hemispheric scale as they pass below. The first of these Applications Technology Satellites (ATS-1) was launched December 6, 1966, into a synchronous orbit to hang over the equatorial Pacific Ocean.

With the operational satellite system, ESSA is able to keep a close watch on tropical storms and hurricanes. Air-traffic control over the routes in the United States has become more efficient with the periodic cloud "snapshots," which controllers are now able to obtain from the satellite as it passes overhead. The cameras aloft also observe ice formation in Hudson Bay, the Gulf of St. Lawrence, and the Great Lakes. They monitor icebergs calving off the Greenland and Antarctic ice caps.

The advent of the tall towers has opened an era of the economic exploitation of space, and its rewards may outweigh the entire cost of developing space technology.

Chapter 11

Gemini

THE CRUISE OF THE "MOLLY BROWN"

It was Tuesday morning, March 23, 1965. Nearly two years had passed since Gordon Cooper's 34-hour flight in Faith 7 had wrapped up Project Mercury. Now, Project Gemini, phase two of the manned-space-flight program, was ready to launch the first crew in a two-seater Gemini spacecraft.

At 4:40 a.m., Gus Grissom and John W. Young were awakened by Deke Slayton, the astronaut crew chief. After a brief physical checkup, which found them fit to fly, they sat down to a preflight breakfast of tomato juice, cantaloupe, porterhouse steak, eggs, and toast.

Grissom, the command pilot, had christened the spacecraft the "Molly Brown," after the musical comedy *The Unsinkable Molly Brown*. The title (not the plot) recalled his narrow escape from drowning after his suborbital flight in 1961, when his Mercury vehicle, Liberty Bell 7, sank in the Atlantic Ocean. It also brought to the space program a more whimsical and appealing attitude toward the naming of flight vehicles than the unctuous symbolism NASA employed in Project Mercury in christening vehicles "Freedom," "Faith," or "Friendship," with the numeral seven referring to the original seven astronauts. Whimsy, however, never sat well with the self-conscious NASA directorate, whose members had to think about what Congress might say, nor with the agency's script-writers and image-makers, who had to think of what history might say. Nothing officially could be done about "Molly Brown," which was Gemini 3, after the press picked it up, but thereafter Gemini craft were designated only by numbers.

At 6:45 a.m., the pilots swung themselves feet-first into the spacecraft, at the 100-foot level of the service tower, the Titan 2 ICBM launch vehicle standing below. The morning was graced with a clear, golden sunrise, which lighted the sky to a pale blue. When the Complex 19 service tower rolled back, the rocket and spacecraft, 109 feet tall, became silhouetted against the morning sky like a giant finger pointing out man's fate. In place of the familiar Atlas, the more powerful Titan 2 would boost the Gemini crews on their space missions. It was a two-stage rocket with a total thrust of 530,000 pounds, or about 165,000 pounds more than Atlas delivered. Gemini was bigger and heavier than Mercury, and it had engines which would enable it to do what no other manned space vehicle had done before—change its orbit.

Conceived as an intermediate step between Mercury and Apollo, the Gemini program got under way December 7, 1961, when NASA awarded a contract to McDonnell to develop a two-man spacecraft that could be maneuvered in orbit to make a rendezvous and dock with another vehicle. This technique was essential for Project Apollo, which had adopted the lunar-orbit-rendezvous mode of landing men on the moon. In this mode, it will be recalled, two of the three Apollo astronauts would descend to the surface in a Lunar Module or Bug, leaving the third in the Apollo spacecraft, which would continue in orbit around the moon at an altitude of 48 miles. For the return to earth, the two men on the moon would blast off in the Bug and make rendezvous and dock with Apollo, casting the Bug adrift after they reentered the mother ship.

While NASA directors were dubious about musical-comedy titles, they knew that they could rely implicitly on classical symbolism. Project Gemini had been named for the Gemini (twins), Castor and Pollux, sons of Zeus, who became enshrined as twin suns in the constellation of Gemini. Shaped like Mercury, the conical Gemini vehicle contained half again as much cabin space and weighed more than twice as much. Its total weight at liftoff was 8360 pounds. It was 18 feet and 5 inches long and 10 feet in diameter at the base of its adapter section. The spacecraft consisted of two sections. The capsule where the crew rode was called the reentry section. Behind it, skirting out like the mouth of a bell, was the adapter section, which contained the new fuel-cell batteries, an array of 16 jets comprising the Orbit Attitude Maneuvering System (OAMS) and four solid-fuel retrorockets for reentry. Just before retrofire, the adapter section would be cut loose, and just after retrofire, the retrorockets would be jettisoned, too. The reentry module, with its own landing system and attitude-control jets, would then begin the long arc to the earth.

At supersonic speed in the atmosphere, the reentry capsule had lift, and could fly a little. Its center of gravity had been offset so that the amount of lift could be controlled as it hurtled through the atmosphere on its downward arc. Control was managed by rolling the vehicle. Theoretically, the crew could land the craft at any point within an area of 28,000 square miles. By rolling upside down, the pilots could increase the lift and lengthen the descent arc about 200 miles. By flying "heads up," they could shorten the arc by 200 miles. By applying continuous roll they could eliminate the lift vector and allow the spacecraft to follow a ballistic path down. The roll or "bank angles" were indicated to the crew by a computer about the size of a shoebox. It could predict the flight path at any given moment. When the craft reached a point where a ballistic trajectory would carry it to the impact point, the computer would signal the crew to establish a continuous roll to equalize lift and drag. When the craft was slowed to subsonic speed, the lift effect vanished.

The Gemini spacecraft was equipped with dual hatches, one for each astronaut, so that either or both crewmen could climb out of the vehicle on his own side This design turned out to be ideal for extravehicular activity, which had been considered as a possibility late in the program. However, five days before the launch of the Molly Brown, the Russians upstaged the American program dramatically with the 10-minute space float of Lieutenant Colonel Aleksey A. Leonov on the flight of Voskhod II. The feat spurred NASA to institute spacewalking earlier in the program than the directors had intended.

The rendezvous and docking target of Gemini was an Agena rocket, equipped with a docking collar into which the cylindrical nose of the Gemini would fit and become latched. Docked with Agena, Gemini in effect acquired another launch vehicle, one capable of propelling it into an orbit many times higher than any manned spacecraft had ever gone before. Operationally, this was the manner in which a manned spacecraft might be propelled on a lunar journey with a rocket powerful enough to do the job. But this form of Earth Orbit Rendezvous was not to be exploited in the Gemini program as a means of extending the range of manned space flight. That was left to Apollo.

Except for a false indication of a leak in the oxidizer line of the Titan 2, causing a 29-minute delay, the countdown was surprisingly trouble-free. The Molly Brown lifted off Pad 19 at 9:24 a.m. on twin pillars of translucent, violet smoke which made Titan's exhaust look thin and wispy compared to the fuming billows of the old Atlas. Up she went, straight and true, toward orbit. Young, who had not flown before, had a moment to look out the window and see the unforgettable vision of the earth as a planet, which is the initiation of an astronaut. He was genuinely awed, even though he knew beforehand what it would look like.

A member of the second group of nine astronauts selected in September 1962, John Young was 34 years old and a lieutenant commander in the Navy. He had entered the Navy after being graduated with a bachelor of science degree in aeronautical engineering from the Georgia Institute of Technology and had served as a Navy test pilot and program manager for the F4H weapons-system project. In 1962, he set the world "time to climb" record in a Navy F4B fighter aircraft. He married a Georgia girl, Barbara White of Savannah. At the time of the flight, their daughter, Sandy, was 8 years old and their son, John, who was 6, was just getting over the chicken pox. Barbara Young and the children watched the liftoff on television at home, near the Manned Spacecraft Center, Houston. "That's it," she said breathlessly as she saw the Titan 2 lift off Pad 19.

Grissom was talking to Gordo Cooper, the capsule communicator at the Cape. Each was assuring the other that all was well and there wasn't a thing to worry about. There wasn't.

"THERE ARE NO WORDS . . ."

The first stage had fired properly. The second ignited as the first fell away and began to steer the spacecraft into orbit.

"Steering is good from here," Cooper said. Grissom said he thought so, too. "Roger, Molly Brown, you are go from here."

"Roger," said Grissom. "Molly Brown is go."

As the spacecraft approached the Canary Islands, Grissom reported it had developed a tendency to yaw to the left. He was able to counter this by using the thrusters to correct the ship's attitude. Later, he discovered that the leftward drift had been caused by water being ejected into space by a water boiler in the cooling system. It was the first and the least of the many malfunctions that were to plague all ten manned flights of Project Gemini.

It was clear over Australia. Grissom and Young looked down to see the lights of Perth twinkling below them. Over Texas, near the end of their first revolution around the earth, the astronauts performed the first orbit change of the space age. Grissom fired the 100-pound OAMS thrusters for 74 seconds in the direction of flight, slowing the craft down about 50 feet a second. The braking, slight as it was, dropped the apogee or high point of the Molly Brown's orbit from 139 to 105 miles. It had the effect of circularizing the orbit, and was the kind of maneuver that would later be required for rendezvous with another orbiting vehicle. According to the rules of orbital mechanics, the pilot could change the apogee of his orbit by thrusting at perigee, the low point, 180 degrees away. He could alter his perigee by thrusting at apogee. Over Texas, the Molly Brown was nearing perigee, the lowest point of her original, oval orbit.

On the second revolution, Grissom fired the big thrusters to make a change in the plane of the orbit, amounting to one-fiftieth of a degree—another test of the vehicle's capability to achieve rendezvous.

For the most part, the conversation between the spacecraft and ground was terse. It lacked the sense of wonder and awe which had made the Mercury flights so colorful. In Mercury, the astronauts had discovered what it was like to fly in orbit. In Gemini, they concentrated on the mechanics of moving around in the new environment. And this was work, not adventure. The dialogue between ship and ground had all the zest of a couple of cab drivers reporting to the dispatcher. Cooper asked the crew to look for the second stage of the Titan on the second revolution over Florida, inasmuch as the stage had been tracked by radar, drifting along behind the Molly Brown.

"Any success on contact with the booster?" he queried as the ship sailed over Florida.

"No," said Grissom, "we were facing the wrong way."

"See anything over the West Coast?"

"Well, we saw a part of California and what I imagine was Arizona."

"Okay, Molly Brown," said Cooper.

West of Hawaii, on the third revolution near apogee, Grissom fired the OAMS thrusters in retrograde for 1 minute and 49 seconds. The braking slowed the craft by 96 feet a second, dropping perigee from 98 to 52 miles. At the new low point of the orbit, the spacecraft would eventually reenter the atmosphere and land without firing its four retrorockets, but this effect was not tested. Twelve minutes after the maneuver, Grissom jettisoned the adapter section and fired the retrorockets.

Gemini Control at the Cape instructed him to bank left 45 degrees and then reverse the bank angle to the right 55 degrees. He carried out the maneuvers, but the computer showed that the Molly Brown would land about 25 miles short of the impact point in the Atlantic Ocean, where the aircraft carrier *Intrepid* waited. This figure proved optimistic. The Molly Brown floated down on her ringsail parachute 58 miles short of the impact point, about 9½ miles from the Coast Guard cutter *Diligence,* which had been stationed 50 miles uprange from the carrier for just such a contingency.

Neither Grissom nor Young was able to account for the landing error, a serious undershoot. However, flight director Christopher Kraft suggested that the bank-angle commands may have been garbled.

Young was enthusiastic about the flight. "It was incredible. It was unbelievable," he said. "There are no words in the English language to describe the beauty of it. Man! I was impressed!"

C'MON IN, ED . . .

The flight of the Molly Brown, last of the named spaceships, proved that the United States had developed a maneuverable space vehicle which apparently could rendezvous with another in orbit. Space-agency executives were well pleased with its performance. A new aura of confidence, bordering on boldness, began to appear, in contrast to the caution which had characterized Mercury.

Addressing the Fifth National Conference on the Peaceful Uses of Space at St. Louis, May 26, 1965, George E. Mueller, associate administrator for manned space flight, characterized the Gemini 3 mission as an "historic milestone." Rendezvous and docking were scheduled for 1966. During 1965, he said, the program objectives were to determine the effects of weightlessness on long-duration flights, lasting up to 14 days, and to develop man's "capability to step out into the nothingness of space and do effective work." This referred to NASA's intention of duplicating Leonov's spacewalk of the previous March 18 as soon as possible.

The Russians had not overlooked any opportunity to exploit Leonov's feat. They brought a color film of the event and showed it at midnight May 5, at the Chicago meeting of the American Astronautical Society in the Conrad Hilton Hotel. The film made a deep impression on the 700 engineers and scientists who saw it and who could appreciate the technology which the spacewalk represented. The 20-minute document showed the cosmonaut climbing out of the Voskhod through a hatch, resembling one in a submarine. He emerged like an overfed butterfly in his bulky spacesuit. The figure of the spacesuited human became clearly defined against the sun's glare and the blackness of space. Background music had been dubbed in. It was Russian folk music, very lilting and cheerful, reminiscent of a sleigh ride through snowy woods, and it served to enhance the strangeness of the bulky figure slowly somersaulting, head over boots, in space.

Dr. Oleg Gazenko, the Soviet medical specialist who used the film to illustrate his lecture (or gave his lecture as a device to show the film), explained that the camera on the spacecraft was providing the cosmonaut with up-and-down orientation. Holding on to Voskod with both hands at first Leonov reached out with one hand, like a swimmer poised to shove off in an unknown sea, and then let go. With a little push, he began to drift away from the ship. He started to tumble and when he lost sight of the camera, he lost his orientation, according to Gazenko. Then he was truly lost, said the Moscow physician. Leonov came around in a forward roll, and saw the earth, the sun, and once more the ship. A tether held him to the ship, so that he could not drift away. He carried his own oxygen in tanks on his back, like a diver. As he described his sensations, Leonov's voice sounded high-pitched and rapid. For a long moment, the camera showed him silhouetted against the sun, alone in a vast emptiness. No stars appeared in the violet dark, only the white-suited man and the greenish ship. Sometimes above, sometimes below, sometimes to one side, the earth slid by, the surface obscured by white puffs of clouds. During the ten minutes Leonov floated, he crossed the Eurasian continent. He reported that he could see the Black Sea coast of the Crimea when he came out of the hatch. He floated across the Ural Mountains and the Siberian steppes. As he reeled himself in and crawled back into the ship, his last sight below was of the Yenisei River. He covered about 2500 to 2900 miles outside the spacecraft.

NASA reacted to this spectacular feat by moving up the launch of Gemini 4 to June 3, 1965, more than a month ahead of schedule, and rushing preparations for an American space float.

As initially planned, the flight of Gemini 4 was scheduled to last four days "to evaluate the effects of prolonged exposure of the flight crew to the space environment in preparation for flights of longer duration." The decision by NASA to add an extravehicular activity to the flight plan was simply a conditioned response to the Russian stimulus. However, the space agency spokesmen explained that extravehicular activity, or EVA as it came to be called routinely, was planned in 1964, when development of the EVA suit and life-support equipment was started. Nevertheless, when the flight of Gemini 4 was first programmed, there was still no Eva suit. One was adapted from the flight suit designed for Gemini. It consisted of a cotton undergarment, a cotton comfort layer, a rubber pressure layer, a layer of link net, seven layers of aluminized mylar insulation which were interspaced with six layers of dacron, a thick felt layer (to absorb micrometeorites), and a tough, white, reflective outer layer of nylon. The suit was designed to create a microenvironment in which a man could live in space. It pressurized his skin so that the internal pressures of his body would not rupture his blood vessels, it retained his body heat so that he would not be frozen stiff, and it enabled him to breathe and get rid of some of his excess body moisture.

The crew of Gemini 4 consisted of two Air Force test pilots, Major James A. McDivitt, 35, a Korean War combat veteran, and Major Edward H. White II, 34, a graduate of the United States Military Academy with a master of science degree in aeronautical engineering from the University of Michigan. McDivitt was command pilot, and White, pilot.

White was slated to perform the EVA if it was to be done on that flight. No one could be sure that the space walk would be included until May 24, when qualification tests of all the EVA equipment were completed. White had rehearsed the art of space floating in an altitude chamber, which simulated the partial vacuum at 180,000 feet. However, the actual EVA exercise would be taken at 528,000 feet. In all, White spent 60 test hours in the EVA suit. He rehearsed climbing out and back into the spacecraft 110 times before the flight. Even then, the Gemini flight directorate would not commit itself to the EVA experiment until after Gemini 4 was launched.

White's EVA suit weighed 34 pounds, 10 pounds more than the ordinary Gemini flight suit. It was fabricated by David Clark Company, Worcester, Massachusetts. A tether or "umbilical cord" one inch in diameter kept the astronaut attached to the spacecraft. Wrapped in plastic tape coated with gold, the tether included a half-inch-wide, flat, nylon line capable of holding one-half ton, a silicone-rubber oxygen hose one-quarter of an inch in diameter, four electrical leads, and a communications lead. It was 25 feet long and weighed 9 pounds on the ground. While the umbilical cord assured the space floater of a supply of oxygen from the spacecraft, a chest pack 13 inches long, 6 inches wide, and 2 inches deep provided emergency oxygen and a means of regulating the suit pressure. White also would carry a means of propulsion which technicians at the Manned Spacecraft Center called the "Zot" gun. It was a twin-barreled jet shaped like the handle bars on a bicycle and it enabled the space man to propel himself from one point to another by squirting jets of oxygen in the direction opposite to the one he wished to go. Officially, the Zot gun was called the Hand Held Maneuvering Unit, or HHMU. It weighed 7 pounds on the ground and was rigged to carry a camera. "Lashed up" at the Manned Spacecraft Center, the Zot gun proved the best maneuvering device NASA brought into play in the Gemini program. The only trouble with the early model was that it did not hold enough gas.

Delayed by electrical trouble in the erector tower, the launch of Gemini 4 was executed an hour and 16 minutes late at 10:16 a.m. on June 3, 1965. As soon as the spacecraft was freed from the second stage of the Titan 2 booster, McDivitt fired the rear OAMS thrusters to add 5 feet per second to his velocity and pull away from the following rocket. Then he turned the capsule around so that it rode blunt end forward, pilots facing to the rear.

The stage was tagging along behind them, tumbling. It carried a flasher beacon to make it more visible. First objective of the four-day flight was to make a rendezvous with the 27-foot cylinder on the second revolution by flying within 20 feet of it. However, McDivitt found that the spent stage kept drifting away from him during the first revolution. He kept on thrusting now and then to stay within 500 feet of it, but this procedure began to use up a great deal of fuel. "It's taking a lot more fuel than we anticipated," he told the Guaymas, Mexico, tracking station. "We're just holding our own and it's farther out than we want. Do you want us to close with it, or save fuel?"

Flight director Kraft told McDivitt to conserve fuel. "We're more concerned with the lifetime [of the flight] than with matching the booster," he said. "I guess we'll scrub it [the rendezvous].'"

"We've used 50 per cent of our OAMS fuel," McDivitt said.

"Knock it off," advised Grissom at the Cape. "It's not worth it."

Ed White was then advised to prepare for his swim in space, which the flight plan called for him to execute on the second revolution as Gemini 4 passed over the United States. There was a major constraint on the EVA, however, and the late launching did nothing to ease it: flight directors wanted White back inside the spacecraft before it entered darkness. But it took the crewmen so long to go through the 40-item EVA check list that they decided to postpone the EVA until the third revolution. This reduced the daylight time for the exercise by 90 minutes. It meant that White would be performing the spacewalk toward mid-afternoon, Eastern Standard Time, and would have to get back inside by the time he reached Bermuda in order to complete it in full daylight. Kraft did not want him "fooling around in the dark" and did not want the Gemini hatch open on the night side. He wanted to keep hazards at the minimum, since no one knew what to expect in this new operation.

"It sounds like you've been awfully busy," said Gus as Gemini 4 came over Florida north of the Cape at the end of the second revolution and sailed down the Atlantic Missile Range.*

"Yeah, I finally got a chance to look out the window," replied McDivitt. "It's really nice."

Four hours after liftoff, as Gemini 4 passed over the Carnarvon, Australia, tracking station, McDivitt reported that "all hands" were ready for the EVA adventure. He depressurized the cabin from 5 to 2 pounds per square inch to test suits. When he and White were satisfied that the suits were holding, McDivitt flushed the remaining oxygen out of the cabin. In this respect Gemini 4 differed from Voskhod 2, which maintained its near-normal atmosphere of oxygen and nitrogen at a pressure close to that at sea level during EVA. In order to leave the Voskhod, Leonov had to squeeze through a submarine-style pressure chamber and, once in space, he breathed pure oxygen, like a diver.

In Gemini 4, egress to space was simpler. When all was ready, White simply unlatched the hatch over his head. He

* The Atlantic Missile Range became officially designated as the Air Force Eastern Test Range on May 15, 1964. The old name for this 6000-mile "shooting gallery" has persisted, however.

was startled when it suddenly popped up three or four inches as a result of pressure from excess oxygen being vented by his suit. McDivitt reached over and held the hatch from flying open, grasping a lanyard used to pull it down for closing. There was some danger of the hatch being damaged if it was allowed to swing back rapidly. White climbed up on his seat and mounted the outside camera, then took a short series of pictures with it while he stood in the seat, his head and torso protruding into space through the open hatch. McDivitt reported that all was well and Kraft told White to go ahead and climb out. Gemini 4 was passing north of Hawaii when he propelled himself away from the hatch firing the Zot gun.

White moved down the center line of the spacecraft under the propulsive force of the gun. He found he could move in yaw (right or left) and in pitch (up or down) adequately with the gun. "If you can control your pitch and your yaw and translate fore and aft, you can actually get from point A to B," he said later. The influence of his tether was to pull him up above the spacecraft. It was not attached to the vehicle in the best position and tended to pivot White toward the adapter section of the spacecraft, where he did not want to go. The Zot gun quickly ran out of fuel and White had to maneuver himself by the tether only. He worked his way out to the nose of the spacecraft with the idea of determining whether he could hold on to it while McDivitt maneuvered the spacecraft by firing the OAMS thrusters. "But . . . I took one look at the stub antenna which was our connection with radio back to earth and I felt that this wasn't any place to play around with and so I didn't do any work around the nose of the spacecraft." Instead, he pushed off from the vehicle. He found that with the tether influencing all his movements it was difficult to move in the direction he planned to go unless the tether was perpendicular to the surface of the vehicle. At an angle to the surface, the tether caused him to roll when he pushed off. White pulled himself back by means of the tether and pulled on it to anchor his feet on the metal skin of the craft. In this way he could walk across the ship until the angle of the tether changed and he found his feet going out from under him.

One aspect of the EVA worried him at the time. He had been told that his conversation with McDivitt and the ground would be carried live on radio and television. He wondered,

"What do you say to 194,000,000 people when you're looking down at them from space?" He decided it was the wiser policy to say nothing to them but to carry on as a test pilot and confine his remarks to McDivitt. That turned out to be a splendid decision since White was able to communicate spontaneously and unself-consciously the feelings of wonder and delight he experienced while floating free in orbit.

"Hey," he said, "I'm looking right down and it looks like we're coming up on the coast of California. I'm going in a slow rotation to the right. There is absolutely no disorientation associated with this thing [as Leonov was reported to have experienced]. There's no difficulty in recontacting the spacecraft. I'm very thankful in having the experience to be first. Right now, I'm actually walking across the top of the spacecraft. I'm on top of the window."

"Hey," protested McDivitt inside, "you just smeared up my windshield, you dirty dog. You see how it's all smeared up there?"

"Yep," acknowledged White, tumbling slowly, head over boots.

"Ed," continued McDivitt, "I don't even know exactly where we are but it looks like we're about over Texas. Hey, Gus, we're right over Houston."

"We're looking right down on Houston," said White, a native of San Antonio.

"Yeah," said McDivitt. "That's Galveston Bay down there."

White said, "The sun in space is not blinding. It's quite nice." He was talking to McDivitt, and the two conferred a few moments. While doing so, they could not receive ground communication. McDivitt told White he had better see if Houston wanted anything. Houston said, "The flight director wants for White to get back in."

Neither crewman aloft seemed to hear that instruction. White continued to float and roll like a happy porpoise in slow motion. He was having the time of his life. Grissom's voice came up from the Cape calling, "Gemini 4, Gemini 4 . . ."

"Hey, Gus," said McDivitt. "This is Jim. Got any messages for us?"

Gemini 4 was passing over Florida, between the Cape and Miami, and Ed White had floated all the way across the

United States, from sea to sea. Starting near Hawaii, he had been outside the capsule for 20 minutes, doubling Leonov's time out.

"Gemini 4, Gemini," called Grissom. "Get back in."

"Okay," one of the crewmen responded. There was a silence. White, still cavorting outside, said, "Hey, this is fun."

"Well, come on back in, Ed," urged McDivitt.

"I don't want to come back to you," said White. "But I'm coming."

"You said that three and a half or four days ago, Buddy," said McDivitt.

"I'm coming," said White.

"You've got four minutes to Bermuda," said Grissom.

"I'm trying to," said White.

"Whoops. Take it easy, now," said McDivitt.

"The hand hold on that spacecraft is fantastic," said White. "Aren't you going to hold my hand?"

"Ed," said McDivitt sharply, "come on in here."

"I'll open the door and come through there," said White. (He was kidding. The hatch was open.)

"Come on, Ed. Let's get back in here before it gets dark," McDivitt urged.

"It's the saddest moment of my life," said White.

"Are you getting him back in?" asked Grissom impatiently.

"Yeah," said McDivitt. "He's standing in the seat now and his legs are down below the instrument panel."

White sat down and began to pull the hatch shut, but it resisted. It took the combined efforts of both pilots, pulling hard on the lanyard, to get it shut because of the pressure of exhaust gases vented by their spacesuits. White had to use both hands to operate the bar-and-lever latching mechanism and McDivitt had "to pull like the devil" on the lanyard to hold the hatch tightly shut so that it would latch securely.

White later explained that "the conversation that you heard between Jim and me coming back in was that I was disassembling the camera. I had certain things I had to put up—it took me about five minutes to get them all put up." [1]

The spacewalk demonstrated without question that a trained man can maneuver in orbit, free of the capsule, and this opened up a new horizon of human activity in the interplanetary medium. It made the assembly in orbit of a space station from prefabricated parts a reasonable enterprise.

There was, however, one hint of problems to come. In mounting the movie camera on a bracket behind the door, White had to exert himself and McDivitt cautioned, "Hey, you're starting to breathe pretty hard." White said he did not believe he was tiring himself too much at the time, "but I did realize that I was putting out energy for the first time during the EVA exercise." [2]

The remainder of the four-day flight was comparatively uneventful. The astronauts reported two unidentified orbiting objects, one quite large, which they saw on their 38th revolution over India. The objects were believed to be pieces of debris from American or Russian launch vehicles—space junk. The crewmen were brought in by ground control on a ballistic-landing trajectory after their computer failed. They splashed down at 12:12 p.m. Monday, June 7, 1965, on their 63rd revolution, 390 miles east of Cape Kennedy. They were 50 miles short of the planned impact point. However, within 38 minutes of splashdown, they were picked up by a Navy helicopter and flown to the aircraft carrier *Wasp*. They climbed out of the flying machine and, without displaying a trace of vertigo, nausea or weakness, walked along a red carpet which the grinning Marines had unrolled for them.

The space agency was delighted with the results of Gemini 4, which succeeded on all counts except for the rendezvous effort with the second stage of the Titan. Next on the spaceflight agenda was the mission of Gemini 5—a flight of eight days. If it was successful, it would break the Soviet orbital flight record of five days (81 revolutions) established by Lieutenant Colonel Valery F. Bykovksy in Vostok V in June 1963.

EIGHT DAYS IN AUGUST

Gemini 5, with Cooper, promoted to lieutenant colonel, and Navy Commander Charles Conrad, Jr., in the cockpit, was boosted off Pad 19 at 9 a.m. August 21, 1965. Its major objective was to test the performance of men and machines in orbit for eight days.

Conrad, 35, was graduated from Princeton University with a bachelor of science degree in aeronautical engineering. He served as a Navy flight instructor and had been in the second group of astronauts selected in September 1962. He and his wife, the former Jane DuBose of Uvalde, Texas, have four boys: Peter, Thomas, Andrew, and Christopher.

On the first revolution, trouble appeared in the fuel-cell battery which was being used in manned flight for the first time. The battery produces electricity by the reaction of hydrogen and oxygen gases, which are forced into stacks of cells under pressure. A by-product of the reaction is water. No sooner had the flight settled into its first orbit than Cooper reported that pressure was falling rapidly in the fuel-cell oxygen supply from 800 pounds per square inch, considered optimum, to about 70 pounds. The pressure, which was necessary to move gaseous oxygen out of the tank where it was chilled and stored in liquid form, was built up simply by heating the tank with an electrical wire, so that some of the oxygen would vaporize and expand out of the container through a connecting line to the battery. As the oxygen pressure dropped, it seemed likely that the flow of oxygen into the fuel cells would be reduced to the point where the battery could not operate. If that happened, the crew would have to switch to the spacecraft's reentry-system storage batteries, and land.

The fuel-cell battery consisted of two sections, each housed in a cylinder 2 feet long and 1 foot in diameter. Together, they produced 2 kilowatts of electricity at a peak of 26½ volts. Each section contained three stacks of 32 cells each, connected in series. The electrochemical reaction took place in the cell, where, in the presence of a catalyst, the hydrogen atom would release an electron to the anode (positive terminal). Hydrogen ions (nuclei) remaining would migrate through a membrane separating hydrogen from oxygen gases into the oxygen side, where they would combine with oxygen, and with electrons passing through an external circuit, to produce water.

According to the manual, it was necessary to maintain a pressure of 800 pounds per square inch to keep this reaction going. The fall in pressure indicated a faulty heating element in the fuel-cell battery tank where 46 pounds of oxygen were stored, but since the tank was in the adapter section, beyond the reach of the crew, nothing could be done about it. At Gemini Control, flight director Kraft realized he was facing a major crisis. Should he order the astronauts down at the end of their third revolution, so that they could land near the prime recovery zone, or allow them to continue the flight as long as the fuel-cell battery continued to put out electricity? The success of the previous two flights had provided Gemini Control with confidence in the spacecraft, and Kraft knew the crewmen about as well as he knew himself. He was determined to keep Gemini 5 aloft as long as it was reasonably safe. Kraft ordered Cooper to "power down" while experts on the ground studied the problem. NASA engineers conferred with experts of the General Electric Company, the manufacturer, who believed that the battery would continue operating even at reduced oxygen pressure. It all depended on whether the pressure stabilized at some level where power would still be produced. After several hours, the oxygen pressure seemed to stabilize at 60 pounds per square inch. Even though that was only one-twelfth normal pressure, the battery continued to make electricity. On Sunday, August 22, the oxygen pressure rose to 82 pounds and prospects for the mission brightened.

Meanwhile, shortly after the problem had appeared, Cooper deployed a small device called a "radar evaluation pod." He and Conrad were scheduled to practice rendezvous with it. The pod was actually a radar transponder—amplifying and reflecting radar signals beamed to it, so that it was readily detectable by the onboard radar set. The rendezvous maneuver was postponed while the life expectancy of the fuel-cell battery looked doubtful. When it became apparent that the battery would continue to provide enough power, even at reduced oxygen pressure, for the rendezvous maneuvers, the pod's batteries had become exhausted. Sunday morning, as Gemini 5 was flying over Australia, the Carnarvon radar station reported that the pod had drifted some 75 miles away from the spacecraft. Dead as a radar mirror, the pod was no longer usable as a rendezvous target.

New troubles related to the fuel cells appeared late Sunday.

THE PHANTOM AGENA

Because the fuel-cell battery had been operating at low power for 25 hours, it had not used the expected amount of hydrogen, and pressure in the tank began to build up. Like oxygen, the hydrogen had been stored as a superchilled fluid, but at a temperature 126 degrees colder than liquid oxygen. Heat from the ship was vaporizing the hydrogen, and pressure building up in the tank caused it to vent some of the gas. The venting had the effect of a jet and soon began to disturb the attitude of the spacecraft. In order to deal with this problem, Kraft instructed Cooper to bring the fuel cells up to full power, which would use up the gaseous hydrogen. To everyone's astonishment, the battery performed perfectly at full power even though the oxygen-tank pressure remained about one-tenth of what the manual said it should be. With all the power they needed, Cooper and Conrad put the spacecraft through a series of orbit changes, aiming at a hypothetical rendezvous with a nonexistent target, represented by chart coordinates. The hypothetical target was referred to as the "phantom Agena." Even though the crew used more fuel for the exercise than planned, it looked successful on paper. If there had been an Agena there, Cooper and Conrad could have made a rendezvous with it.

By Tuesday, August 24, it was apparent that the fuel cells would continue working even on a trickle of oxygen, but another problem appeared: a fuel shortage was threatened. The next day, the hydrogen tank began to vent again, like an overheated tea kettle, causing the ship to execute a tumbling roll.

As the tumbling rates increased, the crew had to reduce them by applying thrust—which used more precious fuel.

Another problem appeared late Wednesday. The fuel cells were producing more water than expected, as the by-product of electricity generation. The by-product water, undrinkable because it was contaminated by the chemical catalyst, was piped into a rear tank. In order to save space, the crew's drinking water was also stored in the tank but in a bladder, which prevented it from being contaminated by the fuel-cell water. The storage arrangement was fiendishly ingenious. As Cooper and Conrad drank the water from the bladder, it would decrease in volume and thus provide more room in the tank for the water from the fuel cells. However, because the fuel cells were making more water than expected, the tank was filling faster than the bladder was being emptied. The only outlet for water was through the urinary tracts of the astronauts, and the drinking water from the bladder eventually was urinated into a container in the cockpit. Under this waterworks arrangement in Gemini 5, there was only one solution to the water problem created by the fuel cells. That was for the crew to drink heartily, and Dr. Berry, the flight surgeon, kept exhorting them to do so every chance he had.

On Thursday, August 26, Cooper and Conrad were hailed by Gemini Control as they passed over Texas for exceeding the five-day flight of Bykovsky in Vostok V. Cooper acknowledged the salute philosophically and assured Dr. Berry he was drinking all the water he could hold. He passed on the news to Kraft that No. 7 and 8 thrusters, which yaw the ship to the left, had just conked out. Later in the day, Conrad added that No. 3 and 4 thrusters which yaw the ship to the right were showing signs of failure.

But there was good news from Dr. Berry. Both astronauts were in fine physical shape, he reported, probably because of all that water they were drinking in the race to keep ahead of the fuel cells. The records of the flight of Gemini 5 would show that the human components were holding up better than the mechanical ones. Still, Gemini 5 flew on and on around the earth, toward an eight-day record. As long as the major systems continued to function, Kraft saw no reason to end the flight early.

Once more, Kraft told Cooper to cut down the power, this time in order to reduce water production by the fuel cells. When Cooper powered down, a new problem appeared. The spacecraft became cooler and those maneuvering thrusters which were still working became sluggish. Cooper feared they might freeze up altogether, rendering the OAMS thrusters useless for further flight. In that case, the crew would have to prepare for reentry, using the reentry controls in the nose of the capsule for attitude control.

On Friday, August 27, the pilots allowed Gemini 5 to drift in orbit as they took turns sleeping. Under the jet of the venting fuel-cell hydrogen tank, the ship tumbled slowly. Gemini controllers had estimated that by Friday the venting should stop. But because the cells were being used at low power, to avoid water problems, less hydrogen had been expelled from the tank than expected and the tank continued to vent gas. There was no way to beat the fuel-cell problem without generating a new one.

By cutting power to reduce water, the crew protracted the hydrogen venting which threatened premature exhaustion of the OAMS fuel. The dilemma may one day appear in sagas of the Early Space Age as "Gemini's Choice." Neither alternative was a good one.

On Saturday, August 28, the seventh day of the mission, it was agreed aloft and on the ground to power up, come hell or high water. As soon as the crew did so, the hydrogen venting stopped and the ship was quickly stabilized. There were sighs of relief in the tracking network around the world.

Since early Friday, Cooper and Conrad had been listening to snatches of Dixieland music, broadcast to them over high-frequency radio as a test of reception in that mode. McDivitt, capsule communicator at Gemini Control, was the disk jockey. In their spaceship, with its near-to-overflowing water tank, its jammed thrusters, and its balky battery, the sound of music was immensely reassuring. Cooper loved Dixieland and Conrad didn't mind it. As music to orbit by, it sounded bathetic, but the idiom was right, and it helped to take everyone's mind off the trouble upstairs.

At 7:55 a.m. Sunday, August 29, Cooper and Conrad splashed down in Gemini 5 about 414 miles southwest of Bermuda and 80 miles from the aircraft carrier *Lake Champlain,* which was on station waiting for them. They had flown 120 revolutions of the earth in just 65 minutes short of eight days. Bearded and feeling "crummy" in their flight suits, the 38-year-old Cooper and 35-year-old Conrad were able to walk away from their spacecraft after the longest period of confinement in orbital flight by human beings up to that time. It was only a short walk but an historic one. They climbed stiffly out of the spacecraft after Navy frogmen fastened a flotation collar around it and they walked around the collar.

The two had faithfully carried out 16 scientific experiments, including geologic, oceanographic, and meteorological observations, and had verified Cooper's claim that it was possible to see the wakes of ships, thin columns of smoke, aircraft contrails, and even aircraft, from orbital altitude. But the most significant experiment was the physical condition of the astronauts themselves. The eight-day flight had produced no physiological problems. Gemini 5 proved the critical point that a round-trip voyage to the moon was well within the physical capability of trained men. That in itself was new knowledge in the world.

THE RENDEZVOUS

Having demonstrated the feasibility of a lunar trip, the corps of astronauts was ready for the next step in Project Gemini: rendezvous with an Agena target rocket. Although the first rendezvous try had been set for early in 1966, the program was going so well that the Agena rendezvous was programed for the flight of Gemini 6 in 1965. The mission was scheduled to lift off Pad 19 on October 25, with Walter M. Schirra, Jr., 42, promoted to Navy captain, as command pilot, and Thomas P. Stafford, 35, an Air Force major with a bachelor of science degree from the Naval Academy, as pilot. Stafford was married to a home-town sweetheart, Faye L. Shoemaker of Weatherford, Oklahoma. Their two daughters, Dianne and Karin, were then 11 and 9 years old. Stafford was an expert on performance flight testing, and was the co-author of two books in the field. He had served as chief of the Performance Branch of the Air Force Research Pilot School at Edwards Air Force Base, California.

The chase of the Agena rocket by Gemini required a double launching. The Agena was launched first from Pad 12 by an Atlas ICBM at 10 a.m. on October 25 toward a 185-mile-high circular orbit. Two miles from the Atlas pad, Schirra and Stafford waited in Gemini 6 atop their Titan-2 launch vehicle to commence the six-hour chase of the Agena 101 minutes after it was boosted.

The Agena target vehicle was a modified version of the Air Force Agena D, and was similar to the second-stage rockets which had propelled Ranger to the moon and Mariners to Venus and Mars. The rocket was 26 feet long and 5 feet in diameter and weighed 7000 pounds, almost as much as Gemini, when fueled and in orbit. Agena had two propulsions systems. The primary one was able to develop 16,000 pounds of thrust and a secondary one delivered 400 pounds maximum for attitude control. The fuel used was UDMH (unsymmetrical dimethylhydrazine), and the oxidizer was IRFNA (inhibited red fuming nitric acid) in the main propulsion system and MON (mixed oxides of nitrogen) in the secondary system.

When the Gemini had made rendezvous and then had docked with Agena, the pilots could control the Agena engines and use them to propel the combined vehicle system in a variety of maneuvers. The Agena main propulsion system could add about 2368 feet per second to the orbital velocity of the joined vehicles (25,000 feet per second). It would enable the astronauts to reach a considerably higher altitude than men had flown before. A more powerful rocket, capable of adding 11,000 feet per second to Gemini's orbital velocity, would have made it possible for the crew to fly around the moon. That, however, was not the object of the Gemini rendezvous and docking program. Its function was simply to prepare flight crews for rendezvous and docking in Project Apollo.

To achieve a rendezvous with the Agena, the Gemini 6 crew had to make a series of precise changes in their initial orbit in order to match the orbit of the target at a point quite near it. Although the exercise followed the principles of Newtonian physics, which every high-school student is supposed to know but rarely does, the whole concept of rendezvous appeared alien. There was no perceptible counterpart in human experience. The quirks of orbital mechanics, as in the case of communications satellites, had to be learned. To overtake a vehicle ahead, one had to slow down, and to slow down in orbital flight, relative to the ground, one had to speed up. In orbits near the earth, in this respect, vehicles travel faster than in orbits farther away. It was difficult for many laymen to visualize that while a higher orbit made the vehicle go slower, it required more energy to achieve. Hence, there was the apparent contradiction of speeding up to slow down. Conversely, in order to speed up the revolution of the spacecraft around the earth, the astronaut must apply braking force and drop to a lower orbit. The difference in orbital speed from higher to lower orbits is the result of gravity, which accelerates the ship faster when it is nearer the earth than when it is farther away.

Thus, it was impossible to chase Agena simply by thrusting after it if it was any distance away. For, by using the thrusters, the crew would add energy to the spacecraft and increase the size of its orbit, in which it would then be moving more slowly than the target. In order to catch up with a vehicle ahead of it, the spacecraft had to travel for a time in a lower orbit, and then translate into the orbit of the target on an interception course. Only at short range, under a mile or two, was it possible to thrust toward the target and reach it.

Schirra and Stafford sat silent in Gemini 6 as the Atlas thundered upward, bearing the Agena target vehicle toward orbit. They were to wait until the Agena completed one revolution of the earth and was passing over the Cape before they would be launched. In Project Gemini, the term "revolution" rather than "orbit" came into use to describe the number of times a spacecraft was moving around the earth. Revolution was different from orbit because it was referenced to a particular point on earth. While it took about 90 minutes for a space vehicle to complete one orbit in space, it took six minutes longer for the vehicle to reach the point on the earth's surface it started from. The reason for this was that in the time the vehicle had been flying around the planet, the planet itself was rotating and the point of departure on the surface had moved 22.5 degrees to the east. Thus, it took 96 minutes for the vehicle to make a complete revolution of the earth and reach the point from which it started. An orbit was 360 degrees in space, but a revolution required the spacecraft to travel that additional 22.5 degrees in order to circumnavigate the rotating planet.

While Agena circled the earth at 185 miles altitude, Gemini would be launched into an elliptical orbit 100 by 168 miles. At the time of the Gemini launch, Agena would be 1,050 miles ahead, but, because Gemini was in a lower orbit, the spacecraft would be moving faster around the earth and would overtake Agena. On the second revolution, the crew would fire the rockets at second apogee, a maneuver which would increase the altitude of perigee from 100 to 134 miles. This would reduce the catch-up rate. Then the crew would fire the rockets at third apogee to raise perigee to 168 miles, circularizing their orbit at that altitude, only 17 miles below that of Agena. In the new orbit, Gemini would approach to within 161 miles of Agena. On the fourth revolution, as Gemini approached to within 39 miles of the target, the crew would fire the thrusters to add 32 feet a second to the spacecraft's velocity. Gemini 6 would then glide up to the 185 circular orbit slightly ahead of Agena and the crew would reduce the velocity difference between the vehicles to less than one foot a second. That would be rendezvous. They would then dock with Agena at the speed at which one puts his car into the garage.

The 10 a.m. launching of the Agena followed a flawless countdown. After the vehicle vanished into the heavens, there was a brief moment of silence at the crowded Cape press site as more than 500 news correspondents waited for word that Agena was in orbit. The word never came. Agena vanished between Bermuda and Antigua, and radars saw a number of pieces on its track, an indication that Agena had blown up when its main propulsion engine ignited to boost it into orbit after it separated from the Atlas.

Nevertheless, the radar could have been seeing "chaff," and the tracking network kept on looking for Agena. It was only after 45 minutes, when the Carnarvon, Australia, station radioed back "no joy" (no contact) that Gemini control scrubbed the mission and ordered Schirra and Stafford to dismount from their perches in Gemini 6.

There was no joy in Capesville, it was said, for Agena had struck out. But there was always another day.

RENDEZVOUS 7–6

It came in December. While the Air Force Systems Command launched an investigation to find out what had gone wrong with Agena, the space agency went ahead with preparations to launch Gemini 7 on a long-duration flight lasting 14 days. Since Gemini 6 and its Titan 2 rocket were still available, and an Agena target rocket was not, NASA made the bold decision to carry out a rendezvous with two manned vehicles, using Gemini 6 to chase Gemini 7.

Accordingly, Gemini 7 was launched on its 14-day mission at 2:30 p.m. EST Saturday, December 4, 1965, with Air Force Major Frank Borman, command pilot, and Navy Commander James A. Lovell, Jr., pilot. Both astronauts were 37 years old and had been selected in the second group of astronauts in the fall of 1962. Borman was a 1950 graduate of the United States Military Academy who joined the Air Force. In 1957, he received a master of science degree in aeronautical engineering from the California Institute of Technology and served as an instructor in thermodynamics and fluid mechanics at West Point. He was a graduate of the Air Force Aerospace Research Pilots School and an instructor there. He had married the former Susan Bugbee of Tucson, and had two sons, Frederick, 14, and Edwin, 12.

Lovell, born in Cleveland, was a graduate of the Naval Academy and a test pilot at the Naval Air Test Center, Patuxent River, Maryland. He was also a graduate of the Aviation Safety School of the University of California. He was married to the former Marilyn Gerlach of Milwaukee. At the time of the flight, they had three children, Barbara Lynn, 12, James A., 10, and Susan Kay, 7, and were expecting their fourth.*

As soon as Pad 19 had cooled, technicians were swarming over it to prepare for the launching of Gemini 6. The flight of Gemini 7 settled down into a routine. Borman had four electroencephalograph electrodes pasted to his scalp so that ground stations could monitor changes in his brain-wave pattern during working, rest, and sleep periods. One of the objectives of the experiment was to correlate depth of sleep with pilot performance. Another was to compare brain-wave patterns which Borman exhibited during weightlessness with those made previously on the ground. The experiment could not be completed because on the third day of the flight Borman accidentally pulled the electrodes loose. The analysis of the brain waves recorded showed that deep sleep on the first "night" in orbit was limited to about one-half hour, and most of the sleep period was spent in a drowsy state. On the second "night," sleep began quickly after a brief period of drowsiness and rapidly became deep. It then lightened over a period of 90 minutes in "a typically normal cyclic fashion." [3] "Three further such cycles ensued before the astronaut awakened through a drowsy period." The records did not indicate whether the astronaut was dreaming and he did not recall dreams.

On Monday, the third day of the flight, astronaut Elliott See, the capsule communicator at Houston, read the day's news to the crew. He advised Borman that his two sons were settling down in school after watching Dad lift off Saturday afternoon. Flight director Kraft advised that the supplies of

* Jeffrey Carl arrived January 14, 1966.

consumbales—food, water, oxygen, and hydrogen—were good for 20 days. No need to hurry back. Lovell asked if his wife, Marilyn, was still expecting. "You'll be the first to know if she's not," replied Elliott See.

NOT ON SUNDAY

The countdown for the launch of Gemini 6 began on Saturday, December 11, when the Gemini 7 crew had been in orbit a week, and Schirra and Stafford once more climbed into their capsule early Sunday morning. Fundamentalist radio preachers in Florida condemned the launch attempt as a desecration of the sabbath, but no one paid any attention to them. This time, the "friendly target vehicle" was Gemini 7, flying overhead. At 9:45 a.m., the Titan 2 engines ignited and Schirra called out that the spacecraft clock had started. There was a gust of violet smoke and then the Titan engines shut down. Observers were stunned. In the seconds that followed, Schirra was on the verge of ejecting himself and Stafford from the capsule, which was equipped with aircraft-type ejection seats in the event of a fire or explosion on the pad or during liftoff. However, the panel lights showed the quick-witted command pilot that the Titan engines were not building thrust and in a millisecond decision Schirra decided that the right thing to do was sit tight. He was right. There was no danger of any explosion or fire.

After he and Stafford again dismounted from Gemini, technicians discovered that the immediate cause of the engine abort was the premature dropping of an electrical plug at the base of the rocket. Later, technicians discovered also that a plastic dust cover was obstructing the oxidizer-inlet line of a gas generator in the first stage. It would have caused the engines to shut down a fraction of a second later if the plug had not fallen out. Gemini 6 simply was not fated to get off the pad on Sunday.

Gemini 6 was made ready for a third launch attempt on Wednesday, December 15. In preparation for it, Borman thrust Gemini 7 into target orbit, which was nearly circular at altitudes between 185.9 and 187.3 miles. On the third try Gemini 6 was boosted smartly off Pad 19 by a flawlessly functioning Titan 2 at 8:37 a.m. Wednesday as Borman and

Lovell passed over the Cape on their 164th revolution. When No. 6 was injected into orbit, it trailed No. 7 by 1380 miles.

A crucial test of the Gemini program was at hand. Gemini 6 entered an orbit shaped like the capital letter "O" with perigee at 100.5 miles near the Cape and apogee at 165.6 miles over the Indian Ocean. When Borman heard the orbit figures, he exclaimed, "Wonderful!" It was not as precise as the flight plan called for, but it was good enough to make rendezvous a fairly straightforward exercise.

Schirra and Stafford began overtaking Borman and Lovell rapidly, since the No. 6 crew was flying the lower or "inside" orbit. Schirra made his first orbit change near perigee south of New Orleans on the second revolution when he fired the aft thrusters to add 14 feet a second to Gemini's velocity. The effect was to raise apogee slightly to 168 miles. Over the Indian Ocean some 45 minutes later at apogee, he burned the aft engines again, this time for 1 minute and 17 seconds. Velocity increased by 60.8 a second. The effect of this maneuver raised perigee by 34 miles to 134.5 miles. The third maneuver moved the Gemini 6 orbital plane eight-hundredths of a degree north to match that of No. 7.

"How are we doing?" Schirra called down as Gemini 6 passed over Texas. "Just great," replied flight director Kraft.

With a "tweek" burn—a small push—to adjust apogee more precisely, Schirra and Stafford continued to gain on Gemini 7 on their third revolution. As they approached to within 230 miles, Schirra announced he had "locked on" to the target with the ship's radar. He then fired his aft thrusters over the Indian Ocean so that perigee would be raised to 168 miles, circularizing the orbit. However, the circularization was slightly lopsided, with dimensions of 165 and 170 miles. At Gemini Control, flight technicians computed a terminal maneuver for the No. 6 crew which would enable the vehicle to intercept its target on the fourth revolution over the Pacific Ocean.

As Schirra began the terminal maneuver, the two vehicles were only 37 miles apart. At 23 miles, each crew could see the other spacecraft coming nearer. Schirra then boosted No. 6 into the orbit of No. 7. At 2:30 p.m. December 15, 1965, Stafford called out, "One hundred and twenty feet apart and fitting." The first orbital rendezvous between two manned vehicles had been accomplished.

Controllers at Gemini Control were standing as the rendezvous neared completion and at Stafford's report they broke out little American flags. It was an historic moment. Schirra eased No. 6 to within one foot of No. 7 and the two crews peered at one another. There was some movement in Gemini 6 and presently a sign appeared at Schirra's window which said, "Beat Army." Schirra flew No. 6 entirely around No. 7 and then settled into a formation flight with No. 7. After four revolutions, he widened his orbit by raising apogee and dropping perigee. Then both crews relaxed and dozed in their sleep period, secure from collision by the immutable laws of orbital mechanics.

Suddenly, Gemini controllers were alerted by a message from Schirra: "This is Gemini 6. We have an object, looks like a satellite, going from north to south, up in polar orbit, He's in a very low trajectory . . . looks like he may be going to reenter pretty soon. Stand by. Looks like he's trying to signal us."

The message threw Mission Control into a state of strained alert until the faint, off-key strains of someone playing "Jingle Bells" on the harmonica came over the Gemini network. It was Schirra whipping up a little Christmas spirit, with Stafford jingling some small bells. It wasn't a prank, Schirra said later, but merely a diversion to relieve the tension and to remind the folks down below of the Christmas season. "But I think we convinced Chris Kraft and many of the people on the flight-control team that we did, in fact, have an unidentified flying object up there," Schirra said.

Thursday morning, December 16, Schirra and Stafford waved a "so long" to Borman and Lovell and fired retrorockets to reenter. The two vehicles had flown along together for 20 hours and 22 minutes after rendezvous. Schirra steered No. 6 by bank angles to an impact point only 13 miles west of the aircraft carrier *Wasp*. It was the first successful controlled landing in Project Gemini.

Upstairs, Borman and Lovell continued on their appointed rounds, feeling a little more "itchy and crummy" every day. A fading fuel-cell battery threatened to force them down on Friday, December 17, but since it was putting out enough power to meet requirements, Kraft allowed the crew to finish the mission.

Borman and Lovell early in the flight had gained permission to take off their flight suits and work in their underwear, becoming the first men to fly so informally in space. In Voskhod I, the Russians had achieved a shirt-sleeve environment, but in Gemini 7, the astronauts had gone them one better with an underwear environment. It was a remarkable tribute to the spacecraft manufacturer.

The longest of all the space voyages of man ended on December 18, 1965, with Borman's precise landing in the Atlantic Ocean at 8:05 a.m. EST 695 miles east southeast of Cape Kennedy and only 11 miles from the *Wasp*. He and Lovell had flown 5,716,900 miles, 20 times the distance to the moon, in 330 hours and 35 minutes—just 5 hours and 25 minutes under 14 days. The astronauts had donned their flight suits before retrofire and now found the garmets bulky and uncomfortable as they climbed out of the floating spacecraft to be whisked away to the deck of the *Wasp* by helicopter. On deck, they walked with slow strides, as though making their way through a swamp. Their legs felt heavy after two weeks of weightlessness, but they could walk.

And now, critical data were in. American spacemen could survive at least 14 days in weightless flight quite handily and had learned the art of rendezvous. There still remained the elusive problem of docking with Agena. That was next.

TAILSPIN

Gemini-Agena rendezvous and docking were accomplished brilliantly at 6:15 p.m. EST March 16, 1966 by Neil A. Armstrong, 35, a civilian test pilot who had worked for NASA on the X-15 rocket airplane, and Air Force Major David R. Scott, 33, who held master of science and engineering degrees from the Massachusetts Institute of Technology.

Gemini 8 was launched at 11:41 a.m. from Cape Kennedy, 101 minutes after the target Agena reached orbit without mishap. After they circularized their initial oval orbit and altered its plane by one-half a degree, Armstrong and Scott found they had closed the distance between Gemini 8 and Agena from 1050 miles at Gemini liftoff to 170 miles. They secured a radar "lock" on the 26-foot target rocket at a range of 158 miles and sighted it at 76 miles. Six hours after liftoff, while flying over the Pacific Ocean, Armstrong advised the

Hawaii tracking station that they were 150 feet from Agena and matching velocities at 25,365.9 feet per second. Armstrong carefully moved the Gemini at three-quarters-of-a-mile-per-hour relative speed to within two feet of Agena's docking collar. He notified the flight controller Keith K. Kundel on the tracking ship *Rose Knot Victor*, that all was ready for docking. As the two vehicles passed over the South Atlantic Ocean, Kundel gave Armstrong a "go" to dock. The command pilot eased the conical nose of the bottle-shaped spacecraft into Agena's docking collar, the collar latched, and the final objective of Project Gemini was accomplished.

"It was a real smoothie," Armstrong reported.

The novelty of being docked with the powerful rocket was exhilarating, in the ship and on the ground. For a few minutes, the docked vehicles moved smoothly on Gemini's fifth revolution, without any tendency to change attitude. Armstrong took control of Agena's secondary propulsion system, used for steering, and fired its jets to make a 90-degree turn. They found that the maneuver took 16 seconds. The joined flight continued across the Atlantic Ocean, Africa, and the Indian Ocean.

At 6:42 p.m., 27 minutes after docking, Armstrong and Scott noticed they were starting to roll and swing as they passed over southern China. The rate built up rapidly until they were rolling at about 36 degrees a second. Quickly, they sent corrective commands to Agena's attitude-control system. After three minutes, they reduced the gyrations to a point where they believed they had regained control, but four minutes later, the roll and yaw motions began again. This time, they were more violent and the crew feared their vessel would be torn apart. Assuming that the gyrations were caused by some malfunction in Agena's secondary propulsion system, causing it to fire wildly, the astronauts decided they had better undock as soon as they could reduce the rock and roll so that it could be done safely. But no sooner had they backed off from Agena than they began to spin faster than ever, at the rate of one revolution a second. The pilots then knew they were in trouble and that the problem was in the Gemini. At this time, Gemini and Agena were passing over the coast of China and approaching the tracking ship, *Coastal Sentry Quebec*, stationed south of Japan. Houston called the ship and the *Coastal Sentry* advised that the spacecraft had undocked from Agena. The ship called the crew.

Armstrong reported, "Well, we consider this problem serious. We're toppling end over end but we are disengaged from Agena. It's a roll or nothing. We can't turn anything off."

It appeared to the flight directors at Houston that Gemini had gone out of control. The *Coastal Sentry* advised Houston that "He's in a roll and he can't stop it. His OAMS regulating pressure is down to zero." The crew came on again: "We're in a violent left roll here at the present time. We apparently have a roll of a stuck hand controller."

The *Coastal Sentry* asked Houston Flight Control, "Did you copy that, Flight? They seem to have a stuck thruster. We can't seem to get any valid data here. He seems to be in a pretty violent tumble."

Gemini 8 was gyrating so violently that scrambled antenna patterns made transmission exceedingly fragmentary. Voices cut in and out with garble. At length, Armstrong reported, "We are regaining control of the spacecraft slowly in RCS [reentry control system] direct."

Relief spread through the control room, and the controllers knew the flight was over. The pilots had resorted to use of the reentry control system in the nose of the Gemini to stop its plunging and rolling, which apparently had been caused by a defective thruster in the OAMS engines. RCS was designed for use only after Gemini had jettisoned the adapter section which contained the OAMS engines and was preparing for retrofire. It was a firm rule at the Manned Spacecraft Center that once the RCS was activated, the spacecraft had to land as soon as possible, since there was no other control system on the vehicle to resort to in event the RCS malfunctioned.

As Gemini 8 came within range of the Hawaii tracking station, Armstrong reported that he still had no control over the thrusters in the OAMS array of thrusters, but was slowly bringing the Gemini under control with the RCS. Hawaii told the crew to prepare for landing in an emergency area of the Pacific on the seventh revolution. "Get into retro attitude [blunt end forward] as soon as possible," Hawaii advised. Houston called Hawaii and suggested, "You might ask the crew if they have any idea where the Agena is from them. That would give us a hack on the [Agena] orbit."

"Do you have any idea of the Agena position at present?" Hawaii asked Gemini 8.

"We saw it about ten minutes ago," said Scott. "It looked to be a mile or so underneath us."

Hawaii then asked the crew to relate the sequence of events in the emergency. Armstrong began relating that he had the OAMS off in the spacecraft and had just turned on a tape recorder Then his voice faded out as Gemini 8 passed beyond range of the station. Since Gemini passed over South America as they began their sixth revolution, far to the south of the tracking ships, the story remained incomplete. When they returned to Hawaii on their sixth revolution, Armstrong and Scott reported that they had regained yaw control in the OAMS and once more were able to maneuver the ship with that system instead of using the RCS up forward. They wanted to continue the flight. Scott had been scheduled to take a space walk. But Flight Control, Houston, said no.

Armstrong brought Gemini 8 down into the Western Pacific Ocean at 10:23 p.m. EST 500 miles west of Okinawa and 725 miles south of Japan within sight of circling rescue aircraft.

After Air Force swimmers attached a flotation collar to the space vehicle to keep it afloat, Armstrong and Scott calmly ate their suppers of dehydrated food as the vehicle rocked gently on the sunlit Pacific. They were picked up by a Navy destroyer, the U.S.S. *Mason*, and were taken to Okinawa. From there, they were flown back to Cape Kennedy, where they told the whole story of their venture in the debriefing.

For reasons which have never seemed credible, NASA withheld the voice tape report of the pilot's dilemma after the gyrations began. Newsmen at the Manned Spacecraft Center protested vigorously and the agency finally played the tapes the following day, March 17. The NASA spokesman, Paul Haney, explained that the tapes had been withheld because it was feared that the "high voice levels" of the crewmen might give newsmen the wrong impression. However, there was nothing in the tapes to suggest that either crewman was near hysteria, as the excuse for withholding the tapes implied. Both Armstrong and Scott sounded as though they were certainly alarmed at their situation, but it was clear that neither considered himself in mortal danger.

It is more plausible that NASA withheld the tapes until officials in Washington could determine the gravity of the situation. At the time the linked vehicles went out of control, they were in radio range of the Chinese mainland. A radio signal, by accident or design, might have activated the remote controls of the Agena secondary propulsion system to start the tailspin. However, as it became apparent later that the Gemini OAMS was acting up, and when the astronauts required control of their vehicles, there was no longer any dire reason for withholding the tapes. Later, a transient short circuit in the Gemini OAMS was announced officially as the cause of the trouble. But since the OAMS was jettisoned with the adapter before reentry, no one will ever know how that happened.

The withholding of the tapes was a clear act of suppressing news at a time when it was most urgent for the news-media correspondents to have all data possible. Its effect was to prevent the public from learning the full measure of the crisis until after the crisis was resolved. Even though the civilian space agency is mandated by law to make full disclosure, it often balks at doing so, even to Congress, and seeks to prevent its contractors from doing so. This is the tune bureaucracy plays to protect itself against criticism and public awareness of its mistakes, inefficiencies, and ineptitudes, and NASA plays the tune as often as any bureaucracy. Fortunately this time, the story had a happy ending. That would not always be the case, as we shall see later.

THE ANGRY ALLIGATOR

The flight of the next Gemini seemed to be unable to extricate itself from a pall of grief and trouble. The prime crewmen, Elliott M. See, Jr., and Charles A. Bassett II, were killed in an airplane crash at St. Louis February 28, 1966. The backup crew, Stafford and Lieutenant Commander Eugene A. Cernan, 32, of the Navy, entered the Gemini 9 spacecraft on the morning of May 17, 1966, only to climb out again when the target Agena's failure to achieve orbit caused the mission to be scrubbed. This time, the fault was found in the Atlas guidance system, rather than in the Agena. It had happened to Stafford before, on the first attempt to launch Gemini 6, and he observed philosophically, "You really can't get your hopes up until that Agena comes across the States."

A native of Chicago, Cernan had received a bachelor of science degree in electrical engineering from Purdue University and a master of science degree in aeronautical engineering from the Naval Post Graduate School at Monterey, California. One of the "third generation" astronauts selected in October 1963, he was married to the former Barbara Atchley of Houston. They were the parents of a daughter, Teresa Dawn.

Unable to make another Agena ready promptly, NASA brought out a makeshift target called the Augmented Target Docking Adapter (ATDA). This contraption had been lashed up as a backup target for Gemini 8 out of Gemini parts by McDonnell and used the Gemini stabilization system. It was 12 feet long and 5 feet in diameter and weighed 2400 pounds at launch atop the Atlas ICBM. It was by no stretch of the imagination a rocket. It was simply something to dock with.

The ATDA was boosted by the Atlas into a roughly 185-mile circular orbit shortly after 10 a.m. June 1, 1966. Stafford and Cernan lay back on their couches, awaiting their turn. Three minutes before Gemini 9 was to lift off, trouble appeared in the ground link to the spacecraft computer, preventing the controllers from programing the computer with last-minute data. The launch had to be postponed until June 3, when Gemini at last lifted off Pad 19 at 8:39 a.m.

As Stafford and Cernan approached the ATDA on the third revolution over New Guinea, the pilots saw that the shroud protecting the nose of the vehicle during flight up through the atmosphere had failed to come off. From one mile away, Stafford could see the jaws of the shroud had opened only halfway and had become jammed in that position. He reported, "We have a weird-looking machine here. Both the clam shells of the nose cone are still on but they are open wide. The jaws are like an alligator's jaw that is open at 25 to 30 degrees. It looks like an angry alligator out here rotating around." Docking was not possible with the shroud still attached to the ATDA, and the best use the crew could make of the "angry alligator" was as a rendezvous target. At one point, Stafford maneuvered Gemini 9 to within three inches of the "angry alligator" so that Cernan could photograph the shroud and its wires up close.

Early on the morning of June 5, Cernan prepared for his spacewalk. It was programed to be a long one, lasting 2 hours and 25 minutes, during which time he would test a 166-pound Air Force maneuvering unit he would carry on his back. With the AMU (Astronaut Maneuvering Unit), he could move 40 feet from the nose of the spacecraft. At 10:22 a.m., Cernan climbed outside and looked down. He saw a city he believed was Los Angeles. Looking slowly around him, he moved to the adapter skirt at the rear of the Gemini, where the AMU was stored. He made his way to the bulky maneuvering unit, saw that it appeared to be in order, and began to struggle into the straps, which fitted over his shoulders so that the unit rode on his back like a knapsack. In zero gravity, the effort was exhausting and he noticed that he was breathing rapidly, exhaling fog into his helmet visor. The fogging began to limit his vision. The suit air-conditioner was not handling the humidity. Presently, Cernan was unable to see through the visor. He had been trying to connect the oxygen hoses and electrical plugs between his suit and the AMU. This seemed inordinately difficult. Hoping his faceplate would clear, Cernan rested, but the plate continued to be fogged over. At best, he was able to achieve only 40-per-cent vision through it. He abandoned his struggles with the AMU on advice from the command pilot and made his way forward, to the hatch. The movement once more increased the faceplate fogging. He was constrained to peering through a clear space directly in front of his nose.

At 11:48 a.m., when it was obvious that the equipment was not going to allow him to perform any purposeful activity outside, Cerman climbed back into his seat. He had been outside 1 hour and 26 minutes most of which was spent in frustrated efforts to maneuver and don the AMU. Officially, the Manned Spacecraft Center logged the time the crew spent in extravehicular conditions at 2 hours and 5 minutes, the time the hatch was open.

Stafford flew Gemini 9 to a precise landing in the Atlantic Ocean June 6, less than two miles from the *Wasp*. He and Cernan elected to remain in the spacecraft until the carrier came alongside and hoisted the vehicle to her flight deck. The first attempt to do work in EVA had shown that it would not be easy.

A CHANGE IN LUCK

Two failures with the Agena target vehicle and the ridiculous episode of the "angry alligator" began to erode the confidence of many observers, including this one, in the technical management of the Gemini program. The successful rendezvous of Gemini 6 and 7 and the rendezvous and docking of Gemini 8 with the Agena were marred by malfunctions in equipment which all but defeated the major flight objectives.

With the flight of Gemini 10, however, the bad luck, or bad management, which had plagued the project vanished. The final three Gemini flights—Nos. 10, 11, and 12—were all executed with brilliance and polish that made everyone look good. More significantly, they demonstrated that the art of rendezvous and docking in earth orbit had been mastered by United States astronauts.

Gemini 10, with Commander John Young as command pilot and Air Force Major Michael Collins, 35, as pilot, lifted off Pad 19 at 5:20 p.m. EST July 18, 1966. It was by far the most daring manned flight of the space age, but it was executed so smoothly that, except for an episode when the astronauts were blinded by an eye irritant during an EVA mission, it appeared to be a routine exercise.

This time the astronauts had two targets, their own Agena 10, which had been orbited successfully 101 minutes earlier and was 920 miles ahead, and the Agena left in orbit from the Gemini 8 flight of March 16. It was 575 miles behind as Young and Collins were injected into their initial orbit by the Titan 2.

One of the "third generation" astronauts selected in 1963, Collins was a graduate of the United States Military Academy who had joined the Air Force. He had been experimental flight test officer at the Air Force Flight Test Center, Edwards Air Force Base, California. He was married to the former Patricia M. Finnegan of Boston, and had three children, Kathleen, 7, Ann, 5, and Michael, 3½.

Five hours and 21 minutes after liftoff, Young reported that Gemini 10 was ready for docking with its Agena. All thruster systems aboard the spacecraft and the rocket were thoroughly checked to avoid another tailspin before Young was given a "go" to dock. This time, there was no trouble. The astronauts began to perform a series of maneuvers to try out the Agena's propulsion systems. On the first principal maneuver, Young fired the Agena's primary propulsion system to boost the joined vehicles into a new orbit with an apogee of 458 miles, a new altitude record.

"Mike threw the switch and it was really something," said Young. "We had a negative one *g* and were driven forward in the cockpit. We got a tremendous thrill on our way out to apogee." The crew was driven forward when the Agena blasted because they were seated facing the big rocket in docking position. Remaining in the high orbit overnight, the crew then fired the Agena in retrogade July 19 to come down to an apogee of 236.6 miles, with a 182-mile perigee.

They attempted to circularize the orbit at 236.6 miles by means of the Agena secondary propulsion, but achieved only a 240-by-236.6-mile orbit. In the new orbit, the linked vehicles were near the orbit of the Gemini 8 Agena, with which Young planned to attempt a rendezvous later in the flight. Meanwhile, both astronauts opened their hatches and Collins stood up on his seat to take photographs of stars in ultraviolet light for Professor Karl Henize, an atronomer at Northwestern University, who became so intrigued with the prospect of doing astronomy in space that he later applied for astronaut training himself and was selected in 1967. Since ultraviolet light is screened out by earth's atmosphere or by a window, the photographs had to be taken in the "open" outside the atmosphere. During this stint, on the 15th revolution, an irritant found its way into the oxygen system and acted on both astronauts like tear gas. In a few moments, both were blinded by tears. Unable to continue the photography assignment, Collins sat down and he and Young slammed the hatches shut and repressurized the cabin.

"It smelled like lithium hydroxide," Collins reported. "Neither John nor I could do anything so I came back in." Lithium hydroxide was used along with charcoal to remove carbon dioxide and odors in the environmental-control system of the vessel. With the hatches closed and new oxygen in the cabin, the crewmen recovered. The nature of the irritant was not determined because it was gone when the vehicle landed, but Dr. Berry, the chief flight surgeon, believed an antifogging mixture used to keep faceplates clear of breath moisture might have been involved.

Gemini 10 remained docked with Agena for 39 hours. Then Young backed it away. On the third day of the flight he steered the spacecraft into a rendezvous with Agena but did not attempt to dock with it. While Young held the spacecraft to within three feet of the Agena, Collins opened his hatch, climbed out, and, secured by an umbilical line, propelled himself with the Zot gun across the gulf between the two vehicles to the Agena. The act demonstrated that the United States had the capability of inspecting any space vehicle in orbit at will. From the metal hide of the Agena, Collins plucked a canister of microorganisms which had been subjected to space conditions since March 16. The organisms included T-1 bacteriophage, some of which survived, it was reported later. Similar organisms had survived 30-hour exposure to high-altitude conditions on balloon flights and 16 hours on Gemini 9. Collins also picked up a micrometeoroid collector from the fantail of the Gemini on his tour outside.

Fuel was running low and Young suggested that Collins curtail the EVA, planned for 55 minutes, to 35 minutes. Somehow, Collins mislaid his Hasselblad camera which he had been using to take pictures of the earth. There was one plate holder in it. Also, in the confusion of his reentering the spacecraft, the flight plan, a booklet about as thick as a suburban telephone directory, disappeared, and the micrometeoroid package vanished, too. They are probably still in orbit. Despite these minor mishaps, the flight went well enough to restore confidence in the successful outcome of the program. Young brought Gemini 10 down from an altitude of 232 miles July 21 to a pinpoint landing in the Atlantic.

M EQUALS 1

Gemini 11, with Charles (Pete) Conrad as command pilot and Lieutenant Commander Richard F. Gordon, Jr., 36, pilot, was boosted off Pad 19 at 10:42 a.m. EST on September 12, 1966, just 97 minutes after its Agena target rocket was launched into orbit. Conrad, who had flown with Cooper on Gemini 5, maneuvered his vessel into a first-orbit rendezvous with Agena over Hawaii 84 minutes after leaving the pad. Ten minutes later, he docked with the Agena over Texas. It was the crowning achievement of Project Gemini.

Gordon, who had won the Bendix Trophy Race from Los Angeles to New York in 1961, setting a new speed record of 869.74 miles an hour and a transcontinental record of 2 hours and 47 minutes, said he would have doffed his helmet, if he could have, to the piloting skill of his Gemini partner. They sighted the Agena ahead of them over Madagascar after crossing the Atlantic after liftoff, and rapidly moved toward the rocket's orbit in a series of maneuvers. West of Hawaii, Conrad told Gordon, "Hey, I've got to fly her down a mite." He braked with the forward thrusters. "Okay," said Gordon, "I'm going to give you a range-rate reading. Boy, is that thing bright." The sunshine was blinding off the metal skin of the 26-foot Agena. Gordon called out the closing rates, like a lookout sounding for shoals. "Twenty-five feet a second," he said, referring to the relative velocities of the two vehicles. "Twenty. Fifteen. . . ." Conrad's voice came clearly over the radio. "We are station keeping and looking at the TDA [target docking adapter]," he announced. "Tell Mr. Kraft would he believe it, M equals one." ("M equals one" meant rendezvous on the first revolution.)

"Roger," said John Young at Flight Control, Houston. "Outstanding. And he believes it." Conrad then docked with Agena 11.

On Tuesday morning, September 13, Gordon climbed out of the spacecraft as it crossed the West Coast south of San Diego on its 15th revolution to begin two hours of planned extravehicular activity. He looked down and saw that he was floating over Houston. Lightning was flashing down there, he said, "like a stop light."

Gordon had several assigned tasks. The principal one was to hitch a 100-foot nylon rope from the Gemini to a bar on the Agena, thus tying the two vehicles together. The purpose of this exercise was to determine whether the tethered vehicles could be stabilized by an effect called gravity gradient after they undocked.

The chest-pack dehumidifier Gordon was wearing proved unequal to the task of handling the perspiration generated by his exertions. Like Cernan, Gordon found himself blinded with perspiration. Panting and trying to see through his fogged faceplate, Gordon sat down on the Agena, facing

Gemini. Conrad called out over their intercom, "Ride em, cowboy. How you doing?"

"I'm breathing hard," said Gordon. He certainly was. His heavy, labored gasps could be heard through the world-wide communications hook-up, at the rate of 40 a minute. His heart rate registered 162 beats a minute, compared with Conrad's 120 at the time. The EVA was overtaxing him. After 44 minutes of EVA, Conrad told Gordon to come back in. "Listen," he told Houston, "I just called Dick back in. We are repressurizing the cabin right now. He got so hot and sweaty he couldn't see."

"I know how it is," sympathized John Young, the capsule communicator down below. "When you get where you can't see, you've got to close the lid."

Gordon had succeeded in moving back to the fantail of the Gemini adapter section to get a cosmic-ray film pack which was fastened there and bring it back to the cabin for analysis after the landing. He had completed the chore of linking Gemini and Agena with the 100-foot tether. Although he had also intended to wipe the spacecraft windows with a rag, so that the dirt which accumulates on them during liftoff and flight could be analyzed, he was called in before he could carry out that task.

Weary from the day's work, the astronauts ate their condensed rations, drank as much water as they could hold, and began their sleep period. In the early morning hours of Wednesday, September 14, Gordon fired the Agena's main engine over the Canary Islands. The blast added 918 feet per second to the velocity of the joined vehicles. "It's going . . . it's going," Conrad called, as Gemini-Agena climbed to an apogee of 850.5 miles, a new altitude record over Australia. Perigee remained at 180 miles. From the highest perch on which men have ever sat, the astronauts could see a panorama of 17,200,000 square miles of the earth's surface. "Fantastic, fantastic," said Conrad, looking down at Australia, Borneo, and Southeast Asia spread out below him. The air was quite clear that day. Gordon photographed the scene with a Hasselblad camera.

After two revolutions in the elongated orbit, Conrad fired the Agena engine as a brake, dropping apogee back down to 180 miles. The orbit was now circular. Gordon then opened the hatch and stood up on the seat to take star photographs in ultraviolet light for Henize.

When he completed the photo assignment, Gordon closed the hatch. Conrad then undocked the Gemini from Agena, backing off nearly to the end of the 100-foot tether. This exercise was designed to determine whether two space vehicles could be stabilized when tied together if they were aligned vertically with the center of the earth. Theory held that in such an alignment a gravitational gradient would be set up.* That is, the end of the vehicle nearer the earth would be subject to greater gravitational force than the end farther away. Every particle of matter in the universe, says Newton's law of universal gravitation, attracts every other particle with a force which is directly proportional to the product of their masses and inversely proportional to the square of the distance between them. Applied to Gemini-Agena 11, it meant that when the two vehicles were pointed at the earth, the greater force exerted on the portion of the vehicle system nearer the center of the earth would tend to stabilize the system and damp out random oscillations. It was a free way of stabilizing attitude, if it worked.

Conrad and Gordon began the experiment on their 32nd revolution and the effects astonished them. In short order, they found themselves spinning around like a yo-yo. "Boy, this is really wild," Conrad said. "The tether has a great big bow in it." He eased the Gemini outward to tighten the line, but it seemed to be impossible to straighten it. "It's like the Agena and I have a skip rope between us," he said. "It looks like we're skipping rope with it. We've got a weird phenomenon here. The tether is doing a spin. That's the reason I can't get to the end of it. I can't get it straight."

John Young called up from Houston, "The consensus here, Pete, is that you will never get the spin out of the tether."

Conrad decided to try the second part of the experiment, since it appeared that the gravity-gradient part of it was not going to work. By gently firing the Gemini thrusters, he started the tethered vehicles rotating around each other,

* A suggestion that gravity-gradient stability could be achieved in orbit by two objects of approximately equal weight connected together "like a dumbbell" was made by James B. Fisk, president of Bell Laboratories, in an address May 24, 1961, to the American Iron & Steel Institute, New York.

dumbbell fashion. "This is no easy job," he commented. "Well, we, uh, have started. Hey, this is not going to work. As a matter of fact . . . well, I guess we'll wait and see."

Physical forces no one had warned the crew about continued. Suddenly, it looked as though the tether was going taut. "Hang on," cried Conrad, "here goes the jerk. We can't feel it, but, by golly, we're not oscillating." The yo-yo effect vanished. The tether, which could stand a pull of 1000 pounds, had indeed gone taut. But no jerk had been felt, which astonished Conrad because there had been a great deal of slack.

"He's yawing 30 degrees to either side of us," said Conrad, referring to Agena. "We're rolling. The tether is holding. We've got good, steady tension and it looks like we've got a good spin going."

Now it appeared that the Gemini was revolving around the Agena. Except for a slight roll, the Gemini oscillations were damping down. The joined vehicles were stabilizing as they slowly rotated around each other at the rate of 360 degrees in nine minutes.

Young suggested that somebody lose a pencil in the spacecraft to see if artificial gravity had been set up by the rotary motion. Gordon reported that a camera which had been floating in the cabin had moved back against a bulkhead. At Houston, it was calculated that the rotation was producing a force equal to 1.5-thousandths of a gravity. It was the first time "artificial gravity" had been generated.

After two hours of experimenting with the effects of tethered rotation, Gordon pressed a firing stud that expelled the tether from Gemini. The last he saw of the tether, which seemed to have a will of its own, it was slowly winding around the Agena "like a Christmas present." Thursday morning. Conrad landed Gemini 11 in the Atlantic Ocean, 700 miles east southeast of Cape Kennedy.

FOUR DAYS VACATION WITH PAY

The experiences of Gordon and Cernan in trying to perform useful tasks outside the spacecraft demonstrated that extravehicular motions consumed more energy than anyone had expected. Another attempt to investigate the problem was planned on the flight of Gemini 12, the final one of the Gemini program. Lovell was command pilot and Air Force Major Edwin E. (Buzz) Aldrin, Jr., 36, a doctor of science from the Massachusetts Institute of Technology, performed the space walk.

Gemini 12 was lifted at 2:46 p.m. on November 11, 1966, about 98 minutes after its target Agena was launched. Four hours and 13 minutes after the Gemini launching, Lovell caught the Agena on the third revolution over the Pacific Ocean and docked with it. However, a defective turbopump appeared in the rocket. This problem made it inadvisable for the crew to use the Agena propulsion system to boost them to a planned apogee of 460 miles. Instead, they moved the joined vehicles into an orbit which would enable them to photograph an eclipse of the sun on their tenth revolution, Saturday morning, November 12, as they passed south of the Galapagos Islands. The Manned Spacecraft Center reported later that the pictures were not successful.

Aldrin opened his hatch, stood up in his seat, and took photographs of stars in ultraviolet light on the night side of the orbit and ground photographs on the day side for 2 hours and 18 minutes. The scholarly astronaut was delighted with the view, and Lovell said, "What did I tell you, Buzz? It's four days of vacation with pay to see the world."

On Sunday morning, November 13, Aldrin climbed out of the spacecraft to tackle the problem of working in free fall. He had been briefed thoroughly by Gene Cernan and Dick Gordon, and he had practiced his EVA maneuvers for months in the swimming pool. Moving carefully and deliberately, weighing every motion, he managed to keep cool. This time, his 42-pound, chest-pack dehumidifier kept him dry. He was careful to avoid overloading its air-conditioning system. He found a 100-foot nylon rope which had been attached to the Agena before it was launched and tied it to the Gemini. He turned on a movie camera to record all of his motions. Then he wiped the windshields of the Gemini which were smudged with thruster exhaust and aerosols picked up in the flight through the atmosphere. He observed that the windshields seemed to be as dirty inside as out. Later investigation showed that contaminants seeped in between the inner and outer windshield panels.

Then the world's first orbital spaceworker pulled himself to the fantail of the Gemini adapter, where he performed the work assignments in the flight plan. These consisted of plugging and unplugging electrical and fluid connectors, unscrewing bolts and screwing them in again, and manipulating hooks and rings. Aldrin used a combination of foot restraints and body tethers installed in the adapter. He said they were important fixtures to prevent his body from moving when he used his forearms, wrists, hands, and, occasionally, shoulders. The restraints also prevented the muscles from working against each other, as they did if the astronaut tried to hold himself steady in the weightless state. In contrast to the struggles of Cernan and Gordon, Aldrin crawled over the Gemini like a human fly. Monitoring the astronaut's performance at Gemini Control, Dr. Berry reported that Aldrin displayed no unusual stress. Aldrin's only complaint was that his left foot showed a tendency to float upward when only his right foot was secured in a stirrup restraint. "That's the Cernan effect," advised Lovell. Cernan had noted the problem on Gemini 9.

After Aldrin had worked a while, Lovell suggested, "Buzz, why don't you rest two minutes because I want to change a tape and this is a good time to change it."

"All right," said Aldrin. "But I'm getting a slightly warm posterior. The sun is setting very agreeably. It's great from a visibility standpoint, but it's not too good from a heating standpoint. So it's getting a little warm."

"Roger," said Lovell. "That's the sunning effect."

Aldrin said he had no trouble positioning his body when he used the foot restraints. He worked his way around the world for 2 hours and 9 minutes, describing his experiences as the spacecraft flitted from one tracking station to another.

For a while, he was talking to the station at Kano, Nigeria. Then Houston called, "One minute to loss of signal. You'll be in Tananarive [Madagascar] in nine minutes." In range of Tananarive, Aldrin continued his recital of the details of his work.

At the end of his EVA period, Aldrin gathered together his equipment like a careful plumber, unlatched the camera from its mounting, shoved everything inside the hatch, and climbed in himself. The cabin was repressurized with oxygen and the astronauts relaxed.

They had proved that man could perform useful work in orbital free fall if he knew how. He had to be trained for it, however, and Aldrin was.

Besides a spacewalk record, Aldrin left behind him in orbit two canvas pennants. One said, "Nov. 11—Vets Day" in honor of the day Gemini 12 was launched. The other said, "Go, Army—Beat Navy." It was the West Pointer's reply to the sign Schirra displayed at the rendezvous of Gemini 6 and 7 saying, "Beat Army." With a camera, a flight plan, and now these signs drifting about, space was becoming a souvenir-hunter's paradise. It already was a junk yard, with hundreds of pieces of rockets and miscellaneous hardware whizzing around in orbit.

After a rest, the crew of Gemini 12 prepared to attempt the gravity-gradient experiment. Lovell undocked from the Agena and carefully maneuvered so that Gemini was above the rocket. Both vehicles pointed downward. At the full extension of the 100-foot tether, the 19-foot Gemini and the 26-foot Agena would have made a linked system 145 feet long. A stabilizing increase in gravitational force at the bottom of a system that long might be expected. However, like Conrad in Gemini 11, Lovell found himself at the mercy of the unpredictable forces which kept the nylon tether from becoming taut and the spacecraft yo-yoing. At one point, the tether looped and threatened to foul Gemini's radio antenna. Both Gemini and Agena were rolling, and Lovell reported that the motion was difficult to control. However, in the end, mind prevailed over matter, and the theory began to prove itself. The gravity gradient began to take effect. After three hours, Lovell reported that both Gemini and Agena had achieved stability.

On Monday, November 14, Aldrin performed a third EVA, standing up in his seat with the hatch open for an hour. When he closed the hatch again, he was the undisputed EVA champion of the world, with 5 hours and 38 minutes of extravehicular activity. He spent one hour's more time outside his ship than John Glenn had spent inside on the first Mercury flight.

The Gemini 12 crew's main chore on Monday was to take part in an experiment with France to observe a yellow cloud of sodium vapor. The vapor was to be injected into the high atmosphere as a wind tracer by a French sounding rocket, launched from France's Hammaguir Launch Facility in Algeria. Lovell and Aldrin would attempt to photograph the cloud as they passed over the Sahara desert on their 40th revolution. Against the yellow desert, however, neither astronaut could see the yellow cloud. The French fired a second rocket to produce a second cloud, but when the Americans came around again, they still could not see it.

As the flight progressed in its third day, three of Gemini's 16 thrusters and one of the six fuel-cell battery units failed. However, William C. Schneider, the mission director, decided to keep Gemini 12 aloft the full four days planned.

The flight of Gemini 12 ended Tuesday, November 15, and with it ended Project Gemini. The spacecraft landed by autopilot at 2:21 p.m. EST 730 miles southeast of Cape Kennedy within sight of the aircraft carrier *Wasp*. Both space travelers were in fine condition. And the directors at Houston ignited the customary cigars.

"To go to the moon," said Robert R. Gilruth, director of the Manned Spacecraft Center, at the post-landing conference, "we had to learn how to operate in space. We had to learn how to rendezvous and dock, to light off large propulsion systems in space, to work outside the spacecraft, endure long-duration missions, and how to make precise landings from orbit. In the ten manned flights in 18 months, we did all the things we had to do as the prelude to Apollo."

Chapter 12

Fire in the Cockpit

Exhilarated by the success of Project Gemini, NASA moved promptly toward the first manned flight of the three-couch Apollo spacecraft, the vehicle designed to put Americans on the moon. As the Gemini program ended late in 1966, the crew of the first Apollo, which bore the factory number 012, was deeply engrossed in training for an orbital flight that might last as long as 14 days. Gus Grissom, veteran of Mercury and Gemini, was the command pilot. Ed White, the first American spacewalker, was the senior pilot. The third crewman was personable, 31-year-old Roger B. Chaffee, a lieutenant commander in the Navy and graduate of Purdue University in aeronautical engineering. He occupied the right-hand seat as the pilot, White sat in the middle, and Gus occupied the left-hand seat.

The Apollo spaceship was by far the most complex vehicle ever built in the United States. Its development made excruciating demands on the state of the art and on the quality of American industrial performance. For that reason, it was clearly a symbol of the best of American technology, and NASA officialdom was quite proud of it, as was the contractor, North American Aviation, Inc. In the evolution of space vehicles, it was as far removed from the Gemini two-seater as Gemini was from Mercury.

Apollo basically consisted of two sections: the cone-shaped Command Module (cabin) where the crew would ride, and the cylindrical Service Module attached to its base, containing the main propulsion engine with 21,500 pounds of thrust to drive the spaceship home from lunar orbit, as well as attitude-control thrusters, fuel-cell batteries, and portions of the environmental-control system that was designed to keep the cabin at a shirt-sleeve temperature. Joined together, the modules were 34 feet long and weighed 67,000 pounds at launch. The cylindrical Service Module was 12 feet, 10 inches in diameter, as was the Command Module at its base. A launch-escape tower, similar to the one used on Mercury, was affixed to the nose of the Command Module. It contained a torpedo-shaped solid-fuel rocket capable of lifting the cabin instantly off the Saturn booster to an altitude of 3000 feet so that parachutes could be deployed for a soft, sea landing in the event of fire or explosion of the Saturn fuel during countdown on the pad.

On the return from the moon, the Service Module would be detached. Only the Command Module, looking like a child's top, would reenter the atmosphere, blunt end forward, and land on the sea on its three striped parachutes. The Command Module, standing 11.7 feet high, was constructed of two shells with two layers of microquartz fiber insulation between them. The inner shell comprising the pressure vessel, in which the crew was housed, was made of aluminum honeycomb bonded between aluminum alloy sheets. The outer shell was fabricated of stainless-steel honeycomb, braced between stainless-steel sheets.

The entire Command Module was coated with an ablative heat shield to ward off heating by atmospheric friction of up to 5000 degrees Fahrenheit during reentry at lunar-return velocity of 7 miles a second—2 miles a second faster than the reentry speed of the orbiting Mercury and Gemini spacecraft.

Essentially, the Command Module cabin was designed as cockpit, office, laboratory, radio station, kitchen, bedroom, bathroom, and den, according to the contractor. Test equipment was provided to enable the astronauts to check difficulties in any of the spacecraft systems. The crew could move about to some extent in the cabin, but during most of the flight the astronauts would be confined to their couches, constructed of aluminum and titanium and padded with nylon webbing encased in plastic.

On February 26, 1966, NASA had had an encouraging success with Apollo on an unmanned, suborbital test launch, 5500 miles down the Atlantic Missile Range. The completed spacecraft was boosted by an uprated Saturn 1 and it landed on parachutes 200 miles off Ascension Island at the tip of Africa.

During the flight, the Apollo's main engine was fired, shut off, and fired a second time. However, the engine showed a drop in fuel pressure after it had run for 100 seconds and its performance was rated only as "fair." As a first test, however, the vehicle's systems generally worked well enough to satisfy the NASA Apollo directorate. George E. Mueller, associate administrator for manned space flight, termed the flight "the successful first step of the manned Apollo program."

It had been the expectation in the space agency to move directly into Apollo as soon as Gemini was over, and there were indications after this test that the launch of Grissom, White, and Chaffee might come as early as December 1966. However, the NASA directorate had programed the manned test for the first quarter of 1967, and problems which cropped up in the environmental-control system of the manned vehicle, No. 012, ruled out an earlier launching.

THE CHILDE APOLLO

In preparation for the commencement of Apollo manned test flights in earth orbit, the Manned Spacecraft Center at Houston staged a two-day briefing in December 1966 for news-media correspondents in the Center's plush auditorium. The sessions were conducted on a note of high optimism, relieved only by a gust of dry Hoosier humor from Grissom. In response to a question, Gus observed that he would consider the lunar mission a success "if we all get back."

Joseph F. Shea, the Apollo program manager at Houston, compared the effort of going to the moon as a process of growth, like that of a child. In the progression from Mercury to Gemini, performance criteria became more stringent, and, in Apollo, the stringency seemed to go up by an order of magnitude. At least 20,000 failures had been logged in Apollo during the six years of its development, Shea said. More than 200 individual failures had been logged against 100 components in the Environment Control System alone, he said.

"Apollo," he explained, "requires a higher level of maturity than Gemini. We are going through the maturation process now. If you looked at any one of the Gemini flights, you would find a small but significant number of problems in important systems. Even the last flight had some fuel-cell problems and several thrusters out. I think you could recognize what is acceptable for earth orbit where you can get back home in a relatively short period of time—in effect, push a button and you're down. But for the lunar program, when it's three or four days before you can actually get back, that class of problems which was acceptable in Gemini would not be acceptable in Apollo. And if we noted them in the earth-orbital part of the mission, they would cause an abort. So in effect, Apollo has a requirement for hardware maturity that is significantly higher than the similar requirements we had in Gemini. It is kind of like watching children grow up. We are going through that maturing process on Apollo now. My feeling is we are somewhere in the middle of adolescence."

This process of maturation had begun in 1960 when Apollo was conceived as an earth-orbital extension of Mercury. "Childe" Apollo had begun to grow after President Kennedy committed the nation to a manned lunar landing.

At the time when NASA decided to use the Apollo design in a lunar vehicle, the decision had not been made on the mode of the flight and landing. It will be recalled that, while the spacecraft design was being developed during 1961, the space agency was still considering the giant Nova, with 12 million pounds of thrust in its first stage, as the booster to launch Apollo directly to the lunar surface. Then, as it became apparent that such a huge launch vehicle could not be built in time to accomplish the landing by the end of the decade, and the smaller Saturn 5 was designated as the moon booster, the question of earth-orbit rendezvous versus lunar-orbit rendezvous became the subject of an extensive and sometimes bitter debate.

During this period, however, both the Command and Service Modules were already under contract at North American

Aviation's Space and Information Systems Division at Downey, California.

The initial design of Apollo did not provide for the lunar-orbit-rendezvous mode. In this mode, as described in Chapter 6, two astronauts would leave the Apollo as it orbited the moon and crawl into the attached lunar-landing module through a docking tunnel. They would then detach the Lunar Module from the Apollo, descend to the lunar surface, and carry out the landing mission.

On the return trip, they would lift off the moon by means of the 3500 pounds of thrust ascent engine. After making rendezvous and docking with the orbiting Apollo, the lunar explorers would crawl back into the Command Module cabin through the tunnel, seal the hatches, and then detach the Lunar Module ascent stage, leaving it in lunar orbit when Apollo fired its main propulsion engine for the return to earth.

The first models of the Apollo developed at North American did not contain the docking tunnel, and consequently could not be used for the lunar landing. However, because they would require all other systems that the lunar vehicles would need, they were developed as Block 1 vehicles for flights in earth orbit. As Shea described it, "Block 1 is sort of a legacy of the problems associated with getting the modes defined in the first place." Spacecraft No. 012, in which Grissom, White, and Chaffee were to fly, was a Block 1 vehicle. A more advanced model, Block 2, would go to the moon.

THE INFLUENCE PEDDLERS

Following President Kennedy's call for the lunar landing, Project Apollo had become the greatest nonmilitary contract prize in history. Five major aerospace companies entered the lists, eagerly seeking the distinction of building what would be the first manned interplanetary vehicle. They were General Dynamics, Astronautics Division, in conjunction with AVCO; General Electric, Missile and Space Vehicle Department, in conjunction with Douglas Aircraft, Grumman Aircraft and Space Technology Laboratories; McDonnell Aircraft, in conjunction with Lockheed, Hughes Aircraft, and the Vought Astronautics Division of Ling-Temco-Vought; Martin; and North American Aviation.

Five proposals were submitted early in October 1961, after design studies had been made for NASA under contract by the Martin Company. The proposals were submitted for detailed study and evaluation to 190 panels of experts in the space agency, each panel reviewing the plans for specialized systems. The panels formed what NASA called a "Source Evaluation Board." The fundamental mission of the Board was to advise the NASA directorate which of the proposals represented the best chance of success.

On November 24, 1961, the Board made its report to the Administrator, James E. Webb, and to his deputies, Robert C. Seamans, Jr., and the late Hugh Dryden, who died in 1965. The board ranked the Martin Company highest in over-all ratings for technical approach to the Apollo project, technical qualifications, and business qualifications. On the scale the Board used, Martin was rated 6.9, General Dynamics, 6.6, North American, 6.6, General Electric, 6.4, and McDonnell Aircraft, 6.4. In areas of technical approach and qualifications only, the Board rated the Martin Company 6.1 in first place and North American 5.9 in second.

In spite of this rating by the Source Evaluation Board, Webb, Seamans, and Dryden selected North American. It was their judgment that North American "had by far the greatest technical competence," according to a memorandum Seamans provided the Senate Committee on Aeronautical and Space Sciences.

The NASA administrators agreed that Martin had the best technical approach to Apollo, but they said this might have been the result of Martin's prior experience in preparing the Apollo design feasibility study.

Webb, Dryden, and Seamans concluded that because of "the very great engineering complexities involved in the Apollo project," the exceptional competence which they imputed to North American—largely on the basis of its development of the X-15 rocket airplane—"had to be considered as an overriding consideration in selecting a contractor," according to the Seamans memorandum.

The administrators concluded that the key personnel at North American "were better than Martin's by a significant degree"—again quoting Seamans. This judgment was to be seemingly refuted early in 1967 when North American hired two of Martin's top executives.

Neither Webb nor any other NASA official made public in 1961 the reasons for North American's selection, and none of the contractors, who are accustomed to surprises from their main customer, the United States government, made any outcry at the administrators' decision to override the experts. NASA announced the selection of North American as Apollo contractor on November 28, 1961, and estimated that the initial phase of the contract would cost in excess of $400 million. (By the end of 1966, the value of North American's contract with NASA on Project Apollo amounted to $2,218,000,000 for 31 flight vehicles, 23 engineering-test models, and 19 full-scale mockups, along with an adapter for the Lunar Module, mission simulators, trainers, and tracking and ground-support equipment.)

The system of awarding large government contracts came to public notice nearly two years after North American's selection, when a civil lawsuit was filed in the United States District Court of the District of Columbia. It alleged that the Secretary of the Senate Democratic Majority, Robert G. (Bobby) Baker, had represented that he could get government contracts for North American Aviation.

The plaintiff in the suit was the Capitol Vending Company which operated automatic vending machines dispensing cigarettes, candy, soft drinks, and food in several government buildings and other establishments. Named as defendants in addition to Baker were his law partner, Ernest C. Tucker, and business partner Fred B. Black, Jr. Well known in Washington, Black was North American Aviation's capital consultant and was retained by the company at a fee of $168,000 a year.

Capitol's president, Ralph Hill, complained that Baker, Tucker, and Black helped him win a vending-machine concession from a North American Aviation subcontractor and then had Capitol ousted when Hill refused to sell his business to them. The subcontractor was Melpar, Inc., of Falls Church, Virginia, a subsidiary of Westinghouse Air Brake.

Melpar at the time held a $150,000 antenna subcontract on Apollo and was doing $100 million worth of work on circuit boards for North American's Autonetics Division on the Minuteman 1 ICBM.

Hill stated in the suit that "Robert Baker as Secretary of the Majority [in the Senate] was able to and did represent to defendant Fred Black that he was in a position to assist in securing contracts for North American Aviation, Inc."

The suit stated that Baker, Tucker, and Black had majority control of a vending-machine company, incorporated in Delaware in the fall of 1961, called Serv-U, Inc.

"As partial return for the services performed by Fred Black and Robert G. Baker," the plaintiff continued, "North American entered into an agreement to permit Serv-U to install vending machines in its plants in California. Fred Black and Baker assisted in securing contracts between Melpar and North American. In partial return, Melpar has agreed to enter into an agreement with a vending-machine operation in which the defendants, Fred Black and Robert G. Baker, have a financial interest."

These allegations concerning a highly placed, influential, and popular employee of the Senate plunged official Washington, which reacts to even the flimsiest gossip, into a dither. Senator John J. Williams, the conscience of Congress, called on the Senate floor for an investigation of Bobby Baker and his extracurricular activities. He was joined by Senator Mike Mansfield, and the Senate Committee on Rules and Administration began an investigation in October 1963. Baker resigned his Senate post.

So began the "Bobby Baker case" on Capitol Hill. The Senate Rules Committee investigation failed to unearth what services Baker and Black had performed for North American that induced the company to grant Serv-U the vending-machine concession. But the committee did ascertain that North American signed a contract with Serv-U on January 30, 1962, three months after winning the Apollo award, and became Serv-U's first customer. Until the advent of Serv-U, the main vending-machine concession at North American's plants had been held by Canteen Corporation, an old, established firm.

The Rules Committee found that Serv-U was immediately successful. Its contract, starting at the Los Angeles plant of North American, soon was extended to cover the company's Space and Information Division at Downey, California, where Apollo and the S-2 second stage of the Saturn 5 moon rocket were being fabricated, and to the Rocketdyne Division, which was making the H-1 engines for the uprated Saturn 5 and the J-2 hydrogen engines for both the S-2 and S-4B upper stages of the Saturn 5. In all, within a year after it entered the vending-machine field, Serv-U was supplying vending-machine services to more than 40,000 employees of North American. According to a statement by Serv-U officers on a mortgage-loan application in Baltimore, Serv-U's net annual income was $720,000 to $840,000 a year. Most of this net came from the contract with North American, but some of it was derived from a later arrangement with Northrop Corporation, which had a $39,685,000 subcontract with North American to develop the Apollo parachute landing system. The Serv-U contract with Northrop had been arranged with the aid of Black. As an aside, the Senate committee investigation disclosed that two of Baker's partners in Serv-U were operators of a Las Vegas gambling casino under surveillance by the Federal Bureau of Investigation.

It was in 1962 that Baker and Black moved Capitol Vending into Melpar, which Black also represented as a Washington consultant. Hill asserted that his arrangement with Melpar was canceled after he refused to sell out to Baker and Black.

Questioned by reporters about his influence in getting government contracts, Baker responded, "What possible influence could I have? Who would I talk to?" There were a number of people on Capitol Hill Baker could have talked to. He was the protégé in the Senate of Lyndon B. Johnson when Johnson headed the Senate Democratic majority. He was "like a son" to Senator Robert S. Kerr, chairman until his death in 1963 of the Senate Committee on Aeronautical and Space Sciences. Baker had borrowed money to buy stock in Kerr's company, the Kerr-McGee Oil Company, a powerful independent which Kerr and a partner, Dean McGee, had organized in 1936. The assistant to the president of Kerr-McGee Oil and one of its directors was James E. Webb, whom Kerr had recommended for the post of NASA Administrator. Webb agreed to divest himself of his holdings in Kerr-McGee and also in McDonnell Aircraft when he took the NASA post in order to avoid possible conflict of interest. He was also a director of the Fidelity National Bank & Trust Company of Oklahoma City, owned in part by the Kerr family. Consequently, Baker was on good speaking terms with Webb. In addition, Baker had invested in land in Orange County, Florida, near Orlando, owned by Senator George Smathers and his brother, Ben Smathers, and had engaged in real-estate transactions with the Secretary of Commerce Luther H. Hodges, when Hodges was governor of North Carolina.

There was, in fact, scarcely anyone on Capitol Hill that Baker couldn't talk to or didn't know. He had befriended Black in 1959 and soon became an habitué of the two-room suite Black maintained at the fashionable Sheraton-Carlton Hotel as a base of operations. The relationship between Baker, Black, and North American Aviation officials is dramatized in recordings of conversations in the suite made by the FBI, which planted a microphone there during an income-tax investigation of Black.*

According to the FBI's eavesdropping log, on March 14, 1963, Baker and Black were in the suite along with a crowd of people. A babble of voices, "both male and female," could be heard. The log says:

"Bobby Baker reenters room. Incoming call. Black said he wanted Taylor to come up and meet Baker. Lee Taylor entered. [Leland R. Taylor was vice-president of North American Aviation in charge of public relations.) Introduced to Baker. Taylor begins discussion of charges of conflict of interest against Black. . . . Baker said, 'Hello, Mr. Director. Congratulations.' Leland Taylor had just been appointed a director of North American. Man said, 'Hello, Bobby, how are you?' "

In another portion of the FBI log, Black was recorded April 4, 1963, as saying on the telephone, "Bobby, just talked to Ed." This person was identified by the FBI as Edward Levinson, part owner of the Fremont Hotel and gambling casino in Las Vegas, Nevada, and a partner with Baker and Black in Serv-U and in a banking enterprise.

According to the log, "Black told Levinson that business was good and that he had swung a $120,000,000 to $130,000,000 deal for North American that day. He said he

* Black's subsequent conviction for income-tax evasion was thrown out by the United States Supreme Court because evidence obtained by concealed microphones is inadmissible. The court ordered a retrial. The FBI recordings were subsequently made a part of the court record.

was with Bobby 'a lot today' and will be talking more with Bobby."

The log reported that Black then dialed the telephone and asked for extension 2995. It said that FB (Fred Black) advised Bobby to call Lee Taylor "later when he was alone" and get his reaction to something which the FBI could not decipher.

On April 18, 1963, the FBI log states, "Black spoke to Bobby and said he worked for him today—as he [Black] did a job for Atwood [the president of North American Aviation] today."

After Senator Kerr died of a coronary occlusion on January 1, 1963, the log shows that Black told Kerr's partner, Dean McGee, on February 11: "Two things. First of all, since the Old Man [Kerr] died, this fellow Webb has gotten weaker and weaker where the State of Oklahoma is concerned. We sent them several things before the Senator died. Okay. We got them back and got an okay on one-third of what we wanted to put there. He's just not going to do anything for us. I'm getting concerned about a few things in Oklahoma City itself. NASA is not helping us. When the Senator was alive, he'd be helping. I want you to know North American and F. Black aren't backing up one inch. He's pushing for North American with McGee."

On another occasion, the FBI picked up a conversation between Baker and McGee. Baker discussed friends who owned two banks and a textile business. He wanted to talk about the possibility of forming a holding company.

The fact was that at the time the Apollo contract was let, Bobby Baker's influence on Capitol Hill could hardly be exaggerated. Lyndon Johnson, then the Vice-President and chairman of the Space Council, had relied implicitly on him. He referred to Baker as his "strong right arm . . . the last man I see before I go to bed at night, the first man I see in the morning."[1] Senator Smathers referred to Baker as the "third" senator from South Carolina, where Bobby was reared, the son of a mail carrier in Pickens.

During the Senate Rules Committee investigation of Baker, John L. Atwood, president of North American Aviation, was called as a witness on February 18, 1964, to discuss the vending-machine contract. He testified that Black had advised him in the fall of 1961 that Baker was interested in the concession. Atwood said he knew that Black was a friend of Bobby Baker, but he said he did not know that Black and Baker were partners in the Serv-U Company or owned 57 per cent of its stock. Black merely had said that "friends of Baker" had a vending-machine company that was "very efficient" and merely had urged that North American consider it.

Atwood related that he then summoned his senior vice-president for administration, J. L. Smithson. "I told him that I had received a strong recommendation that there was an excellent, well-managed company in the vending business that might represent some advantage to North American AID," he said. AID was a corporate unit set up within the company to handle vending-machine and food service arrangements for the employees.

Lennox P. McLendon, the Senate Rules Committee general counsel, asked Atwood if he recalled telling Smithson and Leland Taylor "that you wanted them to give consideration to a new vending company in which Robert Baker had an interest."

A.—I remember talking to both of them. I cannot recall whether they were together or not.

Q.—Did you *in substance* tell them that you wanted them to give consideration to a new vending company in which Baker had an interest?

A.—I asked them to check it. I said it had been referred by Baker to Mr. Black. I think I said this. It is my recollection.

Leland Taylor was called to the witness chair and testified, "Mr. Atwood called me into his office on some other matters. During the discussion, he mentioned that Fred Black had spoken to him about a vending company and that he had suggested that we give them an opportunity to make a proposal on some of the vending operations of the company."

The action that North American's AID took on the Baker-Black approach is recorded in the corporation's minutes of December 20, 1961, thus: "Mr. W. H. Cann called attention to the fact that practically all of the corporation's [AID] income is derived from vending operations carried on by Canteen Company. He stated that although the performance of Canteen Company had been completely satisfactory, consideration should be given to the introduction of a competitive vending company in one of the facilities administered by the corporation."

Focused on the general business activities of Bobby Baker, the Rules Committee investigation did not go into the circumstances in which the administrators of NASA had overruled their own experts to award the Apollo contract to North American.

Hill's suit against Baker et al., which had triggered the Baker scandal, dragged on for two years. During that time, Hill's attorney, David Carliner, attempted to subpoena Webb as witness. But the NASA administrator averred that he had more important matters to attend to and refused to accept the summons on the grounds of executive privilege. The suit, seeking $300,000 damages, was finally settled out of court.

Hill's allegation that Black had influenced North American to employ Melpar as subcontractor later was confirmed by Atwood in another lawsuit. This one was brought by Black against North American in the District Court, after North American fired him as a result of the Baker probe. In his answer to Black's complaint that North American had broken a long-term contract with its Washington consultant, Atwood stated that Black had brought Melpar to North American as a subcontractor without disclosing that he was also on Melpar's payroll as consultant. In answer to interrogatories by North American, Black stated that he served two other Apollo subcontractors in the same role. They were the AVCO Corporation, Wilmington, Massachusetts, which had a $36,663,000 subcontract with North American for the Command Module heat shield, and the Aeronca Manufacturing Company, Middletown, Ohio, which had a $13,351,000 subcontract for the stainless-steel honeycomb panels of the spacecraft's outer shell.

THE PHILLIPS REPORT

By 1964, as the relationship between Bobby Baker, Fred Black, and North American was becoming clear to Washington insiders, Project Apollo was moving ahead at Downey without any apparent complications. All reports were optimistic. Associate NASA Administrator Mueller radiated this confidence in an address to an Engineers' Week dinner at the

Biltmore Hotel in Los Angeles on February 21, 1964. He estimated that the number of engineers and scientists at work on the Gemini and Apollo programs was reaching 45,000, about 2.8 per cent of the national employment in aerospace. Mueller had succeeded D. Brainerd Holmes as director of manned space flight, following Holmes' resignation from that post at the end of 1963. "With respect to technology," Mueller told the engineers, "we have examined the program, subsystem by subsystem, and have found all of the knotty questions involved in advancing the state of the art are yielding to hard work. We know of no technological problem that would prevent our accomplishing the Apollo program in this decade."

This note of confidence was reiterated in reports to Congress. In the spring of 1966, Mueller's office advised the House Subcommittee on Manned Space Flight that Apollo schedules were being met. "The subsystems schedules are being met and greater assurance for continued orderly delivery of these subsystems can be expected under the incentive contracts for these subsystems," Mueller's office stated in a written report.[2]

Behind this façade that all was well, however, the internal organization in NASA headquarters was aware that the Apollo program was sick. The Apollo program director at NASA headquarters, Major General Samuel C. Phillips, submitted a report in December 1965 that North American was slipping badly on the spacecraft and that the lunar-landing schedule was in serious jeopardy. NASA administrators did not reveal this situation to either the House or Senate space committee during subsequent appearances at budget hearings in 1966.

Phillips sent a letter to Atwood dated December 19, 1965, stating that he was dissatisfied with the progress on Apollo and also on the S-2 second stage of the Saturn 5 rocket, which North American also was building. His dissatisfaction, Phillips said, was based on findings of a NASA inspection team which showed that key performance milestones in testing and in hardware delivery had slipped continuously at North American. "Degradation in hardware performance and increasing costs" were also cited. Delays in the deliveries of the S-2 had caused a year's delay in the development of the entire flight article. Even though the Apollo program was revised because of delays, the S-2 flight stage still lagged behind schedule. Moreover, the Phillips report added, "the S-2 cost picture . . . has been essentially a series of cost escalations with a bow wave of peak costs advancing steadily throughout the program life."

North American's estimate of the cost of ten S-2 stages had more than tripled, NASA's Apollo program manager stated, and the S-2 was still plagued with technical problems. "Welding difficulties, insulation bonding, continued redesign as a result of component failures during qualification, are indicative of insufficient aggressive pursuit of technical resolutions during the early phases of the program," the report stated.

Aside from the S-2, there had also been a "history of slippages in meeting key milestones in the development of the Apollo Command and Service Modules." The first manned spacecraft 012 had slipped more than one year, in spite of the fact that program schedules had been revised to accommodate the problems that North American was encountering in building the vehicle. Ground testing had been delayed from three to nine months in less than two years. Moreover, the report said, "technical problems with electrical power capacity, Service Module propulsion, structural integrity, weight growth, etc., have yet to be resolved."

Phillips observed that the problems seemed not to be the result of an inadequate work force. To the contrary, he and his inspectors believed that both the S-2 stage and the Apollo Command and Service Module construction programs were overmanned and "could be done—and done better—with fewer people."

Finally, the report stated: "Delayed and compromised ground and qualification test programs give us serious concern that fully qualified flight vehicles will not be available to support the lunar-landing program. NAA's [North American Aviation] inability to meet spacecraft contract use deliveries has caused rescheduling of the total Apollo program. There is little confidence that NAA will meet its schedule and performance commitments with the funds available for this portion of the Apollo program. There is no evidence of current improvement in NAA's management of these programs of the magnitude required to give confidence that NAA performance will improve at the rate required to meet established Apollo program objectives."

Phillips reported that he found that corporate-level interest in the Apollo and S-2 programs at North American was "passive." He said he had observed that the main area of corporate interest "appears to be in the Space and Information Division's financial outlook and in their cost estimating and proposal effort." He said, "This does not relieve the corporation of its responsibility and accountability to NASA for results. . . .

"The right actions now," Phillips advised, "can result in substantial improvement of position in both programs in the relatively near future."

This alarming situation in the Apollo program was not revealed publicly by the space agency. A year later, at the Apollo program briefing at Houston, Shea told two hundred news-media correspondents that "we stayed on the planned [development] curve or actually ahead of the planned curve" in building spacecraft 012, which was then at Cape Kennedy being prepared for the first manned flight.

"LET'S GET OUT. . . ."

Spacecraft 012 was born at Downey in August 1964. The basic structure was completed in September 1965, and the subsystems were assembled and installed in the Command Module between September 1965 and March 1966. Following a "Customer Acceptance Readiness Review" which was conducted by NASA at the factory during the summer of 1966, NASA issued a certificate of flight worthiness and authorized the spacecraft to be shipped to the John F. Kennedy Space Center (KSC). The certificate listed open items of work to be completed at the Center but did not state how many there were. Later, it was found that 113 "Significant Engineering Orders" had not been accomplished at the time the Command Module was delivered to NASA.

The amount of effort and rework required at the Kennedy Space Center to put spacecraft 012 in shape for flight was greater than that required for the first Gemini spacecraft. This in itself was not considered disturbing, because Apollo was a more complex vehicle than Gemini. However, the engineers at Kennedy Space Center inferred from the amount of work to be done on the spacecraft that the design, qualification, and fabrication process had not been completed adequately prior to its shipment to KSC.[3]

Nevertheless, a design certification review was held at NASA headquarters during September and October 1966, under the chairmanship of Mueller and the Board issued a design certification document on October 7, 1966, which approved the design as flightworthy, pending satisfactory resolution of the listed open (incompleted) items.

At this point, it seemed likely that the manned flight might be set for late December or January. During tests at KSC in mid-October, however, the oxygen-system regulator failed. It was removed and redesigned. Later that month, the entire Environmental Control Unit was removed from the spacecraft and returned to the factory for a redesign in the cooling system.

In the meantime, the propellant tank in the Service Module of spacecraft 017 ruptured during a test at Downey. Vehicle 017 was to fly on the first Saturn 5 in an unmanned test during 1967. It was found that the 017 tank had been weakened by pressure tests and, since similar tests had been applied to 012, the Apollo directorate ordered the latter vehicle removed from the altitude chamber and disassembled for inspection. The tank of 012 was found to be satisfactory. But in the meantime, a leak developed in the new water-glycol evaporator which had been installed in the Environmental Control System to solve an overheating problem. The unit had to be returned to factory for repairs, and came back to KSC on December 14, 1966. Manned and unmanned tests of spacecraft 012 were run at sea-level pressure and at simulated high altitude in the altitude chamber on December 29 and 30. Grissom, White, and Chaffee reported that everything worked fine.

Spacecraft 012 was then mounted atop uprated Saturn 1 booster No. 204 at Launch Complex 34. The combined Apollo-Saturn 204 vehicle stood 224 feet tall on the 430-foot-diameter concrete launch pad. It was enclosed by a mobile service tower 310 feet high, weighing 3500 tons. About five hours before the launching, this huge structure would roll back about 680 feet from the rocket on steel rails.

Within the service tower there were seven working levels

above ground and eight movable platforms. These enabled technicians to reach any part of the rocket or spacecraft during countdown and checkout. The tower was like an office building, swarming with employees. who wore helmet-like hard hats on their heads and organization insignia on their shirts. It was as busy as a hive 24 hours a day, with technicians, engineers, mechanics, electricians, and inspectors riding up and down the elevator from one level to another. The hatchway entrance to Apollo was reached on Level 8 through a chamber on an extendable arm called the "White Room." A thousand feet away was the steel and concrete "blockhouse" control center.

At 7:55 a.m. on January 27, 1967, KSC technicians began an all-day "plugs out" test of spacecraft 012. "Plugs out" meant simply that the vehicle's systems would be put through a simulated countdown and launch, during which the power supply would be switched from an external source, the Complex, to an internal source, the spacecraft. The main internal power supply for the spacecraft were its fuel-cell batteries in the Service Module. However, these were not to be used during this test. They were to be simulated by storage batteries outside the vehicle.

The crew, suited up as though ready for an actual flight, entered the spacecraft at 1 p.m. Grissom squirmed in first to take the left-hand seat farthest from the hatch; then White, who took the middle couch; and finally Chaffee. After a brief period of adjustment, the hatch was closed.

The hatch actually consisted of three doors. The outermost, called the boost-protective-cover hatch was a part of the covering which shielded the Command Module during the launching. The middle door was termed the "ablative hatch" because it became a part of the heat shield of the vehicle when the boost-cover hatch was jettisoned after ascent through the atmosphere. The third, or inner hatch, sealed the pressure vessel in which the astronauts rode.

Grissom noticed an odor in his spacesuit oxygen supply. The countdown was held while the oxygen was sampled. No impurities were found and the countdown was resumed, just as though an actual launching was to take place. Beneath the spacecraft, however, the Saturn booster and its S-IVB second stage were unfueled—though the escape tower above the Apollo had a live, torpedo-shaped rocket in it.

The astronauts watched the progress of the countdown on the instrument consoles during the afternoon, talking to the blockhouse over the spacecraft communications system. The cabin had been purged with oxygen, which had been kept at a pressure of 16.7 pounds a square inch, about two pounds higher than normal sea-level atmosphere. The purpose of this standard procedure on all American space vehicles was to seal the inner hatch against a leak. Following the launch, as the outside atmospheric pressure dropped, the inside pressure would be reduced by bleeding the oxygen to space, down to about 5 pounds per square inch. Against zero pressure outside, that sealed the hatch tightly.

There was no indication during that long January afternoon beside the warm sea that none of the crew would ever emerge from the pressurized cabin alive.

Shortly after 5 p.m., communications trouble began between the spacecraft and the blockhouse. The countdown was held at 5:40 p.m. to enable technicians to find out what the trouble was. The problem seemed to be a continuously live microphone in the cabin, which the crew could not turn off. Then new problems appeared in the entire communications network linking the spacecraft with various ground stations in the service tower as well as the blockhouse. When this trouble developed, the countdown had reached the point where the switch from external to internal power was to be made. It was never made.

At 6:30 p.m. there was a surge in alternating-current voltage, and four seconds after 6:31 p.m., a voice from the spacecraft came over the faulty communications system crying, "Hey," or "Fire." Hearers were not sure which word was said or whether both words were said. Some of them believed the word or words were uttered by Grissom. Two seconds later a voice from the spacecraft was heard saying "I've" or "we've." Again, the men with headsets on outside the craft could not identify what was being said. Then there was more: "Got a fire in the cockpit." That came through clearly and the listeners interpreted the message as "We've got a fire in the cockpit." The voice sounded like Chaffee's. For another 6.8 seconds, there was no further word from the cockpit. Then another garbled transmission came through. Listeners heard it as "They're fighting a bad fire . . . let's get out . . . open 'er up," or "We've got a bad fire . . . let's get

. . . we're burning up," or "I'm reporting a bad fire . . . I'm getting out." The transmission ended with a cry of pain. The voice again appeared to be Chaffee's.

As soon as he heard Chaffee say "fire in the cockpit," Donald O. Babbitt, North American engineer who was the Pad Leader, ordered his lead mechanic, James D. Gleaves, to "get 'em out of there." As he turned to a communications box on his left, Babbitt saw flame erupting from the base of the spacecraft. The aluminum-walled pressure vessel had split like a melon. Flames and smoke were pouring into the insulation space between the pressure vessel and the heat-shielded exterior wall of the spacecraft and escaping through access ports into levels 7 and 8 of the service tower.

Babbitt was struck by the concussion of the erupting flames and gases and hurled against the communications box. He struggled to regain his balance, and his first thought, then, as he later recalled, was to get out, but he didn't. He raced across the White Room to a communications technician known as the "elevator talker" and instructed him to tell Robert Moser, the test supervisor, "we're on fire."

The burst of flame and the sounds of the pressure vessel rupturing under high pressures generated by the fire caused technicians to believe that the entire Command Module was about to blow up. Many of them fled to the elevator. Babbitt said that he, Gleaves, and technicians Jerry W. Hawkins and Stephen B. Clemmons grabbed the only carbon-dioxide bottles available and returned to the White Room to remove the hatches. The outer boost-protective hatch was not fully latched, but it was necessary to insert a special tool into a slot in order to get a hand hold on it.

The White Room was filling with dense, dark smoke from the interior of the Command Module and from secondary fires which the eruption of smoke and flame from the spacecraft had set off on level 8. Some of the technicians were able to find gas masks, but others could not and were driven away from the spacecraft by the smoke. Some of the gas masks could not be made to work. Even those that did work were insufficient protection against the dense smoke. They were designed for use in toxic rather than dense-smoke atmospheres.

Other than these inadequate articles, there was no emergency fire equipment in the service tower. It was as though no one had ever anticipated a fire in the spacecraft, or even imagined such a contingency.

Meanwhile, visibility in the White Room was close to zero. Blinded by smoke, Babbitt and four men worked on the outer hatch, by touch part of the time. The technicians were concerned that the heat from the spacecraft would ignite the solid-fuel rocket in the launch escape tower overhead. If that happened, they might all be killed and the pad would be wrecked. With this grim possibility hanging over them, they nevertheless worked on, struggling with the hatch, taking turns backing out to gulp breaths of fresh air and then returning to the smoky White Room. The nightmare seemed to have no ending. The outer hatch came off. As one man rushed out of the smoke for air, he passed the hatch tool to a replacement who came in to continue the struggle with the ungainly mechanism. The second hatch was opened, and then there was only the inner one left. This had to be unlatched and lowered to the floor of the spacecraft.

Hawkins, Clemmons, and L. D. Reece, a North American systems technician, worked on the inner hatch. Hawkins used the hatch tool to unlatch it while Clemmons tried to get a hand hold on it. When the hatch was undogged, it could not be lowered out of the way because of an obstruction. It would go only part way down. Intense heat and smoke came out of the opening, but no flames were visible inside. Hawkins tried to see into the interior. He called the crewmen. There was no answer. Gleaves, who had been forced away by smoke, returned after catching his breath and saw the hatch was part way down. He kicked it. The hatch fell farther down toward the floor. Gleaves took a flashlight from his tool box and peered into the dark smoke. He could see nothing except the faint glow of lights, which were still burning inside. They seemed like candles in the smoke.

Babbitt looked in. He could see two crewmen but could not identify them because they were in their suits and their helmet visors were closed. Backing out of the cabin to catch his breath, Babbitt peered inside again. This time he could see more clearly. He saw White lying on his back with his arms over his head as though reaching for the hatch. A figure, which he believed from its position in the craft to be Grissom, was lying with one arm reaching for the hatch under White's arm. He could not see Chaffee.

"My observation at the time of hatch removal," Babbitt told investigators, "was that the flight crew were dead and that the destruction inside the Command Module was considerable."

Anxious not to break security regulations, Babbitt did not report what he had seen and what his conclusions were over the entire communications network. He simply advised the ground that he would not at that time describe what he had seen. He then descended the elevator and made his report below. At the base of the tower, he met two NASA physicians and a Pan American Airways physician * and advised them that he believed the crew was dead. During this period, no emergency effort was mounted to remove the astronauts from the cabin and determine by direct examination whether they were dead or unconscious.

It took approximately five minutes from the first report of fire in the spacecraft at 6:31 p.m. to get the hatches open at 6:36 p.m. Because they were not equipped to venture into the cabin, filled with heat and smoke, the technicians who had dared death from the possibility of escape-tower ignition could not bring the bodies out and attempt resuscitation. The three doctors did not reach the White Room until nine minutes after the hatches were open, at 6:45 p.m. The spacecraft was still smoldering and the smoke was too thick to enable them to examine the crew without breathing apparatus.

"After a quick evaluation," the accident-investigation report states, "it was decided nothing could be gained by attempting immediate egress and resuscitation. It was evident that the crew had not survived the heat, smoke, and thermal burns."

Later, it was estimated by investigators that the crew lost consciousness between 15 and 30 seconds after one of the suits failed and introduced toxic gases into all the suit-breathing loops in lieu of oxygen.

Autopsies were performed the next day, January 28, at the Cape Kennedy Air Force Station by three physicians from the Armed Forces Institute of Pathology. They were Colonel Edward H. Johnston, Army, Commander Charles J. Stahl, Navy, and Captain Latimer E. Dunn, Air Force. The cause

* Pan American Airways has the "housekeeping" contract with the Air Force for Cape Kennedy.

of death was found to be asphyxia due to inhalation of toxic gases, principally carbon monoxide. Thermal burns were listed as the "contributory" cause of death.

A belief that the victims literally had been "burned to death" was widespread, but the fact was that the flight suits had provided considerable protection and the burns were not as severe as many observers believed.

A separate statement on this subject was issued by Dr. Berry, chief of the Manned Spacecraft Center Medical Department and head flight surgeon in the program. The statement, issued February 27, 1967, disagreed with the finding that burns were a contributory cause of death. Berry stated that "further toxicology studies" confirmed the cause of death as asphyxiation by carbon monoxide," he said, and "a rapid building up of carbon monoxide could result in unconsciousness in seconds and death very rapidly thereafter." While the crew suffered second- and third-degree burns, "these were not of sufficient magnitude to have caused death."

After the doctors had decided that the men were dead, an attempt had been made to remove the bodies. However, because the intense heat had fused the spacesuits in places to the molten nylon webbing of the couches, it was feared that removal of the bodies would derange vital evidence bearing on the cause of the fire. Consequently, it was decided to leave the bodies in position until photographs had been made of the charred interior. Grissom, White, and Chaffee, still in their spacesuits, were not taken out of spacecraft 012 until 2 a.m. January 28, seven and one-half hours after it is estimated that they perished.

A charred wreck was all that remained of spacecraft 012, the first Apollo that was to fly with a crew. A profound wave of shock and grief swept over the thousands of men and women in Project Apollo, and billowed out into the nation. America's most ambitious program in space had claimed its first victims. It had claimed them in a manner which no one had anticipated. The magnitude of the disaster was stunning.

FOUR MINUTES

Within eight hours of the time the bodies were removed from Apollo 012, NASA assembled an investigating team which was charged with "determining the causes or probable causes of all failures, taking corrective or other actions and submitting written reports of such determinations and actions to the Deputy Administrator."

Appointed to the investigatory body, known officially as the Apollo 204 Review Board, were Floyd L. Thompson, director of the Langley Research Center; Colonel Frank Borman, astronaut; Maxime Faget, director of engineering development at the Manned Spacecraft Center; E. Barton Geer, associate chief of the Flight Vehicles and Systems Division at Langley; George Jeffs, chief engineer on Apollo for North American; Frank A. Long, vice-president for research and advanced studies at Cornell University and a member of the President's Science Advisory Committee; Colonel Charles F. Strang, chief of Missiles and Space Safety Divisions, Air Force Inspector General; George C. White, Jr., director of Reliability and Quality in NASA's Apollo Program Office; and John Williams, director of Spacecraft Operations, Kennedy Space Center.

The composition of the Board, on which six of the nine members including the chairman were NASA personnel and the seventh was the chief engineer for the contractor, triggered widespread criticism. Even the non-NASA members of the Board admitted privately that it certainly looked like an "in house" committee. When Long realized how long the investigation would take, he begged off because of the pressure of his duties at Cornell. He offered his services as consultant, however, and NASA promptly accepted them.

The members of the Board had been selected by Webb and Seamans. They acted under instructions from the White House to get to the bottom of the accident, correct the conditions that caused it, and get Project Apollo back on the rails. Neither the Administrator nor his deputy intended a cover-up. From their point of view, the men who could investigate the Apollo fire most effectively were those who were familiar with the engineering details of the spacecraft. Nothing was further from their minds than a whitewash. Along with the rest of the country, they were shocked and dismayed by the tragedy. Politically, as well as technologically, it contained the seeds of disaster, especially if the Russians should leap ahead in the lunar sweepstakes. Bearing the burdens of intensifying war in Southeast Asia and an unpopular tax increase to finance its escalation, President Johnson was deeply troubled by this sudden collapse of a program that seemed to be going so well. As Vice-President, Johnson had been one of the architects of Project Apollo, and there was a chance the landing could be made during his term in office in 1968, but now, that was gone. He had known Gus Grissom of old, since the Project Mercury days, and he had been delighted with Ed White's spacewalk. Chaffee's brightness and humor would have appealed to him. Now these men were snuffed out like candles in the most improbable, unlikely, and incomprehensible accident anyone could have imagined.

Following the criticism, especially in the influential *Washington Star* and *New York Times*, Seamans removed Jeffs from the Board and asked him to serve as consultant.

In place of Jeffs, Robert W. Van Dolah, director of the Explosive Research Center, Bureau of Mines, Department of the Interior, was asked to serve on the Board, and he did.

The Board submitted its report to Webb on April 5, 1967. It surmised that the fire had begun near Grissom's couch, and probably resulted from a very hot electrical arcing, caused by a short circuit. There was so much evidence of arcing after the fire that the one which started it could not be determined. However, a wire bundle which had been carelessly placed where it could be bruised by an equipment hatch door was suspected as the probable point of origin of the fatal electrical arc. Once ignited, the fire burned in three stages in the rich atmosphere of pure oxygen at 16.7-pounds-per-square-inch pressure. The first stage produced a violently hot fire which lasted about 15 seconds after Chaffee sounded the alarm. The flames moved rapidly from the left of Grissom's couch all the way across the 12-foot cabin to Chaffee's couch, traveling along a nylon net which had been strung under the couches to prevent articles from falling into equipment areas during ground testing. The Raschel netting, as it was called, was supposed to be highly fire-resistant. So were the Velcro fasteners used throughout the cabin and Polyurethane plastifoam pads, installed to protect the struts and hatch during emergency egress by the astronauts. These plastic materials were fire-resistant in a normal atmosphere of oxygen and nitrogen, but in pure oxygen at high pressure they burned like gasoline. The flames rose vertically and then spread out across the cabin ceiling. The Raschel netting provided not only fuel but also firebrands of burning, molten nylon. A wall of flames extended along the left side of the cabin, preventing Grissom from reaching the valve which would have vented the cabin's high-pressure oxygen to the outside. Pressure inside the capsule mounted with incredible rapidity, and 15 seconds after the fire started, the inner wall of the capsule ruptured at 29 pounds per square inch.

The fire then entered a second stage as flames rushed through the cabin toward the break in the pressure vessel, pushed by violent convection of gases. The swirling of this whirlpool scattered firebrands throughout the cabin and spread the fire to every part of the interior. Heat became so intense that it melted the soldered connections of the aluminum oxygen lines in the Environment Control System, and additional oxygen poured into the cabin to enhance the burning. About five or six seconds after the inner cabin wall split, the oxygen pressure dropped to the point where it would no longer support combustion, and the fire began to smolder. Heavy smoke formed and soot fell all over the interior. It was during this final period of the fire that carbon monoxide was formed. With their suit-breathing loops breached by the fire, the astronauts were quickly overcome. The Board estimated that the Command Module atmosphere was lethal 24 seconds after the fire started. The hatches were opened 4 minutes and 36 seconds later.[4] The medical report estimated that consciousness was lost between 15 and 30 seconds after the first suit failed, and "chances of resuscitation decreased rapidly thereafter and were irrevocably lost in four minutes." It may therefore be considered probable that chances of resuscitating one or more of the astronauts had been lost by only 36 seconds.

However, because of the heat, smoke, and lack of equipment, it was not humanly possible for the rescuers even to attempt resuscitation promptly. Therein lies the essential tragedy of this fatal accident. If a trained rescue team, properly equipped, had been stationed near the White Room, the fire need not have been fatal to all three, or to any of the crew. The three astronauts were not burned to death in the fire; the report shows clearly they were asphyxiated after the pressure vessel burst and the fire died out.

The medical report tells the whole story. "Hatches were

opened at approximately 6:36 p.m. and no signs of life were detected. Three physicians viewed the suited bodies at approximately 6:45 p.m. and decided that resuscitation efforts would be to no avail."

MEA CULPA

The report of the Board showed that the Apollo spacecraft was a death trap under the conditions of the tests, and that the astronauts could not have escaped the fire once it started. The conditions leading to the disaster were: a sealed cabin, pressurized with an oxygen atmosphere; an extensive distribution of combustible materials in the cabin; vulnerable wiring carrying spacecraft power; vulnerable plumbing carrying a combustible and corrosive coolant; inadequate provisions for the crew to escape; and inadequate provisions for rescue or medical assistance.

"The Apollo team," the Board stated, "failed to give adequate attention to certain mundane but equally vital questions of crew safety. The Board's investigation revealed many deficiences in design and engineering, manufacture, and quality control."

Specifically, the Board found, the Environmental Control System had a history of many removals and failures, and that the corrosive glycol coolant leaked at solder joints. The coolant was "both corrosive and combustible." There were deficiencies in design, manufacture, installation, rework, and quality control in the electrical wiring. No vibration test had been made of the spacecraft, even though it was going to carry men atop the most powerful rocket in the world, the uprated Saturn 1. No design features for fire protection had been incorporated in the craft.

"These deficiencies created an unnecessarily hazardous condition and their continuation would imperil any future Apollo operations."

The test of the spacecraft on the ground, with a 100-percent-oxygen atmosphere under high pressure in the cabin, represented a serious fire hazard when flammable materials were present. The report noted that the hazard was not recognized. There were no emergency fire, rescue, and medical teams present, and the design of the service tower hindered emergency rescue work.

Once the fire broke out, the crew had nothing with which to extinguish it except a water pistol, used for drinking in flight. Because of the rapid build-up of pressure in the cabin, the inward-swinging inner hatch could not be opened; excess pressure in the cabin kept it tightly sealed. The report said, "The crew was never capable of effecting emergency egress because of the pressurization before rupture and their loss of consciousness soon after rupture."

The Board made it clear that the hatch would have to be changed so that a crew could get out of the spacecraft quickly. NASA subsequently made the change to a single hatch, which would be secure in space yet easy and quick to open on the ground. An explosive hatch had been considered on Apollo, but, ironically, had been discarded by NASA and North American because of Grissom's experience in the Mercury spacecraft Liberty Bell 7 when the spacecraft explosive hatch detonated prematurely soon after it came down in the sea from suborbital flight. Grissom had to scramble out and narrowly escaped drowning when his suit filled with water.

In addition to the fact that 113 significant engineering orders had not been accomplished when spacecraft 012 was delivered to NASA, an additional 623 engineering orders—all calling for changes—were issued after delivery. Of these, 22 were not recorded as having been made. Noncertified equipment items—i.e., items which have not been approved for safety or reliability—were installed in the Command Module at the time of the test. There were discrepancies between North American Aviation and NASA specifications on flammable materials. It developed at Congressional hearings that the flammability of materials which burned had not been tested in a pure-oxygen atmosphere.

The release of the report to the public May 7, 1967, was followed immediately by a hearing, called by Representative Olin E. Teague, chairman of the House Subcommittee on NASA Oversight. The title of the committee seemed singularly appropriate to the subject matter. Failure to recognize the hazard of the test in spacecraft 012 was, perhaps, the greatest oversight of NASA's career. "We cannot bring back these brave men who gave their lives to the program," said

Teague, "but we believe our inquiry will help others avoid their fate."

It was a solemn moment for James E. Webb. "I am not happy to be here today," he said. "If any man in this room wants to ask for whom the Apollo bell tolls, it tolls for every astronaut and scientist who will lose his life on some lonely hill on the moon or on Mars."

John L. Atwood, president and chairman of the board of North American Aviation, said the tragedy had been "appalling and shocking to us at North American." He cited the management's credo of "personal responsibility in daily effort." The initials of that motto spelled PRIDE, he pointed out. There was another acronym at the plant, PEP. It stood for a Performance Evaluation Program the company designed to motivate the employees to higher standards of performance. Even so, there had been ". . . deficiencies," Atwood admitted. "We recognize that while we have made every effort to avoid any deficiency, some deficiencies did in fact exist. That these discrepancies have not been specifically identified with the accident does not diminish our resolve to correct them."

Ironically, a 584-page survey on "Space Flight Emergencies and Space Flight Safety," which had been in preparation for a year by the subcommittee's staff, was published a month after the tragic accident. The study analyzed every conceivable kind of accident that could happen in flight—except for a fire in the spacecraft on the pad. Apparently, no one had even considered that possibility.

Webb added a note of optimism: "The Apollo Review Board has filed a report that includes many serious criticisms of both NASA and industry. We will take our part of the blame for what we have done or left undone, but I believe this committee can have confidence that NASA and its contractors have the capability to overcome every deficiency. . . . Whatever our faults, we are an able-bodied team."

Dale D. Myers, the Apollo program manager at North American, a hard-driving, plain-speaking engineer, recited a litany of mild frustration from the contractor's point of view. North American, he said, had not received the go-ahead for the lunar-landing concept until 1962, and had to revise the humidity-control system and define the advanced Block 2 spacecraft in 1964. In 1966, one NASA-inspired modification after another added 70 per cent more wiring and required overlays, splicing, and other changes. These made it impossible to complete and check out the vehicle before it left the factory, Myers said. "We were not successful in checking out before the shipment to Florida because of last-minute changes." As far as noncertified items in the spacecraft were concerned, he added, there were only four. He did not identify them but said they would have been certified later and had nothing to do with the fire.

Somewhere in the 20 miles of wiring in Apollo 012 there had been an electrical arc which had started the fire in the pure-oxygen atmosphere. While the point of ignition could not positively be identified, the Review Board inclined toward the belief that a wire bundle which was subject to chafing by a storage-compartment door under Grissom's couch was the most likely source of the arc.

Myers' testimony before the Subcommittee on NASA Oversight recalled a remark that Joseph Shea, the Apollo manager at the Manned Spacecraft Center, had made at the December news briefing. He had quoted a Russian engineering proverb saying "The better is the enemy of the good." The moral that Shea drew from it was that one cannot continue to revise designs and hardware in a development and meet a schedule. Design revisions in Apollo throughout its developmental life, together with the squeeze of NASA's schedule, had generated considerable pressure on the contractor, Myers said.

Had the pace of the program, dictated by the deadline for accomplishment of the mission that President Kennedy had imposed in 1961, contributed to the carelessness, the slipshod workmanship, and curiously detached management which the report had found? Floyd Thompson, the Apollo Review Board chairman, stated that he did not believe the pace of the program had any bearing on the accident. Moreover, the over-all design of the vehicle, including the adoption of the pure-oxygen atmosphere, appeared to the Board to be satisfactory. Apollo would take men to lunar orbit if it could be made to function properly. Moreover, in spite of the outcry against the hazards of a pure-oxygen atmosphere, it would be retained.

While the use of a diluent gas, such as nitrogen or helium, would reduce the risk of fire, Thompson agreed, it would introduce other risks. One of these was the difficulty of maintaining a healthy mixture in a two-gas system. In Thompson's opinion, the United States simply had not developed a two-gas system technology for spacecraft, in contrast to the Russians, who had used a two-gas, oxygen-nitrogen system since Vostok I. In the opinion of Thompson and the Board, the route toward reducing fire hazard lay in the direction of reducing the likelihood of igniting a fire and elminating materials that would burn. Someday, a two-gas system might be developed for space vehicles, but not now.

The aspect of the accident most difficult to understand for both House and Senate space-committee members was why everyone connected with NASA and the contractor was so ill-prepared for it. Colonel Frank Borman, the commander of Gemini 7, who served on the Board, answered that point concisely. "None of us really placed any stock or gave any concern about a fire in the spacecraft," he said. "There was no undue concern about the hazards of these conditions. I can say for the crew that was killed—they did not think it unduly hazardous, or they would not have entered the spacecraft. I don't believe any of us recognized that test as hazardous. While silk-scarf attitudes are attributed to us, in the final analysis we're professionals and we will not accept undue risks. No one was concerned about the time it took for egress. There was an emergency egress exercise planned at the conclusion of the test. We were primarily concerned about the booster and hypergolic [ignite on contact] fuels in the Service Module.

"I accept my share of the blame," Borman concluded.

Thompson remarked that "if we could have gotten the victims much sooner to doctors, we might have saved them."

But Dr. Berry asserted that when the hatches finally were open, the crew could not have been saved. It was too late.

Both the House and Senate committee members demanded that Webb produce the Phillips report, which had detailed deficiencies at North American in 1965. Webb stalled, urging that the disclosure of such reports, which he regarded as confidential, would "kill the goose that laid the golden egg." He did not make clear what he meant by this, but his remark was generally interpreted to mean that disclosure would discourage frankness in future reports. Members of the House subcommittee demanded to know what NASA had done to remedy conditions at North American in the 13 months between the Phillips report and the fire. It appeared that nothing had been done.

Senator Clinton P. Anderson, chairman of the Senate space comittee, told Webb bluntly that members of his committee were not satisfied with Webb's explanation of follow-up action taken on the Phillips report. Webb had taken the position that the severe criticism in the report was typical of the manner in which NASA dealt with all of its contractors and did not necessarily mean that North American was any worse than the others. In fact, Webb stated, "North American's work in a large area is good. We could not go unless it was. The manufacturing job done by the company is much better than they were given credit for."

It was unnerving to some spectators to hear the head of the space agency defend the operations of a contractor whose carelessness had contributed to the worst tragedy of the space age.

Webb sent a letter to Senator Anderson saying he had tried to think of ways through which the committee could reestablish the confidence in NASA "it formerly had." "I have tried to find some way," his letter said, "this could be done without violating the basic commitments we have made to individuals and companies to regard information given as confidential." No one on either the Senate or House committee knew of such a policy, requiring confidentiality in the relationship between NASA and its contractors.

In the opinion of Senator Margaret Chase Smith, Webb should have brought the difficulties that General Phillips described at North American to the attention of the committee since the project represented 25 per cent of the NASA budget. Why did NASA select North American over Martin in the first place? she demanded.

Webb carefully explained that the agency administrators felt that Martin's higher rating in 1961 by the Source Evaluation Board was due to experience Martin had gained in making a preliminary Apollo design study for NASA—a study in which North American had not participated. "We discovered that the Board had inadequately applied numerical equations to the element of experience," he said, and in addition, North American's cost estimates had been much lower than Martin's. For those reasons, he said, he, Dryden, and Seamans overruled the 190 panels of the Source Evaluation Board.

Webb gave the Senate Committee the confidential report by Seamans on the selection of North American which asserted that while Martin appeared to have the best technical approach to Apollo, "North American had by far the greatest technical competence." Because of the "very great engineering complexities involved in the Apollo project," the report stated, the exceptional competence which the NASA administrators believed North American to possess "had to be considered as an overriding consideration in selection of a contractor."

The Congressional hearings blew themselves out in May without discovering what had really happened in the Apollo program, and without determining why it had failed. The blame for the tragedy seemed to fall on anonymous, faceless electricians, who wired badly and occasionally left wrench sockets in a wiring nest. One such metal socket had been discovered in spacecraft 012 during the investigation.

Shea, who had radiated such confidence that all was well in December, was transferred from the Manned Spacecraft Center to Apollo headquarters in Washington, and shortly thereafter he resigned to enter private industry. North American management was shaken up. Harrison A. Storms, president of the Space and Information Division, resigned. Atwood hired William B. Bergen, president of Martin-Marietta, to take over Storms' post, and Martin's launch operations director, Bastian (Buzz) Hello, as vice-president of North American for launch operations. So was answered the NASA administrators' contention that North American had "by far" greater technical competence to handle the "very great engineering complexities" of Project Apollo.

On June 15, 1967, nearly five months after the fire, NASA made public a report showing that there were 109 malfunctions and deficiencies in a second Apollo spacecraft, No. 017, which was planned to fly unmanned atop the first Saturn 5. The defects included cut wires which made the launch escape system inoperative. Other defects crippled the guidance system, the water supply, the flight instruments, the electrical system, two tape recorders, the reaction control system, the main propulsion engine, the radio, the parachute-landing system, and a flasher beacon designed to make the craft visible at night in case of an emergency landing on the sea.

The battery powering the beacon had exceeded its lifetime. Moreover, it had been installed upside down, so that a connector would be struck by one of the parachutes when it came out of its canister during descent Two replacement batteries shipped to the Kennedy Space Center were also found to be exhausted, and their dates of manufacture were unreadable. A third replacement battery had a short circuit.

Spacecraft 017 had been accepted by NASA on December 14, 1966. Shea had told the December news conference that "it came through as a very clean spacecraft.

"There were only two or three items that had to be fixed on it," he had said, "and by and large I think as far as the checkout in the factory is concerned we've got the project reasonably well hacked."

On January 31, 1968, more than a year after the accident, the Senate space committee issued a report on its Apollo fire investigation The report merely reiterated the general feeling of committee members that "overconfidence" and "complacency" exhibited by NASA in the Apollo program had contributed to the disaster. This well-worn criticism however, did not satisfy all the members.

In a separate statement, Senator Walter F. Mondale of Minnesota, a member of the Democratic majority, said, "NASA's performance—the evasiveness, the lack of candor, the patronizing attitude exhibited toward Congress, the refusal to respond fully and forthrightly to legitimate Congressional inquiries and the solicitous concern for corporate sensitivities at a time of national tragedy—can only produce a loss of Congressional and public confidence in NASA problems."

Two Republican members, Edward W. Brooke of Massachusetts and Charles H. Percy of Illinois, called for further discussion of the space agency's capability of achieving the lunar landing in this decade. "In our opinion," they said, "a delay of the landing into the next decade brought about in the interest of safety or as a result of efforts to avoid excessive costs that might develop in holding to the present schedule would in no way be a political or a technical disaster."

However, the Senate report was a pale anticlimax. By the time it was made public, Project Apollo was back on the rails again, accelerating toward the Kennedy deadline, and much of the rhetoric of condemnation had been drowned out by the thunder of the greatest rocket men had ever launched—the Saturn 5.

Chapter 13

Saturn 501: The Big Shot

Twenty-two years had passed since the last of the V-2 rockets had been launched from Germany in Operation Backfire. In that interval, American technology had produced a rocket 150 times as powerful as the old Vengeance weapon. It was the Saturn 5, designed to propel three astronauts in a 45-ton Apollo spaceship, including the lunar-landing module, to the moon.

On the morning of November 9, 1967, a crew of 450 men prepared to launch the Saturn 5, carrying an operational but unmanned Apollo and a boilerplate lunar module, on its first test flight. The first of the great Saturns, No. 501, stood in its mobile launch tower on Pad A of Complex 39 at the John F. Kennedy Space Center. The lunar launch complex is slightly the northwest of the old military pads on Cape Kennedy. In the morning sunlight, the rocket gleamed a dazzling white, which was enhanced by the contrast of black bands where the three stages, the instrument unit, and the spacecraft were joined and of vertical, black stripes on the booster, where "UNITED STATES" was painted in red block letters. With the shrouded Apollo spacecraft and its escape tower on top, Saturn 501 stood 364 feet tall. It was 62 feet taller than the Statue of Liberty (including the pedestal) and only 34 feet shorter than the South Tower of the Wrigley Building in

Chicago. Fully fueled, the lunar launch vehicle weighed 6,200,000 pounds. It was the biggest and the heaviest structure man had ever attempted to launch, and it probably was the last big chemical rocket that the United States would develop in this century.

At the end of 1967, the Saturn 5 was the ultimate product of rocket technology. It represented the climax of an era

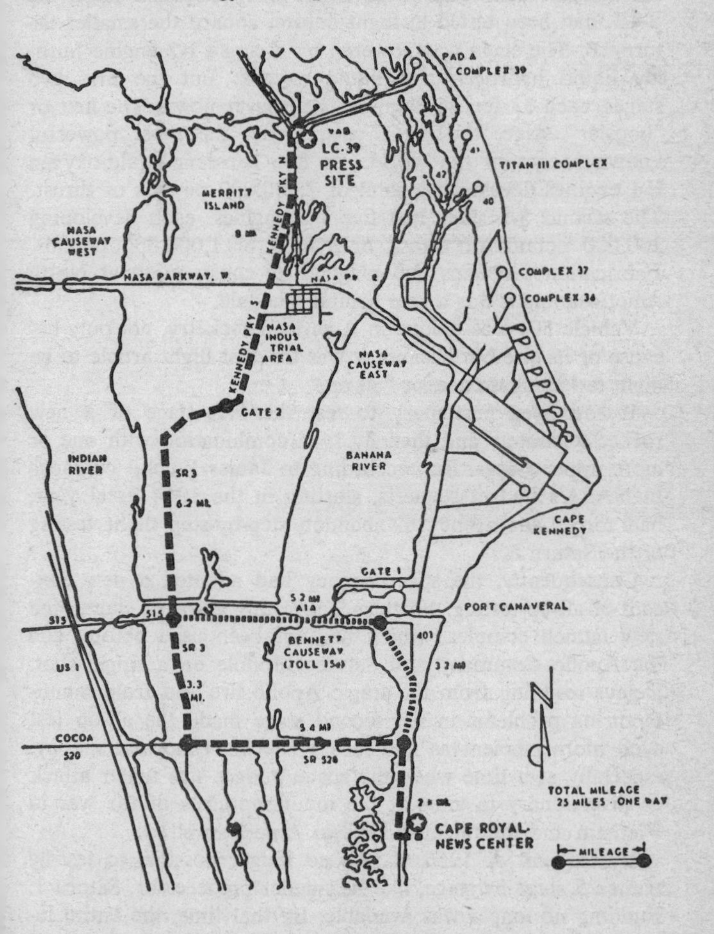

Cape Kennedy.

of development which had begun with V-2. Developing 8,700,000 pounds of thrust in three stages, the Saturn 5 could hurl 11 V-2 war rockets into orbit around the earth or propel four of them to the moon.

On the cool and windy morning of November 9, the ultimate rocket was being counted down for launching to answer the ultimate question: would it fly? Only the third stage, the S-4B, had been tested in flight before, aboard the smaller Saturn-1B. The stage was powered by a single J-2 engine burning liquid hydrogen and liquid oxygen. But the first two stages, each 33 feet in diameter, had never flown. The first or "booster" stage, the S-1C, was, alone, the most powerful known rocket in the world. Its five kerosene-liquid-oxygen F-1 engines developed a total of 7,500,000 pounds of thrust. The second S-2 stage had five J-2 engines, each developing 200,000 pounds of thrust, or a total of 1,000,000 pounds. Fabrication problems had plagued its construction at North American and it was a year behind schedule.

Vehicle 501 was unique in American rocketry, not only because of its size but because it was the first flight article to be flight tested all at once, or "all up."

It had been customary to test the first stage of a new rocket separately and then fly it in combination with one or more upper stages. But, according to James Webb,[1] cutbacks in NASA's budget requests, starting in the 1964 fiscal year, had forced the agency to abandon step-by-step flight testing of the Saturn 5.

Consequently, the space agency had adopted a new concept of all-up testing, which meant testing all of its stages, the new launch complex, which had not been used before, and the Apollo Command and Service Module on a single shot. Delays resulting from the tragic Apollo fire and from manufacturing problems in the second stage made the all-up test even more efficient as a means of saving time and money, especially at a time when the space budget was under attack as diversionary in view of the mounting costs of the war in Vietnam and the racial turmoil in American cities.

By the end of 1966, it was no longer possible to test-fly Saturn 5 stage by stage, like its smaller predecessor, Saturn 1. Funding no longer was available. By that time, the entire lunar-landing program had been geared to all-up testing, and achieving the manned landing in the Kennedy time frame depended on its success. The United States thus was committed to the all-up testing idea—committed actually to go for broke on the initial flight test of a new vehicle with 2,000,000 functional parts.

While the risk of failure in all-up testing seemed to be inordinately high, it was modified somewhat by the development of computerized ground testing and checkout systems. These monitored both rocket and spacecraft continuously. Theoretically, they made it possible to detect breakdowns and malfunctions as rapidly as they occurred and pinpoint them immediately. The automated checkout of both the Saturn and Apollo systems was continuous during their development. That is why the space agency's failure to perceive and act upon the electrical difficulties which the Review Board believed were the cause of the Apollo fire seems so bizarre.

In Apollo, project officials had had a blind spot to the perils of an oxygen-fed fire in the spacecraft. Would a similar blind spot in Saturn 5 testing and checkout cause another catastrophe? Explosions of the Vanguard and Centaur rockets on test flights had been bad enough, but a blowup of Saturn 5 would be a disaster. North American engineers had calculated that it would produce a fireball 3000 feet in diameter.

Since the Apollo fire, and the Congressional hearings which merely produced yapping at NASA's heels, the fortunes of the space agency had dropped to the lowest point since its creation. War in Asia and rebellion at home left Congress with little enthusiasm for space adventure, which the mechanical success of the Gemini program had made to appear routine anyway. Reduced by $420,000,000, the 1968-fiscal-year budget of $4.6 billion which was finally approved for NASA late in the year allowed the agency to do little other than press on to the lunar landing. Project Voyager, programed earlier in the year to land a life-detection capsule on Mars in 1973, was wiped out as a line item in the budget. Other less costly planetary reconnaissance projects were shelved.

Under severe fiscal contraint, NASA could afford to purchase and launch only nine Saturn 5 rockets in the 1968 and 1969 fiscal years instead of 15 as originally planned. The other six could be flown only after the turn of the decade. If the manned landing was to be achieved before 1970, it would have to be done on the ninth Saturn 5 flight. During the spring of 1967, Webb had advised the House Subcommittee on NASA Oversight that the moon landing would be attempted on the 11th flight of the Saturn 5. However, this earliest opportunity was advanced to the ninth flight after budget trimming forced the space agency to postpone two of the planned Saturn 5 flights until after 1970. This kind of juggling was purely theoretical in any case, since no one could be sure in advance how many flights it would take to "man rate" the big rocket for the lunar excursion.

THE 23-DAY COUNTDOWN

In the all-up test of Saturn 5, NASA had been maneuvered by events and by the compulsion to meet the Kennedy deadline into a spectacular gamble. A great deal was riding on the launch of Saturn 501, which NASA decided to call Apollo 4. It was the first of nine launch steps to the moon. If it failed, it was not likely that the landing could be accomplished in this decade.

Of what consequence was such a deadline? The question had become an editorial-page cliché. While President Kennedy had alluded to a landing in this decade seemingly for rhetorical effect, the time frame had almost at once assumed the proportions of a deadline and through the years had become an inseparable aspect of the lunar-landing goal itself. Also, it had developed a function: the Kennedy time frame had served to organize the entire manned space program into a coherent pattern. It had dictated the size and design of the Saturn 5, of the Apollo spacecraft, of the Lunar Module, and of all production, testing, and flight schedules in the Gemini and Apollo programs. The cost structure of Project Apollo was predicated on this schedule. Industrial fabrication plans were made and technicians were employed and steel was ordered and contracts were let on the basis of it.

The Kennedy time frame had, indeed, evolved into a criterion of America's technological potential. The world accepted its achievement, or the failure to achieve it, as a measure of American engineering competence and political persistence. In an era when technological prowess had become a symbol of power, the lunar-landing deadline quickly became a crucial challenge. This was understood on Capitol Hill as the rationale for the $22-billion Apollo Project.*

Since a failure on the first of the nine Saturn shots would make the deadline unattainable, the flight of Saturn 501 was the most decisive since Vanguard in terms of national prestige. In the wings lurked another possibility. In the event of a failure which indicated the need for a major redesign of the launch vehicle, it was considered likely that Apollo would be shelved until after a settlement of the Vietnam war, and the deadline would be irretrievably lost. Flight articles already built would be put in storage. NASA centers would be reduced to stand-by status, with skeleton forces. The engineering and technical teams which provided the brains and sinews of American space capability would be broken up and scattered throughout industry. Project Apollo might be delayed a decade. The cost of losing the gamble on the all-up test of Saturn 501 could be very high. The test could not be allowed to fail. There were 5000 people involved in the checkout of the launch vehicle at Kennedy Space Center and 450 in the launch crew. And every one of them knew it.

On the day of the launch, more than five hundred news-media correspondents were assembled in the race-track-style grandstand of steel and cement which had been erected for spectators at the Kennedy Space Center for the moon shots. The structure seemed symbolic of the wager which the all-up test represented. Three and one-half miles away, Saturn 5 stood beside its mobile launch tower like a giant cartridge. Sunlight flashed through a grove of leaning palm trees to the east of the grandstand. Beyond the rocket sparkled the calm sea. Above the babble of voices and the grinding of buses and automobiles in the parking lot, it was possible to hear the countdown being described over the public-address system. Jack King, the NASA commentator, was saying, "This is Apollo Saturn launch control at T minus 40 minutes and counting. All still is going well with the countdown at this time. We're now in the terminal count and still in process of some final launch-vehicle telemetry checks. We have a completely fueled vehicle on the pad at this time. We're still aiming for the planned liftoff time of 7 a.m. Eastern Standard Time."

* The project had escalated $2 billion in seven years.

It was 6:20 a.m. Incredibily, the count was on time. The first stage had been loaded with 212,200 gallons of kerosene and 346,400 gallions of liquid oxygen. The fuel and oxidizer together weighed 2200 tons. The F-1 engines, once ignited, would burn 15 tons of it every second. In the second stage, there were 267,700 gallons of liquid hydrogen and 87,400 gallons of liquid oxygen. The third stage was loaded with 66,900 gallons of liquid hydrogen and 20,400 gallons of liquid oxygen. More potential energy was locked up in the vehicle than in any other device man had tried to fly—short of the warhead of a thermonuclear bomb.

It had taken more than a year to bring Saturn 5 to this point of readiness at the Kennedy Space Center. In September 1966, the 138-foot first stage had been shipped by barge from Huntsville, Alabama, via the Tennessee, Ohio, and Mississippi Rivers, the Gulf of Mexico, and the Intracostal Waterway into the Banana River estuary, which separates the Space Center on Merritt Island from the Cape. The 81-foot second stage was shipped from North American at Seal Beach, California, by barge via the Panama Canal to NASA's Mississippi Test Center in the Pearl River country. After undergoing static test firings, the S-2 stage was again moved by barge around Florida and into the estuary, arriving at the Center January 20, 1967. The 58.5-foot-long third stage had been flown to Florida August 14, 1966, from the Douglas Aircraft plant at Santa Monica, California, in a specially built freight aircraft called the "Super Guppy." It was a larger version of the "Pregnant Guppy," which had hauled Gemini Spacecraft to the Cape and which had brought the Command and Service Modules of Apollo 017 to the Center from North American's plant at Downey, California, the previous December. The Pregnant Guppy also had delivered the boilerplate Lunar Module to fly on the test, from Grumman Aircraft at Bethpage, New York.

All of these massive components, including the 33-foot instrument unit housing the flight-control instruments, were assembled into the complete rocket during the winter and spring of 1967 at the Center's Vehicle Assembly Building (VAB). This structure, one of the largest buildings in the world, is 525 feet high and has fully developed floor space of 1,500,000 square feet. Overhead cranes can lift the stages and assemble four Saturn 5 launch vehicles, each in its own bay, at the same time. An air-conditioning system of 10,000-tons capacity was installed in the building to maintain an even temperature and control humidity. Maintenance men say that without air-conditioning, clouds form at the top of the building, which is essentially a hollow cube, and it "rains" on the floor. If this sounds like a tall tale, I can say only that the VAB is a tall building, the tallest in the South. Rising above the flat, sandy plain of Meritt Island like a rectangular mountain, its size is curiously deceptive. From a distance, it looks something like a conventional power plant, but as you come nearer, it does not become larger nor appear any closer. It is only when you come within a mile or so and can see automobiles and human figures that you realize the scale.

After being fully assembled, Saturn 501 with Apollo 017 on top was moved in its mobile launch tower directly out of the building on a 3000-ton, diesel-powered tractor, 131 feet long and 114 feet wide. This great crawler, the size of a Delaware River ferry, slowly moved its gigantic load through the great doors of the Vehicle Assembly Building, turned ponderously onto a special road with an 8-foot-deep bed of round stones, and trundled toward Pad A, three and one-half miles distant, at the reckless pace of one mile per hour. As it climbed the 5-per-cent grade to the truncated top of the pad, the tractor deck automatically leveled so that the towering rocket did not tilt. The transfer took 10 hours on August 26, 1967.

The entire flight was planned to last 8 hours, 45 minutes. Saturn 501's first and second stages would boost Apollo toward earth orbit, where the spacecraft would be established at an altitude of 115 miles by a burn of the third stage. Remaining attached to the third stage, the spaceship would coast for two revolutions. On a ground signal, the third-stage J-2 engine would be restarted, as on a lunar mission. However, instead of boosting Apollo to the moon, the J-2 thrust would be limited to injecting the spaceship into a long, looping orbit with an apogee of 10,700 miles. Apollo would then separate from the S-4B stage and use its own Service Module engine to increase the height of the apogee to 11,400 miles. The orbit was a curious one. Its perigee intersected the earth, 41 miles below the surface, so that Apollo would come back into the atmosphere without a retromaneuver. When the spacecraft had passed apogee and started back down, the Service

Module engine would be fired a second time to drive the vehicle into the atmosphere at 25,000 miles an hour, or 36,300 feet a second. The velocity simulated that of a return from the moon and it would test the ability of Apollo's heat shield to withstand reentry friction temperatures of up to 4500 degrees Fahrenheit. Then, if all went well, the spacecraft would arc down for a landing in the Pacific Ocean northwest of Kauai, Hawaii, where recovery ships would be waiting to pick it out of the sea and take it back to California for examination.

The test was unrivaled in its complexity, for it put severe demands on the performance of all lunar flight systems except the Lunar Module. Both the S-4B hydrogen engine and the Apollo engine were required to restart after long periods of idleness.

Expectations of success in the test were muffled in the space agency. George E. Mueller, the associate administrator for manned space flight, advised news-media correspondents at a NASA conference in Washington, October 12, 1967, not to expect too much. "I would anticipate that we would not launch even the first day that we try to launch, but rather will be working out the launch operation for several days, perhaps as long as a week when we do get around to it," he said cautiously. Administrator Webb added, "Well, I tell you, if you plan to go to the Cape you better have your bag packed for at least a week. There is no way anyone can tell when you're going to launch the Saturn 5."

At the time, there was good reason for this pessimism. A rehearsal of the countdown was then in progress at the Kennedy Space Center. Instead of taking three and one-half days as planned, it was taking three weeks. The demonstration was staged to "de-bug" the new launch complex as well as the launch vehicle in advance of the actual countdown leading to a launch. Delays appeared soon after the demonstration was started September 22, 1967. Trouble arose in ground-support equipment, in the first-stage fuel-tank gauge, in the third-stage wiring, in a pressure gauge controlling the flow of gaseous hydrogen, in the Apollo fuel-cell batteries, which had to be changed, and in three computers. By October 4, thirteen days after it started, the 83-hour countdown rehearsal had reached T minus 45 minutes when a computer monitoring the loading of propellant failed. About a half-million gallons of kerosene and liquid oxygen had to be taken out of the first stage.

The countdown was set back to T minus 13 hours and was resumed October 9 when propellant was loaded in the booster. Again, computer trouble appeared and a faulty helium-gas pressure regulator had to be repaired. By the time the countdown reach T minus 5 hours, the launch crew was exhausted. Launch directors called a two-day recess. As the five thousand scientists, engineers, and technicians involved in the vast operation went home, traffic jams materialized on the roads serving the communities of Titusville, Cocoa, Merritt Island, and Cocoa Beach. Everyone knew "they" had stopped counting again.

On October 11, the countdown reached T minus 3 hours and 25 minutes when a battery failed in the second stage. Test directors decided to dump the fuel and oxidizer again and recycle the count back to T minus 13 hours. The next day, it was picked up again and progressed to T minus 5 hours and 5 minutes when another faulty pressure gauge was found. It was replaced. Now the count went down to T minus 17 seconds, when directors saw that liquid-oxygen pressure in the booster was too low to assure proper combustion of the fuel. A faulty pressure gauge was suspected and the count was recycled to T minus 20 minutes. The trouble was not identified, but the count was continued. On the afternoon of October 13, it finally was halted before the count reached T minus 8.9 seconds. That was the point where the five F-1 engines would be ignited. The rocket would be held down by powerful clamps until the engines built up full thrust. At zero, Saturn would go.

The three-week rehearsal countdown had been the longest in the history of American rocketry, and it boded delay and frustration for the actual countdown. After an exhaustive review, the Apollo directorate scheduled the flight test for November 7, but then, as pre-launch preparations fell behind late in October, the test was moved back to November 9. Instead of 83 hours, the countdown was abbreviated to 56 hours and 30 minutes. This included 49 hours of actual working time and 7½ hours of "hold" time which would enable the crew to catch up if they fell behind and had to deal with any emergencies.

Officially, the countdown clock in the control room at Kennedy Space Center began at 10:30 p.m. EST on Monday, November 6. If the count remained on Schedule, Saturn 501 would lift at 7 a.m. Thursday, November 9.

Early Wednesday, an overheated storage battery was found in the second stage. It was replaced without delaying the count. Later that day, an electrical device in the third stage failed to respond to a test. It converted a radio signal from the ground into electrical current which would detonate a package of TNT to blow up the third stage if the rocket went off course. The countdown was halted for two hours Wednesday afternoon while technicians searched for the cause of the failure of the third-stage transmitter to respond to the test signal. No difficulty was found. On later tests, the device worked perfectly. It was assumed that the initial failure was caused by a weak signal, and the countdown was resumed.

One other difficulty appeared Wednesday—a failure in a switching circuit controlling the television monitors—but this was corrected without any countdown delay.

High winds which had been blowing from the north all day Wednesday next appeared as a threat to the launch. Engineers calculated that gusts higher than 32 miles an hour could cause damage to the spacecraft structure during liftoff. However, the weather forecast at midnight Wednesday predicted that while Thursday morning winds would be brisk, they would not reach 32 miles an hour.

The pad crew loaded propellant into the vehicle under the glare of floodlights which illuminated the cloud decks drifting over the launch area. The countdown, as though in defiance of all expectations, continued to run smoothly. Dawn was gray and chilly. A massive steel gantry, looking like an up-ended bridge, was rolled back from the Saturn 5 and its mobile launcher. No. 501 looked ready to go. The commentator was saying:

"This is Apollo Saturn launch control and we are at T minus 2 minutes and counting. T minus 2. We're now beginning to pressurize the tanks within the Saturn 5 vehicle. Now at 1 minute and 40 seconds and counting. Our status board still indicating all is well. T minus 60 seconds and counting. Our status board still shows we're go at this time. T minus 50 seconds and counting. We have transferred to internal power and the transfer is satisfactory. T minus 30 seconds and counting. We'll count down starting from T minus 20. Status is reporting ready for launch, T minus 20, 19, 18, 17, 16, 15,

14, 13, 12, 11, 10, 9 ignition sequence starting, 5, 4 we have ignition. All engines are running. We have liftoff! We have liftoff at 7 a.m. Eastern Standard Time."

A billow of smoky, orange flame, streaked with blinding yellow, spread out beneath the rocket, enveloping Pad A. The five F-1 engines built up thrust quickly. The clamps holding the rocket down flew back and Saturn 501 began to rise like an elevator. The flame formed a thick, round column, flaring out on the ground like a regal red robe the size of a football field. Up went the rocket, clearing the launch tower, piercing the sky. Fires burned all over the launch tower and dense smoke rose from the pedestal. The ground began to shake as though an earthquake had started. On the wind came the rattling roar of a thousand thunders. The shock wave smote observers three and one-half miles away like a great fist. Long, ragged, unbelievably powerful sound waves beat over the Center, as though the earth was splitting in some final cataclysm. Telephone receivers danced in their cradles. Saturn rose like the Washington Monument, moving straight up. The rocket's fiery comet's tail, three times its own length, flared out behind. As the vehicle entered the clouds, the brilliant exhaust was as visible as the sun through light mist. In the concrete and steel control center, plaster dust began falling on the consoles. Wernher von Braun was bellowing, "Go, Baby. Go!" Part of the roof and a window in the Columbia Broadcasting System trailer at the press site began to fall in on commentator Walter Cronkite as he was describing this stupendous scene. He kept on talking while the others in the trailer held the roof and window. The hours of judgment seemed to be at hand.

Saturn 5 raced away, its exhaust dwindling to a sun-bright spark high over the Atlantic Ocean. Control of the vehicle then shifted to the Manned Spacecraft Center at Houston, where Paul Haney, the commentator there, took up the report. "One minute and looking good. Our velocity is now about 2500 feet a second. We are two to three miles downrange. Go—all the way, says the flight director. One minute 50 seconds. Fifteen miles downrange. Two minutes. Inboard engines have cut off at 2 minutes, 18 seconds. Outboard cutoff was 2 minutes, 34 seconds. Fifty miles downrange. The S-2 stage has ignited." At the Cape, spectators cheered and applauded as this milestone in the test was passed.

Velocity rose and the launch escape tower on the spacecraft jettisoned on schedule. The S-2 stage engines cut off at 8 minutes and 40 seconds, the stage fell away, and the third stage J-2 engine ignited five seconds later After burning for 2 minutes and 21 seconds, the S-4B stage shut its engine off. It remained attached to the spacecraft and their velocity was then 25,668 feet per second. Apollo 4 was in orbit.

Saturn 501 had performed brilliantly equaling the wildest expectations of the most optimistic NASA enthusiast. Aloft were the S-4B and the Apollo spaceship, weighing a total of 280,000 pounds—far and away the heaviest load ever placed in orbit. Glynn Lunney, the flight director at Houston, called the attention of Captain Wally Schirra, astronaut, to the weight in orbit. The "package" weighed more than the Mercury-Atlas 6 rocket and spacecraft combined which Schirra had flown five years before. Did anyone then believe it would be possible to heave such a load into orbit? Schirra was impressed. At the Kennedy Space Center, Colonel Gordon Cooper had reacted to the launch with intense excitement. To him and his colleagues in the corps of astronauts, the stunning success of the big shot meant that Project Apollo was back on the rails.

Seventy-five minutes after the launching, microbarographs at Columbia University's Lamont Geological Observatory at Palisades, New York, recorded airwaves produced by the Saturn 5 booster at liftoff. The barograph, which measures air-pressure changes resulting from the passage of sound waves through the atmosphere, records the waves in ink on graph paper on a rotating drum. William Donn, senior research associate and professor of geophysics at City College, New York, said that the appearance of the sound waves on the recording "was quite sensational." The waves kept on coming for 30 minutes. At some points, the time interval between wave crests was five seconds and the average wave was one second long. Donn said that only American and Russian nuclear tests in the atmosphere had produced stronger readings at Lamont Donn added that only two natural events had created stronger air waves than Saturn 5. One was the eruption of the volcano Krakatoa in the East Indies in 1883. The other was the fall of the great Siberian meteorite or comet in 1908.

RECOVERY

As Apollo and the S-4B third stage passed over the Western Coast of the United States on their second revolution, a sequence of events began aboard the rocket to ignite its 200,000-pounds-of-thrust J-2 engine for a second burn. The restarting of the S-4B hydrogen engine after a three-hour coast was the next critical phase of the test. As the vehicles passed over Texas, the Manned Spacecraft Center sent commands to close the valve on the rocket that allowed gaseous hydrogen to escape in order to keep a nominal pressure in the hydrogen tank. Telemetry received at Houston indicated that the valve could not be closed and the pressure in the tank began to fall below the minimum of 31 pounds per square inch required to restart the engine.* For the first time since the launch, controllers at Houston and at the Kennedy Space Center expected failure. The hydrogen-tank pressure had dropped to 27.5 pounds, but the Kennedy controllers reported there was just a chance the engine might start. The joined vehicles were aligned for the burn, which was programed to put the spacecraft into an earth-intersecting orbit with an apogee of 10,710 miles. Apollo was pointed downward at an angle of 40 degrees, but the thrust of the big J-2 engine would at first drive it outward and then the ship would arc back to smash into the atmosphere—if the engine could be restarted.

At 3 hours and 11 minutes into the flight, the big J-2 engine came to life. It burned beautifully at full thrust for five minutes. Apollo then separated from the third stage and began a long climb toward apogee off the southwest coast of Africa. The Service Module engine was fired for 18 seconds. The additional thrust raised the altitude of apogee to 11,234 miles, about 166 miles below the planned peak of 11,400 miles. But another milestone had been passed. Now it remained to be seen whether Apollo's engine would restart as the spacecraft came plunging back five hours later, in order to drive it into the atmosphere at lunar-return speed. There was nothing to do but wait. At Houston, Haney reported that

* NASA later reported that a review of flight data showed that the valve actually had closed.

the concern over these events "is no more and no less than the worries were before the booster lifted off this morning."

The second burn of the Service Module propulsion system on Apollo was scheduled to begin at 3:10 p.m. and last 4 minutes, 20 seconds. It would use the remaining two-thirds of the propellant in the Service Module, reducing the weight of Apollo from 49,729 to 31,600 pounds.

Apollo reached apogee at 12:46 p.m. and began its long descent on an arc that would carry it across the Indian Ocean and halfway across the Pacific toward perigee in the earth's mantle. Ten minutes before the second burn, at 3 p.m., five helicopters sprang off the deck of the aircraft carrier *Bennington,* the prime Apollo recovery ship, which was cruising 600 miles northwest of the island of Kauai. Apollo was then 600 miles high. Paul Haney at Houston reported, "We are coming up on the burn. Mark 8 hours, 10 minutes [3:10 p.m.] and the service-propulsion system is on."

At 3:15 p.m., Houston sent a signal to Apollo 4 to cut off its engine. The spacecraft was rushing toward the atmosphere at 25,000 miles an hour, the fastest a vehicle built for men had ever traveled. There had been 95 engines on Saturn 501 at liftoff and all of them had functioned perfectly. Houston signaled the Command Module, the cabin where astronauts would ride, to separate from the cylindrical Service Module. This was done. Next, radio commands instructed the conical Command Module to turn itself around so that its blunt end was forward. The autopilot executed the maneuver. At 3:19 p.m., the craft encountered the atmosphere at 400,000 feet and began to heat up to 5000 degrees Fahrenheit. Telemetry reports showed that the heat load exceeded an expected maximum of 35,740 British Thermal Units per square foot, and the deceleration load peaked at 8 gravities. Communications were lost with the 6-ton command capsule as its telemetry radio was blacked out by a sheath of ionized air produced by the frictional heat of its descent. The flight director predicted that Apollo would be in the water at 3:36 p.m. Aircraft patrolling northwest of Kauai began to pick up fragmented signals from Apollo as it emerged from blackout. At 3:32 p.m., crewmen on the *Bennington* heard two sonic booms.

"We don't know whether to associate them with an aircraft diving or with the spacecraft," said the commentator. "The *Bennington* . . . there's a cheer . . . and at 8 hours, 33 minutes, 56 seconds the *Bennington* reported they have spacecraft in sight on chutes. In sight on chutes, drifting down . . .

"That report produced a cheer here in the control center and we had a big round of applause. This is Houston.

"The spacecraft is on the water."

HAVE ROCKET, WILL TRAVEL

It was as though a miracle had occurred. The sick program had been made well. Triumph cascaded out of defeat. Hope was rekindled, morale uplifted, and confidence restored in an enterprise which had failed so tragically at the beginning of the year.

By sundown November 9, 1967, in Florida, it had become apparent for the first time that the Apollo mission could be accomplished. As General Phillips said after the launch, "Apollo is on the way to the moon." There was no doubt in the minds of most observers that Saturn 501 could have boosted a three-man crew in their 45-ton spaceship directly to the moon that day.

The flight of Apollo 4 proved that American technology had met one of the toughest challenges in its history. The Apollo launch vehicle was a thundering success. Witnesses seemed to be numbed by the realization that the most complex vehicle system ever created had performed flawlessly the first time. There was no parallel; the event defied adequate description. No one could think of what it was like. No analogies came to mind. No one in the space agency had been able to find the right words or the resounding phrase. Even President Johnson was silent. What one did say was, "By God, it worked." Beyond that, the launch vehicle spoke for itself, in a voice that echoed around the world.

With the successful Saturn 501 test, the United States proved not only that it had mastered the technology of large interplanetary rockets, but also the techniques of launching them and of maneuvering them in space with great precision. The great gamble of all-up testing had been won, and had demonstrated that space work in American industry had reached a new plateau of sophistication.

The Saturn 5 is the rocket that converts all the old dreams

about space travel into vivid prospects, and many of the old fantasies into imminent realities. With this vehicle, the United States has gained access not only to the moon but to the entire solar system. By adding nuclear-powered upper stages to this vehicle it would be possible to explore a large area of the galaxy with instrumented probes. In its present configuration, Saturn 5 brings the moon within the orbit of man, and all of the planets within the reach of automated spacecraft.

The postflight analysis of the telemetered data showed that the performance of all three stages of the huge rocket had been more nearly "nominal" than most of the engineers had dared hope. One anomaly had appeared, however, and it spelled trouble later on, even though it had no effect on the flight. It was a series of vertical oscillations, or shaking, in the first stage. These are called "pogo-ing" because they can be visualized as the up-and-down jumping of a boy on a pogo stick. Pogo-ing was caused by variations in engine thrust. It was a result of vibrational phenomena in the rocket during powered flight and could damage rocket structures if it became severe. The amplitude of the vertical oscillations on the first Saturn 5 flight was relatively small, but it was not hard to visualize the possibility that on future flights heavier shaking could damage soldered joints in the plumbing of the upper stages as well as instruments and structures in the Apollo spacecraft. It could also give astronauts a rough time during the 2 minutes and 34 seconds of first-stage engine firing or shake up a flight crew aboard the Apollo so badly that, even though they suffered no physical injury, the crewmen would not be able to read their instruments, perform navigation and guidance functions, or react promptly to an emergency.

Pogo was not a new problem. It had appeared in the Titan 2 missile at the outset of Project Gemini and had delayed the start of manned flight until the oscillations were greatly reduced. Compared to what could have gone wrong, the pogo problem appeared insignificant indeed, but it left a nagging doubt in the minds of some observers about the prospects of the Saturn 5.

In the glow of Apollo 4's overwhelming success, NASA officials turned to Congress with new confidence for budget support of the lunar landing mission. In spite of over-all budget restrictions, there was never any doubt after Apollo 4 about the continuation of an adequate funding level to complete Project Apollo. That might not have been so if Apollo 4 had failed. Apollo 4 was the first great climax in the lunar flight development program, after the anticlimax of the fire. If Saturn 5 had exhibited a Vanguard-style failure, the entire Apollo program might have collapsed at that point, or, at least, have been stalled. Even Apollo's patron saint, Lyndon B. Johnson, might not have saved it as a top-priority national effort from a catastrophic cutback. The stakes were high in the all-up test gamble on Apollo 4. But it was not the end of the great games of chance at the Kennedy Space Center.

In order to hold new obligational authority for Apollo financing in the 1968 fiscal year at a level sufficient to achieve the lunar landing in 1969, NASA officials were required to cut back post-Apollo space projects in manned flight and unmanned planetary investigation.

While it seemed clear to NASA officials that the American capability for organizing and completing large programs of research and development in space was being challenged by Russia, the urgency of the challenge was being vaporized by the heat of war in Vietnam and social protest at home. National images were changing. The New Frontier had lost its lure. The Great Society had become the Troubled Society, and it was turning its attention inward, as one does in illness, from the contemplation of the cosmic to the contemplation of itself.

NO ROOM FOR FAILURE

In this mood of the nation, Congress approved an appropriation of $4.1 billion for NASA, which fell to $3.695 billion under the across-the-board spending cut to which Johnson agreed in exchange for Congressional approval of an income-tax surtax. Apollo funded at $2.177 billion still remained virtually unscathed, but this amount allowed no margin for failure.

"A manned lunar landing by the end of 1969 depends on success in practically every one of the eight Saturn 5 flights remaining in our operational plan for 1968 and 1969," Webb told the House Committee on Science and Astronautics.[3]

The word "failure" promptly vanished from the lexicon of

NASA. After that, there were no reported failures. There were merely "unfulfilled objectives." And these began to appear in the ensuing flights of Apollo 5 and 6.

First up in 1968 was Apollo 5, the first orbital test of the Lunar Module, the vehicle designed to land two men on the moon. As described in Chapter 6, the American technique for landing on the moon employed Lunar Orbit Rendezvous, wherein the space ship would stand "offshore"—in lunar orbit—while two of its three crewmen descended to the surface in a ferry vehicle originally called LEM, but later abbreviated to LM—an ungainly construction of two segments, or stages, with four, spidery legs and footpads.

The LM has two functions and two powerful engines for fulfilling them. First, it would descend from the orbiting Apollo Command and Service Modules on a precise path designed to settle on a preplanned landing target. Its descent would be governed by a pillar of thrust from its throttleable Descent Propulsion System (DPS). The engine, burning aerozine-50 fuel and nitrogen tetroxide as the oxidizer, could be throttled down from 9710 pounds of thrust to 1050 pounds.

The second function of the LM was to lift the astronauts off the moon and ferry them back up into orbit, where they could rejoin the third man awaiting them in the Apollo. This phase of the operation required the techniques of rendezvous and docking that had been learned in Project Gemini. Once the lunar astronauts were back on board Apollo, they would cast the LM adrift in lunar orbit and fire the Apollo engine to blast out of lunar orbit for a Pacific Ocean splashdown.

In order to carry out the return to orbit, the LM was designed with a second, so-called ascent stage, containing the cabin for the two-man crew. Its single, 3500-pound-thrust engine was not throttleable. It did not have to be. There was nowhere for it to go but up. In leaving the moon, the ascent-stage engine would fire to separate the cabin from the descent stage, which would remain on the moon as the launch platform.

Since 1962, when the decision to go to the moon via Lunar Orbit Rendezvous had been made, the LM had been the most vulnerable part of the Apollo plan. A failure of either of its engines would be fatal for the crew, for if the astronauts survived a crash on the moon, they would perish when their oxygen ran out. There was no plan to rescue them in such an event. The entire lunar mission depended on the flawless performance of this vehicle.

Weighing 16 tons fully fueled, the LM looked like some giant bug from another galaxy, its footpads groping for the moon. It was the first space vessel designed to land on an airless planet, and it could not be tested adequately anywhere but in a vacuum. The machine had been in development since November 1962 by the Grumman Aircraft Engineering Company at Bethpage, Long Island, and by 1968 its cost had reached $1.8 billion. Its development had been months behind schedule, but this was not a pacing item in view of the delay caused by the Apollo fire. Engine and electronic problems had plagued the builders. A backup fuel-injection system had to be designed for the ascent engine to satisfy NASA officials they were leaving nothing undone to guarantee liftoff from the moon.

On the flight of Apollo 4, all components of the lunar transportation system has been tested except the LM. They had flown unmanned and had performed well, and providing a measure of confidence that the day of manned Apollo flight was near.

At the beginning of 1968, the LM was ready for its first test orbit. The flight was designated as Apollo 5. The test vehicle, without legs, had been mounted on Saturn 1B No. 204, encased in a steel cocoon called the Spacecraft-LM Adapter, or SLA for short. The space agency announced that the vehicle would be launched no earlier than January 18, 1968, at Complex 37.

THE FLIGHT OF THE SPIDER

It had been a chilly week flecked with rain at Cape Kennedy. Nearly a year had passed since the hoarse cry "Fire in the spacecraft!" had heralded the worst disaster in space history. Saturn 204 had escaped damage in the oxygen-fed fire which destroyed Apollo 012 and suffocated Grissom, White, and Caffee the year before. Now the launch vehicle was being made ready to boost the LM on its first flight test. Project Apollo was moving on.

The resumption of launch activity did not escape the familiar stops and starts. A power failure blacked out an electronic computer which was monitoring systems aboard the launch vehicle, delaying the start of the 22-hour countdown.

Failures in cooling and electrical systems of the ground equipment at Complex 37 delayed the launch nearly four hours on Monday afternoon, January 22. Then the "small" Saturn's eight Redstone engines ignited in a sunburst of orange and yellow flame driving the LM payload of 31,700 pounds into the darkening January sky with 1,600,000 pounds of thrust. In spite of the fact that it had been parked on the Cape for 18 months, the Saturn rocket worked perfectly. The LM was injected into an oval orbit of 107 by 138 miles. Test conductors settled down to observe its performance.

Four hours after liftoff, the LM's reaction control system was turned on by the flight computer and positioned the vehicle for the first test of the Descent Propulsion System. The engine was to fire for 26 seconds at 10-per-cent thrust, and then to throttle up to nearly full thrust and burn at that level for 12 seconds. The 38-second burn was designed to simulate a partial descent from lunar orbit.

What happened next was a stunning disappointment. The descent engine ignited, ran four seconds, and shut down. Telemetered data showed that the computer had cut it off when it reached only 9.5-per-cent thrust in the time programed for it to achieve 10 percent of full thrust.

The preliminary analysis of the failure showed that the computer was at fault—not the engine. A crew on board could have turned the engine back on. To avoid a repetition of the premature shutdown, the test directors shifted control of the LM's propulsion system from the primary guidance computer to a simpler type of autopilot called the Program Reader Assembly PRA. Unlike the computer, it did not react to feedback from the engine. It simply took control of the descent engine and operated it in accordance with instructions printed on film.

On the second try, at six hours, 10 minutes into the flight, the descent engine fired for 26 seconds at 10-per-cent thrust and 7 seconds at full thrust. Its performance was normal. The engine was fired again for 26 seconds at 10 per cent and two seconds at maximum thrust. Then, as the descent engine was shut down, the Ascent Propulsion System was ignited and boosted the ascent stage away from the descent stage. This "fire in the hole" maneuver, involving the burning of the ascent engine through the descent stage, simulated an abort of a lunar-landing attempt, when the crew would elect to return to the Apollo while en route to the surface.

THE IGNORANT COMPUTER

With growing confidence, the test conductors relinquished PRA control of the vehicle and signaled the LM guidance computer to take over control of vehicle for the second ascent engine burn. Then the fun began. The LM began to dance a jig in space, careening wildly as its attitude-control thrusters began firing seemingly at random.

What had happened? Still, nothing was actually wrong. The queer, electronic logic of the computer had simply asserted itself During PRA control of the LM, the autopilot computer remained in an idling mode and did not "know" that the ascent stage had blasted away from the descent stage. Consequently, it computed the thrust of the Reaction Control System at a level sufficient to move the entire 16-ton mass of both LM stages. rather than simply the upper stage. Considerably more thrust was required to overcome the inertia of the two stages than the inertia of only the ascent stage. The computer's "ignorance" caused the Reaction Control System to operate longer than it was designed to do, and the heat it generated exceeded its temperature limits.

However, the reaction-control thrusters continued to work as they were designed to do under the commands of the autopilot even though they exceeded their specifications. Test observers were pleased about that.

After the gyrations ceased, the spacecraft passed out of radio range of ground stations, and on the next revolution, much of its orbit would be in the southern hemisphere. It was then decided to conduct the second Ascent Propulsion System burn on PRA control. The engine was ignited 7 hours, 44 minutes after liftoff on the LM's fifth revolution as it came within radio range of Hawaii, and burned until its fuel was used up—for about 6 minutes. During the sixth revolution, the ascent stage reentered the atmosphere west of Hawaii and presumably disintegrated.

Before dawn on the morning of January 23, the Apollo directorate delivered a qualified verdict. The flight of Apollo 5 showed that the LM was a "mature" spacecraft, and, with confirming ground tests, was ready to fly men. "It was a very good mission," said General Phillips. "I'm convinced . . . that it did achieve its essential objectives, that it puts us in a very good position to move on into the Apollo program."

In spite of the ambiguities in the test, Project officials were genuinely relieved. It now appeared they had a lunar ferry that would work.

Next up was Apollo 6, the second unmanned test of the Saturn 5 with an Apollo spacecraft as payload. Another moment of truth approached with the countdown for that mission.

POGO RIDES AGAIN

Apollo 6 was conceived as an unmanned flight of 9 hours, 29 minutes duration, designed to test for the first time the capability of the Saturn 5 launch vehicle to boost the Apollo spacecraft to the moon and retest the Command Module heat shield in an entry into the atmosphere at lunar return speed of nearly 25,000 miles per hour.

After being boosted into a 115-mile circular orbit, Apollo and the third (S4B) stage of the Saturn 5 were to coast for two revolutions around the earth, and then, over Cape Kennedy, the S4B's hydrogen-oxygen engine was to be restarted. its 225,000 pounds of thrust would inject the spacecraft into an elliptical orbit with an apogee of 320,000 miles, well beyond the moon.

However, shortly after separating from the third stage, the Apollo was programed to fire its own engine as a retrorocket to slow down the vehicle and reduce its apogee to about 13,820 miles. On its return toward the 115-miles perigee of the orbit, the spacecraft engine was to fire a second time to increase the ship's velocity and drive it into the atmosphere at 24,900 miles, an hour. This would be the final test of the Apollo heat shield before men rode in the vehicle. The Apollo would be recovered in the Pacific Ocean southwest of Hawaii.

At 7 a.m. EST April 4, 1968, the five F-1 engines of the Saturn's first stage ignited with a blinding billow of flame which seemed momentarily to consume the entire Launch Complex The massive vehicle once more began to lift smoothly into the morning sky, as it had on the lift of Apollo 4, the thunder of its 7,500,000 pounds of thrust billowing back over the John F. Kennedy Space Center like the roll of cosmic drums. Again, the ground shook, the steel grandstand at the press site rattled, and a bass throbbing seemed to shake the universe.

Saturn No. 502 lifted beautifully into its trajectory out over the Atlantic, and for 148 seconds of the first-stage burning, another triumph appeared to be in the making. But, unreported at the time, the 363-foot rocket began to shake along its long axis. Again, the vertical oscillations—the Pogo Effect– had appeared. This time it was more severe than on the 501 flight. The violent shaking of the first stage was invisible to observers using optical instruments, but it could be discerned on Control Room instruments at both the Kennedy Space Center on Merritt Island and the Manned Spacecraft Center, near Houston.

After the great F-1 engines cut off and the five liquid hydrogen-oxygen engines of the second (S-II) stage began firing, the problem vanished from the consoles. But not the consequences: new trouble appeared. Pressure dropped in the combustion chamber of the stage's No. 2 engine after 2 minutes, 47 seconds. The engine continued to blast away, however, until pressure fell abruptly at the fuel-oxidizer inlet and the engine cut off after 4 minutes and 21 seconds of operation—1 minute and 47 seconds early. One second later, Engine No. 3 cut off.

At that point, the upper stages of the Saturn and their Apollo payload were some 75 miles above the Atlantic Ocean. The guidance system, sensing a drop off in acceleration, simply programed the remaining three engines of the second stage to fire 29 seconds long in order to reach the velocity at which the autopilot would cut them off. Then the third stage took over the load and propelled itself and the Apollo into orbit, but the guidance and navigation system overcompensated for the early cut-off of the second-stage engines and caused the third-stage engine to burn longer than was needed to make up the deficiency in velocity. Apollo 6 and the third stage entered an elliptical orbit 227.7 by 112.7

miles high instead of the neat, circular orbit of 115 miles that had been planned.

As the S4B and Apollo coasted, controllers carefully checked the fuel in the third stage and determined there was enough left for a second burn. In spite of the extra propellant used in the S4B to inject Apollo 6 into orbit, enough fuel remained to make the translunar injection. As Apollo and the S4B approached Florida at the end of its second revolution, controllers prepared for the restart of the third-stage engine.

"Three hours and 11 minutes into the mission," said the mission commentator at the Manned Spacecraft Center. "Standing by. Our restart maneuver has been initiated. This is Apollo Control Houston. Three hours 12 minutes 35 seconds. Booster reports he looks good at this time. We are about 30 seconds away from our reignition time. Apollo Control Houston standing by. No confirmation on ignition at this time. Standing by. Three hours 14 minutes 52 seconds. We did not have reignition of our third stage engine. We have sent real time command number 71 to separate [the stage] from our vehicle."

The third stage had failed. Houston signaled the Apollo engine to fire, and it responded promptly, burning nearly 7½ minutes and boosting the spacecraft away from the dead third stage to an altitude of 13,893 miles. The long burn of the Service Propulsion System of Apollo used up nearly all its fuel. With only 23 seconds of burning time remaining, Mission Control canceled the second burn, which was to have accelerated the plunge of the Command Module into the atmosphere to test the heat shield. The small amount of fuel remaining would not have increased the velocity by any significant amount. Apollo splashed down in the Pacific Ocean during the afternoon about 600 miles northwest of Hawaii. The mission was as close to failure as Project Apollo had come since the fire, but officials took an optimistic view.

"We did successfully achieve orbit in spite of the malfunction of those two stages," said Mueller at the post-mission news conference. "We would have been able to successfully carry out the—a—lunar mission in this particular instance if the S4B stage had reignited."

Worried and harassed, the Apollo directorate began to search for an explanation of this abrupt reversal in the rising fortunes of Apollo. No one was sure immediately after the debacle what had gone wrong, but there was a growing suspicion that the pogo-ing of the first stage was related to the failures in the other two.

One fact was clear. Saturn 5 was not yet ready to boost Apollo to the moon.

Chapter 14

Genesis Revisited

This vertical shaking of 502 had contributed to the rupture of a fuel line feeding the spark igniter of the No. 2 engine in the second stage. The fuel injector failed, causing a drop in pressure which activated the automatic shut-down system. The No. 3 engine then shut down "in sympathy," not because of any malfunction but because of a wiring error which hooked it into the emergency shut-down circuit of the No. 2 engine. The failure of the single J-2 engine is the third stage to restart in orbit was attributed to a fuel-line rupture, too. However, it appears to have occurred rather late in the first burn of the engine, which established Apollo 6 in orbit. The degree to which the vibrational stress of pogo contributed to the rupture was not established.

THE POGO FIX

When the story of Apollo 6 had been ascertained by flight data and verified by ground tests, the problem was reduced to making the "fixes" that would eliminate pogo and strengthen the J-2 fuel lines. The cure for pogo was to inject helium, an inert gas, into the F-1 engine liquid-oxygen system to dampen oscillations in the line. This would prevent them from resonating with structure vibration. Only two cubic feet of helium were needed for all five F-1 engines. It would be supplied before the launch from a ground tank and replenished during flight from onboard tanks, in which helium normally was stored for operating valves and providing fuel-tank pressure. By mid-July, NASA had assembled "modification" kits which were to be installed on all Saturn 5 first stages if the helium fix worked.

On August 19, 1968, General Phillips announced that Saturn 503 was ready for a manned flight test in December.[1]

THE DECISION ON 503

The decision to fly Saturn 503 manned came at the end of April, just a few weeks after the 502 debacle. The third Saturn 5 had been scheduled early in 1968 to fly unmanned in May, but as the 502 test flight was counted down, officials revealed they might skip a third unmanned test of the big rocket if 502 performed satisfactorily. As we have seen, 502 did not. Nevertheless, the Apollo directorate decided after a flight review at Huntsville to move 503 out of the May test slot and use it after the Apollo 7 flight in autumn—the first manned test of the spacecraft—as the booster for the second manned test of the spacecraft in December. The mission was to become known as Apollo 8. The decision was made on engineering assumptions that the fixes for pogo in the first stage and the J-2 engine failures in the second and third stages would work. But it was made before the proposed fixes could be fully tested.

By eliminating the third, unmanned flight test of Saturn 5, NASA would save an estimated $300 million in flight-development costs leading to the lunar landing. But the economy introduced a more conspicuous element of chance than had been visible before in manned space flight. While the NASA public-affairs apparatus kept projecting an image of conventional, step-by-step, flight development, it was obvious that the caution that had characterized manned space flight development in Mercury and Gemini was vaporizing in the heat of expediency. A new, bolder approach to the lunar program gained ascendancy in the space agency.

Saturn 503 reflected it clearly. Never before had men been

entrusted to a rocket so early in its development. The Saturn 5 had flown only twice—and not well the second time. In his review of United States space activities in 1968, John Glenn had remarked: "In the early days. such a development pace would have been unthinkable. The Redstone rocket had been flown 70 times before it launched Al Shepard on his suborbital flight; the Atlas had flown 91 times before boosting me into orbit and there were 34 launchings of the Titan 2 missile before it lifted Gus Grissom and John Young in Gemini 3." [2]

MISSION C PRIME: LUNAR ORBIT

By the fall of 1968, the Apollo transportation system was in shape, from a psychological point of view if not a technical one, for manned flight testing. The Saturn 5 rocket had been fixed, and even though the fixes had not been flight-tested the Apollo directorate had faith they would work. Moreover, the Apollo spacecraft had reached "maturity," according to its manufacturer, North American Rockwell. The Block II version of the conical Command Module had been made as fire resistant as possible within the weight limitation. All changes from the earlier Block I version, which had burned, were completed. A single, quick-opening hatch replaced the dual hatch which had made the earlier version a death trap. The new hatch could be opened in three seconds. In fireproofing the cabin, the manufacturer had eliminated all nonmetallic materials for which metal or other inorganic substitutes could be found. Beta cloth, a fiberglass material coated with crushed rock, had been used to tie Teflon-wrapped wire bundles, and protective covers had been added at the corners to prevent chafing. Aluminum solder joints were armored. The aluminum junctions of the water glycol (coolant) and oxygen lines were replaced with stainless steel. More than 20 miles of wiring had been rearranged to reduce stress and abrasion. The most effective safety measure, however, was the change in the ship's atmosphere while it stood on the pad during countdown, from pure oxygen at 16 pounds pressure per square inch to a mixture with 60 per cent nitrogen. During powered flight up into orbit, the mixture would be vented gradually to space and replaced with oxygen at 5 pounds per square inch, the approved breathing atmosphere. Some materials which would burn in pure oxygen could not be eliminated entirely in the cabin. However, the introduction of nitrogen while the ship was on the pad, when the cabin pressure had to be somewhat higher than the 14.7 pounds per square inch outside to seal the capsule properly, made the interior reasonably fireproof.

By late summer, therefore, the Saturn 5 was deemed ready. The Apollo spaceship was ready. But the third major component of the earth-to-the-moon transportation system, the Lunar Module, what of it? Alas, it was not ready. Lunar Module No. 3 had been a source of frustration at the Kennedy Space Center since its arrival in June from the manufacturer, the Grumman Aircraft Engineering company. A sequence of electronic problems appeared in the ascent stage, delaying the checkout of the vehicle for its first manned flight test aboard Apollo 8. The rendezvous radar did not lock on to a target properly; faulty switches threw the checkout schedule into confusion It took a whole day to discover why a "glitch" (electrical transient) was produced when one switch was thrown. The trouble was found in two springs controlling the switch pole position. They made electrical contact for a split second when the pole was moved off the center position. In replacing faulty switches, technicians had to replace entire panels in which the switches were mounted. Another trouble source was the electroluminescent lighting of the instrument panel, designed to glow at adjustable levels of intensity. Changing the brilliance caused electronic interference with and introduced errors in an instrument called the Flight Director Indicator, which displays the vehicle's attitude in flight.

By mid-August, it was clear to Apollo directors that LM-3 could not be checked out for a manned test flight in December. The directorate then made another curious decision. Faced with alternatives of flying Saturn 503 without the LM in December or of delaying the Apollo 8 mission until the end of February, when LM-3 would be ready, the directorate decided to fly without the LM in order to hold to the December schedule.

Never before had a prime test article been deleted from a flight because it could not be made ready when the flight was scheduled. In the past, policy had been to delay the flight until the article it was designed to test was ready. Perhaps some precedent for this apparently bizarre priority had been set in Project Gemini when the Agena target vehicle failed and substitute targets had to be found. But in the case of Apollo 8, the decision to fly without the LM seemed pointless.

As initially conceived, Apollo 8 was an earth-orbit test of the LM and the Apollo. The LM crew would practice rendezvous and docking with the Apollo to check out the LM guidance and propulsion systems and rehearse maneuvers to be employed on a lunar landing mission. Eliminating the LM from Apollo 8 would reduce it to a repetition of Apollo 7.

The devision to fly minus the LM made sense only when it was realized that the scheduled December flight of Apollo 8 was the earliest time the United States could exercise an option to fly a crew to lunar orbit—if such a demonstration became politically necessary. The deletion of the LM did not affect the option, except to make it more risky, for the powerful engine on the LM provided a backup propulsion system in case the Apollo engine failed while the ship was in lunar orbit. In such an eventuality, the LM engine could boost Apollo out of lunar orbit on a flight path that would return the crew to earth. Without it, a failure of the Apollo engine would leave the crewmen stranded in orbit around the moon.

What seemed to make the December flight so urgent was a belief in NASA that Russia was preparing to launch a manned spaceship around the moon, possibly early in December. Moreover, a flight to lunar orbit would stand as a memorial to the outgoing administration of Lyndon B. Johnson, one of the architects of Project Apollo.

At the end of the summer, the Apollo flight development schedule called for the first manned LM test on Apollo 9 at the end of February 1969; a second manned LM test with maneuvers out to 4000 miles on Apollo 10 in May; a flight to lunar orbit and test of LM there on Apollo 11 in July, and a possible lunar landing some time later on Apollo 12.

21S FOR ZOND

Apollo 7, the first manned test of the moon ship, slipped noiselessly through September into an October 11 launch date. Meanwhile, the shadow of Russian competition in the lunar venture acquired substance. After the automatic link-up in orbit of two unmanned satellites, Cosmos 186 and 188 in the fall of 1967, the Russians executed a second unmanned rendezvous and docking of Cosmos 212 and 213 in the spring of 1968. These efforts were interpreted in NASA as a clear signal of Russian intention to resume the race to the moon.

That impression was strengthened September 22, 1968, when the Soviet government reported from Moscow (TASS) that a space vessel called Zond 5 had been launched, unmanned, around the moon and had been guided automatically back to Earth to a splashdown in the Indian Ocean. The vehicle had the mass of the Apollo command module. It carried a cargo of turtles, flies, worms, plants, and seeds, which later were examined to determine the effect of radiation beyond the earth's magnetic field on living organisms. The Zond 5 event, later followed by a similar circumlunar flight of Zond 6 which landed in the Soviet Union, made it clear that the Russians had the capability of sending men around the moon. They had obviously mastered the guidance problem for circumlunar navigation, while the American capability for such a trip remained untried. Yet, with the Saturn 5–Apollo system, the United States had more than the simple capability of sending a ship on a circumlunar trajectory that would whip it around the moon and bring it back to Earth—Saturn 5–Apollo could go into orbit around the moon. It appeared that the Zond vehicles lacked a propulsion system to brake their velocity for the drop into lunar orbit and then to boost them out again for the flight back to Earth. Consequently, at the time of the Zond demonstrations, Saturn 5–Apollo had the capability of doing more than the Zond system. And the first opportunity to prove it to the world was Apollo 8. Thus, the Soviet exploits increased pressure to employ the lunar orbit option on Apollo 8.

The Zond 5 event underscored the final warning of Administrator Webb that Congressional stinginess in NASA financing was putting the American space effort in second place. It was the last official warning of this voluble promoter of the American space effort, for on September 16, 1968, Webb announced that he was resigning his post as administrator effective October 7, his sixty-second birthday.

For nearly eight years, the fast-talking Jim Webb had served as NASA's boss and chief evangelist. Now, he was discouraged but still battling. Webb had private reasons for leaving the agency he had run with a rare combination of political sophistication, managerial talent, and red-faced bombast. He had ceaselessly campaigned for space funds with fervor, delivering the great message of the nation's future in space and the Russian challenge at every Chamber of Commerce, Rotary, or Kiwanis meeting he could attend.

In his farewell message, Webb warned that the USSR would "retain" its lead in space technology and exploration for years to come because of repeated budget reduction in the United States program. He gloomily forecast that the chances of landing men on the moon by 1970 were considerably less, because of the budget cuts in 1968, than they had been in 1967.

Webb's poor-mouthing, in the light of hindsight, appears exaggerated. He must have known that Apollo was in a high state of readiness for its first manned flight and that the prospects for a landing in 1969 had never looked so good. Yet, he left the scene in a cloud of pessimism. Mr. Johnson appointed a rather obscure deputy administrator, Thomas O. Paine, a quiet, scholarly metallurgical engineer who had come to NASA from the General Electric Company, as acting administrator.*

By the end of September, it became apparent that the flight crew of Apollo 8, consisting of the Gemini 7 team of Frank Borman and Jim Lovell plus Air Force Major William A. Anders, a 1963 recruit to the corps of astronauts, was training for a flight to lunar orbit.

A STREETCAR NAMED APOLLO

In the first week of October, the countdown began for the launch of Apollo 7.

Two minutes after 11 a.m. EDT on October 11, the eight Redstone engines of the Saturn 1B ignited in a sunburst of orange-yellow flame and lifted Apollo 7 and its crew smoothly into orbit around the earth. Wally Schirra, now a Navy cap-

* Paine was later appointed administrator on March 5, 1969, by President Nixon.

tain, commanded a crew of two "third-generation" astronauts who had not flown in space before. They were Major Donn F. Eisele, 38, a U.S. Naval Academy graduate who had switched to the Air Force, and Walter Cunningham, 36, a civilian scientist-pilot with the rank of major in the Marine Corps Reserve.

In his pre-NASA experience, Eisele had specialized as a project engineer and experimental test pilot at the Air Force Special Weapons Center, Kirtland Air Force Base, New Mexico. Cunningham, who had received his flight training in the Marine Corps, had worked as a research scientist for Rand before joining NASA. As a doctoral candidate in physics at the University of California, Los Angeles, he had developed a magnetometer flown on NASA's first Orbiting Geophysical Observatory.

Apollo 7 was injected into an elliptical orbit of 140 by 173.6 miles attached to the S4B, the second stage on the Saturn 1B. At the end of the first revolution, Schirra separated the Apollo from the S4B, turned the spacecraft around, and nosed it to within four feet of the 28-foot aluminum cone, the Spacecraft-LM Adapter, on the forward end of the rocket. On a lunar flight, this structure would house the Lunar Module during launch and powered flight up through the atmosphere. Four panels at the top of the cone are designed to swing out to free the LM so that the Apollo can dock with it. When both vehicles are joined, the Apollo thrusters were to ease the vehicles away from the S4B, which then would be abandoned. On this flight, however, the aluminum cone was empty except for a docking target with which Schirra could simulate a docking with the LM. This exercise, however, was curtailed by the failure of the panels to open fully, and Schirra declined to risk a collision by bringing Apollo any closer than four feet.

Although Schirra could not complete this transposition maneuver, he was able to demonstrate that the Apollo responded well enough so that it could be executed when the SLA panels were fully open. The spacecraft was then eased away from the S4B. On the second day of the flight, the distance between them had widened to 100 miles. The flight plan called for Schirra to maneuver the Apollo back to within 50 feet of the rocket stage to demonstrate the spacecraft's rendezvous capability.

The mechanical systems worked smoothly at the outset, but "malfunctioning" appeared in the human component. Schirra came down with a head cold, and the first symptoms of upper respiratory infection appeared in his fellow crewmen. These were aggravated by the drying effect of the pure oxygen atmosphere the crew was breathing.

For Schirra's cold, medical science had little to offer. Dr. Berry advised him to take aspirin and a decongestant and to drink plenty of water. The infection, combined with lack of sleep, which had always been a chronic problem, increased the spacecraft commander's irritability. Schirra, age 45, expressed unconcealed impatience with some of the judgments made by younger, less experienced men who were controlling the flight at Houston.

One source of friction was a compact, four-and-one-half-pound television camera, developed by Radio Corporation of America at a cost of about $5 million. The camera was actually an experiment in visual communication between space and ground, but it was to provide an unanticipated bonus in stimulating public interest in the men aloft and their marvelous machine.

The device emitted a 5-watt signal which was beamed to ground stations at Cape Kennedy and Corpus Christi, Texas. Before the flight, the astronauts were not enthusiastic about the camera. It was an extra piece of gear with which they had been saddled for public-relations purposes. Besides, they were not impressed by its initial performance in ground tests.

The Manned Spacecraft Center, however, had promised the television networks an orbital television show Saturday morning, October 12, and the network chiefs had cranked up their enormously complex machinery to telecast it. When the time approached to start the transmission from Apollo 7, however, Schirra announced that he and the crew were too busy with the flight plan to fool with the camera.

Controllers protested that all had been arranged. Schirra explained his position tersely: "You have added two [engine] burns to this flight schedule. You have added a [urine] water dump. And we have a new vehicle up here. I tell you this flight TV will be delayed without further discussion until after the rendezvous with the S4B rocket stage."

Deke Slayton took the microphone at Houston to speak for the Establishment.

"Apollo 7, this is capcom No. 1 [the boss]. All we have agreed to do on this is to flip the switch on. No other activity is associated with TV. I think we are still obligated to do that."

Schirra wouldn't budge. "We do not have the equipment out," he said. "We haven't even eaten at this point. I still have a cold. I refuse to foul up our time lines this way."

"Roger," said Slayton. He knew when to stop pushing the man upstairs. He knew, also, that Schirra was retiring after the mission from active duty as an astronaut. It was the Navy pilot's last space flight and he meant to make it as perfect a match with the flight plan as possible—to realize the always hoped-for, seldom achieved "text-book" flight. Nothing would be allowed to interfere with that goal. Slayton understood what was in his colleague's mind.

On the third day of the flight, Cunningham came down with the cold. Apollo 7 now resembled a flying sick bay. Dr. Berry explained that Cunningham had reported cold symptoms four days before the flight and Eisele three days before. But after the pilots were dosed with antihistimines and decongestants, their symptoms had disappeared and had not reappeared when the medication was stopped. Schirra had shown no preflight symptoms at all.

"If I knew how to cure the common cold," Dr. Berry told a pack of newsmen who were hounding him on the subject, "do you think I would be standing here talking to you gentlemen? I would be sailing on my private yacht!"

Early Monday morning, October 14, trouble appeared in the ship's electrical system, precipitating an engineering conference among NASA and contractor engineers and electronic specialists. An inverter—one of three which transform 28-volt direct current from the fuel cells to 115-volt alternating current—had been going "off line" intermittenly. Something was tripping a circuit breaker. While it required only a push of a button to put the No. 1 inverter back on line, the cause of the disconnect became worrisome.

Just before Apollo 7 crossed the Red Sea on its 39th revolution, a second inverter went out, along with the first one. The master alarm light glowed in the control panel.

"We have the master alarm and no indication as to the cause," Cunningham reported. The scientist-astronaut suspected that something was overloading the inverters, causing an overcurrent trip circuit to disconnect them. While the trip circuit could be quickly closed and the inverters put back to work, the "something" suggested electrical trouble which could only get worse. For the first time it appeared the mission might be brought down early.

The inverters provided alternating current to operate the fuel cell pumps, environmental-control-system glycol pumps, spacesuit compressor and fans in the liquid oxygen and hydrogen tanks serving the fuel cells. An analysis of the problem was carried out over the entire global network serving Apollo communications. It suggested that current surges overloading the inverters appeared when two fans in the liquid-oxygen tank shut off. The fans went on and off automatically. Their function was to circulate the fluid oxygen and keep it at uniform density as it passed over heating elements which partially vaporized it. Vaporization provided pressure to move the oxygen out of the tanks and into the fuel cells, where it combined with hydrogen to produce electricity and drinking water.

When the automatic control shut off the fans, it appeared that the current they were using surged momentarily to the AC (alternating current) bus, or distribution point, tripping the overcurrent circuit which shut off one or more inverters. The anomaly was especially noticeable when little AC current was being used.

"It's not nearly as much of a problem when you're powered up," Pogue told Donn Eisele. "But only when you're powered down. The fix is to keep one of the oxygen fans on while the other is off."

This could be done by manual control, and it worked. After the flight, it was decided that the voltage surges were caused by electrical arcing between the contracts of the motor switch which turned the oxygen tank fans on and off automatically. Even if the in-flight diagnosis was wrong, the fix was right, and the inverter drop-outs ended.

THE WALLY, WALT, AND DONN SHOW

Once these matters were settled, Schirra agreed to do the long-awaited TV show, the first "live" telecast from an American spacecraft in orbit.

As Apollo 7 passed over the southern United States Monday on its 45th revolution, the little TV camera began transmitting, first to the big antenna at Corpus Christi and later to the station at Cape Kennedy. After the usual "snow" and psychedelic swirls, the screen showed a surprisingly clear picture of Schirra, Eisele, and Cunningham in flight coveralls, floating around the cockpit of Apollo 7 in zero *g* like two-legged dolphins. The control panel of Apollo made an impressively complex display in the background. The transmission began at 10:41 a.m. EST and continued for about seven minutes.

The picture was so clear that Stafford, the capsule communicator at Houston, said that by looking at the TV monitor on the wall he could see out of Schirra's window, as Apollo 7 was passing over the Texas coast of the Gulf of Mexico.

Schirra was in top form. He had printed some signs which he held up to the camera, under Stafford's direction. One said, "From the Lovely Apollo Room high above everything" and the other, "Keep those cards and letters coming in, Folks."

The little camera was one of the most successful innovations in photography since Thomas Edison. For the first time, it brought the performers in the national space program into full view of their sponsors, the taxpayers. For the first time, it enabled the citizen on the ground to identify to some degree with the citizen in space.

Apollo 7 continued to perform without serious flaw. Schirra was sold on the spacecraft. "A first-class spacemobile," he called it. "A Cadillac of space," said the ground controllers. By the end of its sixth day in orbit, Apollo had demonstrated its spaceworthiness for a round trip to the moon.

In practicing telescopic tracking of landmarks, the crewmen picked up Hurricane Gladys, which was moving northward in the Gulf of Mexico toward Florida. There was some concern that it might threaten Apollo 8 on the pad at the Kennedy Space Center. Schirra, looking down at the storm, described it as a "doozy . . . a real spinner" with its tight, cloud spiral and central eye.

On the eighth day of the mission, one of three fuel-cell battery stacks began to overheat. A warning light glowed in the cabin. Mission Control advised the crew to turn the stack off until it cooled. That solved the problem. But, as the mission neared its end, controllers began adding chores to the flight plan, and the extra work, along with the colds, which had not abated, triggered several abrasive exchanges between space and ground. One of the extra chores jammed the spacecraft computer because, as crewmen complained, some "idiot" on the ground fed it navigation data which required it to find the square root of negative numbers.

"I want to talk to whoever it was that thought up that little gem," Eisele announced. "That one really got to us."

Schirra on one occasion told the ground, "I have had it up here today and from now on I am going to be an onboard flight director for these updates [changes in plan]. We are not going to accept any new games . . . or do some crazy tests we never heard of before. Each [new] test is going to be reviewed thoroughly before we act on it."

"Roger," came the reply from Mission Control. When the command pilot was in that mood, no one argued with him.

Tired, bearded, and sniffling, the crew of Apollo 7 did sound as though ready to mutiny. What did it mean, the newsmen demanded at a news conference? What were the implications of this war of words? Who runs the flight—the ground or the commander upstairs?

All these questions were disposed of succinctly by Public Affairs Director Paul Haney with the ease of long practice in disposing of queries no one could or would answer.

"Something happens to a guy who grows a beard," explained Haney solemnly. "Next thing you know, he starts to protest."

The matter dissolved in merriment.

But another problem had appeared. The crewmen wanted to reenter the atmosphere without spacesuits and helmets. This would make it easier for them to free their ear passages of mucus from their colds. The passages had to be clear so that air pressure could be equalized quickly on each side of the eardrum during the rapid descent in the atmosphere, when cabin pressure would rise from 5 to 14.7 pounds per square inch in a matter of minutes.

Pilots use the trick of holding the nose to clear the ear passages. Schirra said this could not be done with the helmets on, and if the helmets were not worn, he did not want to wear the flight suit either, because the neck piece would be uncomfortable without the helmet attached.

Deke Slayton objected to leaving the suits off. He pointed out that the stirrup-like foot loops which help restrain the astronauts in their couches during landing were too large for the soft footgear the astronauts wear with the suits off, since the loops were designed for space boots. Sea conditions in the Atlantic Ocean impact zone south of Bermuda promised rough landing. Slayton feared it would be rougher than any in Gemini.

The argument waxed back and forth all day Monday, October 21. Control insisted that the helmets be worn, too, to prevent injury to the face on impact, arguing that the bubble helmets could be removed quickly enough to blow the nose during descent if necessary.

Apollo 7 came home the next morning. Before dawn, along with thousands of persons, I saw the Command Module pass over Houston as a bright star serenely sailing across Texas, after Schirra had fired the Service Propulsion Systems to brake the ship out of orbit 500 miles east of Hawaii and then had separated the Command Module from the Service Module. The crew came in with suits on but helmets off, landing in the Atlantic about 5 miles from aircraft carrier *Essex*. For a few minutes, the Command Module hung upside down in the water in what the manufacturer calls "stable position 2."

Then the crewmen pumped air into inflation bags around the submerged, conical nose of the craft to float it rightside up in "stable position 1." Radio contact was resumed with the outside world as the antenna came out of the water, and word flashed to Houston that all was just fine. Schirra reported the crew was riding a "pink cloud" on the way down, and splashdown was easy.

Apollo, the first of America's interplanetary transports, had proved itself to be as steady as a streetcar.

"We can see the lunar landing now," said General Phillips after the splashdown. "Apollo 7 has given us confidence that we can accomplish it by the end of next year."

Next up was Apollo 8.

OFF FOR THE MOON

The flight plan for the mission to the moon was at once the simplest and riskiest since John Glenn's ride around the earth nearly seven years earlier.

Apollo 8, with Borman, Lovell, and Anders aboard, would be launched at 6:51 a.m. CST, Saturday, December 21. On the first revolution around the earth, attached to the S4B, the crewmen would check the spacecraft systems. If all went well, they would be given a "go" to fire the S4B engine after passing Australia on the second revolution. The boost of the 225,000 pounds of thrust, J-2 engine would add 10,000 feet per second to their orbital velocity of 25,400 feet per second. The additional energy would raise the apogee of their orbit from 118 miles to 240,000 miles.

As they sped away from Earth on a translunar flight path, the astronauts would correct their course along the way so that they would pass the leading edge of the moon (as it moves through space) at a distance of 69 miles from the surface. The point of closest approach was to be known as "pericynthion," after "Cynthia," one of the classical names for the moon. If the launch was on time, Apollo 8 would reach this point at 4 a.m. on December 24.

If the crew did nothing further, Apollo 8 would curve around the moon and head back to Earth on a free return trajectory to land in the Indian Ocean, like Zond 5.

However, the flight plan called for the crew to fire the Service Propulsion System (SPS) engine as a retrobrake, so that Apollo 8 would fall into an orbit around the moon. After 10 revolutions of two hours each, just 69 miles above the surface, the SPS engine would be fired again to accelerate the spacecraft out of lunar orbit and return it to the flight path that would take it home. On the trans-Earth trajectory, the crew would once more make course corrections so that the Command Module, after separating from the Service Module, would enter the atmosphere over the Western Pacific Ocean and splash down southwest of Hawaii in an impact zone where the aircraft carrier *Yorktown* would be waiting.

THE INVADER FROM HONG KONG

Hopefully immunized against the spreading epidemic of Hong Kong flu, which was sweeping across America in December of 1968, the crew of Apollo 8 put in long hours at the Kennedy Space Center studying and rehearsing a flight plan no man had ever attempted to follow before.

The crew was a highly trained and even-tempered one. Borman, the spacecraft commander, was 40 years old. So was Lovell, the command module pilot, who had flown with Borman to a world's record of 14 days in space aboard Gemini 7 at the end of 1965, and who had also been the command pilot of Gemini 12. The "tenderfoot" on the team was 35-year-old Major William A. Anders of the Air Force. He was trained as the Lunar Module pilot, but on this mission, he was going to the moon without a Lunar Module. Anders and his wife, Valerie, were parents of five children, ranging in age from 4 to 11 years old. Like Eisele, Anders had been graduated from the Naval Academy and had switched to the Air Force. He held a Master of Science degree in nuclear engineering from the Air Force Institute of Technology and had been in charge of the nuclear-reactor shielding and radiation-effects program at the Air Force Weapons Laboratory at Kirtland Air Force Base. If anyone was qualified to assess the radiation hazard to which the crew would be exposed when it passed through the Van Allen belts and left the protective shield of the geomagnetic field, it would be Anders.

The final countdown for the flight to the moon began Thursday night, December 19. At Complex 39, Apollo 8, illuminated like the Washington Monument, was visible for miles up and down the coast, a glowing needle against the dark December sky. Early Friday, birdwatchers on the chilly Atlantic beaches reported the spectacle of two 40-year-old gentlemen jogging along the surf with their younger colleague to get in shape for a journey to the moon. Dr. Berry pronounced the crew fit to fly. No couch, no sniffle.

At 2:30 a.m. on Saturday, the crew was awakened for breakfast and suiting up. The van took them to Pad A and they rode up 36 stories of the launch tower in a steel elevator to wiggle feet first into Apollo 8. The countdown droned on, without a hitch. The sun rose in the clear, winter sky, the stars faded and the countdown reached zero. The great engines of Saturn 503 ignited with their earth-shaking roar. As the monster rocket raced upward on its pillar of fire, and began to arc over the Atlantic, Borman advised Mission Control that the ride on 503-Apollo 8 was smooth and easy. Saturn 503 worked perfectly and the moon farers made their first orbit around the earth checking the spacecraft systems.

Down below at Mission Control, Houston, the capsule communicator, Michael Collins, talked with the crew. He had been selected to fly the Apollo 8 mission originally, but was forced to drop out because of surgery to remove a spur on a vertebra.

With unexpected suddenness, as Apollo 8 approached Australia on the second revolution, Collins announced, "You are go for TLI."

TLI meant Trans-Lunar Injection—flight to the moon.

The announcement, reflecting one of history's momentous decisions, was made so casually it sounded routine. Hundreds of persons at the press site at the Kennedy Space Center paid no attention to it. The trip to the moon seemed so unreal that many observers, including this one, had trouble relating this fabulous voyage to the ordinary, conventionalized nomenclature being used to describe it. The only term that tended to swing with the occasion was "pericynthion," the poetic reference to the point of approach nearest the lunar surface.

Apollo on its second revolution was still attached to the Saturn third stage, the S4B. The joined vehicles made a stack 93½ feet long, with a gross weight (at liftoff) of 164 tons. The S4B had used some of its fuel to push itself and Apollo 8 into orbit. Now it remained for the J-2 engine to restart and accelerate this mass to the moon. The last time an attempt had been made to restart the S4B in space—on Apollo 6—it had failed.

But this time the J-2 engine came to life instantly on the firing signal from the Apollo guidance system. Engine ignition took place southwest of Hawaii, where it was still dark. From the 118-mile altitude, the flaming exhaust of the J-2 engine could be seen by predawn risers in the Islands.

"You've got ignition," cried Collins. "You look good. Thrust is good."

"You are solid," added the communicator at the Hawaii tracking.

"Roger," said Borman. "We look good here."

"You are go," Collins repeated. "You look good."

"Thank you, Michael," replied Borman calmly.

The J-2 engine burned for 5 minutes and 10 seconds. It accelerated the joined vehicles from 17,432 miles per hour to 24,259 miles per hour. Apollo 8-S4B swung out and away from the earth in a widening orbit and presently entered the sunlit dawn east of Hawai.

Six hundred miles from Earth, the crew watched the planet recede.

"How does it feel up there?" asked Collins.

"Very good, very good," said Lovell. "Everything is going rather well. It looks just about the same way it did three years ago."

He was thinking of Gemini 7. Borman must have been thinking of it, too, for he acknowledged a call from Houston, saying, "Gemini 8 . . . Correction, Apollo 8."

"Roger, Gemini 8," responded Collins.

From an altitude of 6500 miles, Lovell reported that the earth was assuming the shape of a disk.

"Good show," enthused Collins. "Get a picture of it."

"We are," said Lovell. "Tell Conrad he lost his record." (He was referring to the altitude record of 849 miles reached in Gemini 11 by Conrad and Gordon.) "We have a beautiful view of Florida now. We can see the Cape, just the point. At the same time, we can see Africa. West Africa is beautiful. I can also see Gibraltar at the same time I'm looking at Florida."

Climbing "uphill" to the moon against the gravitational pull of Earth, Apollo 8's velocity slowed relative to the home planet, from the initial 24,259 to 3749 miles per hour in just 24 hours. At 6:38 a.m., CST on Sunday, December 22, Mission Control announced: "We have passed the 100,000-mile mark. We are now 100,738 miles from the earth."

Apollo 8 had sped through the Van Allen radiation belts and cleared the earth's magnetic field extending 40,000 miles out on the sunward side. The crew became the first human beings to emerge from the great umbrella of magnetic force. Apollo 8 had crossed the last frontier of the planet.

To the astronauts, riding backward toward a point in space

where the moon would be 66 hours after liftoff, the frontier was invisible. It was none the less real. A radiation counter indicated the dosage of solar-particle radiation leaking through the cabin walls. It was no higher than that received in an airplane.

On the second day of the moon mission, Houston received a shock. Borman revealed he had been sick with the symptoms of violent gastrointestinal upset. He had nausea, vomiting, diarrhea, headache, fever, and soreness in the roof of the mouth. He reported the illness after he had slept for several hours and had begun to feel better. At first, NASA withheld the information, which was discussed between ship and ground on a "private line"—that is, it was not rebroadcast to news media observers from the control room.

The worst of the illness had seized Borman about 2 a.m. Sunday CST. The condition was not made public by the space agency until 38 minutes past noon that day. By that time, Borman insisted he had recovered from all symptoms and was feeling "fine."

On questioning by Dr. Berry, Borman said that an "uneasiness" akin to nausea was experienced by Lovell and Anders when they squirmed out of their spacesuits and got up from their couches to start moving around the cabin. It lasted only a short time, Borman reported, and all crewmen believed it was a form of motion sickness.

"It was just when you get out of the suits," Borman insisted. "Uneasiness, not nausea, really, but a sort of awareness of motion, like the zero *g* airplane."

He refered to the few seconds of weighlessness which can be experienced in an airplane flown on a toboggan trajectory. This procedure was used in the early training of astronauts. Dr. Berry prescribed medication in the flight kit to control the bowels and reduce nausea if it returned.

BEINGS FROM THE BLUE PLANET

Early Sunday afternoon, the crewmen turned on the midget television camera that had graced the flight of Apollo 7, and a picture filtered down to Earth, showing them milling around the cabin, like larvae in cocoon. It was not flattering.

Borman explained that he was rolling the ship to give the

folks back home, all snug in their living rooms on the Sunday before Christmas, an idea of what it looks like to be halfway between the earth and the moon.

The trouble with this effort was that the telescopic lens of the camera would not pass a picture. The short range lens was substituted, but it showed Earth merely as a whitish blur. Borman then switched to a shot of Lovell, who was squeezing water in a pack of choclolate powder to concoct a chocolate pudding.

Collins advised that Anders' seven-year-old son, Gregory, wanted to know when Daddy would meet up with Santa Claus on the mission.

"Roger," said Anders. "We saw him earlier today and he was heading your way."

Collins agreed to pass the word along.

On Monday, December 23, after minor course corrections, the crew verified the flight path by sighting through the telescopic sextant the stars Altair, Atria, and Antares. Lovell reported having some difficulty sighting the moon, which appeared as a white haze in the sextant optics. At one time, he reported that the moon had a bluish ring around it. This, too, appeared to be a distortion of the optics.

Blinding sunlight, on the other hand, made the ship's scanning telescope, which viewed a much broader field than the sextant, useless in sunward directions. In addition to these illumination problems with the moon and the sun, outgassing of silicone window-sealing material, which decomposed in sunlight, had beclouded two of the five spacecraft windows by depositing an oily film on the inner panes and had reduced visibility in a third.

At noon, Houston time, Apollo 8 was 171,360 miles out from Earth. Its velocity had showed to about 2288 miles an hour. Lovell announced that all hands were feeling good. "We're going to fix it up now so that we all have one more rest period before LOI," he said.

LOI meant Lunar Orbit Insertion—the next big maneuver. In Mission Control, the time posted for LOI was 3:59 a.m., CST, December 24, a moment that no one who took part in the flight in space, at Houston, or at tracking stations around the world, was going to forget.

During Monday afternoon, the crew televised views of Earth from a distance of 201,500 miles. The Blue Planet appeared in a three-quarter phase. The north pole, in darkness, was at the left of the picture while the clouded south pole, at the right, was brightly illuminated. Geographically, Apollo 8 was over South America. The terminator or darkness boundary bisected the Atlantic. Evening had fallen in Europe, where millions of persons watched these fantastic pictures on their home television sets via communications satellite relay. The transmissions were received by, the big NASA antennas at Goldstone, California, and Madrid, Spain.

"What you're seeing, Mike, is the Western Hemisphere," Borman was explaining to Collins, as though to a pupil. "The top over there is the north pole. In the center, just lower to the center, is South America, all the way down to Cape Horn. I can see Baja, California and the southwestern part of the United States. There is a big cloud bank going northeast —covers a lot of the Gulf of Mexico up to the eastern part of the United States."

"Roger," said Collins. "Could you give me some ideas about the colors, and also could you try a slight maneuver. It [the earth] is disappearing. We see about half of it."

Borman rolled the spacecraft and Collins said, "There you go! That's fine. Stop it right there!" Collins now held the long-distance record for TV show direction.

"Okay," said Borman. "For colors, waters are sort of a royal blue. Clouds, of course, are bright white. The reflection off the earth is much greater than the moon. The land areas are generally a brownish, sort of dark brownish to light brown. . . . Many of the borders of the clouds can be seen by the various weather cells. A long band of various cirrus clouds extends from the entrance to the Gulf of Mexico out across the Atlantic. Southern hemisphere is almost completely clouded over and up near the north pole there are quite a few clouds. Southwestern Texas and the southwestern United States are clear."

Borman broke off the travelogue for a moment and then said, "You are looking at yourselves at 180,000 [nautical] miles out in space [about 207,000 statute miles]. Frankly, what I keep imagining is . . . if I am some lonely traveler from another planet what I would think about the earth at this altitude . . . whether I would think it was inhabited or not.

COLLINS: Don't see anybody waving, is that what you are saying?

BORMAN: I was just kind of curious if I would land on the blue or brown part of the earth. You better hope that we land on the blue part.

COLLINS: So we do, Babe.

PERICYNTHION

Shortly after the telecast, Apollo 8 crossed a gravitational-force frontier and entered the predominant gravitational sphere of influence of the moon. The craft began to speed up. The crossing was made at a hypothetical point 38,894 miles from the moon and 202,916 miles from Earth along the flight path. The spaceship was falling toward the moon at 2719 miles an hour.

Hour by hour its velocity increased toward a maximum of 5696 miles an hour at pericynthion. Monday evening, it was quiet aboard Apollo 8. The crewmen were tired and took turns sleeping. At midnight, the astronauts busied themselves checking guidance and navigation equipment and spacecraft systems, and preparing cameras. Their prime photo target was Landing Site No. 1 east of the Crater Maskelyne in the Sea of Tranquillity. It should be just east of the terminator, in early morning sunlight as Apollo 8 came over the moon's eastern horizon heading west.

Mission Control radioed directly to the ship's computer the data required for the Lunar Orbit Insertion burn of the SPS engine, upon which all now depended. Under computer control, the guidance system would put the ship in position for the burn and fire the engine to reduce velocity to about 3657 miles an hour, dropping Apollo 8 into an oval orbit of 69 by 196 miles around the moon.

As the ground loaded the flight computer for the LOI maneuver, Borman checked the new data against the flight instruments. He wanted to make certain there was no error, for a mistake of only 70 miles would plunge them into the moon. The hours rubbed slowly into Tuesday December 24. The Manned Spacecraft Center was a blaze of light on the dark Texas Gulf plain, its grassy campus alive with people hurrying along the walks and drives on urgent business. The

3000 persons involved in flight operations worked steadily at the tasks. Around the world, men in tracking stations, aboard ships in all the seas, and in aircraft listened and waited.

The message "Go for LOI" was received with a terse "Roger," and the ship was positioned for the retro-burn. At 3:59 a.m., according to the computations, Apollo 8 reached pericynthion, 69 miles above the surface of the moon, and passed around its leading edge to the far side, the side that faces away from Earth. The SPS fired for four minutes, six and one-half seconds, but that was not immediately known on Earth, for Apollo 8 was behind the moon.

In Mission Control, where men sat like statues, two clocks were ticking. One showed the time that the first signal from the spacecraft would be received as it cleared the moon on a free return to Earth trajectory. If the engine signal came at that time, it would be known that the SPS had failed to fire and the crew was coming home. The second clock showed the time of signal acquisition if the engine did burn on schedule.

For 19 minutes no electronic message came from Apollo 8. Signal acquisition time in case of engine failure passed. The men in the control room began to relax. At 4:48 a.m., Mission Control announced:

"We're standing by. This is Apollo Control, Houston. Mark. Three minutes from predicted time of acquisition. Standing by. Apollo Control, Houston. Mark. Two minutes Apollo Control, Houston. Mark. One minute from predicted time of acquisition. Apollo Control, Houston, Jerry Carr [Major Gerald P. Carr, USMC] has placed a call. We've heard nothing yet but are standing by. Apollo Control, Hous . . . We are looking at engine data! And it looks good! Tank pressures look good! We've got it! We've got it! Apollo 8 is now in lunar orbit!"

A relieved cheer came out of Mission control. The next voice was that of Lovell, who reported that Apollo 8 was in a lunar orbit of 69.5 by 194.3 miles.

Immediately, Carr warned the crew that telemetry showed that the primary-cooling-system temperature was too high. Anders turned on the ship's secondary evaporator [cooler] and inspected the enviromental control system. He brought up the steam pressure in the evaporator to speed cooling and the temperature in the cooling system began to come down. All hands then turned their attention to the moon.

"Apollo 8, Houston," called Jerry Carr. "What does the ole moon look like from 60 miles? Over."

LOVELL: Okay, Houston. The moon is essentially gray, no color. Looks like plaster of Paris or sort of a grayish, deep sand. We can see quite a bit of detail. The Sea of Fertility doesn't stand out as well here as it does back on Earth. There's not as much contrast between that and the surrounding craters. The craters are all rounded off. There's quite a few of them. Some are newer. Many of them look like—especially the round ones—like they were hit by meteorites or projectiles of some sort. Langrenus is quite a huge crater. Its got a central cone to it. The walls of the crater are terraced, about six or seven different terraces on the way down. And coming up now, the Sea of Fertility and our old friends, Messier and Pickering [craters]. And I can see the rays coming out of Pickering. We're coming up near our P-1 Initial Site, which I'm going to try and see [a landmark ahead of Landing Area No. 1]. Be advised the round window—the hatch window—is completely iced over. We can't use it. Bill and I are sharing the rendezvous window.

HOUSTON: Roger Got any more information on those rays? Over.

LOVELL: Roger. The rays out of Pickering are quite faint from here. There are two different groups going to the left. They don't appear to be any depth at all. Just rays coming out.

CARR: Bill, if you can tear yourself away from that window, we'd like to turn off the secondary evaporator.

ANDERS: Roger. Going off.

LOVELL: Okay. Over to my right, the Pyrenees Mountains are coming up and we're just about over Messier and Pickering right now. Our first initial Point [landmark which would be used by a Lunar Module crew in the descent to Landing Site No. 1] is easily seen from our altitude. We're getting quite a bit of contrast as we approach the terminator. It's very easy to pick out our first initial point and over its mountain we can see the second initial point, the triangular mountain.

ANDERS: And now we're coming up on the craters Columbo and Gutenberg. Very good detail visible. We can see the long, parallel faults of Gaudibert and they run through mare material right into the highland material.

To the pilots, the lunar markers pointing to Landing Site No. 1 were like those of a bombing range, marking the target.

LOVELL: We're directly over our first Initial Point now. It's almost impossible to miss, very easy to pick out, and we can look right over into the second Initial Point. I can see very clearly the five-crater star formation we had on our lunar chart.

During their first two revolutions around the moon, the crewmen were intent on testing their ability to see ground features. On the second revolution, Lovell reported: "We're directly over our favorites, Messier and Pickering, again. The view at this altitude, Houston, is tremendous. There is no trouble picking out features that we learned on the map. We have just passed over the Sea of Fertility and the mare is darker . . . the 'bomb range' has more contrast."

Landmarks on the far side of the moon were identified in the ship-to-earth conversation with the names of men who had helped pave the way to the moon. There were the craters Borman, Lovell, and Anders, which had simply been numbered reference points on the map. It was easier to refer to them by names. Other Lunar Farside craters were named for Gilruth, director of the Manned Spacecraft Center; Debus, director of the Kennedy Space Center; Phillips, Apollo program director, and George Low, the spacecraft program manager. Three small craters nested close to each other on Farside became memorials to Grissom, White, and Chaffee. Two others memorialized the astronauts Charles M. Bassett and Elliott See, who were killed in an airplane crash at St. Louis early in 1966. Craters were named for Webb, for Chris Kraft, flight operations director, and for Tom Paine, acting head of NASA. The naming system was informal and did not include Presidents Kennedy or Johnson. But it signified that the men of Apollo 8 thought of the moon as an American moon.

One crater, a fairly large one, was named for John Aaron, a console operator in Mission Control. It was he who recognized the need to cut in the secondary water boiler to cool the ship down on its first revolution around the moon after the primary cooling system failed to dissipate the heat load in the brilliant sunshine of Lunar Farside. The emergency had eluded the crewmen, who were glued to the two spacecraft windows that had escaped fogging.

On the third revolution, Apollo 8 was given a "go" for its second Lunar Orbit Insertion burn, LOI-2, designed to circularize the orbit at 69 miles. The second burn was also done behind the moon shortly after Apollo 8 vanished from radio view of Earth around the far side at 7:55 a.m. When it reappeared at 8:40 a.m., it was in a new orbit of 69.9 by 71.3 miles. Mission Control called that good enough. Borman reported that the SPS had burned for 11 seconds. Then he excused himself and took a nap.

IN THE BEGINNING . . .

Several hours later, after a breakfast of fruit cocktail, eight bacon squares, cinnamon toast cubes, an orange drink, and cocoa, Borman asked what was going on over there on the bright, blue planet called Earth.

"Your TV program was a big success," replied Mike Collins. "It was viewed by most of the nations of your neighboring planet, Earth. It was carried live all over Europe, including even Moscow and East Berlin. Also in Japan and all of North and Central America and parts of South America. San Diego welcomed home today the Pueblo crew in a big ceremony. They had a pretty rough time of it in the Korean prison. Christmas cease fire is in effect in Vietnam. And if you haven't done your Christmas shopping by now, you better forget it."

"Thank you," said Borman.

"How about your news?" asked Collins.

"Well," said Borman, "we're looking forward to a big burn here shortly."

He meant the SPS firing that would boost them out of lunar orbit and back on the trail to earth.

"Mike," Borman added, "I think I can say it without contradiction—it's been a mighty long, dry spell up here."

It was the first indication of how the crew was reacting to long hours of tense wakefulness and the bleak, alien landscape.

"I guess you can say anything you like without contradiction," said Collins gently.

As Apollo 8 sailed around the moon, crew and ground controllers observed a new phenomenon. Apogee of the orbit tended to increase slightly while perigee tended to decrease. The opposite effect is experienced in Earth orbit, where apogee tends to shrink because of air drag at perigee.

Early Christmas eve, the crew turned the TV camera on Earth and transmitted the picture to its inhabitants.

BORMAN: This is Apollo 8 coming to you live from the moon. . . . We showed you first a view of Earth as we've been watching it for the past 16 hours. Now we're switching so that we can show the moon that we've been flying over at 60 [nautical] miles altitude . . . What we will do now is follow the trail that we've been following all day and take you on through to the lunar sunset.

The moon is a different thing to each of us. I think that each one of us carries his own impression of what he's seen today. My own impression is that it's a vast, lonely, forbidding type existence, a great expanse of nothing, that looks rather like clouds and clouds of pumice stone, and it certainly would not appear to be a very inviting place to live or work. Jim, what have you thought about?

LOVELL: Well, Frank, my thoughts are very similar. The vast loneliness up here of the moon is awe inspiring, and it makes you realize just what you have back there on Earth.

BORMAN: Bill, what do you think?

ANDERS: I think the thing that impressed me the most was the lunar sunrises and sunsets. These in particular bring out the stark nature of the terrain and the long shadows really bring out the relief that is here and hard to see and is very bright. . . .

As Apollo 8 moved across the sunlit eastern edge of the moon, the astronauts described what the camera was seeing, down on the surface. Borman announced that "the crater you see on the horizon is the Sea of Crises." He pointed out a rill which ran along the edge of a small mountain, turning sinously, like an ancient river bed.

"I hope all of you back on Earth can see what we mean when we say that it is a very foreboding horizon, a very dark and unappetizing-looking place," said Borman.

As Apollo 8 approached the terminator, Borman announced that "the crew of Apollo 8 has a message that we would like to send you." It was a reading from the first book of the Bible, Genesis.

ANDERS: In the beginning, God created the heaven and the earth. And the earth was without form, and void; and darkness was upon the face of the deep. And the spirit of God moved upon the face of the waters. And God said, Let there be light; and there was light. And God saw the light, that it was good: and God divided the light from the darkness.

LOVELL: And God called the light Day, and the darkness He called Night. And the evening and the morning were the first day. And God said, Let there be a firmament in the midst of the waters, and let it divide the waters from the waters. And God made the firmament, and divided the waters which were under the firmament from the waters which were above the firmament: and it was so. And God called the firmament Heaven. And the evening and the morning were the second day.

BORMAN: And God said let the waters under the heaven be gathered together in one place and let the dry land appear: and it was so. And God called the dry land Earth; and the gathering together of the waters called he Seas: and God saw that it was good.[3]

And from the crew of Apollo 8, we pause with good night, good luck, a Merry Christmas, and God bless all of you—all of you on the good Earth.

Late Christmas eve, the crew prepared to fire the SPS engine for three minutes and 23 seconds to boost the ship out of lunar orbit with an increase in velocity of 2401 miles per hour and set it on the trans-Earth trajectory for the Blue Planet.

At 11:42 p.m. the ship vanished around Farside and the men in Mission Control settled down to wait. At 12:10 a.m. December 25 the SPS was to fire—but no signal would be received until the spacecraft came out from behind the moon. Would it still be in orbit? Or would it be headed home?

For Chris Kraft, the 37-minute wait for the ship to clear Farside was the longest the flight operations director could remember.

In Mission Control, Astronaut Thomas K. (Ken) Mattingly, capsule communicator, began calling Apollo 8.

Out of the crackle came a faint reply: "Apollo 8, over . . ."

Mission Control erupted in cheering. And Borman was saying, "Roger. Please be informed there is a Santa Claus!"

HOUSTON: That is affirmative. You are the best ones to know.

The Apollo engine had fired precisely as programed. And the first men around the moon were safely on their way home.

Chapter 15

Landfall

In the darkness before dawn of December 27, 1968, Apollo 8 splashed down in the Pacific Ocean about 600 miles northwest of Christmas Island. Three miles away rode the prime recovery ship, the aircraft carrier, U.S.S. *Yorktown.*

Half way around the world, in Houston, controllers at the Manned Spacecraft Center whooped it up with joy and relief. They broke out small American flags and tied them to the consoles. They lighted cigars and filled the room with blue smoke. Even men who had quit smoking for years "lit up" to observe the ritual, which had begun in Project Mercury.

"Hello there, Houston. How you doing?" Borman called on the radio as Apollo 8, its hide blackened by the heat of the fastest entry into the atmosphere ever made by a manned vehicle, rocked on the long, Pacific swells like an illuminated channel buoy. "The moon's not made out of green cheese at all. It's made out of American cheese."

At first light, helicopters churned above the moon ship and swimmers were in the water affixing a flotation collar around it. The crew was transferred quickly by helicopter to the great ship. Then the spacecraft was hoisted aboard like a toy and the *Yorktown* steamed toward Hawaii.

THE TRIUMPH OF THE SQUARES

At Houston, the acting administrator, Tom Paine, addressed a press conference. He was deeply moved but calm. "Here we are this morning at the end of an almost flawless mission," he said. "We feel humble that we were the ones to perform this historic feat . . . It might show the restless students of the world the benefits . . . the triumph of the 'squares' who work with computers and slide rules, of engineering and of science, and of men who read from the Bible on Christmas Eve."

It was a fine morning for Americans. Apollo 8 had given the nation an achievement in which many millions felt vicarious pride. It was the brightest event of an otherwise dismal year and it lifted the mood of the land. In one stroke, Saturn 5-Apollo 8 had demonstrated that the whole of this massive and expensive conglomerate of machinery was greater than the sum of its myriad parts. Here was a transportation system that could take men to the moon, or at least to within 69 miles of it, and bring them home. Now the taxpayers could see what they had purchased for $23,915,900,000 [1].

But the last 69 miles were going to be the hardest—the most hazardous and nervewracking—for no one had ever landed on an airless surface before, or left it, or knew the perils.

Still untested in manned flight was the Lunar Module, the vehicle in which two astronauts would traverse that last 69 miles from lunar orbit down to the surface. In contrast to the Apollo, which had undergone a program of 16 tests before men were entrusted to it, the LM had been tested in space only once (Apollo 5) and then its performance had been ambiguous.

On its second space test, the LM would be manned in Earth orbit. That would be Apollo 9. On its third, the vehicle would be manned in lunar orbit and descend to 50,000 feet from the surface. That would be Apollo 10. On its fourth journey into space, the LM would be used in the attempt to land two men on the moon and lift them back up to lunar orbit, to make rendezvous with the awaiting Apollo for the trip home. That would be Apollo 11, the lunar landing mission.

In the euphoria after Apollo 8, there was speculation that

the Apollo directorate would attempt the moon landing on Apollo 10. The fact was, however, that the directors of Apollo were not inclined to take any risks they didn't have to take. There was no longer any pressure on them to accelerate the flight development program. Politically and competitively, the heat was off. Apollo 8 had given the outgoing Johnson Administration its moment of glory. The mission had signaled also that the landing could be made in July, within the time frame to which President Kennedy had committed the program.

So far as the moon was concerned, Russian competition seemingly had withered away. The Russians had executed a brilliant display of manned spacecraft maneuverability in Earth orbit with a crew transfer between two vessels, Soyuz 4 and 5, in January. The exercise proved that they, too, had mastered rendezvous and docking, but the demonstration was oriented toward the construction of a modular space station, not a landing on a moon. As an act following Apollo 8, it made little impression on the American public.

Apollo 10, therefore, remained a test—the second in a two-part sequence for the LM. Phase One would be the test of the vehicle in an orbit around the earth. Its performance as a spacecraft capable of making rendezvous and docking with Apollo was to be assessed near the earth, where the margin of safety was greater. This test would be made during the ten-day mission of Apollo 9. If all went well, the LM would be flown to lunar orbit with Apollo 10. It would be separated from the mother ship and injected into a mountain-grazing orbit, which is possible around the moon, where there is no atmosphere to slow the vehicle down. At closest approach to the ground, or perilune, the LM would sweep to within 50,000 feet of the surface, in line of sight of Landing Point No. 2 in the Mare Tranquillitatis, so that the LM crew could photograph the site.

In the sequence of missions leading to the landing, Lunar Module operations on Apollo 9 would simulate as closely as possible those on Apollo 10 which, in turn, would simulate those on Apollo 11, except for the landing.

Once this scheme had been adopted, it added up to a sequence of step-by-step tests which seemed to have been brilliantly plotted. There was little to suggest that the flight development program had been largely shaped by expediency, desperation, and not a little luck.

When the flight plan for Apollo 10 had gelled, there was no possibility that it could be switched to a surprise lunar landing at the last minute. Lunar Module 4, to be flown with Apollo 10, would carry extra equipment for elaborate radar tests, the extra weight being offset by reducing to one-half the normal fuel load in the Ascent stage of the LM. Enough fuel would be aboard to enable the Ascent stage, where the crew rides, to maneuver for rendezvous with Apollo, but not to lift the stage off the lunar surface into orbit. LM-4 could land on the moon, but if it did, it could not get off.

THE SOLAR SYSTEM DUMP

Apollo 9, with LM-3 aboard, was rolled out of the Vehicle Assembly Building at Kennedy Space Center and trundled to Launch Complex 39, Pad A, on January 3, 1969. Preparations for a ten-day Earth orbit flight, starting February 28, moved along smoothly until two days before launch. Then the human component in the loop broke down. The crew turned up with colds. The flight was grounded four days.

Manning the mission were two veterans of Gemini, James McDivitt, spacecraft commander, and David Scott, Command Module Pilot. Both were Air Force colonels at this time. The third man, the Lunar Module pilot, was Russell L. Schweickart, 33, a member of the 1963 group of astronaut appointees, who had not yet flown in space. Red-haired, lanky, with electric-blue eyes in an open, boyish face, Schweickart was one of the first of the young, civilian scientists to join the corps of astronauts. When he was accepted for training, he had been working as a research physicist at the Massachusetts Institute of Technology, where he had received a bachelor of science degree in aeronautical engineering and a master of science degree in aeronautics and astronautics, specializing in upper-atmosphere physics. At the outset of Schweickart's career in NASA, his wife, Clare, found it hard to believe that he would ever go to the moon. The idea became more real as time went on. The Schweickarts were the parents of five children: Vicki, 9, Randolph and Russell, both 8, Elin, 7, and Diana, 4.

Apollo 9 lifted smoothly off Pad A on the morning of March 3, 1969, and on the second revolution around the earth, Scott separated the spaceship from the Saturn third stage (S4B), in which the LM was carried. He turned the 24-ton Apollo command and service module around and eased it forward so that its conical front end docked securely with the LM in the S4B. Then, after examining the docking tunnel, a cylindrical passage 18 inches long and 32 inches in diameter through which crewmen could crawl from one vessel to the other, the crew pressurized the LM with oxygen. By activating ejection springs in the S4B, the crew separated the Apollo-LM from the big rocket and then fired the Apollo thrusters to maneuver farther away. Ground control, advised that the crew was in the clear, reignited the rocket's J-2 engine and the S4B rushed away "like a great star disappearing into the distance," the crewmen reported. Later, the rocket engine was fired again to hurl the third stage into an orbit around the sun, where the third stage of the Apollo 8 booster had been sent in December and where three Mariner space probes and miscellaneous machines had been dumped, forming man's first solar junk yard.

"We just sent the S4B hyperbolic and got it out of your way," Houston advised Apollo 9.

On the second day of the flight, McDivitt and Scott swiveled the Apollo engine as they fired it in order to shake the joined Apollo-LM vehicle, a "stack" some 58 feet long. Scott reported that the joined vehicles made up a stable system. The supervehicle also performed nicely in response to both automatic and manual thrust control. In this formation, the flight to the moon would be made on the next mission.

McDivitt and Schweickart crawled through the tunnel into the LM on the third day. Just before he left Apollo, Schweickart was hit by nausea and vomiting. When it subsided, he crawled into the LM, where nausea hit him again. Nevertheless, he was able to perform his piloting duties for nine hours aboard the LM. He and McDivitt switched on the instruments, extended the four-legged, spidery landing gear, checked the data from the flight computer with those from Apollo and from the ground, and fired the Descent Engine for six minutes, enlarging the orbit of the joined machines.

Schweickart was satisfied that the Descent Propulsion System was working properly and reported that the guidance system presented no problems. After several hours in the LM, McDivitt and Schweickart began to think of it as part of a third spacecraft, with engines at either end. Now they were flying it from the Lunar-Module end. Such an operation might be required in case the Apollo engine broke down during insertion of the joined vehicles into lunar orbit. "We would then employ the Descent Engine on the LM to bring us home safely to Earth," Scott explained after the flight.

The main event of the fourth day was a trip outside the spacecraft by Schweickart to test the moon suit and determine the order of difficulty in the transfer of a crewman from one vessel to the other. The Russians had done it for the first time on the Soyuz 4 and 5 missions.

In spite of Schweickart's insistence that his gastrointestinal problem had eased and that he felt fine, Mission Control was uncertain about authorizing him to perform the extravehicular activity. The EVA plan called for him to come out of the LM hatch, holding on to hand rails, and stand on the "porch," a small platform below the hatch and above the ladder the crewmen would use on the moon to descend to the ground from the cabin in the Ascent Stage. Then, Schweickart was to pull himself to the open hatch of the Apollo Command Module, using hand rails and holds, to get in, feet first, and to stand on the couch, with Scott steadying his legs. In that position, the spacecrawler was to rest about 15 minutes. Next, he was to climb out again over the Command Module to the LM, picking up several thermal test packages on the outer skins of both vehicles, then to climb back inside the LM, and to close the hatch and repressurize the vehicle. Scott similarly would close the Apollo hatch and repressurize the Command Module.

GUMDROP AND SPIDER

As a means of distinguishing between the two vessels and among the crewmen in each one, flight planners had resorted to a parody of the military code-name practice. The Apollo was dubbed "Gumdrop," and the Lunar Module was "Spider." Schweickart was to be known during the EVA as "Red Rover." The coding betrayed the influence of Saturday morning television cartoons in the astronauts' households, which were filled with preadolescent cartoon devotees. It lent a lighthearted, if bizarre, touch to the exercise in space. In fact, the whole exercise lends itself to the cartoon narrative mode:

During the morning of March 6, we find Red Rover in Spider getting ready for his daring stunt outside the Lunar Module, 135 miles above the ground. Jim McDivitt, who talks for Spider, is hurrying him up.

SPIDER: This is Spider, here. Just so everybody is familiar, I think we'll do one daylight pass out on the porch.

(This means that Jim McDivitt has made the decision that Rusty Schweickart, alias Red Rover, will do the daring space walk!)

HOUSTON: Roger. Copy, Spider, and we agree with that wholeheartedly.

(What else can they say? After all, they're down here and Spider's up there.)

SPIDER: Roger. (To Red Rover). I'll tell you what we're gonna do. You go on outside, get accustomed to what you are doing and I'll take a couple of pictures of you. You look around and when you look like you're stabilized and you think you can handle something, I'll send out the camera to you.

Now, let us listen in to what Spider and Red Rover are actually doing up there, getting ready for the EVA:

"They cleverly put on that piece of rubber that we've never had on it before. . . ."

"Take it off this side . . . to clean that out when we leave . . . ."

"Okay. Throw that up here."

"Okay. The camera is up there. Put the handle on it."

"How? I can't get that thing screwed in. Look at that!"

"I can't get it out."

"That's supposed to go in that bag over there. Stick that in that bag."

What are Red Rover and Spider up to? Even Gundrop is wondering.

GUMDROP: Rusty, how you feeling?

RED ROVER: Good.

HOUSTON: Spider and Gumdrop, we've got you through Carnarvon [Australia tracking station]. Standing by.

GUMDROP: I'm all set to depress [depressurize the cabin by flushing out the oxygen] whenever you give the word.

SPIDER: Okay, we're all set over here, Dave.

They depressurize and open their hatches, and Red Rover slowly and carefully climbs out of the LM and stands on the porch. Gumdrop says he can see Red Rover's foot. Red Rover asks if Gumdrop can see his toes wiggle. Sure can, says Gumdrop. Then Gumdrop climbs up on the couch and leans out of the Command Module. He looks like a mole with a red helmet on.

Red Rover plants his spaceboots in a pair of metal stirrups on the porch called the "golden slippers." They are painted gold to reflect sunlight so that his feet will remain cool. The stirrups keep him anchored to the LM. Of course, he is secured to it also by a tether so that he cannot drift away.

Red Rover then looks around him and exclaims: "Boy, oh Boy. What a view!"

SPIDER: Isn't that spectacular?

RED ROVER: It really is. There's the moon . . . right over there.

SPIDER: Why don't you say hello to the camera, or something?

RED ROVER: Hello, there, camera. Boy, this is great! There is Baja California. Oh, very pretty.

HOUSTON: Well, folks, you have heard it here live, first hand—the adventures of Red Rover and his friends, Gumdrop and Spider.

THE EYE TEST

Red Rover spent 37½ minutes outside the LM on the porch. He did not climb over to the Apollo hatch. Later, he related, "I was comfortable the whole time, moving up and down the handrail and looking at the control required to transfer from one vehicle to the other. It was a pleasant surprise to me that this was far easier in flight than it had been in any simulations on the ground."

On the fifth day, the LM was unhitched from Apollo and nudged into a diverging orbit which took it 100 miles away from the mother ship. This was the most critical maneuver of the Lunar Module test. It would show whether the LM propulsion and guidance system could bring the vehicle back to make rendezvous and dock with Apollo. If the LM failed, Scott could make the rendezvous with Apollo.

The test was a rehearsal for a similar separation, rendezvous, and docking maneuver around the moon on Apollo 10, where the LM would enter a descending orbit that would swing it down to an altitude of 50,000 feet before it rejoined Apollo at 70 miles altitude. And Apollo 10 was visualized at the time of Apollo 9 as the dress rehearsal for the lunar landing.

The test was carried out in three parts. First, Schweickart burned Spider's powerful Descent Engine to kick the spacecraft in a preliminary divergent orbit that took it 55 miles from Gumdrop. As Spider sailed away, Scott in Gumdrop remarked, "That's a nice-looking machine. That's about all it looks like, too—a machine."

At 55 miles' distance, McDivitt and Schweickart found all systems working well and fired the descent engine again to widen the distance from Gumdrop. Now they were on their own, for if they could not return to Gumdrop or if Gumdrop could not reach them, they could not return to Earth, at least not alive. The LM carried no heat shield, since it was not designed to land on a planet with atmosphere. Once it entered the atmosphere, it would burn up.

As Spider coasted out to 85 miles from Gumdrop, the crew in the Ascent Stage separated it from the Descent Stage. This enabled McDivitt and Schweickart to test the Ascent Engine of the two-stage vehicle. Designed to lift the astronauts off the moon and enable them to maneuver to a rendezvous in lunar orbit with Apollo, all the Ascent Engine had to do on this test was nudge them back into a converging orbit in which they could dock with Gumdrop.

As the two vehicles circled the earth, Spider's crew burned the Ascent Engine and the distance between them began to close. Spider used its small thrusters to make the final orbit adjustment to intercept Gumdrop. As the vessels approached, Scott observed that Spider was upside down relative to Gumdrop.

"Oh, I see you out there, coming in the sunlight," he said. "You're the biggest, friendliest, funniest-looking Spider I've ever seen."

McDivitt handled the docking. He was nearly blinded at times by the sun glare from Gumdrop's metal hide. "It wasn't so much a docking as an eye test," he complained. "Okay, Houston. We've locked up."

He and Schweickart crawled back into Apollo. Then the LM Ascent Stage was cast off. The Apollo 9 mission still had five days to run. The crew practiced navigation and landmark sighting and took hundreds of photographs. But the LM test was over when the final docking was completed, and the weird-looking lunar lander had passed it *cum laude.*

Apollo 9 splashed down on March 13, less than three miles from its prime recovery ship, the U.S.S. *Guadalcanal,* in the Atlantic Ocean north of Puerto Rico. The July lunar-landing attempt now seemed to be assured.

There remained Phase Two of the LM field trials—the test in lunar orbit. Saturn 505, with Apollo 10 on top and LM-4 in its steel cocoon on the third stage, was being prepared for moonflight on the Ides of May.

THE SECOND VOYAGE TO THE MOON

In Apollo 8, man had journeyed to within 69 miles of the surface of the moon. Apollo 10 was designed to bring him to within 50,000 feet of the surface for the final test of the Lunar Module before the first attempt to land.

The second visit of *Homo sapiens* to the moon May 18-26, 1969, was considerably less somber, less pious and more professional in mood (if not in execution) than the first. To Tom Stafford, John Young, and Eugene Cernan, the Apollo 10 crew, the alien contours of the moonscape were fascinating, rather than depressing; challenging, rather than forbidding. In contrast to solemn recitations which had come from Apollo 8 on Christmas Eve, the language of the Apollo crewmen was studded with the conventional visceral, sexual, and profane expressions of men under stress, struggling with balky equipment and intermittently malfunctioning machinery. But it was also illuminated with the delight, the triumph, and the awe of man gazing upon another world.

The crew of Apollo 8 had regarded the moon with foreboding—as alien, inimical to man. The "Ten" crew saw it differently, as a new land to be explored, described, interpreted. To the second team of visitors, the tans, dark browns, and grays of the maria, the weird glows in craters illuminated by earthshine, the jumble of boulders which became visible in sunlight, the steep, canyon-like rilles, and the clustered volcanoes were scenes of elemental beauty.

If the voyage of Apollo 10 was carried out more nearly in the mood of a Magellan than in that of the Ancient Mariner, the difference might be explained by the fact that it was the second time around. But there were also important differences in context.

The Apollo 8 mission had been escalated from Earth orbit to lunar orbit by two mainly political objectives: to nail down a clear American lead in manned space flight before the end of the Democratic administration and to present President Johnson with a lunar triumph before he left the White House. In terms of Apollo flight development, the lunar-orbit mission of Apollo 8 was premature. The crewmen were not so well prepared for it as they might have been if it had come later, even though their training had been intensive. Perhaps no amount of training could have prepared them for what they beheld on their first circuit of the moon at 69 miles.

Apollo 10 was devoid of the motives that had launched Apollo 8 to lunar orbit. The "Ten" mission was a straightforward engineering test of the Lunar Module in the moon's environment. Its crew had been trained for it and also had acquired psychological preparation from the experiences of Borman, Lovell, and Anders. Since the Apollo 10 astronauts were to remain in lunar orbit three times as long, they had been given more training in landmark identification and lunar "geology," or selenology.

The quaint idea of code naming the Apollo command and service modules "Charlie Brown" and the Lunar Module "Snoopy," Charlie's faithful beagle in the "Peanuts" comic strip, injected a note of whimsy into the mission, as had the "Spider and Gumdrop" code of Apollo 9. Finally, the naming of lunar landmarks after NASA people and their wives or after features of the Old West made the moonscape less alien and awesome.

THE WORLD INCREDIBLE

The launch of Apollo 10 was another triumph of the timetable. It went up precisely as scheduled, at 12:49 p.m.

Eastern Daylight Time on Sunday, May 18, from Pad B of Launch Complex 39. Saturn 505 was the fifth moon rocket to get off the pad on time—and the fifth to be launched. As it bore Apollo 10 high over the cloudy Atlantic Ocean, Cernan could be heard exclaiming over the radio, "What a ride!"

"Here's four and one-half *gs*," Stafford reported. "Just like old times. It's beautiful out there."

All the crewmen had been in space before. Stafford, a veteran of Gemini 6 and now a full colonel in the Air Force, and Cernan, a Navy commander, had flown together in Gemini 9 to encounter the "angry alligator"—that malfunctioning docking target—in Earth orbit. John Young, a Navy commander, had made the first manned flight in Project Gemini with Gus Grissom and had flown Gemini 10, with Michael Collins.

On the second revolution, over central Australia, the Saturn third stage (S4B) engine was ignited and burned five minutes, 43 seconds to boost Apollo 10's velocity by 10,438 feet per second and inject both rocket and spacecraft into the translunar trajectory. From Earth orbital velocity of 25,000 feet per second, the speed of Apollo 10 rose steadily. As it reached 27,500 feet per second, the crew rode from night side into a Pacific dawn.

"What a way to watch a sunrise!" exulted Cernan.

At 31,000 feet per second (21,136 miles per hour), Stafford reporting that the still-thrusting S4B engine was producing some vibrations and yawing. But the insertion into the translunar flight path was good—so accurate, in fact, that only one small correction in the course was required to bring Apollo 10 into a 69-mile orbit around the moon.

"Would you believe it," Stafford called to Mission Control, Houston. "The world is starting to fade away."

"We believe it, Tom," replied Air Force Major Charles M. Duke, Jr., the capsule communicator. "You are go here. It was a perfect insertion."

The S4B vibration was one of a series of minor malfunctions which plagued the entire mission, but which did not interfere with its execution. John Young, the Command Module pilot, said, "It felt like it was running rough, at least compared to the Titan." Stafford added, "That TLI (translunar injection) frequency was a little bit too much. We thought sure it was coming unglued."

At an altitude of 4117 miles, Young separated the Apollo spacecraft from the S4B, turned the spacecraft around, and proceeded to dock with the Lunar Module, housed in the forward adapter section of the rocket. Millions of persons watching television that Sunday afternoon saw the approach of the big spacecraft to the rocket, in which the Lunar Module seemed to be folded, like a fetus in the womb. After the separation the crew had begun to televise the scene with the Westinghouse color television camera, an instrument only 17 inches long with a variable-focus, zoom lens.

Looking back at Earth, the volatile Cernan exclaimed, "That world is just incredible. Holy moley! I sure hope we can show it to you. I really do."

Young performed the docking with seeming ease. Spring releases were activated to free the LM from its metal cocoon. The springs pushed the big, but nearly empty, S4B from the greater mass of the joined Apollo–Lunar Module, and the television camera showed the rocket, gleaming brightly in the sun, about 300 feet away. Stafford called down to Houston, "You tell the people who worked on that machine that we sure appreciated it."

Shortly after the LM was pressurized with oxygen, another problem appeared; and it was to cause serious inconvenience later on. Oxygen pressure in the tunnel connecting the docked Apollo and Lunar Modules ruptured a Mylar-Fiberglas insulation pad on the Apollo forward hatch. When both the Apollo and LM hatches were opened to the tunnel, the mass of Mylar bits and Fiberglas blew into both cabins. Eventually, after the crewmen had removed their space suits, the Fiberglas bits worked their way beneath the crew's coveralls and caused them to itch for the remainder of the eight-day mission. Stafford described the insulation particles as "just a little snow."

Apollo-LM moved away from the S4B with a short burn of the Apollo engine and Houston signaled the rocket's automatic control system to vent the remaining propellant in the rocket through the engine. This added 126 feet a second to its velocity and sent it out on a path that would take it past the trailing edge of the moon and, with the acceleration provided by lunar gravity, sling it off into a junk-yard orbit around the sun.

At an altitude of 24,200 miles, the crew produced another

television show for the dwindling home planet, featuring the dwindling home planet. It arrived during prime TV time Sunday evening when the doings of Project Apollo were viewed by a maximum national audience. At that time, Apollo 10 was still moving uphill against the earth's gravitation pull at 8986 miles per hour—a little more than one-third of its initial velocity of almost 25,000 miles an hour at translunar injection. Not until it neared the moon three days later would it breast the top of the "hill" and begin accelerating down the other side under the moon's gravitational influence.

By looking into a small minitor aboard the ship, the crew could see in black and white the picture that the color camera was transmitting to Earth. Cernan called down to Houston, "Charlie, if you see this, it's going to be out of this world, literally." Stafford reported that out the window he was looking "right at the good, old U.S. of A." He could see the Rocky Mountains "sticking out" and Baja California, a conspicuous land configuration from space which every orbital crew has pegged. "Can't tell whether you have any fog in Los Angeles or not, but Alaska is pretty much fogged in," he said. The resolution of the small camera was "fantastic," in the words of the crewmen who watched the picture in the four-pound monitor. Observers in the control room at Houston agreed. The colors were true—the oceans were blue, land masses were brownish, and the great ice sheet of Antarctica, where the austral autumn was well advanced, still gleamed in the waning sunshine.

The color camera employed a rotating color wheel with red, blue, and green filters, which passed in front of the camera's image tube as the wheel spun at 600 revolutions a minute. The camera thus transmitted separated red, blue, and green images to NASA receiving stations at Madrid, Spain, and Goldstone, California. These were combined to produce a single color picture by conversion equipment at the Manned Spacecraft Center at the rate of 30 frames a second, a rate compatible with the standard rate for commercial television. The color wheel, or field sequential system, used by Westinghouse in the camera, was developed in the Columbia Broadcasting System laboratories about 25 years ago, but it did not go beyond the experimental stage in the 1940s and 1950s because the images it produced were not compatible with the black and white system. Although the focusing frequently was not as sharp as in conventional television camera work, the color transmissions I saw during the flight at the Manned Spacecraft Center were magnificent. They opened a new dimension in visual communications. And the colors were true—judged by the descriptions of the astronauts.

The flight proceeded uneventfully except for an error in chlorinating the drinking water which caused chlorine to accumulate in some water Stafford drank. He complained vigorously.

"Tom," called Mission Control solicitously, "did that water taste—could you taste any chlorine at all in that water when you first started using it?"

"You bet your sweet bippy," retorted Stafford. "It's gotten lots better but there was chlorine in it to start with."

THE HOT DOG MODE

Twelve hours out from Earth, the crewmen retired, leaving the control of the ship's attitude to the autopilot. In order to equalize the sun's heat around the dual space vehicles, the Apollo-LM "stack" was maneuvered so that it stood on its tail perpendicular to the flight line between Earth and moon. Its small reaction-control-system thrusters then were fired to roll the stack, rotisserie style, at about the rate of one revolution an hour. Throughout the night, the 58-foot double spacecraft turned slowly like a hotdog on a spit before a distant fire.

From time to time, however, Apollo-LM tended to pitch or yaw out of the rotisserie alignment. Whenever it moved 20 degrees to one side or the other, the autopilot would fire a thruster to restore the planned barbecue attitude so that the heating of vehicular skin would be uniform. Each time the thrusters fired, a shuddering went through the entire spacecraft structure and awakened the crew.

Stafford complained that it felt like a "dull thud and the whole stack vibrates in a dance of three cycles . . . kind of a boom . . . rum . . . rum . . . rum." It was a minor discomfort compared with the upper respiratory infections, the nausea, vomiting, and dizziness, experienced by crewmen on the previous manned Apollo flights. The Apollo 10 crew escaped the head colds and nausea which had plagued the earlier flights and did not even complain of the dizziness their predecessors on Apollo 8 had described when they got up from their couches. For the first time on an Apollo mission, everyone felt hale and hearty for the entire trip. That in itself was a major advance in the technique of interplanetary voyaging.

By "reveille" Monday morning, Apollo 10 was 100,000 miles out from Earth, moving moonward at 3927 miles an hour, or less than one-sixth of the velocity at which it had begun the journey. The ship was still climbing uphill, against Earth's pull, and would continue to decelerate until it fell into the gravitational clutch of the moon about 2:40 p.m. EDT, Wednesday, May 21, at a distance of 38,640 miles from the center of the moon. So, at least, prophesied the calculations of Philip Shaffer, Mission Control's imperturbably precise, flight-dynamics officer.

Shortly after midmorning, the crewmen issued a historic weather report as they looked back at the receding earth. Portugal and western Spain were clear but eastern Spain along the Mediterranean was under clouds. Italy was clear south of Rome. Sicily, Sardinia, and Corsica were partly cloudy to cloudy. Greece was clear and Turkey had "very scattered clouds." Bulgaria was clear with partly scattered clouds, but the rest of Europe was mostly hidden by clouds. North of the Black Sea, a large part of Russia was clear, but the rest was cloudy. Arabia was clear. So were Israel and Jordan. Libya and Egypt were clear except for a cloud strip over the center of Egypt running from Saudi Arabia across the Sinai Peninsula. Africa was clear "pretty much" to the south, except for the Cape, where South Africa was cloudy. In the early morning light, parts of Africa looked "like velvet."

"That's your morning weather report from about 100,000 miles," Cernan advised. It was one of the most significant contributions of Apollo 10 to space development, inasmuch as it demonstrated how effectively a manned orbital station could perform meteorological observations.

Perhaps a man aloft could see no more than a weather satellite camera, but he could interpret in real time, and he could learn from what he saw. Apollo 8 had made an unanticipated and fascinating contribution to the study of Mars in the photos its crew took from midway between Earth and moon of the Sahara Desert; the photos of the ground texture in the Sahara bore a strong resemblance to that of the Martian surface in the Mariner 4 photographs.

Mission Control observed; "Looks like Ole Charlie Brown [the Apollo spacecraft] is motoring right along in good shape there. And Snoopy [LM] is hanging in there real well too."

Dr. Willard Hawkins, flight surgeon, advised that the radiation exposure to which the crew had been subjected in passing through the Van Allen belts and beyond the geomagnetic field was the equivalent of about three chest X-rays. Happily, Stafford and Young turned on a casette tape player and serenaded the ground with a tinny rendition of "Up, Up, and Away."

"That was really beautiful, you guys," encouraged Charles Duke on capsule communicator duty. "Y'all been practicin' a lot, we can tell that."

"It's Tom and John on the guitar and the three of us singing," explained Cernan. "We had trouble stowing the bass drum aboard, though."

At noon, Stafford reported that he could see a glistening speck in the distance and wondered if it could be the S4B, nearly 1800 miles away.

A DEMONSTRATION OF ZERO G

The only course correction on the flight to the moon was made Monday, 26 hours, 32 minutes, and 56 seconds after lift-off. It was a seven-second burn of the Apollo engine adding 48.7 feet per second to the velocity of the dual spacecraft. The burn moved the vehicle slightly out of its flight plane to the right to reduce its inclination to the lunar equator when it reached the moon. Charlie Brown and Snoopy were then about 124,000 miles from Earth.

For the first time on the flight, the crew saw the moon, a thin crescent, but looking larger, Cernan said, than he had ever seen it before. Like Apollo 8, this mission had been launched at the new moon so that Landing Site 2 on its eastern face near the equator would be illuminated by morning sun for good light-dark contrast when the Lunar Module buzzed it Thursday at 50,000 feet. Moon sightings during the outbound flight were infrequent. Much of the time, the astronauts were riding backward and when their windows were in line of sight of the crescent moon the view was often washed out by sunlight.

On Monday afternoon the crew sent back another television show of the earth, featuring South America and the Andes Mountains, then switched the camera inside to illustrate what they were doing in the cabin—which was incomprehensible to the average viewer. "We didn't get a chance to shave this morning before this show," Cernan apologized. "I hope that doesn't bother anybody."

At Houston, Charles Duke said that picture definition inside the cabin was so good he could read Cernan's wrist watch, which looked like 4:00 or 4:05 p.m.

"It's 1605 [4:05 p.m.] Cape time," Cernan confirmed.

Inside the cabin, Stafford was standing (or floating) right side up while beside him Young was working on star sightings upside down. The scene was so casual it seemed unreal. Cernan raised the question: Which one was really right side up? He reversed their positions, relative to the viewers, by turning the camera around. Once more right side up, Stafford put his hand on or under the head of the upside-down Young and bobbed the Command Module pilot up and down like an oversized ten-pin. It was an historic demonstration of the nature of weightlessness in free fall.

The drinking water problem persisted. This time, it was the presence of hydrogen gas in the water, which gave the crewmen gastrointestinal gas when they drank it. This was the water manufactured by the three fuel cells as a byproduct of their generation of electricity by the mixing of hydrogen and oxygen. Not all the surplus hydrogen combined with oxygen to make water and hydrogen remained in the water as a free gas, creating bubbles great and small.

Apollo Control had worked out a procedure for separating the gas from the water. A quantity would be poured into a plastic bag and the astronaut would whirl the bag around. It was theorized that this exercise would separate the gas bubbles from the water. But the theory did not work.

"What happens," Stafford explained, "is that we start off with a bag full of water and bubbles. Little, bitty bubbles. And we end up with a bag full of water and great, big bubbles. There is no way to separate the bubbles from the water that I can see."

"Did you try spinning it the other way?" Mission Control

asked. The experts at Houston promised to analyze the problem, but somehow the answer eluded them.

THE SHUDDERING SPACECRAFT

Overnight Monday, the double spacecraft was jockeyed into a stable attitude in the rotisserie mode so that its yawing and nodding did not reach the 20 degree limit when the thrusters would fire to bring the stack into position again. All was peace and quiet aboard Apollo 10. In the morning, Stafford reported that even though the thrusters had not fired once, "every so often the whole stack just gives a little shudder." Contractor engineers believed the shuddering of the structure was caused by sloshing of the fuel in the Lunar Module's Ascent Engine tanks, which were not filled on this mission since that engine would not be needed to lift the Ascent Stage of the LM from the lunar ground. Controllers advised the crew that the Apollo 9 astronauts had mentioned that morning that they, too, had experienced the "shudders" on their flight, when the LM was out in front.

Reaching a stable flight position was an undramatic but important event on the flight. It bore out engineering predictions that the Apollo-LM stack could fly to the moon and home again in a stable attitude without the constant use of the thrusters to maintain the proper barbecue roll attitude. This was a milestone in the development of interplanetary travel, for it saved thruster fuel which on missions longer than a mere jaunt to the moon would be critical.

Once more, the S4B blinked into sight and the computer in Mission Control said it was by then some 4000 miles away. Tuesday evening, Young advised the night-shift capsule communicator, Air Force Major Joe Henry Engle, that both Earth and moon looked the same size.

"The moon is in the left window and the earth in the right window," Young said. "You can see the moon just as the sun sets behind the right window [the ship was continuing to rotate slowly to keep it cool]. There's a period of time there—less than a minute—when you can see the moon. It's practically a new moon. It's only a sliver from where we are."

"I bet that's a pretty good sight from there, too," said Engle. "Right, John?"

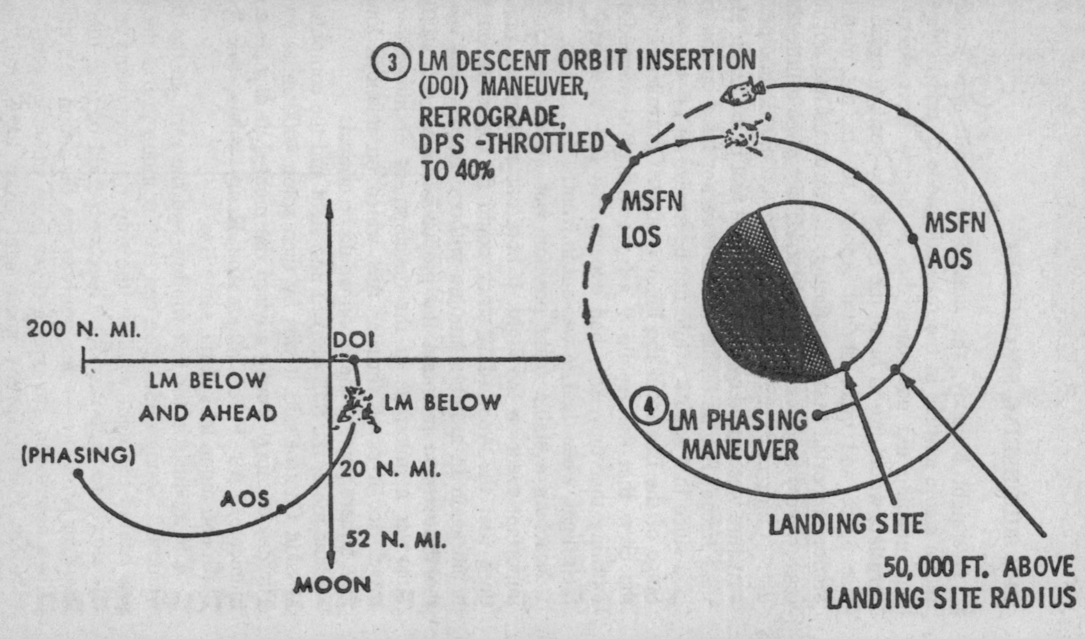
LUNAR MODULE DESCENT ORBIT INSERTION
3 LM DESCENT ORBIT INSERTION (DOI) MANEUVER, RETROGRADE, DPS -THROTTLED TO 40%
MSFN LOS
MSFN AOS
4 LM PHASING MANEUVER
LANDING SITE
50,000 FT. ABOVE LANDING SITE RADIUS
200 N. MI.
DOI
LM BELOW AND AHEAD
LM BELOW
(PHASING)
20 N. MI.
AOS
52 N. MI.
MOON

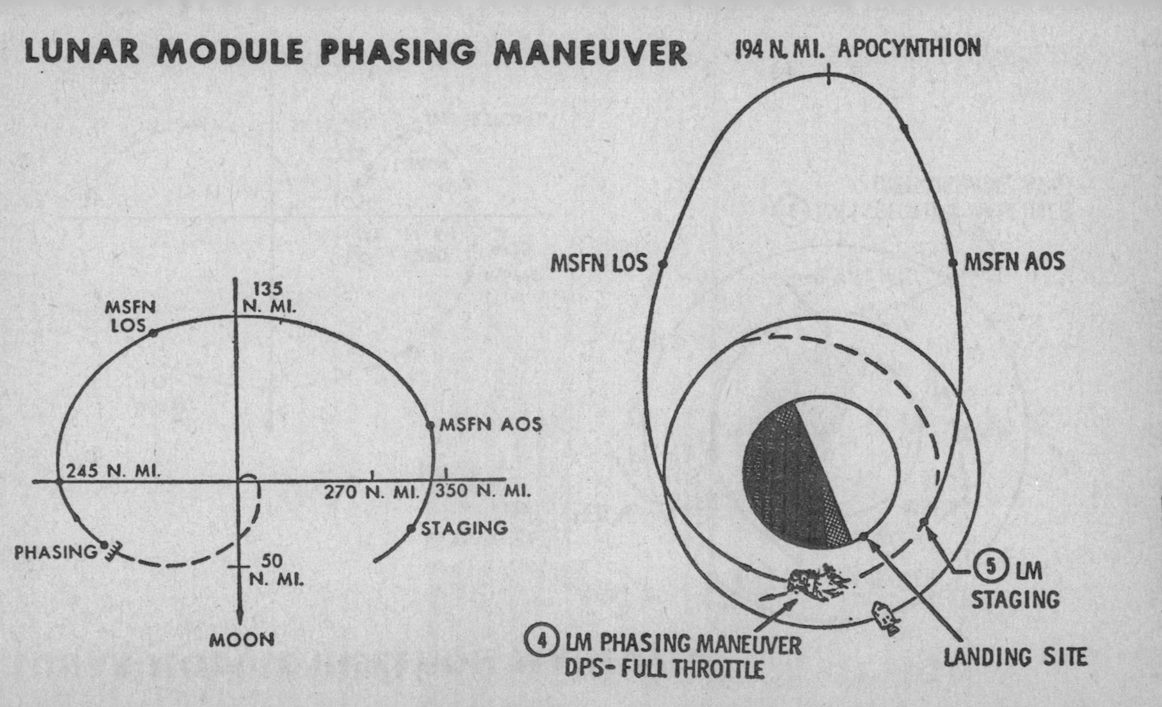
LUNAR MODULE PHASING MANEUVER
194 N. MI. APOCYNTHION
MSFN LOS
MSFN AOS
135 N. MI.
MSFN LOS
MSFN AOS
245 N. MI.
270 N. MI.
350 N. MI.
STAGING
PHASING
50 N. MI.
MOON
5 LM STAGING
4 LM PHASING MANEUVER DPS- FULL THROTTLE
LANDING SITE

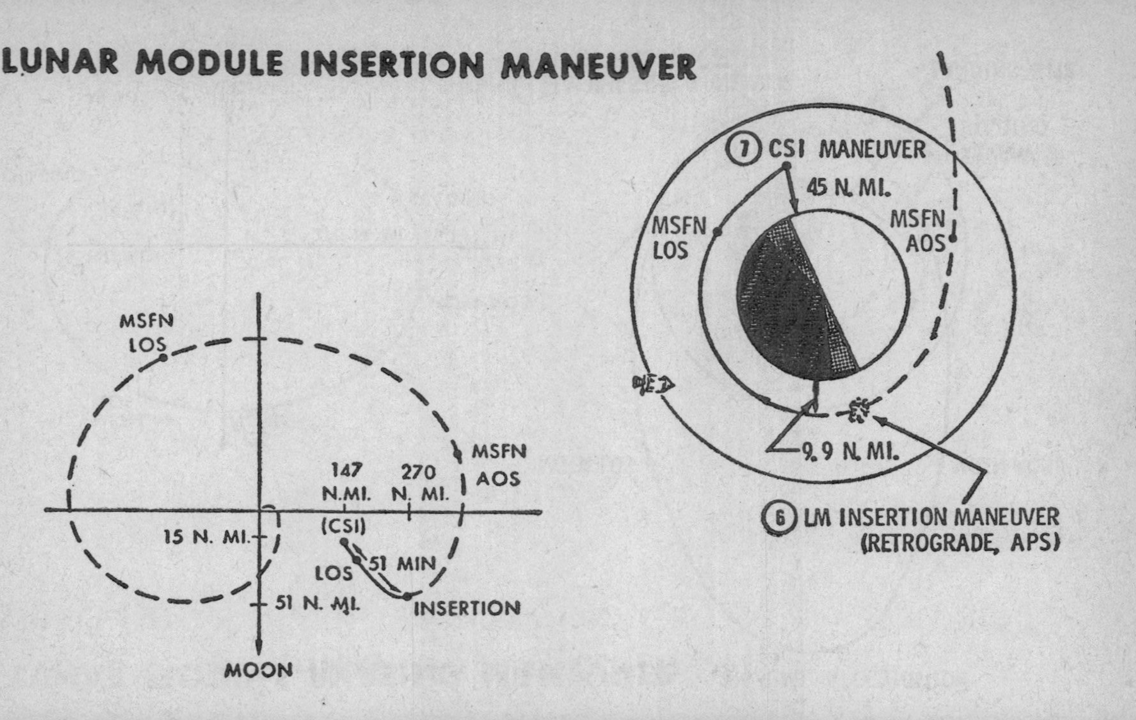
LUNAR MODULE INSERTION MANEUVER
MSFN LOS
147 N.MI.
270 N. MI.
MSFN AOS
15 N. MI.
(CSI)
51 MIN
LOS
51 N. MI.
INSERTION
MOON
7 CSI MANEUVER
45 N. MI.
MSFN LOS
MSFN AOS
9.9 N. MI.
6 LM INSERTION MANEUVER (RETROGRADE, APS)

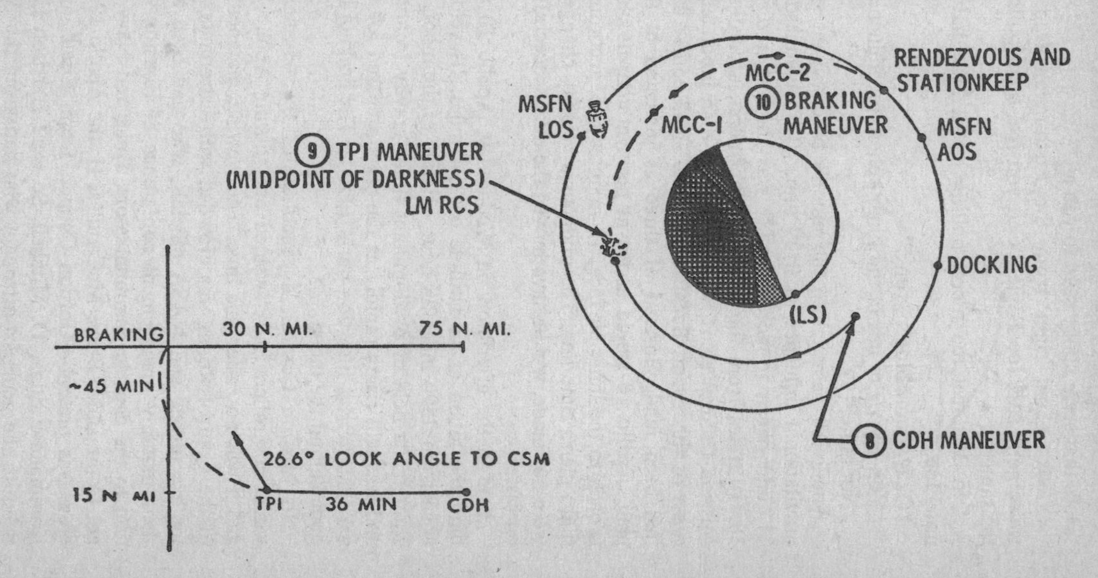
LUNAR MODULE CONSTANT
DIFFERENTIAL HEIGHT AND TERMINAL PHASE MANEUVERS
RENDEZVOUS AND STATIONKEEP
MCC-2
10 BRAKING MANEUVER
MSFN LOS
MCC-1
MSFN AOS
9 TPI MANEUVER (MIDPOINT OF DARKNESS) LM RCS
DOCKING
(LS)
8 CDH MANEUVER
BRAKING
30 N. MI.
75 N. MI.
~45 MIN
26.6° LOOK ANGLE TO CSM
15 N MI
TPI
36 MIN
CDH

"The moon looks as big as the earth does," said Young. "Does that seem about right to you all?"

"Looks about right from the earth-moon transit graph we've got," said Joe Engle. "Can you feel them pulling about the same?"

"No. We feel the moon pulling just a little harder right now, Joe."

"Okay. Something's wrong."

"You say we're not in the lunar sphere [of influence] yet?"

"Not quite."

Apollo 10 passed into the moon's predominant sphere of gravitational influence at 2:39 a.m. EDT, Wednesday, May 21, and began to accelerate. It was then at a distance of 38,710 miles from the center of the moon, according to the Mission Control chart, slightly farther than predicted for the onset of this event. Because only one small course correction had been made, instead of three as indicated in the flight plan, Apollo 10 would arrive at pericynthion, its closest approach to the lunar surface, 11 minutes late, according to the flight plan's time line. The midcourse velocity changes, which were not made, would have moved the dual spacecraft along a bit faster.

Late in the afternoon on Wednesday, Apollo 10 went behind the moon at an altitude of 294 miles. Its velocity had risen from 2590 miles an hour, relative to the moon, when it entered the field of predominant lunar gravity to about 5300 and was still rising rapidly as the crew prepared for the critical Lunar Orbit Insertion burn [LOI-1]. The braking is always done behind the moon.

In Mission Control, a repetition of the 45-minute wait which controllers had "sweated out" during Apollo 8 began, and it was no easier this time. The viewing chamber adjacent to the Control Room was crowded with astronauts, contractor executives, and NASA officials, who had been attending their monthly management conference at Houston. In addition to the capsule communicators, Duke and Navy Lieutenant Commander Bruce McCandless II, the Apollo 10 back-up crew was present—Gordon Cooper, Donn Eisele, and Navy Commander Edgar D. Mitchell. So were Harrison H. (Jack) Schmitt, the geologist-astronaut who had given the Apollo 10 crew some instruction in lunar formations; Rusty Schweickart, the Apollo 9 Lunar Module pilot; Deke Slayton, astronaut crew chief; Chris Kraft, director of flight operations; Gilruth, the Manned Spacecraft Center Director; General Phillips, Apollo Program director; George Hage, mission director; Debus, the Kennedy Space Center director; von Braun, Marshall Space Flight Center director; George Mueller, associate NASA administrator for Manned Space Flight; Lou Evans, president of Grumman Aircraft Engineering Corporation, the Lunar Module prime contractor, and his chief engineer, Joseph Gavin; and William Bergen, president of the Space Division of North American Rockwell, maker of the Apollo spacecraft. The audience was quiet as the minutes oozed away toward another moment of truth.

THE MOON IS BROWN AND GRAY

The LOI burn was scheduled at 4:44 p.m. at an altitude of 113 miles. The 20,500 pounds of thrust Apollo engine was to have burned 5 minutes, 54 seconds, braking the ship's velocity from 5675 miles an hour, to which it had been accelerated by lunar gravity, to about 3240 miles an hour and dropping the dual spacecraft into a longish oval orbit around the moon.

At 5:18 p.m., a sigh went up from controllers and the VIP observers. This was the time when a signal would have been received from the vessel if it had failed to go into lunar orbit and was headed on an Earth-return trajectory instead. There were still 11 minutes and 45 seconds to wait if Apollo 10 was in the planned orbit.

"We are six minutes away from the time we should be hearing from Apollo 10," reported a public affairs spokesman. "Three minutes, fifty-six seconds away. We are waiting. Mark. Two minutes. Everyone here is watching displays and listening. Thirty seconds. Ten seconds. We have acquisition of signal! We are getting data!"

HOUSTON: Hello, Apollo 10. Houston. Over.

APOLLO 10: Roger, Houston. Apollo 10 can tell the world that we have arrived.

The dual spacecraft, code-named Charlie Brown and Snoopy, was speeding along an equatorial track in its initial orbit of 196 by 68.6 miles. It followed the trail which Apollo 8 had blazed in December. Below unfolded a landscape all brown and gray, and the crew stared at it, fascinated, as they began to cross the face of the moon we can see, riding backward, with their heads down.

"Hello, Houston," Cernan called. "You'd have to see this planet to believe it."

"We have our student geologists here overlooking the surface and they'll report in a minute," Stafford said.

Some of the landmarks the crew called out as they identified lunar features on their charts had never before appeared on maps of the moon. They crossed the Mare Smythii (Smith's Sea) named after an English admiral-astronomer who died more than a century ago and reported a large gouge with the nonclassical, nonscientific title of "Ghost Crater." About 375 miles to the west, Stafford announced: "We've Langrenus now out here. Very beautiful sharp peak right in the center."

Langrenus is a vast, walled plain, about 84 miles in diameter, named for a 17th century Belgian selenographer. The peak Stafford saw was probably the one that rises to 3000 feet from a twin-peaked central mountain group. About 240 miles to the west, Cernan announced: "We've seen Big Taruntius. You'll be glad to know we're walking right up our LM chart. We're looking at Messier and Messier A, and Secchi ought to be coming up. And that's Secchi."

"Big" Taruntius was a crater 36 miles in diameter filled with interior hills and ridges, surrounded by a sheer, 3000-foot wall. It was named after a Roman philosopher, circa 86 B.C. Messier, named for an 18th century French astronomer, is a deep, small crater only 9 miles across. The American astronomer William Pickering had reported transient events there, now believed to be emissions of gases, and not far away is another small, deep crater named for him.

FRONTIERSVILLE

Beyond the crater Secchi, which bore the name of a 19th century Italian astronomer, the lunar voyagers saw a northwesterly running, rocky spine, identified on their charts as Gemini Ridge. From the eastern limb of the moon to Landing Site No. 2 in the Mare Tranquillitatis at 23 degrees, 38 minutes, 51 seconds east longitude, the distance appeared to be 1250 miles. At the dual spacecraft's velocity of 5346 feet per second, or 3640 miles an hour at this point in the orbit, the distance was covered in less than 21 minutes.

Beyond Secchi appeared a newly named feature, Apollo Rille, a long, narrow cleft. To the north of it, there was a canyon-like structure identified on the charts as Dry Gulch. These landmarks had shown up in the Lunar Orbiter 5 reconnaissance photos and in those taken by the Apollo 8 crew. While new, selenographic nomenclature was informal, designed only for temporary identification of landmarks, it lent a bacon-grease and woodsmoke flavor to this alien landscape in the tradition of the American frontier.

It was evident that the strangeness of the moonscape had been tempered by the frontier-style place names the Apollo directorate was using. This nomenclature seemed to lift the Apollo 10 crew's morale. At least, it put their journey into a more familiar and acceptable tradition of exploration—in contrast to the voyage of Apollo 8, which had sailed over the edge of the world to gaze upon what seemed to the crew as a primeval phase of creation.

Westward into the lunar morning sped Apollo 10. Lost Crater, Star Crater, and the Twin Craters, bisected by Apollo Ridge, passed beneath. To the south, the Pyrenees, a chain of mountains rising no higher than 10,500 feet above the surface, could be seen. Lost Basin appeared, then Low Mesa, Lonesome Mesa, and Bear Mountain.

Houston called: "Apollo 10—if you're near Site 1, if you get a chance, could you comment on the volcanic cones on the highlands south of the track? Over."

Illuminated by the morning sun, Landing Site 1 lay just to the north of the Bear Mountain region and the crater Maskelyne A. About 62 miles to the west; the great crater Maskelyne, named for a 19th century English astronomer in the "old style" of Earth-based, telescopic observation, was surrounded by "new style" landmarks called Snake Ridge, Wash Basin, Boot Hill, Duke Island, Cape Bruce, Jack's Basin, Barbara's Mesa, and Bob's Bend.

North and west of Maskelyne ran Smoke Ridge and to the south appeared the great surface gash called Sidewinder Rille. Diamondback Rille, another cleft feature, appeared to the north. West of these features, Apollo 10 crossed the terminator, the day-night boundary, and sped on into the lunar night.

The landscape below, however, was well illuminated by Earth light. Landing Site No. 2, the goal of the next flight, was clearly visible in pre-dawn Earth shine. It would be illuminated by the rising sun and highly photographable the next day when Stafford and Cernan were scheduled to descend to 50,000 feet over the Tranquil Sea.

NASA ROAD 1

South of the track, Stafford saw Theophilus, a plain 60 miles or so in diameter, ringed by a steep, sharp wall 15,000 feet high. "It's got a very pronounced central peak which is not nearly as high as the rim and it's got a little rim crater just on the inside which is very easily distinguishable," he reported.

"Roger," said Houston. "We've located you on a map. Go ahead, Tom."

"Yeah. Well, I'm right over Maskelyne and Maskelyne B now, leading right into Landing Site 2. And Sidewinder Rille and Diamondback Rille stand out just tremendous here. We're just about to cross the terminator. Boy, that's really something there. Don't see why fish aren't dumped down that creek. Those rilles are something else again."

The astronauts were reflecting a recent viewpoint that the rilles, especially the serpentine ones, might have originated as river valleys. The implication of this point of view, of course, was that the moon once had a hydrosphere and possibly, for a brief time, an atmosphere.

Schmitt, the geologist-astronaut at Houston, had instructed the crew to look for evidence of piled-up rocks or levee material along the sides of the rilles. This would further support the water-erosion theory of their formation. But Stafford and Cernan reported that the rille walls appeared clean and sharp.

The landmarked trail to Landing Site 2 was easy to follow, Stafford said. It was just like NASA Road 1—the dual-lane divided highway bordering the Manned Spacecraft Center. North and west of Landing Site 2 stood a man-made monument, Surveyor 5, which had landed on the morning of September 10, 1967, had radioed more than 18,000 pictures back to Earth, and had made the first chemical analysis of the lunar soil showing it to be similar in composition to terrestrial basalt.

North of Site 2 lay the "old style" crater Sabine E, and to the south was Moltke, with Last Ridge directly below. Chain Gulch, Worm Rille, and Chuck Hole came into view. Southwest of the site was a region called the "Oklahoma Hills" on the chart. It was there that Stafford reported seeing volcanoes, and Cernan supported his identification of these formations as volcanic. The crewmen also advised their geology mentor, Jack Schmitt, that both Diamondback and Sidewinder Rilles resembled Canyon Diablo, the meteor crater near Winslow, Arizona. They had sharp, steep walls without any build-up or levees along the side. Even the rille called "U.S. 1" seemed to have vertical edges, at least from an altitude of 69 miles. Stafford and Cernan promised a more detailed description when they came down to 50,000 feet on Thursday, May 22, for the final test of Snoopy, the lunar module.

Cernan jolted listeners with a report that the center of some of the lighter craters, illuminated in earthshine, "glow as if they're lit by radioactivity—they just glow in this very low, dim light." Schmitt suggested that Cernan might have seen vagrant sunbeams reflected by smooth rocks.

As Apollo 10 moved across the lunar night, Houston asked the crewmen to look some 475 miles to the north of their track and observe the crater Aristarchus, where "transient events" had been reported on Monday and Tuesday. The reports came from amateur and professional astronomers in California, New Mexico, and Spain who are members of the volunteer Lunar International Observers Network (LION). A "transient event" is a brief change in the appearance of a lunar feature, a darkening, a glow, or a brightening, which some astronomers attribute to emission of gases from the interior of the moon. Such events have been reported for centuries—some by witnesses claiming to have seen them with the naked eye.

LUNAR SNOW STORM

Aristarchus was close to the crew's northern horizon, but Stafford located it. Nothing unusual was visible there. Apollo 10 passed out of the Ocean of Storms and into more rugged country, then vanished from Earth radio view as it "rounded the corner" to Far Side to begin its second revolution. On this pass, Stafford said he could "really see some boulders" at the bottom of the crater, Messier-A. Landing Site 2 appeared to be somewhat rougher than expected in the dawnlight. Looking rearward, toward the rising sun, the crew reported seeing the solar corona, "and it's really beautiful."

On Wednesday night, as the dual space vehicle began its third revolution, the Apollo engine was burned again behind the moon for 14 seconds to subtract another 139 feet per second from the ship's velocity and to circularize the orbit at nearly 69 miles. Cernan then removed the Command Module hatch and crawled into the tunnel linking the Apollo with the LM. The tunnel was filled with bits and pieces of Mylar and Fiberglas from the broken hatch insulation pad. It was a mess, Cernan reported.

"It looks like cotton," Stafford told Houston. "It tastes like Fiberglas." In the Lunar Module, Cernan checked the communications system and, when he crawled back into the Apollo, he reported that the LM cabin was filled with insulation debris.

"Would you believe we've been living in what you might call snow for three days and we found out where the rest of it is," Cernan told Mission Control. "It's in our good friend Snoopy."

In addition to causing the crewmen to itch, the insulation "snowstorm" was blamed for the apparent failure of the tunnel's oxygen vent valve. The valve seemed to be clogged or stuck when the crew attempted to depressurize the tunnel Thursday morning by dumping the oxygen to space. The depressurization was necessary to check the hatch seals on both Lunar Module and Apollo Command Module before the two separated.

After a conference with Houston, Stafford and Cernan, fully rigged in their spacesuits, opened the LM hatch to the tunnel and vented the interior of the LM to reduce oxygen pressure in both the LM cabin and the tunnel—from 5 to 3.5 pounds. Then they closed the LM hatch and repressurized the LM cabin to 5 pounds per square inch. There was no leakage across either the LM or Apollo hatches into the partially depressurized tunnel. The effects of the clogged tunnel vent valve had been circumvented, so to speak. But there was more.

Continued gas pressure in the tunnel seemed to act like a lubricant in the mechanical linkage between the Lunar and Apollo Command Modules, so that every time a thruster fired on the Apollo to maintain the rotisserie roll, minute sliding would occur between the joined vehicles. When Snoopy, the Lunar Module, had slipped three degrees out of alignment with Charlie Brown, the Apollo Command Module, Houston became concerned. The twisting or bias threatened to damage the latching mechanism which secured the vessels at docking. If the latches were damaged or broken, the LM could not re-dock with the Apollo after making the excursion to 50,000 feet, and Stafford and Cernan would have to go outside the LM in order to get back into the Apollo. This mode of transfer from one vessel to the other presented a risk which Mission Control did not wish to take.

THE MOUNTAIN-GRAZING ORBIT

Extra-vehicular crew transfer had been accomplished by Russians on their twin Soyuz spacecraft flights in January. On Apollo 9, Rusty Schweickart had been scheduled to step off the LM "porch" and climb to the Apollo hatch, but that exercise was not carried out. The Soviet transfer was done by cosmonauts securely tethered to docked vehicles. Any attempt by astronauts to swing from one undocked vehicle to another, irrespective of how close they could be brought together, was breaking "new ground" in manned space flight. Houston had no desire to attempt it on this mission.

The sliding of the two spacecraft, therefore, threatened to force Mission Control to scrub Snoopy's mountain-grazing descent to 50,000 feet. Tension increased in Mission Control as the bias between Snoopy and Charlie Brown increased to 3½ degrees.

Charles Duke called up to crew: "We're concerned about this yaw bias in the LM and apparent slippage of the docking range. We'd like you to disable and to keep disabled all roll jets until after undocking. We will not maneuver to undocking attitude. Just hold what you've got."

The joined vehicles were nearing the western limb around

which they would soon disappear to Farside. "Okay, Charlie Brown and Snoop," Duke called. "Three minutes to going over the hill. You're 'go' for undocking and we'll see you around the other side."

"Roger," said Stafford and Cernan in Snoopy and John Young in Charlie Brown.

"Snoopy—correction," Duke came back on the air quickly. "If it's apparent that the LM interface has slipped around to about 6 degrees, do not undock and let's come around again and look at it. Over."

"Roger," said John Young in the spaceship Charlie Brown.

Flight Director Glynn Lunney told Duke: "Charlie, let them know it's three and one-half [degrees] now."

"Roger," said Duke. "Your yaw bias right now—the slippage—is 3½ degrees right now. So about double what you've got—if it goes that far—do not undock."

At this point, radio communication was lost as Snoopy and Charlie Brown turned the corner. The Control Room settled down to another 45-minute wait. When the vehicles finally reappeared, they were 30 to 40 feet apart. Stafford reported to Houston that all was well and "I hope we can get back on the nominal after that insulation kind of goofed us up."

Young fired Charlie Brown's thrusters to move away from Snoopy. There was a flurry of excitement when Snoopy's radar seemed to be inoperative, but the problem was solved by a power-switch adjustment.

"Keep up the good work, boys," Young called to his colleagues in Snoopy. "You will never know how big this thing gets when there ain't nobody in here but one guy."

"You will never know how small it looks when you are as far as we are," replied Gene Cernan.

"Snoop and Charlie Brown—we see you separating on the big tube," interposed Duke.

SNOOPY: See you, John. Have a good time while we're gone, babe. Don't get lonesome out there, John. And don't accept any TEI updates! [that is—don't return to Earth without us].

Once more, a critical engine burn—the one which would decelerate Snoopy into a new orbit with a low point or pericynthion at 50,000 feet—was done behind the moon.

EARTHRISE

Snoopy's descent engine was burned as a brake for 59 seconds at 10-per-cent throttle for the first 15 seconds and at 40 per cent for the remaining 44. It slowed the vessel's velocity by 71.3 feet per second. From an altitude of 70.3 miles at the time of burn, Snoopy's new orbit would bring it to 9.6 miles, or 50,688 feet, of the lunar surface at pericynthion at a point about 282 miles from Landing Site 2. Actually, on its first, low orbit pass, Snoopy went more than 3000 feet lower. Then the ship rose toward apocynthion—the high point in the orbit—at 70.3 miles behind the moon.

Moving out in front of Charlie Brown, Snoopy began its long dive toward the Mare Tranquillitatis at 4:35 p.m. Thursday, while Charlie Brown remained sedately in its orbit of 70.3 and 66.1 miles. When Charlie Brown reappeared around the moon, John Young reported that Snoopy's descent orbit insertion burn looked good. "They are down there among the rocks, rambling about the boulders," he said.

HOUSTON: Roger, Charlie Brown.

YOUNG: They just saw Earthrise. They say they are looking at the horizon now.

SNOOPY: Hello, Houston. This is Snoopy.

HOUSTON: Roger, Snoopy. Go ahead.

SNOOPY: We is down among the boulders, Charlie.

HOUSTON: Roger. I hear you are weaving your way up the Freeway. Can you give me a postburn report?

The burn was on time, Snoopy reported, and Snoopy's AGS (Abort Guidance System) was indicating a pericynthion of 9.8 miles. Stafford described Earthrise as "magnificent." He added: "You can also tell Jack Schmitt that there are enough boulders around here to fill up Galveston Bay."

As Snoopy passed over Smyth's Sea (Mare Smythii) on the eastern limb, Cernan reported that it was "a fantastic sight." He reported many signs of volcanic activity in the past. Craters showed a pure white near the edges and black at the bottom. It was like crossing a black and gray sea, "a beautiful sight."

"It looks like we're getting so close all you have to do is put your tail wheel down and we're there," said Cernan.

"Bet it looks like they're really hauling the mail," Houston told Charlie Brown after Young sighted Snoopy in his optics.

"Surprisingly enough, Charlie," Stafford came in, "it really doesn't look like we're moving too fast down here. It's a very nice, pleasant pace."

A few minutes later, Stafford came back: "Here it comes. I've got Diamondback. . . . Diamondback Rille is very easy to see. These rilles look like they may be as much as a couple of hundred feet deep and very smooth. The surface actually looks very smooth, like a very wet clay, but smooth with the exception of the bigger craters. The best description I can give of these rilles is of a dry . . . a dry desert out in New Mexico or Arizona. Okay. Here we're coming on the site."

The landing area, about two miles square, actually lay five to six miles north of Snoopy's orbital track, instead of directly under it as planned. While the "miss distance" was not too far to be corrected by the LM's descent propulsion system, it represented an error in navigation which could not be explained immediately. Stafford realized that if he had been attempting to land, the miss distance would have presented a serious problem. In a landing attempt requiring excessive fuel in the powered descent, a miss distance of this magnitude could force the crew to abort the landing if there was insufficient fuel to reach level ground.

At this point, communications were interrupted between Snoopy and Houston. Charlie Brown passed the word that Houston wanted Snoopy to get in the proper attitude for the phasing burn, the maneuver that would start Snoopy heading back to rendezvous with Charlie Brown.

CHARLIE BROWN: Snoop, Charlie Brown. Do you read?

Snoopy suddenly returned to full, live communication and a voice was saying, edged with frustration: "You know, this God-damned filter has failed on me. My Hasselblad [camera] just failed. Oh, I tell you, man, that's something. Oh, look at that. . . . If I don't have any more . . ."

HOUSTON: Snoop, Houston. We're reading you now. We're counting eight minutes to the burn. Over.

SNOOPY: Roger. We're going to phasing attitude.

The phasing burn was designed to change the orbit to an elongated loop with a peak altitude of 218.5 miles behind the moon. The pericynthion over the Mare Tranquillitatis would be only slightly higher than during the first pass, at 13.5 miles.

Ten minutes after passing Landing Site 2, which Stafford and Cernan had tried desperately to photograph, Snoopy's descent engine was fired again, 26 seconds at 10-per-cent throttle and 16 seconds at full throttle, to boost velocity by 195.4 feet per second. Snoop then began a long climb to high apocynthion behind the moon. In this "dwell" orbit, Snoopy would allow Charlie Brown, still in its roughly circular flight path, to move out ahead so that it would be in the proper position for interception and rendezvous.

As both vehicles passed the terminator, Snoopy's flashing light became highly visible to Young in Charlie Brown, about 150 miles away and apparently Charlie Brown appeared momentarily to Stafford and Cernan, for a voice emanated from Snoopy saying: "That's unbelievable, isn't it? It's like a thing out there on a string where you can touch it. Where the hell are we going?"

INTERCEPTION

CHARLIE BROWN: You're there, you're there!

SNOOPY: Man, are we there! This one happened so good, it's just pathetic [meaning the burn was good].

On the fourteenth revolution of the moon, Snoopy came around trailing Charlie Brown. Stafford and Cernan got ready for the big test—staging the LM in lunar orbit, which means separating the ascent stage from the descent stage, and then using the ascent engine as a brake to reduce apocynthion from 218.5 to 51.7 miles, thus returning to an orbit from which Snoopy could intercept Charlie Brown. Houston advised Snoopy that its ascent engine looked ready to go.

The ascent engine, delivering 3500 pounds of thrust, could not be throttled and Cernan said that he expected it "may give us a kick."

The Snoopy pilots quickly separated from the descent stage and Houston reported seeing it on radar, but all was not well in lunar orbit. Snoopy began to gyrate in roll.

"Something is wrong with that gyro," said a voice from Snoopy. "Roll is 180 [degrees] and pitch is 233 . . ."

HOUSTON: Snoopy, Houston. We show you close to gimbal lock [radar lock on Charlie Brown].

This was not supposed to happen at staging. Snoopy was to remain in a stable attitude. Instead it was rolling and pitching as though searching frantically with its radar for its friend, Charlie Brown, so that the radar could lock on to Charlie's transponder.

SNOOPY: Something went wild during that staging.

CHARLIE BROWN: How is the staging?

HOUSTON: They're staging. Had a wild gyration, but they got it under control.

Immediately after separation from the descent stage, the ascent stage had rolled for about 30 seconds under the control of the Abort Guidance System. Stafford halted the roll by over-riding the AGS and taking manual control of the vessel. As the ship was stabilized, Cernan remarked: "Is the AGS in inertial, Tom? Okay, that's good. I don't know what the hell that was, Babe." Neither did Houston—immediately. Later, Glynn Lunney suggested that the gyration seemed to be the result of a switch in the wrong position on the Abort Guidance System. Instead of "Attitude Hold," where it was supposed to be, the switch may have been placed in the "Auto" position. This would have turned the guidance of the ascent stage over to the computer which had been programmed to point Snoopy's radar antenna toward Charlie Brown and keep it there. It was Houston's theory that the mispositioned switch was the cause of the alarm. But Stafford did not buy the theory. He concluded that the Abort Guidance System, the secondary guidance system being used for the maneuver, had simply malfunctioned until he took control of the ship away from it.

Two minutes after passing Diamondback Rille, Cernan counted down and Snoopy's ascent engine burned for 15.2 seconds, braking velocity by 207 feet a second and dropping apocynthion to the planned 51.7 miles. It was a few minutes after 8 p.m., and Thursday, May 22, was proving to be the longest day in Project Apollo up until now. Yet, despite the difficulties and frustrations of the mission, the flight was moving on schedule.

Rounding the moon to Farside, Snoopy's 100-pounds-of-thrust reaction control system (RCS) engines fired for 32 seconds. They raised the velocity by 50.5 feet a second and the altitude of pericynthion on the other side of the moon from 13.5 to 51.7 miles. An hour later, on the 15th revolution, a brief RCS burn increasing velocity by only 3.4 feet per second trimmed Snoopy's path at a constant height of 17 miles below that of Charlie Brown.

Once more the crew looked down on Landing Site 2 and reported that 25 to 30 per cent of the area seemed clear for landing. Stafford commented: "However, if you come down in the wrong area and you don't have the hover time [enough fuel left to hover], you are going to have to shove off [stage the LM and return in the ascent stage to Apollo upstairs].

In its lower orbit, Snoopy rapidly overtook Charlie Brown. When the vessels were 44 miles apart, Snoopy's RCS thrusters were fired for 15.6 seconds to boost the velocity by 24.1 feet per second and send Snoopy scurrying on an intercept course with Charlie Brown. When the two disappeared around Farside on their 16th revolution, they were only 26 miles apart. And when they reappeared 45 minutes later, a voice from Snoopy was announcing:

"Snoopy and Charlie Brown are hugging each other. We is back home."

On the morning of May 24, as Apollo 10 climbed out of lunar orbit on the way back to Earth, Stafford, gazing back at the receding moon, remarked: "It's still a beautiful view. In fact, just looking at it, you recollect you've come a long way, so just imagine where we're going to go in a few years Over."

HOUSTON: Roger, that, Tom.

APOLLO 11: THE THIRD VOYAGE

Apollo 10 splashed down in the Pacific Ocean east of Pago Pago shortly before dawn in that part of the world on May 26. Lookouts on the prime recovery ship, the U.S.S. *Princeton*, saw it coming in after reentry, glowing like a meteorite. The glow faded, the parachutes opened and Apollo 10 sank slowly toward the waves, landing lights blinking.

Two and one-half weeks later, General Phillips announced that the first attempt to land men on the moon, the Apollo 11 mission, would be launched July 16 as planned. Equipment malfunctions and anomalies which had plagued the Apollo 10 flight had been reviewed. Not all were understood at that time, but none was considered a constraint to the July flight. Moreover, the crew was ready. The civilian astronaut, Neil A. Armstrong, 38, was commander. Air Force Lieutenant Colonel Michael Collins, 38, and Colonel Edwin E. Aldrin, Jr., 39, were Command Module and Lunar Module pilots respectively.

Collins was to remain in the Apollo orbiting the moon at 69 miles while Armstrong and Aldrin would descend in the LM. They were scheduled to touch down at Site 2 in the Mare Tranquillitatis at 4:23 p.m. Eastern Daylight Time, July 20, 4 days, 6 hours and 51 minutes after liftoff from Cape Kennedy.

As Chris Kraft, the director of flight operations at the Manned Spacecraft Center, had promised after Apollo 8, the first order of business for the first men on the moon was rest and a check of the LM systems. Armstrong and Aldrin would remain in the LM for nearly 10 hours.

The first man to climb out of the LM and set foot on the moon would be Armstrong, who, with David Scott as copilot, had executed the first docking in orbit with an Agena rocket on the flight of Gemini 8. During the war in Korea, Armstrong was a Naval aviator and flew 78 combat missions. In 1955, he was graduated from Purdue University with a bachelor's degree in aeronautical engineering. He became a career test pilot for the old NACA and then for its successor, NASA, before joining the corps of astronauts in 1962. Armstrong had flown the X-15 to more than 200,000 feet altitude, he had also flown the X-1 rocket airplane and had tested just about every fighter aircraft in the United States inventory in the late 1950s.

He is an expert on gliders, a gold-badge holder of the Federation Aeronautique Internationale; an Associate Fellow of the Society of Experimental Test Pilots and of the American Institute of Aeronautics and Astronautics, and a member of the Soaring Society of America. Armstrong is a pilot's pilot, a firm, quiet man one inch under six feet tall, weighing 165 pounds, with light hair and blue eyes, giving the impression of unshakable confidence. He is married to the former Janet Shearon of Evanston, Illinois, and at the time of the Apollo 11 flight their two sons, Eric and Mark, were 12 and 6 years old respectively.

At 2:17 a.m. July 21, Armstrong, accoutered in his bulky moon suit and portable life-support pack, was to back out of the LM Ascent Stage and start climbing down the ladder. At the second rung, he was to pull a D-ring to deploy a television camera housed in a pod on the Descent Stage. Turned on, the camera would telecast a knothole view of Armstrong climbing down the remaining rungs of the ladder and taking his bearings on the moon. He would stand first on one foot to check his balance, then move around on both feet, testing the action of his suit. Aldrin would photograph him from the LM window. Then Aldrin would pass out the Hasselblad camera to Armstrong on a conveyor belt and Armstrong would photograph Aldrin descending the ladder.

While Aldrin takes his bearings, Armstrong would take the television camera out of the pod along with its collapsible tripod and set it up about 30 feet from the LM. The camera would make a continuous record of the astronauts' activities in its field of view and transmit it to Earth as irrefutable documentation. Aldrin would extract from the pod a roll of aluminum foil on a staff. Pushing the staff into the ground, he would raise the aluminum foil on it, like a sail on a mast, and hook the foil to the top, turning it sunward to trap particles of the solar wind. Later, it would be rolled up and stowed back aboard the Ascent Stage for the return to Earth, where the microscopic tracks left by the solar wind could be studied.

Both astronauts would then inspect the LM for leaks and take note of the extent to which its descent-engine plume had burned or scattered the soil. Next, Aldrin would lift two important pieces of scientific equipment out of the pod with the LM's rope and pulley conveyor: a 4-pound seismometer and an 11-pound laser mirror (moon weight). Housed in an 11-inch cylinder 15 inches high, the seismometer consists of a suspended weight which would jiggle if the ground shook from a quake or meteorite impact. A radio powered by solar cells would transmit the frequency and intensity of the jiggles to Earth on command from an Earth station. Designed to operate two years, the apparatus was expected to send data which would reveal clues to the moon's internal structure, and show whether it is internally active like the earth.

The laser mirror consisting of 100 corner-reflecting prisms of synthetic silica was designed, like a highway-sign reflector, to turn a light beam back to its source. In this case, the beam would be a laser emission either from Cloudcroft Air Force Base, New Mexico, or from Hawaii, making a round trip of nearly one-half million miles. With the mirror on the moon, it would be possible to make Earth-moon measurements with an uncertainty of only 33 inches—to detect hitherto unmeasurable variations in the rotation and orbits of both Earth and moon. These might confirm or refute a theory that the radius of the moon's orbit is increasing and that the force of gravity linking the Earth-moon system is weakening.

Aldrin would set up the seismometer about 70 feet from the LM and open its solar-cell panels. The instrument would start operating immediately. One of the first "events" it would transmit to Earth would be the liftoff of the LM Ascent Stage. Armstrong would deploy the laser mirror about 10 feet away from the moonquake detector.

Having performed these tasks, the two would then begin collecting soil and rock samples, picking them up with tongs and shovel and dumping them into plastic bags. About 50 pounds of rocks would be collected, including samples taken at a depth of 10 to 12 inches. Armstrong would add to this collection a 2-pound sample he was to bag shortly after he alighted on the moon and carry in a suit pouch against the contingency of an early departure. If the lunar astronauts had to leave hurriedly because of a landslide or a leak in the LM, at least they would have the contingency sample to bring home.

After stowing all samples in two vacuum-sealed containers, Aldrin would climb up into the LM and Armstrong would pass up the samples to him on the conveyor. Then Armstrong also would get back into the LM, close the hatch, and pressurize the cabin, giving the crewmen an opportunity to return to the LM oxygen supply. Having made this switch, they would depressurize the cabin again, open the hatch and throw out on the ground their overshoes, backpacks, a still camera, a lunar tool tether, and spacesuit connector covers. These items, along with the TV camera on its tripod, the seismometer, the laser mirror, and an American flag which the astronauts would set up on a metal staff, complete with crossbar to keep the flag unfurled in the solar wind, would constitute a monument to man's arrival on the moon. It was consonant with the old tradition of leaving trash in the explorers' wake, although never before had anyone abandoned a $70,000 television camera.

NO COFFEE BREAK

The extravehicular activity (EVA) would be limited to 2 hours and 40 minutes, a short day's work under the open, airless sky although the portable life-support system was designed to operate for 4 hours. Earlier in the program, two EVA periods had been planned, with a rest period inside the LM between to resupply oxygen, water, and batteries in the life-support backpacks. This plan had been modified by second thoughts. No one knew how a man in a bulky moon suit and backpack would fare in one-sixth Earth's gravity. On the Earth, it was possible only to approximate this condition in a swimming tank or in an airplane during a low-*g* or zero-*g* dive. It might be easy to walk around on the moon. It might be terribly difficult. It was better, Apollo directors reasoned, not to be too ambitious the first time out of the cradle.

After 2 hours and 27 minutes on the moon, Armstrong and Aldrin would blast off in the Ascent Stage, using the Descent Stage as the launch pad. Following the same orbital maneuvers as in Apollo 10, they would make rendezvous and dock with the Apollo 11 Command Module at 5:32 p.m. July 21, after reentering it, and would jettison the LM by sending it off into solar orbit.

The crew would then fire the Apollo engine at 56 minutes after midnight, July 22, for the return to Earth. Splashdown was scheduled at 12:51 p.m. EDT, July 24, in the Pacific Ocean southwest of Johnston Island after a flight of 8 days, 3 hours, and 19 minutes.

The recovery operation for Apollo 11 was governed by a quarantine procedure to prevent the possible contamination of Earth's biosphere by lunar organisms, if any. The quarantine had been worked out in elaborate detail by an Interagency Committee on Back Contamination, composed of members from the United States Public Health Service, the United States Department of Agriculture and of the Interior, the National Academy of Sciences, and NASA. The quarantine was not based on any theory of life on the moon but simply took into consideration the fact that while life on the moon seemed improbable no one could possibly be sure. It was an insurance policy, purchased with the philosophy that although the risk of back contamination from the moon was very low, the value of what was being insured, the biosphere of Earth, was incalculably high.

The quarantine was designed primarily to shield terrestrial life from any volatile epidemic. The astronauts were to be held in strict biological isolation from the biosphere for a period of 21 days after the liftoff from the moon, and the lunar samples would be isolated for 50 to 80 days. The crew could, however, appear on television.

After splashdown, the crew would exchange its spacesuits for specially designed biological-isolation garments to serve as barriers between them and the environment. Breathing masks would filter the air they exhale.

Similar garments would be worn by Navy swimmers attaching flotation gear to the spacecraft, but their breathing masks would filter the air they inhale. The moon men would then be flown by helicopter to the deck of the prime recovery ship, the aircraft carrier *Hornet,* and then be transferred to a sealed van about the size of a large house trailer called the Mobile Quarantine Facility. In this isolation unit, the crew, with the doctors and aides who were to join them in the quarantine, would be transported by the carrier to Hawaii and flown from there by C-141 cargo airplane to Ellington Air Force Base, near the Manned Spacecraft Center, Houston. The last leg of the journey would be made by the returning heroes in the sealed trailer, hauled by truck to the Center's $8,000,000 Lunar Receiving Laboratory. There the first men on the moon and the man who waited for them in lunar orbit would finish out the 21-day quarantine in quarters furnished in General Services Administration style like a second-rate motel. They would undergo debriefing and innumerable biological tests, play Ping-pong, read, and watch television until released to their families and the official pomp and ceremony awaiting them.

The recovery procedure was a modification of an earlier plan which called for the crew to remain in the Apollo until the carrier came alongside and hoisted it up on the deck, when the astronauts would scramble into a quarantine facility through a plastic tunnel linking spacecraft and trailer hatches. This plan was abandoned after the boat crane lifting Apollo 9 out of the sea broke down and dumped the capsule into the water. The crew had previously been transferred to the ship by helicopter. If they had remained in the capsule, they might have been injured.

CONTAMINATION COMPROMISE

Under the new plan, which required the Apollo hatch to be opened to the atmosphere while the capsule rocked in the sea, it was possible that if some vagrant lunar microorganisms had infiltrated the spacecraft they might escape into the atmosphere. The areas around the Apollo hatch where the astronauts would egress and the rubber life rafts where they would await helicopter pickup would be swabbed down with a disinfectant by Navy swimmers. The astronauts would apply a disinfectant rinse to each other's biological insulation garments.

Even these energetic and photogenic precautions to prevent the back contamination of the earth did not alter the character of the new procedure as a compromise, based on levels of probability. There was the extremely low probability of living organisms on the moon and there was the much higher probability of an accident to the Apollo capsule while the astronauts were in it during its transfer to the deck of the carrier. It was easier for the Apollo directorate to believe that a boat crane might break down, it seemed, than that there might be pathogenic organisms on an airless, apparently waterless moon.

No compromises were made, however, in the operation of the Lunar Receiving Laboratory, covering 83,000 square feet of floor space in Building 37 on the campus of the Manned Spacecraft Center. The crew and the samples they returned would be confined in this extraterrestrial quarantine station under uniquely aseptic conditions. Nothing brought from the moon could escape to the outside. Even the atmosphere in the building was sterilized before being exhausted outside the building.

SAMPLING THE MOON

The LRL was more than an Ellis Island for space travelers on the Gulf Plain of Texas. It was the first depot for the analysis of extraterrestrial material behind what the laboratory staff believed were impenetrable biological barriers.

In the biomedical section, lunar soil samples would be exposed to a dozen different culture media and to germ-free mice, Japanese quail with a rapid reproductive cycle, cockroaches, shellfish, and 33 species of plants. Primates were not included among the experimental animals as that order of mammals was ably represented by the astronauts in crew quarantine.

It was planned to fly the lunar soil samples directly from Johnston Island to Ellington Air Force Base so that analysis could begin at the LRL no later than 27 hours after splashdown. A small sample was to be examined in the underground radiation laboratory for short-lived radionuclides, produced by cosmic-ray bombardment of the moon. Of the 50 pounds of rocks and soil to be returned, only 5 per cent was allocated to biological testing. About 40 per cent would be subjected to optical, mineralogic, petrographic, chemical, magnetic, radiation, X-ray diffraction, and fluorescence tests. After the 50- to 80-day quarantine period, it would be distributed to investigators throughout the United States and in Canada, England, Germany, Japan, Finland, and Switzerland.

From the lunar-soil analysis, the investigators expected to reach several general conclusions about the landing-site surface, including a determination of whether it was basalt, as indicated by the Turkevich experiment on Surveyor 5. Detailed analysis would reveal much about the origin and evolution of the moon, and, indirectly, about the origin and development of the terrestrial planets in the solar system. Any evidence of living organisms, existing or fossil, would be the biological discovery of the century—provided the absence of terrestrial contamination could be established. The conditions under which the lunar soil was to be retrieved, transported, and analyzed should go a long way toward dispelling the contamination issue, which for years has beclouded claims of finding prelife organic compounds and fossil organic residues in meteorites.

THE CASE OF THE REGRESSING NODE

All these preparations for the greatest adventure, man's first step out of the terrestrial cradle into his cosmic neighborhood, contributed to NASA's unshakable posture of confidence that Apollo 11 would take its crewmen to the Sea of

Tranquillity and bring them back to the Pacific Ocean. All the manned Apollo flights had met their objectives with relatively little trouble. On Apollo 10, however, there had been seven equipment malfunctions and anomalies, and one serious error in navigation. Only one of the malfunctions threatened to interfere with a flight objective. This was the failure of the oxygen dump system in the LM-Apollo tunnel, initially attributed to clogging by insulation from the rupture of the insulation pad on the Apollo forward hatch. Later analysis of the problem showed that the dump line actually had been jammed by a wrong fitting.

The other problems were the vibration of the S4B rocket at translunar injection, an oxidizer leak in the Command Module's Reaction Control System, difficulties with the primary evaporator for emergency cooling after it dried out during the launch, overheating of two of Apollo's three fuel cells during the return from the moon, and the gyration of the LM Ascent Stage as it separated from the Descent Stage.

The gyration was still not explained when General Phillips announced that Apollo 11 would go on schedule. By that time, the Manned Spacecraft Center was investigating a theory that something had gone haywire in the circuitry of the Abort Guidance System switch. That made more sense to Stafford and Cernan than the earlier notion that one of them had inadvertently put the switch in the wrong position.

However, the question raised by the anomaly was moot so far as the Apollo 11 mission was concerned. When they lifted off the moon, Armstrong and Aldrin would be using the Primary Guidance and Navigation System, not the secondary AGS.

The insulation pad on the forward hatch of the Command Module was removed from the Apollo 11 machine. The S4B vibrations were still not understood when the decision was made to launch Apollo 11 in July, but the Apollo 10 flight review had established that the oscillations had not exceeded safety limits. Special instruments were being installed in the Saturn third stage to see what was causing the shaking—but they would not prevent it if it recurred on Apollo 11 translunar injection.

The navigation error which had caused Stafford and Cernan to miss Landing Site 2 by five to six miles had uncomfortable implications. It was the result not of carelessness but of lack of knowledge about the gravitational lumps and bumps in the moon and their effects on low, near-equatorial orbits. The site is two-thirds of a lunar degree north of the equator. To bring the LM directly over it, the flight plan inclined the Apollo orbit slightly, about 1.3 degrees, to the equator.

Because of the moon's rotation, at about one degree every two hours, the point called the node, where the orbit crossed the equator, regressed—that is, appeared to move backward along the flight path. If Apollo 10 passed directly over Landing Site 2 on one revolution, the site would have regressed out of the orbital plane on the next and succeeding revolutions. Because of the orbit's inclination, the site would not remain directly beneath, but would appear to move to one side with the regression of the equatorial node.

This regression had been observed on the Lunar Orbiter flights and the flight of Apollo 8. Consequently, guidance men at Houston cranked data into the Apollo 10 guidance computer to adjust the orbit so that the LM would pass over Site 2 on the 13th revolution when it made its first, low pass.

Evidently, this adjustment would have worked beautifully if nodal regression had taken place under what the engineers assumed were the laws of physics. But the nodal regression did not take place as calculated and the effect of adjusting the orbit was to put the LM five to six miles out of plane with the site.

Several explanations have been advanced for the phenomenon of the apparent nonregression of the node on Apollo 10. One expert suggested that gravitation effects produced accelerations on the spacecraft of which the guidance experts at Houston were unaware or did not expect. He believed that, instead of being inertial, with the moon rotating under it, the Apollo 10 orbit had been somehow "dragged along" by the moon's rotation. If so, no one at Houston seemed to know why.

In any case, it was decided that the Apollo 11 orbit would be inclined precisely as Apollo 10's had been, but this time, the apparent lack of nodal regression would be taken into account. Houston advised that at the proper time, the orbit would be "biased" to put the LM in the proper descent orbit plane to bring it down on Landing Site 2. One could not help gain an impression from this that the lunar landing was going to be flown by the seat of the pants.

The danger of missing the site was the likelihood that nothing but rough, potholed, boulder-strewn country would be within the LM's landing range. The descent engine could correct an out-of-plane descent for a distance of six miles. Beyond that, as Stafford had warned, the crew would have to find a level place before the fuel gave out or hustle back upstairs and abandon the landing. The crew might live to try again some day, but another crew would make the next try. The commander would face a choice between a treacherous landing that might cost him and his pilot their lives or an abort of the mission and loss of the fame, fortune, and place in history of being first on the moon.

One source of the mysterious accelerations that were putting bends and twists in lunar orbits were dense concentrations of mass believed to be under the lunar surface in the circular "seas." Since gravitational acceleration is proportional to mass, these structures presented an invisible obstacle course to pilots trying to steer a rocket ship passing near them into a landing ellipse three and one-half miles long by one and one-half miles across.

Large and denser than the surrounding material, the structures had become known as "mascons," for mass concentrations. Aside from creating navigation hazards, the mascons had fascinating implications. Twelve of them had been detected from their effects on the flight paths of Lunar Orbiters and Apollo 8 by William L. Sjogren, Paul M. Muller, and Peter Gottlieb of the Jet Propulsion Laboratory. Muller and Sjogren deduced the existence of six mascons beneath the great ringed maria called Imbrium, Serenitatis, Crisium, Nectaris, Humorum, and Aestatum from Lunar Orbiter 5 tracking data in 1967-1968. Six additional mascons were later reported by Sjogren and Gottlieb to be located in the crater Grimaldi, the Mare Humboldtianum, the Mare Orientale on the western limb of the moon, the Mare Smythii on the eastern limb, and two unnamed maria, one at 27 degrees east longitude and 5 degrees south latitude—not far from Landing Site 2. The mascons betrayed their presence by distorting the orbits of spacecraft passing over or near them, and, because of their density, they could be mapped as regions of higher gravity than surrounding areas of the moon.

The mascons were credited with producing a peculiar "yo-yo" variation in the orbits of Apollo 8 and 10. On each revolution of the moon, the apolune, or high point, of these orbits would rise, while the perilune, or low point, would drop. In Earth orbit, the reverse occurs: the apogee, or the high point, drops because the vehicle loses energy from contact with residual atmosphere at perigee, the low point. There is no atmosphere around the moon to affect orbits, but the mascons change them dramatically. On its third revolution, the Apollo 10 orbit was 70.3 miles apolune by 69 miles perilune. On the 25th revolution, apolune had risen to 77.1 miles—nearly seven miles higher than on the third revolution—while perilune had dropped to 62.5 miles, six and one-half miles lower than on the third revolution.

All the mascons detected so far appear to lie beneath the ringed seas, some so ancient as to be nearly obliterated. Professor Urey has observed that the gravitational anomaly produced by the mascon in the Mare Imbrium might be accounted for by a flat, circular slab of chondritic meteorite material 2.5 miles thick, 416 miles in diameter and 6.2 miles below the surface. In this case, the fact that it remained near the moon's surface and did not settle deep within it indicates that the moon is more rigid than the earth and has a lower temperature.[2]

It would seem from these estimates that the mascons may be large buried meteorites, or planetoids, which crashed into the moon and made the circular seas. But another interpretation has been advanced by Muller, who has speculated that the mascons are accumulations of lunar sediments which might have been deposited by water in deep basins.[3] This theory supposes the existence of a primeval hydrosphere, an idea which has gained some support from Lunar Orbiter and Apollo photos of the rilles, sinuous clefts resembling ancient river or lake beds.

COUNTDOWN

The countdown for the landing of men on the moon began at 8 p.m., EDT Thursday, July 10, on a hot, cloudy evening. An international brigade of more than 3000 news media correspondents began filling the motels in Cocoa Beach and Titusville. The next day, Armstrong, Aldrin, and Collins underwent exhaustive physical examinations. They were in excellent condition, the flight surgeon reported. There was no sign of infection and Dr. Berry was keeping the crew in partial quarantine in the hope of preventing any from developing on the flight. This precaution kept the astronauts isolated from everyone except a handful of medical and technical people with whom they had to work. The exclusion policy was even applied to President Nixon. He backed out of his acceptance of an invitation to dine with the crew on the night before the launch after Dr. Berry publicly warned against it. This bit of medical advice irritated the administrative hierarchy in NASA which was anxious to bring the President as close to the program as possible and angered Frank Borman who had extended the invitation to Mr. Nixon on behalf of Armstrong and the crew. On his return from a goodwill mission to Russia, Borman criticized Berry, asserting that the advice should have been given privately—if at all. It was obvious, Borman said, that the giver failed to understand psychology inasmuch as a visit from the Commander-in-Chief would have been more important to the crew than the risk of any respiratory infection it might have involved.

Dr. Berry held his ground. The pre-launch dinner invitation to the President had been issued without consultation with him—and he was responsible for the physical well being of a flight crew, not Borman nor the public affairs people. All too fresh in the physician's memory was the violent, flu-type infection that had stricken Borman, himself, on the moonward flight of Apollo 8; the head colds that had made the crew of Apollo 7 miserable for their entire flight, and the bout of nausea that had overtaken Rusty Schweickart on Apollo 9 during the critical test of the Lunar Module.

The medical reasoning was clear, and Mr. Nixon quickly accepted it. He agreed to be present on the Hornet when the crew returned. Then the situation would be reversed. The crew would be in quarantine to protect mankind. No one knew whether there were microorganisms on the moon or, if so, if they would be pathogenic. Against such a contingency, it was essential to keep the crewmen as free of contagious illness as possible as a means of clarifying to some extent the origin of any infection they might develop in the post-flight quarantine at Houston.

The episode made a brief flurry in the press because it

seemed to illustrate that the right hand of NASA did not understand, or appreciate, what the left hand was doing. This was not simply a bureaucratic schism. It represented a fundamental difference in point of view between the engineers and astronauts, or "Rover Boys," on the one hand, and the physicians and biologists, or "Nervous Nellies," on the other. To many of the "Rover Boys" the hazards of radiation and long-term effects of free fall or weightlessness had never seemed real. The possibility of "bugs" on the moon seemed downright ridiculous. The only "bugs" the "Rover Boys" were concerned about were the malfunctions in the spacecraft systems. Borman expressed the "Rover Boys" view nicely when he told a press conference at the Kennedy Space Center before the launch that while in Russia he had gained the impression that the cosmonauts "have suffered rather more than we have from the medical community." That is, he added, the Soviet medical approach to space flight has been more conservative.

About 175 miles east of Cape Kennedy, a flotilla of Russian naval vessels appeared July 11. They were dispersed over an area of 12 square miles and heading generally down the Atlantic toward Havana—ostensibly to pay a port call July 20 at Cuba's invitation. The flotilla consisted of a guided missile cruiser, two guided missile destroyers, two submarines and a submarine tender. The U.S.S. *Thomas J. Gary*, a U. S. Navy destroyer based at Key West, was shadowing the Russian warships.

From the nature of their armament, the Soviet vessels appeared to be fully instrumented to track Apollo 11 from launch to orbit, and to observe the performance of each of the three stages of the Saturn 5 rocket. The U. S. Navy doggedly refused to speculate about this possibility. Ordinarily, the Russians use instrumented fishing trawlers off the Florida coast to monitor the launches of missiles and rockets from Cape Kennedy. The presence of such a formidable array of ships was not only a possible tribute to Apollo 11, but also a display of interest in a test of the U. S. Navy's Poseidon missile that was launched July 9 down the Atlantic Missile Range.

A clearer Soviet response to the flight of Apollo 11 seemed to appear July 12 when Tass announced that a vehicle called Luna 15 had been launched to the moon. It followed rumors that the Russians had been preparing to send to the moon a robot lander that would scoop up a sample of soil and fly back to Russia with it. The timing of the launch suggested that the Soviets expected that such a feat, if successful, would upstage the American manned lunar landing. So ran the speculation. The Russians did not identify the mission of the craft and merely advised Frank Borman in response to his query that its orbit around the moon would not interfere with the published orbit of Apollo 11.

Officially, the Soviets did not disclose the mission of Luna 15 as it flew on a minimum energy trajectory toward the moon. Borman and Wernher Von Braun believed that the vehicle was a lander that would scoop up a soil sample and fly back to the Soviet Union with it. If so, it would provide heavy ammunition to the critics of manned space flight in the American scientific community and in Congress. In the past, the term "Luna" had described a surveyor-type soft-lander or a lunar orbiter. Was this one a lunar boomerang? No one in the West seemed to know.

The Apollo 11 crew sat for a final news conference Monday night, July 14. It was conducted on a television circuit that enabled a panel of newsmen to interview the astronauts in crew quarters on Merritt Island from the NASA News Center in the city of Cape Canaveral, about 10 miles away.

Except for Collins, the Apollo crewmen appeared withdrawn and volunteered little information about their feelings, their hopes or fears.

Collins said he had a complaint. So far as the landing was concerned, he was odd man out. He would be orbiting the moon in the Apollo Command and Service Module while Armstrong and Aldrin were on the surface. And while a terrestrial audience would be able to watch them on television, Collins could not. There was no TV receiver in his cabin—only the color camera and its monitor.

"I guess I'm one of the few who won't be able to see what's going on down there. I want you to save the tapes," he told Walter Cronkite of CBS, a member of the news panel.

Wednesday, July 16, the day of the launch of Apollo 11 came bright and hot. The Cape area was jammed with sightseers, observers, very important persons, news media representatives, technicians, and thousands upon thousands of children, who would inherit the moon.

The VIP list was led by former President Lyndon B. Johnson and Lady Bird; Vice President and Mrs. Spiro T. Agnew;

Charles A. Lindbergh; most of the President's cabinet officers and commission chairmen, 69 ambassadors of foreign governments, 100 foreign science ministers, attaches and military experts; 19 governors, 40 mayors, 275 leaders of American commerce and industry, and nearly one-third of both houses of Congress.

Outside the west gate of the Kennedy Space Center, another group of citizens had mobilized under the aegis of the Southern Christian Leadership Conference to stage a Poor People's Demonstration on the occasion of the lunar launching. Their purpose was to dramatize the plight of the underfed in a country spending $24 billion in eight years to send men to the moon.

"We came here to dramatize a national problem and not to oppose this great event," said Hosea Williams, one of the demonstration directors. "Our purpose in coming here is to protest the inability of our congressmen and senators to establish humane priorities. We have spent $24 billion to explore outer space. If America put that much into feeding the poor and hungry, then poverty and hunger would be gone from the face of America today."

The Rev. Ralph Abernathy, president of the Conference, appeared Wednesday morning with the demonstrators, some of whom had spent the night along the roadside of U. S. highway 1 near Titusville, singing freedom songs, chanting "I may be black, I may be poor, but I am somebody," and praying.

As Apollo 11 was counted down toward liftoff, NASA officials escorted the demonstrators inside the Center to a point of vantage from which they could view the launch.

All segments of American society seemed to be represented at the point of departure of men for the moon. All were in their places. The poor were in one place; the rich and powerful in another. Outside the gates of the huge Center, along the causeways and on the beaches, were an estimated 1,000,000 persons, who were neither rich nor poor. All eyes were focused on the distant rocket, the symbol of mankind's struggle up from the cradle. Never before, perhaps, have so many people been so highly aware of the coming of a great event.

Armstrong, Aldrin, and Collins breakfasted at 5 a.m., struggled into their moon suits at 5:30 a.m., and rode to the pad in their white van at 6:26 a.m. The routing was carried out punctually. At 6:52 a.m., they were strapped onto their couches in the Apollo spaceship called " Columbia." The Lunar Module, "Eagle," was stowed underneath them in a steel cocoon at the head of the Saturn third stage.

In all respects, except one, the launch of Apollo 11 was the same as those of Apollo 8 and 9. Apollo 11 was a mission of destiny, and it drew as great an audience as perhaps has ever assembled anywhere for a single event.

At 9:32 a.m. July 16, 1969, precisely on schedule, Saturn 506's first stage engines ignited in blinding orange flame and black billows of smoke, and began to rise, like an elevator, with Apollo 11 on top. The ground shook. The massive vehicle cleared the launch tower and split the sky with a rattling roar. Man was on his way to the planets.

The thunder of the five F-1 engines in the first stage was heard around the world. It was the answer to the call of John F. Kennedy in May, 1961 for a national commitment to land a man on the moon and bring him home safely before this decade was out. Mr. Kennedy had speculated that he might not be alive when the appointment on the moon was kept, his brother-in-law, R. Sargent Shriver, was quoted as saying by the Miami *Herald*. Shriver, U. S. Ambassador to France at the time of the launch, was one of the VIP's who came to witness the event. According to the *Herald* account July 17, Shriver quoted Mr. Kennedy as telling him and other members of the family: "I firmly expect this commitment to be kept. And if I die before it is, all of you here now just remember when it happens I will be sitting up there in heaven in a rocking chair just like this one, and I'll have a better view of it than anybody."

Apollo 11 and the S4B, the third stage of the Saturn rocket, headed out over the Atlantic Ocean on their first revolution of the earth in a circular orbit of 118.45 miles. Halfway through the second revolution over the central Pacific Ocean, the Apollo-S4B was oriented for the second burn of the S4B's J-2 engine burning for five minutes 47 seconds and accelerating both rocket and spacecraft from 17,427 miles an hour to 24,250 miles an hour, or an increase in velocity from 25,560 feet per second to 35,570 feet per second at S4B engine cutoff.

"Hey, Houston," Neil Armstrong called down after cutoff, "The Saturn gave us a magnificent ride."

"It looks like you are on your way now," Mission Control responded.

Mike Collins performed the transposition and docking maneuver after separating the Apollo from the rocket. He turned the ship around, docking it with the Lunar Module and then triggering the LM release from the rocket so that the joined vessels could pull away from the S4B. He thought he had used "more gas" than necessary.

As the crewmen drew farther away from the earth, they began to talk, for now the flight was novelty. Armstrong remarked: "You might be interested that at my window right now I can observe the entire continent of North America, Alaska, over the Pole, down to the Yucatan Peninsula, Cuba, the northern part of South America—and then I run out of window."

The flight "up" to Lunar Orbit was uneventful. The crew enjoyed sound sleep. No one complained of dizziness or nausea. There were no emergencies or near emergencies on board. Apollo 11 was a streetcar named Columbia and Eagle, headed nonstop for the moon. Man's third voyage to the moon had become routine.

On the morning of July 17, the second day of the mission, Jim Lovell, who had been commander of the flight's back-up crew radioed Apollo 11, from Mission Control: "Is the Commander aboard?"

ARMSTRONG: This is the Commander.

LOVELL: I was a little worried. This is the Back-up Commander, still standing by. You haven't given me the word yet. Are you go?

ARMSTRONG: You've lost your chance to take this one, Jim.

Later, during the morning of July 17, Houston ordered a midcourse correction that required a three second burn of Apollo's engine to reduce pericynthion, the closest approach to the moon, from 201 to 69 miles.

Apollo 11 had no difficulty establishing a stable attitude using the rotisserie roll that Apollo 10 had pioneered. The roll at three revolutions an hour distributed the sun's heat around the dual Apollo-LM spacecraft.

"How does it feel to airborne again, Buzz?" Lovell called up to Aldrin. The two had flown Gemini 12 in 1966.

"Well, I'll tell you," said Aldrin. "I've been having a ball floating around inside here, back and forth, up to one place and back to another."

"It's a lot bigger than the last vehicle Buzz and I were in," said Lovell.

"Oh, yes, it's been nice," Aldrin agreed. "I've been very busy so far. I'm looking forward to taking the afternoon off. I've been cooking, sweeping, and almost sewing, and, you know, the usual little housekeeping things."

The chatter indicated a relaxed crew with everything under control. No malfunction plagued the vessel as on the previous flight.

"Hey, Jim," Buzz Aldrin called down, "I'm looking through the monocular now and, to coin an expression, the view is just beautiful. It's out of this world. I can see all the islands in the Mediterranean. Some larger and smaller islands of Majorca, Sardinia and Corsica. A little haze over the upper Italian peninsula, some cumulus clouds over Greece. The sun is setting on the eastern Mediterranean now. The British Isles are definitely greener in color than the brownish-green that we have in the islands and the peninsula of Spain."

"Right," Lovell agreed, "I understand that the northern Africa-Mediterranean area is fairly clear today, huh?"

"Right," said Aldrin. "I see a bunch of roads with cars driving up and down."

At the time Aldrin made this report, Apollo 11 was about 127,000 miles from Earth. The astronaut was looking at the region with a monocular of low magnification.

The scenes of Earth continued to intrigue the crew. Collins told Houston later that day—the second day of the mission: "I've got the world in my window." Using the monocular, he said he could see from Seattle all the way down to Tierra Del Fuego at the tip of South America.

From nearly seven miles a second at the beginning of translunar injection, the gravitation pull of the earth had slowed down the spacecraft to less than a mile a second.

In the afternoon of the second and the third days of voyage, the crew pointed the lens of Apollo's color television camera at the earth, and a lovely blue-green half-earth appeared on terrestrial TV screens with the eastern Atlantic and Europe in darkness. Of considerably high resolution were the color pictures of the crewmen working in the cabin and explaining their activities. These television tours were long ones, lasting up to one hour and 36 minutes of continuous photography. They gave the viewers a feeling of reality about the voyage that had seemed unreal at the outset to so many people watching the launch at Cape Kennedy.

In the longest telecast ever sent from space up to that time, millions of Earthlings saw what the interior of the Apollo Command Module, the docking tunnel connecting it to the LM, and the interior of the LM actually looked like. At first Armstrong and then Collins took the world below on a television tour. The Apollo color camera worked marvelously well, transmitting rich color in authentic tones over a distance of 175,000 miles.

From time to time there was some confusion about who was doing what.

"We see lots of arms," commented Charlie Duke, Capsule Communicator on duty at Mission Control.

"The only problem, Charlie," explained Collins, "is these TV stage hands don't know where to stand."

Armstrong and Aldrin cleared the connecting tunnel of the probe and drogue assembly, which are used in docking to guide the latching mechanism into place as the Command and Lunar Modules come together. The LM hatch then became visible. Then he opened the hatch and a light came on in the lunar vessel.

"Just like a refrigerator when you open the door," Collins said. "The light goes on."

The picture was so clear that Duke commented: "We can read the markings on the instruments for the Glycol (the coolant) pressure quantity and you can read the scale on the Eight Ball (flight director attitude indicator which shows the attitude of the craft in relation to the earth or moon)."

"The restraints in here," said a voice, "do a pretty good job of pulling my pants down."

"Roger," said Duke, "we haven't quite got that before the 50 million TV audience yet."

Collins pointed the camera out of the window. On the TV screen at Houston, the blue-green earth swam into view, vibrating like jello.

"If that's not the earth," said Duke, "we're in trouble."

"It is," he was assured. "Just with more cloud bands."

The flight proceeded calmly. A faulty oxygen flow indicator that had been noted earlier in the flight was written off as useless. Houston, however, could monitor the oxygen flow, so that the malfunction was not a problem.

Apollo 11 entered the predominant gravitational field of the moon and began accelerating from the low point of its uphill climb from Earth, 2040 miles per hour, until it reached 5225 miles an hour by Saturday noon.

The crewmen then prepared to fire the Apollo engine to brake velocity by about 1977 miles an hour (2917 feet per second) in order to fall into lunar orbit. The vessel rounded the western limb of the moon and disappeared from radio view. On the Farside, at 1:22 p.m., the Apollo engine ignited and burned five minutes and 57 seconds, about five seconds short of the planned burn. The engine, however, was cut off by the guidance computer which had a will of its own.

When Apollo 11 reappeared, the tracking data at Mission Control showed that it had entered an orbit around the moon of 70.4 by 194.9 miles. On their second revolution of the moon in this orbit, the crew turned on the television camera as they sped across the eastern limb at Smyth's Sea. The now familiar lunar landscape along the equatorial band of Nearside began to unfold.

DESCENT TO THE MOON

The insertion of Columbia and Eagle into lunar orbit marked the successful arrival of the third voyage to the environs of the moon. The whole world was now astir with the portent of what was to come. Mr. Nixon had called for a national "day off" on Monday, July 21, closing most federal offices with one, significant exception—the manned spacecraft center in Houston.

Pravda published an unusual amount of detail about the mission and referred to Armstrong as the "Czar" of the ship. The government of West Germany declared Monday to be "Apollo Day" and school children in Bavaria were given a holiday. Pope Paul VI arranged for a special color TV to be installed in his summer residence to watch the flight, and the BBC in London was reported to be considering a special call to Britons in the event the astronauts decided to climb outside their craft earlier than planned after landing.

On their first two revolutions of the moon, the crew seemed to find the view of the surface almost precisely what they had been led to expect by briefing and photography.

On this voyage, Armstrong, Aldrin, and Collins were looking at the landscape critically, as befitted mariners preparing to make a landfall.

Armstrong observed that the LM showed a tendency to dip down toward the moon during the first orbit, and suggested the phenomenon might have something to do with the mascons.

Apollo 11 cleared Smyth's Sea on the eastern limb of the moon and Collins picked up the first landmark by which the accuracy of the flight path could be established. Landmark Alpha was a bright crater in the floor of the Foaming Sea at 65 east longitude about 790 miles east of Landing Site 2.

The double spacecraft had been guided into a flight path with an inclination of only 1.25 degrees to the lunar equator—the same as the Apollo 10 orbit. By duplicating the orbital parameters on Apollo 10, the guidance team at the manned spacecraft center hoped to avoid the navigation error that had caused Stafford and Cernan to fly five to six miles south of the landing site on their low pass in May. The guidance officials knew what had been wrong with their navigation on Apollo 10, but admitted they did not know enough about gravitational anomalies of the moon to understand why.

To the west of Alpha, Apollo 11 passed over the Sea of Fertility and the crewmen turned their television camera on Langrenus that stood out stark and massive to the south of the track.

Beneath the dual spacecraft lay a crater named Webb, a marker that told the crew they were six minutes away from starting their 50,000-foot powered descent during which the LM descent engine would fire continuously to brake their fall to the ground.

The crater was fairly flat at the bottom and there were three smaller impact craters there and in the west wall.

Langrenus with its remarkable central peak appeared a greenish-yellow on the television screen, but the crewmen said it was a montage of tan and gray scenes.

"The Sea of Fertility doesn't look very fertile to me," said Armstrong. "I don't know who named it."

Aldrin, a Doctor of Science from the Massachusetts Institute of Technology, had the answer. "Well, it may have been named by the gentleman this crater was named after, Langrenus. Langrenus was a cartographer to the King of Spain and made one of the early, reasonably accurate maps of the moon."

"Roger," responded Bruce McCandless, the capsule communicator. "That is very interesting."

"At least it sounds better for our purposes than the Sea of Crises," said a voice from the spacecraft.

"Amen to that," replied the capcom.

The crater Secchi came into view as the dual spacecraft sped westward with an orbital velocity of 3654 miles an hour. In seven minutes from that point, the crew was due to pass out of the lunar morning and into the dusk of Earthshine.

Below, the track leading to the Landing Site 2 spread out before them like a road map. Westward lay the scar of Apollo Rille, the escarpment called Apollo Ridge, Twin Craters Ridge, Smokey Basin, Mount Marilyn, Weatherford, Cape Venus, Lost Basin and then the circular craters Censorinus W and Censorinus T.

Beyond, they looked at Low Mesa, Lonesome Mesa, Bear Mountain, Barbara's Mesa, Cape Bruce, Boot Hill, Wash Basin, Sidewinder, and Diamondback Rilles, dwarfed by the massive crater Maskelyne to the north of their track.

Finally, they came to Last Ridge and the landing site itself. Powered descent would begin at a triangular shaped mountain called Mount Maryland, 303.8 miles east of the site.

In the meantime, Mission Control studied an anomaly in the performance of the Apollo engine during Lunar Orbit Insertion. Pressure in one of the two engine banks, either one of which could be used, had fallen faster and lower than expected.

The reason for this could not be determined immediately, but the dayshift flight director, Clifford Charlesworth, asserted that the pressure drop, while mystifying, presented no discernible hazard. Nevertheless, the second Lunar Orbit Insertion burn would be made, he ordered, with the second engine bank.

On previous Apollo lunar flights, the second LOI burn was designed to circularize the orbit. But experience had shown that the curious lunar gravitational bumps speedily converted a circular orbit into an oval one. Consequently, the flight had been planned to put Apollo 11 into a slightly oval orbit which the lunar gravitational field would tend to make circular.

The 17 second burn came off on schedule behind the moon, reducing the orbit to 61.7 by 74.5 miles on the afternoon of July 19.

Apollo 11 was ready to begin the final phase of its mission

the next day. On Sunday, July 20, the Lunar Module, Eagle, would separate from the mothership, Columbia, and go down to the Sea of Tranquility. The crewmen retired early Saturday night and the lights never dimmed at the manned spacecraft center.

Early Sunday, Armstrong and Aldrin began to prepare the Lunar Module for the descent to the moon. They spent most of the morning checking it out and testing its communications with the ground.

At 1:47 p.m., as Apollo 11 was on its 13th revolution, Eagle was undocked from the Apollo Command and Service Module and Collins was alone in Columbia.

"I think you've got a fine looking flying machine, there, Eagle," Collins observed. "Despite the fact you're upside down."

"Well," responded Eagle, "somebody is upside down."

Collins burned the small thrusters on Columbia to move the big Apollo Command and Service Module away from Eagle at about two and one-half feet a second. This would put Eagle about 1100 feet behind Columbia when Eagle began its descent on the next revolution.

As both vehicles rounded the western limb of the moon to begin the 14th revolution, Mission Controllers made a final check of all systems on Eagle and gave the crew a "go" for descent. The descent burn, using the vehicle's powerful descent engine, was made behind the moon, out of radio range of Earth. At Mission Control, the controllers stood by their consoles to await the reappearance of Columbia and Eagle and learn the outcome of the burn.

Columbia was first to appear around the eastern limb of the moon, and Collins reported:

"Listen, Babe. Everything's going just swimmingly. Beautiful."

Then Eagle came into radio view around the moon.

"Eagle, Houston. We're waiting for your burn report," Mission Control called.

"Roger," said Armstrong. "The burn was on time."

His report showed that Eagle's descent engine had fired for 29.8 seconds to brake the vessel's velocity by 76.4 feet per second.

Eagle in its new orbit began moving down to a perilune, or low point, of 9.4 miles above the surface. When it reached

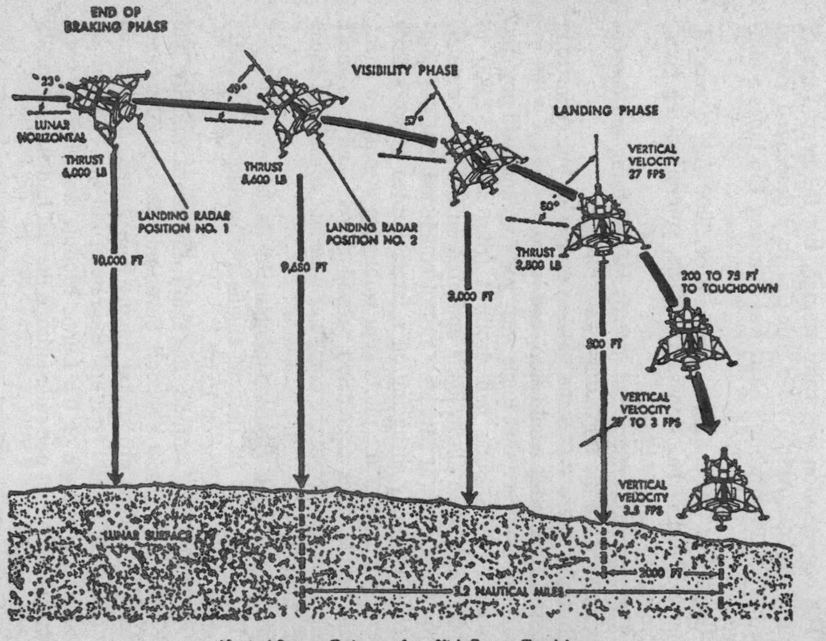

Nominal Descent Trajectory from High Gate to Touchdown

that point, the engine would ignite to take it down the rest of way to Landing Site 2.

From time to time, Mission Control asked Armstrong to yaw the vessel so that the high gain antenna would be aligned with Earth.

The conversation was calm enough, but the heart rates of Armstrong and Aldrin were rising as they descended toward the surface. Armstrong's heart was pumping 156 times a minute.

Eagle was descending at the rate of 2900 feet a second. Charles Duke at Houston said: "Eagle, you're looking great. Coming up—nine minutes."

The guidance system had already started firing the descent engine to brake the ship's downward plunge.

Initially, there was some yawing of Eagle, but Armstrong advised Houston it would stop, and it did.

Houston advised that Eagle was "go" for continued powered descent. The altitude had dropped from about 52,000 to 40,000 feet when alarm lights appeared on Eagle's computer. A quick analysis at Mission Control concluded that the computer was simply shrugging off a larger burden of work than it could handle as the pilots kept calling up displays to check their course. This conclusion proved correct and two more program alarms from the computer were not considered cause for concern.

At 27,000 feet, Houston announced Eagle was still "go."

Eagle reported its descent engine performance was better than in the simulator on Earth.

At 21,000 feet altitude, the ship's downward velocity had dropped to 1200 feet a second. Seven minutes and 30 seconds after the descent propulsion engine had ignited, the altitude had fallen to 16,300 feet, and the rate of fall was 760 feet per second. The plunge of the Lunar Module was slowing comfortably. At about one mile altitude the fall had slowed to 129 feet a second.

Houston announced: "We are now in the approach phase. Altitude 4200 feet."

A moment later, Mission Control advised: "You're go for landing."

"Roger, understand," said Eagle. "Go for landing."

Another computer alarm flashed on Eagle's console and was acknowledged by Houston. The ship's computer seemed to be complaining mightily that it was being overworked. But so was everyone in Mission Control and the descent was fine.

Or so it appeared. However, from the time of the start of the descent orbit burn, it had become increasingly apparent that the descent trajectory had become biased about three miles west of the landing site. Armstrong advised Houston he would be landing "long" from the site—down range.

Nevertheless, Houston kept advising that the descent looked good. The guidance team did not seem to be aware that Eagle was going to overshoot its landing target.

"We're go," said Eagle. "Hang tight. We're go. Two thousand feet."

Aldrin called out the descent rates and angles: "Thirty-five degrees. Seven hundred fifty (feet). Coming down at 23 (feet per second). Six hundred feet, down at 19 (feet per second).

"Five hundred forty feet, down at 15. Four hundred feet, down at nine. Three hundred thirty feet, three and one-half down. We're pegged on horizontal velocity seven-five feet. Things look good."

HOUSTON: Sixty seconds.

EAGLE: Lights on. Down two and one-half. Forward. Forward. Good. Forty feet, down two and one-half. Picking up some dust. Thirty feet, two and one-half down. Faint shadow. Four forward. Drifting to the right a little.

Armstrong moved the descending vehicle to the west to avoid a crater about the size of a football field, filled with big rocks.

Eagle had overshot the landing site touchdown point by about four miles and was hovering near the fringes of a rocky area called the "Cat's Paw."

Eagle reported: "Drifting right."

Then the contact light went on as the extended "feelers" in Eagle's pods made contact with the ground.

EAGLE: Okay. Engine stop. Modes control, both auto; descent engine command override off. Engine arm, off.

HOUSTON: We copy you down, Eagle.

It was 4:17 p.m. plus 42 seconds, Eastern Daylight Time. Man was on the moon.

During the applause and cheers which broke out in the news center, and briefly, in Mission Control, Armstrong's first words after the landing were lost. They were:

"Houston, Tranquility Base, here. The Eagle has landed."

HOUSTON: Roger, Tranquility. We copy you on the ground. You've got a bunch of guys about to turn blue. We're breathing again. Thanks a lot.

TRANQUILITY BASE

In the theater-sized auditorium of the NASA news center, on the campus of the Manned Spacecraft Center at Houston, the television lights went up like flares and a bespectacled engineer in a dark suit appeared in the glare, his face flushed, his hands clasped like those of an opera singer. He was Thomas O. Paine, who had come up through NASA hierarchy to become administrator. Without preamble, he said:

"Immediately after the lunar touchdown, I called the White House from Mission Control and gave the following report to the President, 'Mr. President, it is my honor on behalf of the entire NASA team to report to you that the Eagle has landed in the Sea of Tranquility and our astronauts are safe and looking forward to starting the exploration of the moon.' "

He paused for a moment, and then said in a more conversational tone: "We have clearly entered a new era. I think it means a different thing to each one of us. The voices that we hear coming back from these brave men on the surface of the moon are still hard to believe, and yet it's true. I think that this success is something that has raised the spirits around the world and it has caused us to pause and ponder its meaning which only history in the final analysis will reveal to us."

There was a ring of sincerity about these remarks which made them more moving, in the context of the wonder and awe of what had happened, than any contrived statement. Here was a plain speaking man, without bombast, without subterfuge, who reacted to the advent of men on the moon with humility and gratitude. Paine added: "It is significant, I think, that two days ago, the Soviet Union, at our request, passed to us valuable orbital parameters for the Luna 15 mission in an unprecedented cooperative move. I think it bodes well for the future of mankind's exploration of our sister planet."

The request, made by Frank Borman, had been honored promptly. Moreover, the Russians continued to advise NASA of changes in Luna 15's orbit.

Shortly after Paine concluded a press conference, Mission

Control announced it had been decided to advance by three hours the time when first Armstrong and then Aldrin would prepare to climb out of the LM. Dr. Berry explained that the crew was so well rested and also so keyed up that the rest period that had been planned after touchdown was unnecessary.

At 10:51 p.m. EDT, Armstrong slowly backed out of Eagle's hatch and, directed by Aldrin, planted both feet on Eagle's porch, a platform just below the hatch.

"Move, here, roll to the left," Aldrin instructed, since it was difficult for Armstrong in his extravehicular mobility unit (EMU) to see his feet. The EMU weighed 183 pounds (Earth weight) including a liquid cooling garment, portable life support system, or backpack, oxygen purge system and visor assembly. It provided life support—oxygen, temperature control and water—for a period of four hours outside the LM, but was not to be used on this mission for more than two hours and forty minutes.

"Okay," said Aldrin, "now you're clear. You're lined up on the platform. Put your left foot to the right a little bit. Okay, that's good. Now roll left."

Armstrong collected his dirt sample bags. Aldrin warned: "Okay. Not quite squared away. Roll right a little. Now you're even."

ARMSTRONG: That's okay?

ALDRIN: That's good. You've got plenty of room for your left (foot).

ARMSTRONG: How'm I doing?

ALDRIN: You're doing fine.

Armstrong then opened the equipment pod (the Modularized Equipment Stowage Assembly) on the LM Descent Stage, exposing the lens of the automatic television camera. Aldrin turned the camera on and suddenly on the television screen at Mission Control the lunar scene flashed on the screen.

The forward strut of the LM came sharply into focus, and then into view came the creature from another planet, looking like a diver with a huge pack on his back, gingerly climbing down the LM ladder.

The dimness contributed to this fantastic sight of an alien structure in light and shadows, all steeply contrasted, in the lunar scene. In the shadow of the LM, which towered up 22 feet 11 inches from the surface, it was luminously dark, and in the sunlight, brilliantly white.

"Man, we're getting a picture," called the Capsule Communicator, Bruce McCandless.

"Oh. You got a good picture, huh," queried Aldrin.

"There's a great deal of contrast in it. Currently, it's upside down on our monitor but we can make out a fair amount of detail."

The picture was quickly rectified at Houston.

At 10:54 p.m., Armstrong reached the foot of the ladder. There was a drop of three feet to the ground. He paused on the lowest rung and said:

"I'm at the foot of the ladder. The LM foot pads are only depressed in the surface about one or two inches, although the surface appears to be very, very, fine grained, as you get close to it. It's almost like a powder. It's very fine.

"I'm going to step off the LM now."

Armstrong seemed to float down to the surface. He said: "That's one small step for man, one giant leap for mankind."

A moment later, he added: "The surface is fine and powdery. I can pick it up loosely with my toe. It does adhere in fine layers like powdered charcoal to the sole and sides of my boots. I can only go in a small fraction of an inch. Maybe an eighth of an inch. But I can see the footprints of my boots and the treads in the fine, sandy particles. There seems to be no difficulty in moving around as we suspected. It's even perhaps easier than the simulations at one-sixth *g* that we performed on the ground. It's actually no trouble to walk around."

Armstrong also noted that the descent engine blast did not leave a crater of any size and that the site appeared to be quite level. Then he told Aldrin that he was ready to receive the still camera on the clothesline-style conveyor line the two had rigged after opening the hatch.

When he received the camera and unhooked it from the conveyor line, Armstrong announced he was going to "step out and take some of my first pictures." Then he put the camera down and used a small scoop to get a contingency sample of surface soil and rocks, about two pounds.

"This is very interesting," said Armstrong as he shoved the long-handled scoop into the loose topsoil. "It's a very soft surface but here and there where I plug with the contingency sample collector I run into a very hard surface. But it appears to be very cohesive material of the same sort."

Armstrong put the material into a bag and stowed the bag in a leg pocket of his moon suit. Now, if an emergency came up requiring the moon men to leave in a hurry, at least they would be able to bring that sample back to Earth.

Armstrong then removed the solar wind experiment from the equipment pod, pushed its staff in the ground and unrolled a sheet of aluminum foil like a home movie screen to catch the wind that blows between the worlds, the wind from the sun. When the excursion outside the LM ended, Armstrong would roll up the sheet of foil and return it to Eagle. It would be brought back to Earth where the nature and energies of the sub-nuclear particles it had caught would be analyzed. The wind blew directly on the lunar surface, but it never reached the surface of the earth because it was deflected by the geomagnetic field.

Aldrin, looking out the window of the LM, remarked that the moonscape "looks beautiful from here."

"It has a stark beauty all of its own," Armstrong agreed. "It's like much of the high desert of the United States." He told Houston: "Be advised that a lot of the rock samples out here, the hard rock samples, have what appear to be vesicles (holes) in the surface. Also, I am looking at one now that appears to have some sort of phenocryst (a crystal embedded in fine-grained material)."

Aldrin then backed out of the Lunar Module hatch as Armstrong watched him from the ground. The main problem in getting out through the hatch, Armstrong had found, was clearing the bulky back pack, or PLSS (portable life support system).

"Okay," said Armstrong, "your PLSS is—looks like it is clearing okay. Now drop your PLSS down. There you go. You're clear. About an inch clearance on top of your PLSS."

"Now," said Aldrin, "I want to go back and partially close the hatch, making sure not to lock it on my way out."

"A good thought," commented Armstrong below.

"Okay, I'm on the top step and I can look down over the landing gear pads. It's a very simple matter to hop down from one step to the next."

"Yes," agreed Armstrong. "I found it to be very comfortable

and walking is also very comfortable. You've got three more steps and then a long one."

Holding on to the fourth rung up, with both hands, Aldrin eased himself down to the ground.

"A little more," Armstrong guided. "About another inch. There, you got it. That's a good step. About a three-footer."

"Beautiful, beautiful," said Aldrin, gazing around the landscape.

"Isn't that something?" said Armstrong. "Magnificent sight down here."

This dialogue, coming from two bulky figures who could be seen shambling anthropoidally around their monstrous looking spaceship, sounded more than faintly implausible. It sounded like the conversation of a couple of real estate men inspecting a subdivision. It was not the style of conversation for which several generations of science fiction had prepared the world. It was not even in the heroic tradition of exploration, where one planted the flag, waved a sword, spoke noble words for the ages. Armstrong, the test pilot, and Aldrin, a Doctor of Science from the Massachusetts Institute of Technology, shuffled about the landing site, inspecting their ship, and viewing the landscape, like two gentlemen out for a Sunday evening stroll. Their heartbeats were fairly rapid, but their voices were cool. It was significant that these two gentlemen found the primeval scene beautiful and magnificent, for that indicated that men—or these men—were ready to leave the cradle of the earth and investigate the immediate solar neighborhood.

Armstrong removed the television camera and a tripod from the equipment bay, and set up the camera about 100 feet away, so that its lens could pan the activities of both astronauts near the LM.

The one concession to the old tradition was the erection of a nylon American flag on an aluminum pole with a metal arm to hold the flag extended, inasmuch as there was no air on the moon to make it wave. There had been talk in the space agency about planting the flag of the United Nations on the moon, but that was not the will of Congress. Once the flag had been positioned, the astronauts saluted it and went about their tasks. A flag on the moon is a strange sight. It never moves.

While the place where Eagle had landed appeared fairly

level, the ship was not quite vertical. It yawed to the south some 13 degrees and tilted back about 4½ degrees. In this attitude (posture), however, it was as stable as on the launch pad at Cape Kennedy, for the landing gear had been designed to support the ship at greater angles than these. Aldrin noticed that one of the landing pod probes, designed to "feel" the surface ahead of touchdown, had been broken off and bent back.

While inspecting the landing gear, Aldrin noticed that the surface of the rocks was powdery—an effect that has been anticipated because of erosion by the solar wind. When Houston asked him to repeat that statement, however, Aldrin phrased it another way. He said the rocks looked "slippery."

"You got to be careful that you are leaning in the direction you want to go," said Aldrin. "Otherwise . . . in other words, you have to cross your foot over to stay underneath where your center of mass is. Neil, didn't I say we might see some purple rocks?"

"Find the purple rocks?" asked Armstrong.

"Yes. They are small, sparkly." The rest of Aldrin's remark was garbled, but it was understood at Mission Control that the purple rocks had been collected.

"I would take at first they are some kind of biotite," said Aldrin. "We'll leave that to the lunar analysis." Biotite is a brownish mica that occurs in either igneous or metamorphic rocks. Geologists at the manned spacecraft center suggested that if the rock was biotite it must have been formed under heat and pressure.

It was while Armstrong and Aldrin were setting up the flag that Mike Collins in the good ship, Columbia, appeared over the lunar horizon flying a patient orbit.

"Houston," called Collins. "Columbia in high gain. Over."

HOUSTON: Columbia. This is Houston reading you loud and clear. Over.

COLUMBIA: Yes. This is history. How's it going?

Collins had been on the other side of the moon during the time Armstrong and Aldrin had been out on the surface.

HOUSTON: The EVA (extravehicular activity) is progressing beautifully. I believe they are setting up the flag, now.

COLUMBIA: Great.

HOUSTON: I guess you're about the only person around that doesn't have TV coverage of the scene.

COLUMBIA: That's right. That's all right, I don't mind a bit. How is the quality of the TV?

HOUSTON: It's beautiful, Mike. Really is. They've got the flag up and you can see the Stars and Stripes on the lunar surface.

COLUMBIA: Beautiful. Just beautiful.

Meanwhile, down at Tranquility Base, Aldrin began to evaluate the problem of lunar surface locomotion. He had discovered one had to keep track of one's center of mass and place one's feet directly under it.

"It takes about two or three paces to make sure you've got your feet underneath you," he said. "A kangaroo hop does work, but it seems that your forward ability is not quite as good."

Armstrong counseled that the kangaroo hop did, indeed, work on the moon, but he preferred to put one foot after another in the usual way, and he demonstrated. Aldrin showed what the kangaroo hop looked like. It looked like Nijinsky in a space suit.

This discussion was interrupted by Houston where Bruce McCandless was calling.

"Neil and Buzz," he said, "the President of the United States is in his office now and would like to say a few words to you."

"That would be an honor," said Armstrong.

"Go ahead, Mr. President," said Houston.

MR. NIXON: Neil and Buzz, I am talking to you by telephone from the Oval Room at the White House. And this certainly has to be the most historic telephone call ever made. I just can't tell you how proud we all are of what you have done. For every American, this has to be the proudest day of our lives. And for people all over the world, I am sure they, too, join with Americans in recognizing what a feat this is. Because of what you have done, the heavens have become a part of man's world. And as you talk to us from the Sea of Tranquility it inspires us to double our efforts to bring peace and tranquility to Earth. For one priceless moment, in the whole history of man, all the people on this earth are truly one. One in their pride in what you have done. And one in our prayers that you will return safely to Earth.

ARMSTRONG: Thank you, Mr. President. It's a great honor and privilege for us to be here representing not only the United States but men of peace of all nations, and with interest and curiosity and a vision for the future. It's an honor for us to be able to participate here today.

MR. NIXON: And thank you very much and I look forward —all of us look forward to seeing you on the *Hornet* on Thursday.

ARMSTRONG: Thank you.

ALDRIN: And I look forward to that very much, sir.

Strangely enough, one expected to hear some applause from an audience, somewhere, to complete the illusion of an interplanetary Rotary Club luncheon program, but there was only silence then between the worlds.

There was no one on the moon but Buzz, Neil and Mike, who was orbiting in Columbia without a television set and hence pretty much out of the picture, and Luna 15, which the British expected to land at any time. The moon men went back to work.

Armstrong unlimbered his long-handled scoop to begin the collection of 30 pounds of bulk soil and rock samples, which would be sealed in a vacuum container for shipment to the lunar receiving laboratory at Houston. While he was gathering rocks and soil, Aldrin photographed Eagle from different angles. Upstairs, as Columbia passed over at an altitude of about 70 miles, Collins said he could see a small, white object on the southwest wall of a crater but doubted that was Eagle. It was probably a boulder reflecting sunlight.

The moon men began to realize after the first hour that they were working more slowly than expected. Houston advised them they were a half hour behind the work schedule. Now, time began to run out. Armstrong and Aldrin worked together to set up the seismometer and laser reflector experiments. They put the laser reflector, a mirror that would return a laser beam from Earth back to its point of origin, on a low ridge, and the seismometer at some distance away. Soon after the seismometer's solar panels were opened to the sun, to power its transmitter, seismic signals were received at Houston. But they were not moonquakes. The quake detector was so sensitive it was picking up the footfalls of the moon men.

Armstrong reported that boulders nearby looked like basalt with about 2 per cent white minerals—crystals. Then he amended his earlier statement about seeing vesicular rocks. The holes in the rocks, on close inspection, he said, looked like tiny meteoroid impacts, little impact craters made by missiles the size of BB shot.

There were only 10 minutes for Armstrong to collect 14 documented samples of soil and rock. The plan was to describe each sample by noting its appearance, consistency and the place where it was found. The commentary would be recorded in Houston. During this exercise, geologists at the manned spacecraft center could question the moon men or make suggestions through the capsule communicator.

But in spite of a 15 minute extension of the period outside, there was not enough time to complete this more meticulous field work. Aldrin barely had time to get core samples—dirt packed into the hole of a tube, like a length of pipe, which he pushed and hammered into the ground to a depth of eight to nine inches. When extracted from the tube, the core would present a cross section of lunar soil structure and composition along its length.

One hour and 50 minutes after Armstrong climbed out of the LM, Houston sounded the warning that only a few minutes remained in the work schedule.

Aldrin was still driving the core tube into the ground with a hammer.

"I hope you're watching how hard I have to hit this into the ground to the tune of about five inches, Houston," he said.

"Roger," said Houston.

"It almost looks wet," said Aldrin. At Houston, a geologist attached to the manned spacecraft center, interpreted this to mean finely grained.

Houston then began to hurry the crewmen to pick up the core samples and the solar wind foil and hustle back into the LM. Time and suit consumables—oxygen and water—were running out. There was no chance to document the core samples and the moon men hurriedly filled a second canister with 20 pounds of rocks and soil, also without documentation.

"Buzz, this is Houston," said the capsule communicator. "It's about time for you to start your extravehicular activity close out."

At that point, both astronauts had been living off their portable life support suits for two hours and 25 minutes. Armstrong kept busy picking up more rocks.

"I'm picking up several pieces of really vesicular rock out here now," he called.

There was no time to collect environmental samples by exposing open bottles to lunar vacuum and then sealing them in the hope that analysis back home would reveal some trace of gases near the surface.

There was only time left to gather up the core samples and a close-up camera that the astronauts had used intermittently to take pictures of rocks and soil formations.

"Roger, Neil and Buzz," urged the Capsule Communicator, Bruce McCandless, "Let's press on with getting the close-up camera magazine and the samples return container. We're running a little low on time. Head on up the ladder, Buzz."

The moon men returned to the LM cabin, unstringing the line conveyor. Armstrong had been on the moon's surface a few minutes more than two hours, and Aldrin, about 20 minutes less.

"Okay," said Aldrin. "The hatch is closed and latched."

The LM cabin was quickly repressurized. Armstrong and Aldrin then transferred their oxygen supply from the portable life support pack to the spacecraft. Still suited up, they depressurized the cabin, opened the hatch again and threw out some refuse: the two portable life support packs, a lithium hydroxide canister and arm rests. Then they closed the hatch and once more repressurized the LM cabin.

Columbia was again passing overhead and Houston briefed Collins on what was going on down on the surface.

"The crew of Tranquility Base is back inside their base and repressurized. Everything went beautifully."

"Hallelujah," said Collins as he sailed by 70 miles up.

"Roger, Tranquility," McCandless told the moon men. "We observed your equipment jettison on TV and the passive seismic experiment reported shocks when each PLSS (life support pack) hit the surface. Over."

"You can't get away with anything any more, can you?" Tranquility replied.

"No, indeed," said Houston.

It was 3:40 a.m. EDT, July 21, and the first men on the moon had completed the first "day's" work on the new frontier.

In the growing light of the lunar morning, the men of Tranquility Base curled up on the floor of the Lunar Module and went to sleep.

RETURN FROM THE MOON

Monday was a holiday in most of the United States but at Tranquility Base it was the most crucial day of the space age. Armstrong and Aldrin prepared to launch Eagle from the moon into an orbit from which they could maneuver to rejoin Columbia. Once more, the tension grew in Mission Control as the crew at Tranquility recited their checks in terse, flat tones. To the controllers, Eagle looked ready to fly. The fact that it was slightly canted as it stood on the lunar ground was of so little consequence that no allowance had to be made for it in the initial upward trajectory.

Shortly after 1 p.m. Monday, July 21, Program 12 was loaded in Eagle's primary guidance and navigation system. Time became compressed into a sequence of conscious moments. Mission Control was saying:

"Eagle, Houston, you are go at 3 minutes. Everything is looking good."

That had become a cliché. Nothing had looked bad on the flight of Apollo 11.

Eagle responded with "Roger," then there was a roaring in the interplanetary radio link as though Eagle's radio was picking up the thrusting of the ascent engine. It was 1:54 p.m. EDT. The Ascent Stage moved off the moon swiftly. It left the Descent Stage, a truncated four-legged, polyhedron standing in Tranquility, a monument in perpetuity to man's first visit to the moon. The monument was complete with a plaque of stainless steel, 9 by 7⅝ inches, bearing the two hemispheres of Earth and the words:

"Here men from the planet Earth first set foot upon the moon, July, 1969 A.D. They came in peace for all mankind." It was signed with the names of President Richard M. Nixon and of Armstrong, Aldrin and Collins.

The greatest show off the earth—21 hours and 37 minutes on the moon—was over. And now the controllers watched their screens as Eagle sped upward to 15,000, 22,000, 32,000 feet and slanted over toward the west.

"We're going right down U.S. 1," called Armstrong. Eagle was on course.

"You're going right down the track," came the exultant voice from Mission Control.

Eagle headed out on a nearly horizontal path passing Sabine, Ritter and lesser craters along the ground track. Seven minutes and 17 seconds after liftoff from the moon, Eagle was once more in orbit. It was moving with a velocity of 3775 miles an hour (5,537 feet per second) in an orbit of 10.4 by 53.7 miles. The little spacecraft rose up and around the moon to begin a series of maneuvers which brought it into line of sight with Columbia two hours later. The rendezvous and docking procedures were the same as those on Apollo 10, and on the 27th revolution of Columbia at 5:35 p.m. EDT the ships were docked. There was a brief moment of anxiety when Eagle's docking attitude slipped out of alignment, but it was quickly corrected.

"I accidentally got off in attitude and then the attitude hold system started firing," said Armstrong. It made Collins think someone was hitting Columbia with a sledge hammer. But the docking was completed otherwise without difficulty. Armstrong and Aldrin began moving their gear, their rock samples, the solar wind experiment, their film packs, their remaining cameras (the TV camera had been left at Tranquility Base) and other equipment to Columbia.

After the moon men had resumed their couches in the Apollo spacecraft, Eagle was undocked and Columbia eased away from it. There was not enough fuel left in Eagle's ascent engine tank to fire the vehicle off into an orbit around the sun, where Snoopy had been consigned in May. Eagle remained in orbit and in time, it was anticipated, the strange perturbations of a low orbit around the moon might bring it down to the surface.

"It's been a mighty fine day," said Houston.

"Boy, you're not kidding," responded Columbia, as it disappeared around the moon on the 27th revolution.

Five hundred miles to the east of Tranquility Base the Soviet entry in the lunar sweepstakes Luna 15, had crashed into the region of the Sea of Crises at a velocity estimated by Jodrell Bank at 300 miles an hour. It had attempted to land, but whether its mission was to have scooped a soil sample for return to earth remained a matter of conjecture. Moscow simply announced that the station had reached the surface and its work had ended.

So ended the great space race of the 20th century, with the nylon flag, 3 by 5 feet, of the winner standing stiffly outstretched near a television camera in the Sea of Tranquility.

The crew rested and prepared for the next moment of truth in the text-book flight, trans-earth injection after midnight. Once more, Mission Control examined the Apollo systems and once more, for the third time in Project Apollo, advised a crew:

"You are go for TEI (Trans-Earth Injection)."

At the beginning of the 31st revolution around the moon, as Apollo 11 rounded farside, the spacecraft's service propulsion engine flared for 2 minutes, 29 seconds at 12:55 a.m. Tuesday. It was a perfect burn, as all propulsion maneuvers had been on this voyage. When Apollo 11 came into radio contact, Charles Duke, the Capsule Communicator, called:

"Hello, Apollo 11. Houston. How did it go? Over."

Fringed with static, Armstrong's voice came across the void with these words: "Time to open up the LRL doors, Charlie."

The LRL is the Lunar Receiving Laboratory at the Manned Spacecraft Center where the crew would be held in quarantine until 21 days had elapsed from their lunar departure.

"Roger," said Duke. "We got you coming home. The LRL is well stocked."

As Apollo 11 sped for splashdown shortly after noon Thursday, the project directorate assembled in the news center auditorium for the victory press conference. There was still one more hurdle to go—reentry into Earth's atmosphere—but it had been accomplished twice before and now it appeared the return would be routine.

Perhaps the man on the platform most deeply moved by the events of lunar landing and the departure from the moon was Robert R. Gilruth, the Manned Spacecraft Center director, who came into NASA at the beginning. "I'm proud to realize," he said, "that we've landed on the moon in this decade and that they are on their way home." Gilruth revealed that less than one half of the consumables—oxygen, electric power and water—had been used up while the moon men were outside the LM. The disclosure indicated the extreme caution with which Mission Control was managing the extravehicular activity.

For Wernher von Braun it was the realization, he said, of an old dream.

"I think the ability for man to walk and actually live on other worlds has virtually assured mankind immortality," he said. To which George Mueller, NASA's normally reserved Director of Manned Space Flight, added enthusiastically: "It seems quite clear that the planets of the solar system are well within our ability to explore both manned and unmanned at this time." To go to the stars, man would have to invent a means of using a higher energy source—nuclear fusion, the reaction which powers the sun. The principle is known, said Mueller, adding: "It requires only invention to make it available."

At 1:38 p.m., EDT Tuesday, July 22, Apollo 11 entered the Predominant sphere of gravitational influence of the earth and began accelerating down the 200,000 miles slope toward the Pacific Ocean.

During the journey to Earth, the crew was advised of elaborate preparations to greet them in New York, Washington and Chicago after their 21-day quarantine ended in August. President Nixon, meanwhile, flew out to the aircraft carrier, U.S.S. *Hornet,* awaiting the moonship in the central Pacific Ocean recovery zone.

Mission Control guided the earthward trajectory of Apollo 11 toward an impact zone 943 miles southwest of Honolulu. As Apollo 11 approached the earth Thursday morning, the crew donned flight suits and helmets and prepared for the shock of entering the atmosphere.

Houston called Apollo 11: "Apollo 11, you are looking mighty fine here. You're clear for landing."

Apollo 11: "Yeah, we appreciate that."

Out of the window, the home planet once more filled the heavens as Apollo 11 approached it at 24,670 miles an hour. The Command Module was separated from the service module and swung around for re-entry attitude. It smashed into the atmosphere at 400,000 feet at 12:35 EDT. The radio blackout following entry seemed unbearably long. When the crew was back in communication again, Apollo 11 was descending through the pre-dawn dimness of the Pacific on its main parachutes.

The heat-scarred ship drifted slowly down to the warm sea from which life had come and evolved to man. In this age, man had kept his appointment on the moon to enter another stage in his evolution and swim in the cosmic sea.

Epilogue:

The Moon and Beyond

From the viewpoint that man is destined to expand beyond the Earth, as he has over the Earth, Project Apollo is one of the critical developments in human evolution. It lifted man out of the terrestrial nursery and set him on the road to the stars. It enlarged his habitat by adding to it another world with untapped resources, a world from which he could gain a better view of the universe, its origin and evolution, than ever before. It opened to man the entire solar system. In a single decade, in which he increased the weight of payloads he could loft into orbit from 30.8 pounds (Explorer I) to 280,000 pounds (Saturn 5–Apollo), he had broken the gravitational bonds of his home planet and pushed his frontier some 238,000 miles beyond.

Although the twentieth century cannot see any more clearly than could the fifteenth century where voyages to their respective New Worlds would lead, I believe that we can identify the lunar landing as an event in the expansion of our species. Apollo was a spectacular advance in transportation technology, which has been a function of biological and social development of *Homo sapiens* since the invention of the wheel. The lunar landing was motivated by the traditional pressures which have caused mankind to spread from the place of its origin as a species to the ends of the earth, from deserts to poles. These were the pressures of intraspecific competition and the compulsion to seek and explore, a built-in behavioral characteristic.

Yet, at the culmination of Project Apollo, the mood of the nation was different than at the beginning. In the eight years since John F. Kennedy had called for the manned lunar landing, a time of troubles had overtaken America. The buoyancy, the confidence that the United States could meet any challenge, the high resolve of 1961—all these had been seriously eroded by the interminable war in Vietnam and a revolutionary ferment to which the war contributed at home.

While the landing on the moon stood out in brilliant contrast to the unsolved social problems in the land, the civilian space program did not escape the hostility of influential intellectuals and militant students toward the military-industrial Establishment. The aerospace component of the Establishment had produced most of the Apollo hardware, and this led to a widespread view in academic circles that Apollo was in reality a quasi-military venture and hence a function of the military-industrial complex. The fact that Apollo received the lion's share of the declining federal budget for nondefense research and development was a focus of criticism for many scientists who believed it was robbing them of federal research support.

Finally, the threat of Russian competition, which had helped launch the program in the beginning, faded rapidly as Apollo approached its goal of the lunar landing. This was evidenced by the indifference of Congress and the public to the Soyuz rendezvous, docking, and crew transfer achievement in January 1969—a "stunt" which the American program had not yet matched.

These factors seemed to account for the lack of any significant opposition to the progressive cutting of the space budget to meet the costs of the war. From an expenditure level of $5.3 billion in the 1967 fiscal year, the NASA budget was reduced to a recommended $3.7 billion in fiscal 1970, at the time of the lunar landing.

In preparing this budget in 1968, the Johnson administration had failed to provide funds for ongoing exploration of the moon after the first three Apollo landings. It also had halted the production of Saturn 5 rockets after the 15 vehicles procured for the Apollo and Apollo Applications were delivered. This meant that after 1972, when the last of the moon rockets was launched, there would be no more. For it takes 42 to 45 months to construct and test the Saturn 5.

So it was that on the day of man's greatest achievement, the conquest of the moon, the United States was busy closing out Project Apollo and probably its lead in manned space flight. More than a quarter of the 280,000 space workers employed in Apollo at the peak of the program, in 1966-1967, had been or were being laid off. America had begun the retreat from the moon before it even arrived. Obsessed by social, racial, and economic stresses, the country had turned its attention inward, like the chronically ill person who becomes progressively indifferent to all but his own complaints.

BEYOND THE MOON

At the recommendation of his science adviser, Lee A. DuBridge, and others, President Nixon had altered the 1970 budget without increasing it. The changes provided $46 million to continue manufacturing Saturn 5 rockets, so that vehicles would become available in 1973 at the rate of three a year, and allocated $79 million to equip the last six Apollo-LM vehicles with lunar-exploration instruments, not provided earlier. This alteration was designed to protect the option to pursue manned space flight after 1972, when the presently funded program ended. It would also make more effective scientific use of the $24 billion Apollo transportation system in 1970-1972.

Starting with Apollo 11, ten landing attempts were scheduled from mid-1969 to 1972. The second, Apollo 12, would aim for the maria on the western side of the moon; the third, for a highland formation, like the Fra Mauro crater in the western highland region; the fourth, for the eastern cratered highlands near Censorinus; the fifth, for the region of the Littrow craters east of the Mare Serenitatis; the sixth for the great crater, Tycho, where Surveyor 7 stood; the seventh, for the volcanic domes of the Marius Hills; the eighth for Schroter's Valley, where luminous gas emissions had been reported by Earth observers for many years; the ninth, for the Hyginus Rille, one of the most prominent faultlike structures on the moon; and the tenth, for Copernicus, where it was hoped that material blown out of the interior might be found.

Apollo 20 was the end of the line. NASA officials talked of extended lunar exploration in the style of the famous Antarctic traverses. Operating from a lunar ground base, astronaut-scientists might one day explore far afield, in tracked vehicles with sealed cabins or in rocket helicopters. But the implementation of this style of lunar exploration was a costly enterprise, too costly, perhaps, if funding stayed at the fiscal 1970 level, unless logistic costs could be reduced dramatically.

However, one such advanced lunar mission was described to the Senate space committee by John E. Naugle, associate NASA administrator for Space Science and Applications.[1] An automated surface vehicle might be landed in the center of the Imbrium Basin and guided by radio control from Earth to the Appenine Mountain front. Near a point called Hadley Rille, scientists from a Lunar Module would meet the surface car, transfer their equipment to it, and drive it on a selenologic traverse, perhaps for hundreds of miles. The investigation might be designed to test a theory that the Imbrium Basin was formed by the impact of a smaller second moon of the earth.

NASA also contemplated the development of a large space station of 10,000 cubic feet, to be assembled in orbit between the earth and moon over a period of several years. The plan was sufficiently vague to allow consideration of such science-fiction refinements as artificial gravity and a swimming pool. Stay time would be reckoned in months and transportation would be provided by reusable shuttle rockets. Here was another project as costly as Apollo and one which might not be possible until the nation had extricated itself from the Vietnam war.

At the time of the first landing, a less grandiose space station had been developed—the Saturn I Workshop. It consisted of the empty hydrogen tank of the S4B, the third stage of the Saturn 5, which McDonnell-Douglas Corporation, St. Louis, was outfitting as a roomy and well-equipped laboratory accommodating a crew of three or four for a period up to 56 days. Appendages of the Workshop included a multiple-docking adapter, which would enable a number of different vehicles to dock with the space station, and an airlock.

The Workshop would serve a number of functions. One of them would be as an operating base for the manned solar observatory or Apollo Telescope Mount, which was to be housed in a modified Lunar Module Ascent Stage and flown in Earth orbit. The Workshop would be orbited about 1971 by a Saturn 1B and a second rocket would then ferry the crew up to it.

For exploration beyond the earth-moon system, NASA was preparing to follow up the 1969 Mariner-Mars fly-by missions with a project to put two Mariner-class spacecraft in orbit around Mars in 1971. The orbiters would carry out photographic and radiometeric reconnaissance at a 1200-mile altitude. In 1973, the Viking Mission to the Red Planet would consist of launching two payloads, each containing a Mariner-class orbiter and a Surveyor-style soft-landing machine, using the Air Force Titan 3-C with a Centaur on top. The soft landers would be designed to settle down on the Martian soil and make a limited chemical analysis as well as to take close-up pictures. Viking was a scaled-down version of Project Voyager, which NASA was forced to abandon when cutbacks in space expenditures were ordered by the White House in the 1967 fiscal year. Voyager was to have carried the "Gulliver," an instrument designed to report the presence of living organisms. Such a device was being considered for the Viking landers.

NASA planned to send automated spacecraft from one end of the solar system to the other in the 1970s. In 1972, Pioneer F would travel to the vicinity of Jupiter, via Mars and the Asteroid Belt, followed by Pioneer G in 1973. An instrumented spacecraft would be launched in the other direction in 1973, on a dual-planet fly-by of Venus and Mercury, using the gravitational field of Venus to sling the probe sunward. Toward the end of the 1970s, a "Grand Tour Mission" would launch a small space vehicle on a path taking it near Jupiter, Saturn, Uranus, and Neptune, when the planets were in alignment between 1976 and 1978. Early studies of the Grand Tour Mission, which would provide the first close-up photographs, magnetic-field measurements, and surface-temperature readings of the major planets, had been made at the Illinois Institute of Technology under a NASA contract in the mid-1960s. These had suggested using Jupiter's gravitational field as a slingshot to accelerate a spacecraft passing near the planet to the outermost reaches of the solar system. By 1969, the use of gravitational energy to move space vehicles along was no longer a new idea. The moon's field had been exploited to "kick" the Saturn S4B stages out of the earth-moon system and into solar orbit during Apollo lunar flights.

This blueprint for space exploration in the 1970s was based largely on the assumption that NASA's annual budget would not fall below a level of $3.8 or $3.9 billion. The "hard fact" was, as Administrator Tom Paine had advised the House space committee on March 4, 1969, that the United States could not make effective advances in space and aeronautics with a budget under that level. NASA's Science and Technology Advisory Committee, under the chairmanship of Charles H. Townes of the University of California, had suggested at the end of 1968 that one-half to one per cent of the Gross National Product appeared to be a reasonable allocation for the civilian space program.[2] With a Gross National Product of $900 billion, this recommendation suggests an outlay for space of from $4.5 to $9 billion a year, compared with the $3.7 billion which the Nixon administration has recommended for 1970.

Except for the super space station, none of NASA's plans for the next decade requires the quantum jump in technology which had to be made in Apollo, and even this project would not be comparable so far as rocket development is concerned. If the United States were to mount a new manned-flight effort comparable in magnitude and order of difficulty to that of Apollo in 1961, it would be nothing less than a manned expedition to Mars. But even a Mars mission would not stretch the state of the art as far as Apollo did. Yet, in the current context of war and domestic unrest, a Mars Project would seem irresponsible. Perhaps this is one measure of the change in national outlook in the 1960s.

What will happen if manned space flight is not resumed after 1972? Will the Soviet Union take up where the United States left off?

Three quarters of a century ago, Ziolkovsky described the earth as the cradle of mankind but remarked that one cannot live forever in the cradle. Keeping Apollo's appointment on the moon has provided the means of leaving it.

There may be stops and restarts in going forward, but there is no going back, and there is no foreseeable end to such a beginning.

Reference Notes

CHAPTER 1: *The Phoenix*

1. Y. I. Perelman, *Tsiolkovsky*. Moscow, 1932.
2. *History of Rocket Technology*, edited by Eugene Emme. Detroit: Wayne State University Press, 1964.
3. Robert H. Goddard, "A Method of Reaching Extreme Altitudes," a paper published by the Smithsonian Institution, Washington, D.C., 1919.
4. Walter R. Dornberger, *V-2*. New York: The Viking Press, 1958.
5. Dieter Huzel, *Peenemünde to Canaveral*. New York: Prentice Hall, 1955.
6. Terence H. O'Brien, *History of the Second World War*. London: Her Majesty's Stationery Office and Longmans, Green & Co., 1955.
7. British Government Registrar, General Returns.
8. Dornberger, *op. cit.*
9. Hearings, Senate Preparedness Subcommittee, Nov. 25, 1957.
10. Huzel, *op. cit.*
11. *Historical Origins, Marshall Space Flight Center*, NASA document.
12. *Ibid.*

CHAPTER 2: *The World-Circling Spaceship*

1. R. Cargill Hall, "Early U.S. Satellite Proposals," in Emme, *op. cit.*
2. Emme, *op. cit.*
3. NASA document, *op. cit.*
4. Report of the Special Committee for the International Geophysical Year (COMSAGI), 1954.
5. The newspaper *Vechernaya Moskva (Evening Moscow)*, Apr. 15, 1955.
6. Tass News Agency, Aug. 2, 1955.
7. Personal interview, Feb. 22, 1968.
8. Hearings, House of Representatives Select Committee on As-

tronautics and Space Exploration on HR 11881, 85th Congress, 2nd session.

9. Hearings, Senate Preparedness Subcommittee, Nov. 25, 1957.
10. *Historical Origins, Marshall Space Flight Center,* Nasa document.
11. John P. Hagen, "Project Vanguard," a paper presented at the American Association for the Advancement of Science annual meeting at Philadelphia, Dec. 27, 1962.
12. Personal interview, June 23, 1967.
13. John B. Medaris, *Countdown for Decision.* New York: G. P. Putnam's Sons, 1960.
14. Associated Press dispatch, Oct. 6, 1957.
15. *Aviation Week,* Oct. 21, 1957.
16. *Aviation Week,* Oct. 28, 1957.
17. *Aviation Week,* Oct. 21, 1957.
18. *Aviation Week,* Oct. 14, 1957.
19. *Ibid.*
20. *Commonweal,* Nov. 8, 1957.
21. John P. Hagen, "The Viking and the Vanguard," in Emme, *op. cit.*

CHAPTER 3: *A Wind Between the Worlds*

1. *Scientific American,* March 1959.
2. *Science* magazine, Feb. 27, 1959.
3. Proceedings of the NASA-University Conference on Space Exploration, Chicago, Nov. 1–3, 1962.
4. *Ibid.*
5. NASA Space Science press briefing, Washington, D.C. Jan. 5, 1967.
6. *Transactions,* American Geophysical Union, 38, 176, 1957.
7. Walter Sullivan, *Assault on the Unknown.* New York: McGraw-Hill, 1961.
8. Journal of Geophysical Research, Vol. 65, No. 9, September 1960.
9. NASA Information Release No. 61-280, Dec. 15, 1961.
10. *Philosophy* magazine, 1919.
11. *Zeitschrift für Astrophysics,* Vol. 29, 1951, p. 116.
12. *Journal of Astrophysics,* Vol. 128, 1958.
13. *Pravda,* Sept. 9, 1961.
14. *Journal of Geophysical Research,* Vol. 71, No. 13, July 1, 1966.
15. *Science,* July 11, 1963.
16. Norman F. Ness, paper delivered to American Association for the Advancement of Science annual meeting at Berkeley, Calif., Jan. 28, 1966.
17. *Journal of Geophysical Research,* Vol. 70, No. 13, July 1, 1965.

CHAPTER 4: *Project Mercury*

1. W. F. Hilton, *Manned Satellites.* Evanston, Ill.: Harper & Row, 1965.

2. Maxime A. Faget and Aleck C. Bond, NASA document, "Technologies of Manned Space Systems."
3. NASA headquarters memo, Feb. 19, 1959.
4. John H. Glenn, Jr., report to COSPAR (Committee on Space Research), 1962.
5. *Astronautics*, February, 1960.
6. *Astronautics*, March, 1960.

CHAPTER 6: *Rendezvous*

1. Police Commissioner Michael J. Murphy to Mayor Robert F. Wagner, according to the *New York Times*, March 2, 1962.
2. Soviet report to COSPAR, 1962.
3. Robert Rosholt, *Administrative History of NASA*, U.S. Government Printing Office, Washington, D.C., 1966.
4. U.S. Information Agency Office of Research and Analysis Report, Oct. 10, 1960.
5. Rosholt, *op. cit.*
6. *Ibid.*
7. Hearings, U.S. House of Representatives Committee on Science and Astronautics on 1962 NASA authorization, Part I, page 362.
8. Hearings, Senate Appropriations Subcommittee, Aug. 10, 1962.
9. Eberhard Rees, speech before the American Rocket Society, Honolulu, Sept. 25, 1961.
10. Special Hearing, House of Representatives Committee on Science and Astronautics, 87th Congress, 2nd Session, Document No. 6.
11. Testimony by James E. Webb before Senate Committee on Aeronautical and Space Sciences, May 9, 1967.
12. Hearings, House of Representatives Committee on Science and Astronautics, Subcommittee on Manned Space Flight, Part 2 (B), 89th Congress, 2nd session.
13. Hearings, House of Representatives Committee on Science and Astronautics, 88th Congress, on 1964 NASA authorization, page 560.
14. *New York Times*, Sept. 12, 1962.
15. *Time*, Sept. 21, 1962.
16. Hearings, House of Representatives Committee on Science and Astronautics, Subcommittee on Manned Space Flight, 88th Congress, 1st session.
17. An English-language version of *Das Marsprojekt* was published by the University of Illinois Press in 1953 as *The Mars Project*.
18. Philip Abelson, speech at the University of Maryland, Apr. 29, 1963.
19. NASA official release, undated.
20. Hearings, House of Representatives Committee on Science and Astronautics Subcommittee on Manned Space Flight,

on 1963 NASA Authorization, Part 2, 88th Congress, 1st session.
21. *Ibid.*
22. *Ibid.*
23. Houston *Chronicle*, Oct. 7, 1962.

CHAPTER 7: *Men in Orbit*

1. *Life*, May 18, 1962.
2. *Results of the Second U.S. Manned Orbital Space Flight*, NASA document.
3. *Ibid.*
4. CBS interview by Walter Cronkite, Sept. 13, 1962.
5. Pilot's Flight Report, *Results of the Fourth U.S. Manned Orbital Space Flight*, NASA document.
6. Gordon Cooper's report at his press conference, May 19, 1963.
7. Los Angeles *Times*, July 2, 1963.
8. NASA Project Mercury Summary, October 1963.

CHAPTER 8: *Mare Nubium*

1. Galileo Galilei, *Nuncius Sidereus*, 1610.
2. Ralph Baldwin, *The Measure of the Moon*. Chicago: University of Chicago Press, 1963.
3. L. V. Berkner and Hugh Odishaw, *Science in Space*. New York: McGraw-Hill, 1961.
4. James E. Conel, Robert L. Kovach, Robert C. Speed, and Alden A. Loomis, Jet Propulsion Laboratory, "Geological Exploration of the Planets," paper presented at the NASA-University Conference on Space Exploration, Chicago, Nov. 1–3, 1962.
5. NASA Release 64-16A.
6. H. L. Nieburg, *In the Name of Science*, Chicago: Quadrangle, 1966.
7. Report of the House of Representatives Committee on Science and Astronautics, Subcommittee on NASA Oversight, June 16, 1964.
8. *Science*, Sept. 4, 1964.
9. Harold C. Urey, Lecture at the University of Wisconsin, Oct. 14, 1964.
10. Personal interview, Oct. 14, 1964.
11. Personal interview, Feb. 23, 1965.
12. Personal interview, Apr. 3, 1964.
13. Walter Sullivan in the *New York Times*, March 1, 1965.
14. Ranger 9 Post-Impact Press Conference, Jet Propulsion Laboratory, March 24, 1965.
15. *Ibid.*
16. *Ibid.*
17. Report, House of Representatives Committee on Science and Astronautics, Subcommittee on NASA Oversight, Oct. 8, 1965. Committee Print Serial J, 89th Congress, 1st session.

18. *Time*, Feb. 18, 1966.
19. *Science*, July 22, 1966.
20. *Science*, March 25, 1966.
21. NASA Final Report on Surveyor 1, Nov. 21, 1966.
22. Surveyor 1 Post-Impact Press Conference, Jet Propulsion Laboratory, June 2, 1966.
23. Surveyor 5 News Conference, Washington, D.C., Sept. 29, 1967.

CHAPTER 9: *No Hiding Place*

1. Hearings, House of Representatives Committee on Science and Astronautics, Subcommittee on Space Sciences, 87th Congress, 2nd session (No. 4).
2. *Ibid.*
3. Hearings, House of Representatives Committee on Science and Astronautics on 1962 NASA Authorization, 87th Congress, 1st session, Part 1.
4. Jet Propulsion Laboratory staff, *Mariner Mission to Venus*, New York: McGraw-Hill, 1963.
5. *Ibid.*
6. *Ibid.*
7. Report to COSPAR meeting, Vienna, 1966.
8. *Science*, Dec. 29, 1967, report of Conway W. Snyder, Jet Propulsion Laboratory.
9. *Science*, Dec. 29, 1967, report of James A. Van Allen et al.
10. Snyder, *op. cit.*
11. A. L. Oparin, *Origin of Life*. New York: Macmillan, 2nd ed., 1958.
12. W. M. Sinton, paper delivered at American Astronautical Society meeting in Denver, June 8, 1963.

CHAPTER 10: *The Tall Towers*

1. American Telephone & Telegraph Co., Background Summary, Project Telstar.
2. John R. Pierce, Lecture at Princeton, N.J., Oct. 14, 1954.
3. James E. Dingman, executive vice-president, AT&T, in letter to Sen. Warren Magnuson, Apr. 19, 1962.
4. *Ibid.*
5. Proceedings, 87th Congress, 2nd session, Vol. 108, No. 146, Aug. 17, 1962.
6. E. Jared Reid, "How Can We Repair an Orbiting Satellite?" in *Satellite Communications Physics*, Bell Telephone Laboratories, June 5, 1963.
7. Annual Report, Goddard Space Flight Center, 1965.
8. *Ibid.*
9. James B. Fisk, address to Iron & Steel Institute, New York, May 24, 1961.
10. *National Academy of Sciences Space Application Study, 1967. Interim Report*, Vol. I.

CHAPTER 11: *Gemini*

1. Gemini 4 Postflight Press Conference, June 11, 1965.
2. *Ibid.*
3. Report of the experiment written by W. R. Adey, R. T. Kado and D. O. Walter, Space Biology Laboratory, Brain Research Institute, University of California. The experiment was managed by Dr. P. M. Kellaway and Dr. R. Maulsby, Methodist Hospital, Houston.

CHAPTER 12: *Fire in the Cockpit*

1. *New Republic*, Feb. 4, 1967.
2. Hearings, House of Representatives, Committee on Science and Astronautics, Subcommittee on Manned Space Flight, 89th Congress, 2nd session, Part 1, on 1967 NASA Authorization.
3. Apollo 204 Review Board Report, Apr. 15, 1967.
4. *Ibid.*

CHAPTER 13: *Saturn 501: The Big Shot*

1. News conference, NASA Headquarters, Washington, D.C., Oct. 5, 1967.
2. Release, Office of Public Information, Columbia University, New York.
3. Statement before the House Committee on Science and Astronautics, February 6, 1968.

CHAPTER 14: *Genesis Revisited*

1. At a NASA news conference in Washington, July 19, 1968.
2. "The Year in Space," by Col. John H. Glenn, Jr. (USMC, Ret.), World Book Year Book, 1969, Field Enterprises Educational Corp.
3. Genesis 1:1–10.

CHAPTER 15: *Landfall*

1. Report to the U. S. House of Representatives' Committee on Science and Astronautics by George Mueller, Associate Administrator for Manned Space Flight, NASA, March, 1969.
2. Harold C. Urey, "The Space Program and Problems of the Origin of the Moon," *Bulletin of the Atomic Scientists*, April, 1969.
3. Paul M. Muller, paper delivered to the American Physical Society symposium on the moon, Apr. 28, 1969.

EPILOGUE: *The Moon and Beyond*

1. Hearings of May 1, 1969.
2. Vol. 1, Report of the Science and Technology Advisory Committee for Manned Space Flight to the Administrator, NASA.

Index

Index includes material through Apollo 10.

Aaron, John, 458
Abelson, Philip, 165, 192
Abercrombie, Jack, 190
Aberdeen Proving Ground, 21
Abort Guidance System (AGS), 495, 498
Abort Sensing Implementation System (ASIS), 112, 113, 114
Ad Hoc Committee on Space, 160
Ad Hoc Committee on Special Capabilities, 40, 41, 42, 44
Adam Project, 107
Adams, Sherman, 44
Adey, W. Ross, 227
Advanced Research Projects Agency (ARPA), 161, 186, 264, 266, 267, 303
Advent communications satellite, 264, 266
Aerobee rocket, 27, 34, 41, 49, 51, 104
Aerojet Engineering Corporation, 28
Aeronca Manufacturing Company, 385
Agena rocket, 105, 121, 238, 239, 249, 330; and Gemini 6, 347, 348, 349, 350; and rendezvous and docking with Gemini spacecraft, 355-357, 362, 363, 364, 365, 366, 369
Air Force, U.S., 28, 29, 33, 36, 40, 44, 48, 59, 77, 105, 106, 170, 209, 222, 267
Air Force Ballistic Missile Division, 112
Air Force Missile Test Center, 60
Air Force School of Aviation Medicine, 106
Air Force Systems Command, 33, 170, 350
Airglow, 219
Albermarle, U.S.S., 81
Aldrin, Edwin E., Jr., 371; spacewalk by, 369-370, 371
Allen, H. Julian, 104
Allen, J. Denton, 292
Alpha radiation, defined, 261
Alphonsus crater, 255
Alsop, Joseph, quoted, 58
Aluminum, on moon, 262
American Association for the Advancement of Science, 165, 166, 285
American Astronautical Society, 333
American Institute of Aeronautics and Astronautics, 98
American Physical Society, 70
American Rocket Society, 6, 9, 106
American Telephone and Telegraph Company, 301, 302, 304, 306, 307, 308, 309, 310, 318
Ames Aeronautical Laboratory, 104
Anders, William A., 440, 448, 449, 452, 453, 456, 457, 458, 461, 473. *See also* Apollo 8 (Saturn 503)
Anderson, Clinton P., 404
Anderson, H. R., 275
Andover Earth Station, for Telstar, 306, 307, 308, 309, 310, 311
Andrews Air Force Base, 155
Apex Predictor, 63, 65, 66
Apollo 012, 373-375, 377, 387, 388, 389, 395, 396, 402, 405, 427; certified by NASA, 388-389; destroyed by fire, 391-395, 397-399, 402; NASA's investigation of failure of, 395-401; Senate space committee's report on, 406
Apollo 017, 389, 405, 406, 415
Apollo 4 (Saturn 501), 408-425 *passim*, 427; countdown for, 412-419; and liftoff, 419; in orbit, 420; and Pogo Effect, 424; record of air waves produced by, 420; recovery of, 421-423

Apollo 5 (Saturn 204), 426, 427-430
Apollo 6 (Saturn 502), 426, 430-433, 434, 450; in orbit, 431-432; and Pogo Effect, 431
Apollo 7, 435, 438, 440-447; head colds of crew, 442, 443; in orbit, 441-447; re-entry of, 447; rendezvous with S-4B, 441, 442; television transmission from, 442, 443, 444
Apollo 8 (Saturn 503), 435, 437, 438, 439, 447, 448, 449-463, 464, 465, 472, 473, 478; in lunar orbit, 456-461; lunar orbit insertion, 456, 459; splashdown, 463; television transmission from, 452, 453, 454, 460; trans-Earth injection, 416; translunar injection, 450-451
Apollo 9 (LM-3, "Gumdrop" and "Spider"), 438, 464, 465, 466, 467-472, 481, 493; separation and docking of lunar module, 470-471; spacewalk, 468, 469, 470, 471
Apollo 10 (LM-4, "Charlie Brown and Snoopy"), 438, 464, 465, 466, 472, 473-499; docking of lunar module, 499; en route to moon, 474-487; gyrations of lunar module, 498, 499; orbit of lunar module, 494-499; separation of lunar module, 494; in lunar orbit, 487-499; lunar orbit insertion, 482, 487; television transmissions from, 475, 476, 477, 480; translunar injection, 474
Apollo 11, 438, 464, 465
Apollo 204 Review Board, 396, 397, 401, 402, 403, 411; report of, 399-401, 402
Apollo Project, 160-182 *passim*, 191, 238, 243, 253, 257, 258, 263, 265, 328, 348, 376, 377, 380, 381, 396, 405, 406, 424, 425, 426, 427, 432, 436, 438, 439, 475, 487, 498; cost of, 165, 181, 186, 379, 413 and *n.*, 464; North American Aviation's contract for, 378, 379, 385; and Phillips report, 386-388; revived through success of Apollo 4, 420, 423, 424; and Vietnam war, 413, 425
Apollo Source Evaluation Board, 180
Applications satellites, 163, 326; *see also* Communications satellites; Meteorological satellites
"Argus" test series, 80
Ariel satellite, 81
Aristarchus crater, 236, 491
Armstrong, Neil A., 355, 356, 357, 358
Army Ballistic Missile Agency (ABMA), 45, 46, 47, 48, 50, 57, 77
Army Corps of Engineers, 120, 187
Army Scientific Advisory Panel, 51
Ascent Propulsion System, 428, 429
Astronaut Maneuvering Unit (AMU), 361
Atlantic Missile Range, 46, 50, 59, 101, 113, 114, 119, 238, 273, 337 and *n.*, 375
Atlas-Agena rocket, 105, 121, 238, 247, 266, 267, 268, 272, 288
Atlas-Centaur rocket, 259, 267, 268
Atlas ICBM, 46, 86, 105, 264, 347, 360
Atomic Energy Commission, 106, 246, 261, 312
Atwood, John L., 383, 384, 385, 386, 401, 405; quoted, 401
Adapter (ATDA), 360
Aurora 7, 198, 211; and ionization blackout, 209, 210; in or-
Augmented Target Docking
bit, 201-206; out of fuel, 206-208; retrofire and re-entry of, 205-210; *see also* Carpenter, Malcolm Scott; Mercury-Atlas 7; Mercury Project
Auroral theory, 82; Argus test in, 79, 80, 81
Automatic stabilization and control system (ASCS), 111, 143, 144, 146, 150, 151, 204, 206, 207, 224, 225
AVCO Corporation, 377, 385
Aviation Medical Acceleration Laboratory, 106
Axtel, Herbert, 20

Babbitt, Donald O., 392, 393, 394
Bacteriophage T-1, space conditions survived by, 364
Baker, Robert G. (Bobby), 379-385 *passim*
Baldwin, Ralph, 254
Banded horizon, observed by Glenn, 139, 140
Barostat (atmospheric-pressure switch), 225
Basalt, in soil of moon, 260, 262
Bassett II, Charles A., 359, 458
Becker, Karl, 7

Bell Telephone Laboratories, 171, 302, 303, 304, 305, 306, 312, 318
Bennington, U.S.S., 422, 423
Benson, Charles, 322
Bergen, William B., 405, 487
Berkner, Lloyd V., 52, 171
Berlin Polytechnic Institute, 7, 8
Berry, Charles, 216, 223, 345, 363, 370, 395, 403, 442, 443, 449, 452
Berry, Ken, 310
"Beyond the Planet Earth" (Ziolkovsky), 5
Biermann, Ludwig F., 87
Black, Fred B., 379-385 *passim*
Blagonravov, A. A., 55, 156
Boeing Aircraft Company, 180, 191, 247
Bonestell, Chesley, 237
Booster engine cutoff (BECO), 134
Borman, Frank, 350, 351, 352, 353, 354, 355, 396, 403, 440, 448 450, 451, 452, 453, 454, 455, 458, 459, 460, 461, 462, 463, 473; quoted, 403
Brain-wave patterns, and weightlessness, 351
Bridge, H. S., 88
British Interplanetary Society, 6
Brooke, Edward W., 406
Brooks, Overton, 163, 173
Brown, Harold, 182
Brown & Root, Inc., 193
Brucker, Wilbur M., 53, 55, 57, 66, 67
Brundage, Percival, 54
Buchheim, Robert W., 40
"Bug," *see* Lunar Excursion Module
Bumper-Wac, 34, 35, 37
Bureau of Aeronautics, Navy's, 27, 33
Bush, Vannevar, 33
Butler, Chris, 60
Buzzbomb, 12, 13, 14, 15
Bykovsky, Valery F., 341, 345

Cabin atmosphere, 108; *see also* Oxygen
California Institute of Technology, 27, 34, 165, 246, 275, 350
Cann, W. H., 384
Canopus, 257, 259, 289, 290, 291
Canopus sensor, in Mariner 4, 289, 290, 291
Cape Canaveral, 25, 35, 36, 45, 46, 49, 50, 51, 57, 59, 60, 88, 90, 94, 96, 100, 101, 113, 114, 115, 132, 192, 263, 303, 314, 318, 325; Mercury Control at, 133, 134, 135, 136, 145, 149, 151, 152, 201, 206, 209, 210, 215, 216, 220, 221
Cape Kennedy, 98, 247, 254, 255, 259, 288, 325, 355, 388, 427, 430, 442, 445; Gemini Control at, 332, 342, 343, 345, 346, 353, 354, 370
Capitol Vending Company, 379, 381
Carliner, David, 385
Carpenter, Malcolm Scott, 103, 109, 123, 131, 198-212 *passim*, 216, 217, 219, 222, 228; "fireflies" observed by, 203, 204; quoted, 201-202, 203, 204, 205, 207, 208, 209; and retrofire and re-entry of Aurora 7, 205-210; and U.S.S. *Intrepid,* 212; *see also* Aurora 7; Mercury-Atlas 7; Mercury Project
Carr, Gerald P., 456, 457
Caucasus Mountains, on moon, 236
Centaur program, 256, 263, 264, 265, 266, 267, 268
Centrifugal force, and gravity, 30, 31
Cernan, Eugene A., 360, 361, 368, 369, 370, 472, 474, 475, 476, 478, 479, 480, 488, 491, 492, 494, 495, 497, 498; spacewalk by, 361
Chaffee, Roger B., 373, 375, 377, 389, 390, 391, 392, 393, 397, 427, 458; death of, 395
Chapman, Sydney, 87
"Charlie Brown," *see* Apollo 10
Cherwell, Lord, 13, 87
Chicago, University of, 88, 92, 94
Christofilos, Nicholas, 79, 80
Chrysler Corporation, 36, 43, 180, 192
Churchill, Winston, 13, 15, 315
Clark, Ray, 190
Clarke, Arthur C., 302
Clavius crater, 235
Clement, G. H., 40
Clemmons, Stephen B., 392, 393
Coastal Sentry, U.S.S., 357
Cocoa Beach, Florida, 45, 64, 100, 101, 102, 103, 119, 123, 188, 321
Cognitum, Mare, 250
Collins, Michael, 362, 363, 364, 450, 451, 454, 455, 459, 460, 474; spacewalk by, 364
Columbia Broadcasting System, 476
Command Module, on Apollo, 175, 374, 376, 377, 387, 388, 390, 392, 394, 398, 400, 410, 414, 422, 426, 432, 436, 439, 447, 448

Committee for Evaluating the Feasibility of Space Rocketry, 27, 28, 29
Committee on Aeronautical and Space Sciences (Senate), 162, 378, 381
Committee on Guided Missiles, 32
Committee on Science and Astronautics (House), 163, 243, 257, 425
Commoner, Barry, 285, 286
Communication by Moon Relay (CMR), 302
Communications Satellite Corporation (Comsat), 316, 317, 325
Communications satellites, 300, 302 *ff.*; active-repeater, 304, 308 *ff.*, 314, 320; longevity of, 305-306, 308, 311, 315; medium-altitude, 314, 317, 318; passive, 303; synchronous, 317, 318, 319, 320; *see also* Relay; Telstar
Compton, Arthur Holly, 74
Comsat (Communications Satellite Corporation), 316, 317, 325
Congreve, William, 6
Conrad, Charles, Jr., 341, 342, 344, 345, 346, 364, 365, 366, 367, 368, 371, 451
Continental drift, 71
Convection-cell theory, of seafloor spreading, 71
Cooper, Leroy Gordon, Jr., 109, 140, 141, 146, 221-229 *passim*, 315, 331, 332, 341-346 *passim*, 420, 482; and observation of fine details from orbital altitude, 227, 346; quoted, 227; and radar evaluation pod, 343; *see also* Faith 7; Gemini 5; Mercury-Atlas 9; Mercury Project
Copernicus crater, 236
Corpus Christi, Texas station, 442, 444
Cosmic-ray detector, 94, 275
Cosmic-ray telescope, 94, 294
Cosmic rays, 73-75, 90, 91, 92, 93, 94, 95, 96, 269, 288; investigated by Van Allen, 61, 68, 69, 70; and trapped radiation, *see* Trapped radiation
Courier 1-B, 303
Crisium, Mare (Sea of Crises), 236, 460
Cronkite, Walter, 419
Cunningham, Walter, 441, 443, 445
Darwin, George, 262
Debus, Kurt H., 24, 46, 59, 62, 63, 65, 188, 189, 458, 487
Deep Space Instrumentation Facility 233, 240, 269, 292
Defense Department, 32, 39, 40, 42, 43, 44, 47, 48, 50, 80, 108, 124, 160, 308
Deimos, 282, 283
Delta rocket, 303, 307, 314, 319, 320, 322, 324
Descent Propulsion System (DPS), Apollo, 426, 428, 492
Diana Project, 302
Diligence, U.S.S., 332
Dirksen, Everett, 309
Docking, of Agena with Gemini spacecraft, 355, 356, 362, 363, 364, 365, 366, 369
Docking tunnel, for Apollo, 377
Donn, William, 420
Dornberger, Walter R., 7, 10, 11, 12, 13, 16, 20, 21, 24, 56
Douglas, William K., 123, 130, 131
Douglas Aircraft Company, 29, 180, 191, 377
Drogue parachute, 110, 153, 208, 225, 228
Drummond, Roscoe, 55
Dryden, Hugh L., 108, 161, 172, 173, 180, 378
Duke, Charles M., 474, 479, 482, 489
Dunn, Latimer E., 394

Earth: convection currents in mantle rock of, 72; magnetic tail of, 98, 99; pear-shaped component of, 71, 72
Earth and Planetary Sciences Laboratory (Southwest Center for Advanced Studies), 84
Earth-orbit rendezvous (EOR), 173, 174, 175, 176, 177, 179, 180, 182, 183, 330
Earth satellites, 25-26, 29-31, 33, 38, 39, 85, 158, 162, 163, 264-265; and gravity, 31; manned, recovery of, 104-105; *see also* Communications satellites; Meteorological satellites
Earth Station, Andover, for Telstar, 307, 308, 309, 310, 311
Echo Project, 303, 304, 306, 325
Eckels, Ann, 72
Edison, Thomas, 445
Edwards Air Force Base, 227, 347, 362
Ehricke, Krafft, 266
Eisele, Donn F., 441, 443, 444, 445, 446, 449, 482
Eisenhower, Dwight D., 21, 39, 47, 54, 56, 57, 58, 67, 106, 108, 158, 159, 160, 162, 163, 303
Engle, Joe Henry, 481, 482

Enos (chimpanzee), 117, 127
Environmental Control System, in Apollo, 375, 389, 398, 399, 444
Environmental Science Services Administration (ESSA), 325, 326
Erosion gauge, 61
Esnault-Pelterie, Robert, 6
Essex, U.S.S., 447
EVA suit, 334, 335
Evans, Lou, 487
Explorer 1, 25, 26, 66-71 *passim*, 85, 186, 189
Explorer 2, 69, 103*n.*
Explorer 3, 69, 70
Explorer 4, 76-78, 80
Explorer 7, 82, 83
Explorer 8, 84
Explorer 10, 89, 90, 98
Explorer 12, 83, 84, 89, 90
Explorer 14, 98
Explorer 18 (IMP 1), 96, 97, 98
Explorer 33, 99, 100
Extravehicular activity (EVA), 334-341 *passim*, 362-371 *passim*, 468, 469, 470, 471, 489; *see also* Spacewalk

F-1 engine, 161, 174, 175, 180, 186, 410, 414, 417, 419, 430, 432 434, 435
Faget, Maxime A., 106, 107, 108, 109, 396
Faith 7, 221, 315, 327; in orbit, 221-226, 228; reentry of, 228-229; *see also* Cooper, Leroy Gordon, Jr.; Mercury-Atlas 9; Mercury Project
Fecunditatus, Mare (Sea of Fertility), 236, 457, 458
Federal Communications Commission (FCC), 304, 317
Fermi Institute of Nuclear Studies, 261
"Fireflies," observation of, 142-143, 144, 145, 147, 203, 204, 216, 217
Fisk, James B., 318, 319, 367*n.*
Flamstead crater, 259
Fleming, William, 167
Flight Director Indicator, in Apollo, 437
Fly-by-wire control mode, 111, 115, 143, 144, 146, 204, 205, 206
Forbush, Scott E., 93
Forbush effect, 93, 94, 95
Forrestal, James E., quoted, 33
Fortune, W. C., 38
Frank, Louis A., 275
Franzgrote, Ernest, 261
Frau im Mond (film), 6
Free fall, 31, 127, 131, 369
Freedom 7, 115
French National Center of Telecommunications Studies, 307
Friendship 7, 117, 122, 124, 125, 131; in orbit, 134-150; pure-oxygen atmosphere used in, 157; and question of landing-bag deployment, 145-147, 148, 149; retrofire and reentry of, 150-154; *see also* Glenn, John Herschel, Jr.; Mercury-Atlas 6; Mercury Project
Froelich, Jack, 64
Fulton, James G., 180, 181
Furnas, Clifford C., 41

Gagarin, Yuri A., 114, 127, 156, 163
Galileo, 234, 235, 269, 292; quoted, 230
Gartlein, C. W., 83
Gavin, James M., 46, 47, 51, 53
Gavin, Joseph, 253, 487
Gazenko, Oleg, 334
Geer, E. Barton, 396
Geiger counter, 25, 61, 68, 69, 70, 76, 275
Geissler, Ernst D., 44, 188
Gemini 3 ("Molly Brown"), 327-333, 436; *see also* Grissom, Virgil I.
Gemini 4, 334-341
Gemini 5, 341-347
Gemini 6, 347-355 *passim*, 359, 474; in rendezvous with Gemini 7, 353, 354, 362
Gemini 7, 350, 351, 352, 353, 354, 355, 449, 451; in rendezvous with Gemini 6, 353, 354, 362
Gemini 8, 355-358, 363; gyrations of, 356-357, 358; and rendezvous and docking with Agena, 355-356, 362
Gemini 9, 359-361, 474; and "angry alligator," 360, 362
Gemini 10, 362-364; Agena docked with, 362, 363, 364
Gemini 11, 362, 364-368, 371; and rendezvous and docking with Agena, 365-366
Gemini 12, 362, 368-372; Agena docked with, 369
Gemini Control, at Cape Kennedy, 332, 342, 343, 345, 346, 353, 354, 370
Gemini Project, 327 *ff.*, 332, 335, 347, 348, 352, 354, 356, 362, 365, 373, 386, 411, 412, 424, 426, 435, 438, 447, 466; end of, 372, 373; term "revolution" used in, 349

General Dynamics Corporation, 126, 264, 266, 377, 378
General Electric Company, 42, 43, 325, 343, 377, 378, 440
Geological Survey, United States, 251
Gerathewohl, Siegfried J., 127-128
German Society for Space Travel (VfR), 6, 7, 8, 9
Germany, rocket development in, 1, 5-20 *passim*
Gilruth, Robert R., 108, 118, 146, 170, 177, 180, 458, 487; quoted, 372
Gleaves, James D., 392, 393
Glenn, John Herschel, Jr., 100, 103, 109, 117-126 *passim*, 127-154 *passim*, 155, 156, 157, 200, 203, 204, 216-226 *passim*, 371, 448; banded horizon observed by, 139-140; "fireflies" observed by, 142-143, 145, 147, 203, 216, 217; and Pres. Kennedy, 143, 155; and question of landing-bag deployment, 145-147, 148, 149; quoted, 436; and retrofire and reentry of Friendship 7, 149-154; *see also* Friendship 7; Mercury-Atlas 6; Mercury Project
Glennan, T. Keith, 108, 109, 160, 161, 167
"Glitch," 222 and *n.*, 223, 437
Goddard, Robert H., 5, 6, 8, 9
Goddard Institute for Space Studies, 73
Goddard Space Flight Center, 72, 75, 84, 91, 95*n.*, 98, 135, 307, 319, 321
Gold, Thomas, 236, 252, 253
Gold-box chemical analyzer, 260-262
Goldstein, Richard M., 280
Goldstone station, 241, 244, 245, 248, 249, 250, 273, 277, 280, 304, 454, 476
Golovin, Nicholas, 168
Golovin Committee, 168, 174
Goodrich, B. F., pressure suit, 122
Goonhilly, parabolic dish antenna at, 307
Gordon, Richard F., Jr., 364, 365, 366, 367, 368, 369, 370, 451; spacewalk by, 365-366
Grau, Dieter, 188
Gravity "artificial," 368; and centrifugal force, 30, 31; and orbital speed, 348
Gravity-gradient stability, 365, 367 and *n.*, 371
Grine, Kenneth, 189
Gringauz, K. I., 88
Grissom, Virgil I., 109, 115, 116, 117, 201, 207, 209, 213, 223, 224, 327, 337, 339, 340, 397, 400; on Gemini 3 ("Molly Brown"), 331-332; and spacecraft 012, 373, 375, 377, 389, 390, 393, 397, 398; as victim of spacecraft fire 395; *see also* Gemini 3 ("Molly Brown")
Group for the Study of Reactive Motion, in Soviet Union, 6
Gruene, Hans F., 63, 188, 190
Grumman Aircraft Engineering Corporation, 184, 185, 253, 377, 427, 436, 458, 474
Guadalcanal, U.S.S., 472
Gueriche crater, 250
Guided Missile Development Division, of Army Ordnance Corps, 43, 45
"Gumdrop," *see* Apollo 9

H-1 engine, 161, 180, 186, 187, 381
Hage, George, 483
Hagen, John P., 48, 49, 54; quoted, 49, 58
Hagerty, James C., 58
Hall, Asaph, 282, 283
Hall, Harvey, 29
Ham (chimpanzee), 114, 127
Hamill, James J., 22
Hand Held Maneuvering Unit (Zot gun), 336, 338, 364
Haney, Paul, 358; quoted, 419, 421-422, 446
Harris, Gordon, 53
Harvard Observatory, 284
Hasselblad camera, 364, 366, 492
Hauesserman, Walter, 62, 188
Hawkins, Jerry W., 392, 393
Hawkins, Willard, 479
Heat shielding, 163, 206, 207
Hechler, Ken, 181
Heimberg, Karl L., 188
Heitsch, John, 212
Helium region, above sensible atmosphere, 84
Hello, Bastian, 405
Henize, Karl, 363, 367
Hess, Victor F., 74, 75
High Altitude Test Vehicle (HATV), 27, 29, 30, 33
Hilburn, Earl D., 245
Hill, Ralph, 379, 380, 381, 385; quoted, 380
Hines, Colin O., 92, 96
Hirsch, André, 6
Hitler, Adolf, 12
Hodges, Luther H., 382
Hoelzer, Helmut, 188
Holaday, William M., 47

Holmdel horn, for Telstar, 306, 309, 310, 311
Holmes, D. Brainerd, 177, 178, 179, 180, 181, 184, 386; quoted, 177-178, 179, 180
Hoover, George, 38
Houston Flight Control (Mission Control), 356, 357, 358, 365, 432, 442, 445, 446, 450, 452, 453, 456, 457, 458, 461, 462, 467, 468, 469, 470, 477, 480, 482, 491, 492, 494, 497, 498, 499
Hueter, Hans, 188
Hughes Aircraft Company, 246, 256, 257, 317, 319, 320, 321, 322, 377
Huntsville, Alabama, 23, 35, 36, 37, 38, 42, 43, 44, 45, 48, 56, 160, 174, 435
Huzel, Dieter, 12, 19, 20, 23; quoted, 19
Hydrogen, liquid, 27, 263, 264, 266, 410, 414
Hydrogen engine, 28, 33, 263, 264, 265
Hydrogen-fusion bomb, 18
Hydrogen layer, above helium region, 84
Hydrogen-peroxide jet system, in Syncom 3, 324
Hydyne, 59, 62

Imbrium, Mare, 236
IMP 1, 96, 97, 98
Indian Ocean tracking ships, 138, 139, 145, 149, 219
Infrared radiometer, 272, 279, 325
Inhibiting Command No. 8, for Ranger 7, 248-249
Injun 1, 83, 96
Injun 2, 120
Intelsat (International Telecommunications Consortium), 317
Intercontinental ballistic missile (ICBM), German, 17
Intercontinental telephone, 300
Intercontinental television, 301
Intermediate-range ballistic missile (IRBM), 44, 45
International Congress on Astronautics, 38, 40, 55
International Geophysical Year (IGY), 2, 38, 39, 40, 44, 71, 78
International Telecommunications Satellite Consortium (Intelsat), 317
International Union of Geodesy and Geophysics, 38
Interplanetary Monitoring Platform, 96
Intrepid, U.S.S., 201, 212, 332
Ionosphere, 82, 295, 296, 300, 301, 302
IOS (Indian Ocean tracking ships), 138, 139, 145, 149, 219

J-2 engine, 174, 175, 180, 265, 381, 410, 415, 421, 434, 435, 448, 450, 451
Jaffe, L. D., 258
James, Jack N., 268
Jastrow, Robert, 73, 95*n.*
Jeffs, George, 396, 397
Jet Propulsion Laboratory (JPL), 27-28, 29, 34, 40, 42, 44, 59, 67, 68, 90, 121, 122, 232, 233, 234, 240, 241, 243, 244, 246, 248, 249, 251, 268, 269, 274, 275, 321; in dispute with NASA, 242-243, 245-246, 247
Johnson, Francis S., 84
Johnson, John, 194
Johnson, Lyndon B., 32, 161-162, 164, 165, 182, 193, 297, 298, 309, 310, 381, 383, 397, 413, 425, 438, 440, 458, 473; quoted, 298, 310, 383
Johnson, Marshall S., 269
Johnston, Edward H., 394
Joint Long Range Missile Proving Ground, 34
Julius Caesar crater, 245
Jungert, Wilhelm, 21
Juno 2 rocket, 77, 82, 86
Jupiter, 259, 275; moons of, 256
Jupiter A, 45
Jupiter C, 25, 37-48 *passim*, 57, 62, 69, 76, 86, 105, 112, 113, 161, 186; *see also* Missile No. 29
Jupiter IRBM, 44, 46, 86
Jura Mountains, on moon, 236

Kaplan, Joseph, 38, 39, 40
Kaplan, L. D., 279*n.*
Kappel, Frederick R., 309
Kármán, Theodor von, 28
"Karst" formations, 254
Karth, Joseph E., 245
Kavanau, Laurence, 168
Kearsarge, U.S.S., 213, 223, 229
Keck, John A., 17, 18
Kefauver, Estes, 47, 316
Kelley, Albert J., 243
Kennedy, John F., 3, 127, 143, 155-165 *passim*, 168, 169, 174, 182, 192, 196, 231, 317, 376, 377, 458, 465; quoted, 196-197
"Kennedy effect," 181-183
Kennedy Space Center (KSC), 192, 195, 388, 389, 390, 406, 408, 413, 414, 416, 417-418,

420, 421, 431, 445, 449, 450, 458
Kennedy time frame, 410, 412
Kepler, Johannes, 317
Kepler crater, 236
Kerr, Robert S., 162, 169, 381, 382, 383
Kesselring, Albert, 11
Khrushchev, Nikita, 52-53; quoted, 51
Killian, James R., 57
King, David S., 172
King, Jack, quoted, 413
Kingsport, U.S.S., 322, 323
Kirtland Air Force Base, 441, 449
Kliore, Arvydas J., 295
Knerr, H. J., 20
Koelle, Heinz H., 188
Kolhorster, Wernher, 74, 75
Kozyrev, N. A., 255
Kraft, Christopher C., Jr., 118, 144, 146, 157, 216, 220, 332, 336, 337, 338, 343, 344, 345, 351, 353, 354, 365, 458, 461, 487
Kuers, Werner, 188
Kuiper, Gerard P., 237, 251, 252, 253, 254, 256; quoted, 256
Kummersdorf rocket-testing ground, 7, 10, 11
Kundel, Keith K., 356

Lake Champlain, U.S.S., 115, 214, 346
Lange, Oswald H., 188
Langley Aeronautical Laboratory, 106, 108, 247
Langrenus plain, 488
Lauritsen, Charles C., 40-41
Lederburg, Joshua, 284
Leibniz Mountains, on moon, 236
Leighton, Robert B., 292, 295, 296, 297, 298; quoted, 296
LeMay, Curtis E., 29, 222
Lemnitzer, Lyman, 53
Leonov, Aleksey A., 329, 333-334, 337, 340
Levinson, Edward, 382
Levy, Edwin Z., 109
Lewis Research Center, 108, 249, 267
Ley, Willy, 6
Liberty Bell 7, 115, 116, 117, 327, 400
Libration, 234
Lindbergh, Charles A., 155
Lindenberg, Hans, 20
Liquid hydrogen, 27, 263, 264, 266, 410, 414
Liquid oxygen, 9, 15, 30, 264, 266, 410, 414
Lithium hydroxide, 363
"Little Joe" booster, 113
Lockheed Aircraft Corporation, 377
Loki rocket, 37, 42
Long, Frank A., 396
Lovelace II, Randolph, 109
Lovell, James A., Jr., 350-351, 352, 353, 354, 355, 369, 370, 371, 440, 448, 451, 452, 453, 456, 457, 458, 460, 461, 473
Low, George M., 167, 458
Lowell, Percival, 283
Luedecke, Alvin R., 246
Luftwaffe, 12, 14, 17
Luna series, 98, 235, 241, 257, 258
Lunar Excursion Module (LEM, LM), 176, 177*n.*, 178, 179, 184, 238, 252, 260, 328, 377, 379, 412, 414, 416, 426, 427, 428, 429, 430, 437, 438, 441, 449, 464, 465, 466, 467, 468, 470, 471, 481, 491, 492, 493; cost of, 427
Lunar International Observers Network (LION), 491
Lunar Orbit Injection (LOI), 453, 455, 456, 459
Lunar Orbit Rendezvous (LOR), 176-184 *passim*, 328, 377, 426
Lunar Orbiter Project, 232, 247
Lunik series, 86
Lunney, Glynn, 420, 494, 498

M-1 engine, 174
McCandless II, Bruce, 482
McClure, Ray, 212
McCracken, Kenneth G., 91
McDivitt, James A., 335-341 *passim*, 346, 466, 467, 468, 469, 471
McDonell Aircraft Company, 106, 108, 109, 110, 115, 162, 192, 328, 360, 378, 381
McElroy, Neil H., 53, 55, 56, 57
McGee, Dean, 381, 383
McGough, John, 188
McIlwain, Carl, 69, 70
McLendon, Lennox P., 384
McMath, Robert R., 41
McVittie, George, 52
Mach, Ernst, 104*n.*
Mach 1, defined, 104*n.*
Madrid station, 454, 476
Magnetic storm, 82
Magnetometer, 89 and *n.*, 94, 270, 293
Magnetopause, 85, 87, 88, 89, 96, 98
Magnetosphere, 81-85, 87, 88, 90, 98
Manned Spacecraft Center, 172, 192, 193, 194, 196, 330, 336, 357, 358, 361, 369, 375, 395,

402, 405, 419, 421, 432, 442, 455, 458, 463, 476, 490
Manned Spaceflight Office, 167, 177
Mansfield, Mike, 380
Manual proportional control mode, 111, 146, 205, 206
Marconi, Guglielmo, 307
Mariner 1, 272
Mariner 2, 90, 244, 247, 272, 273-280, 281, 282, 284, 286; in encounter with Venus, 277-278; solar wind measured by, 275; sun sensors in, 273; telemetry of, 278, 279, 280, 281
Mariner 3, 272, 288
Mariner 4, 90, 286-298, 478; Canopus sensor in, 289, 290, 291; photographs received from, 291, 292, 293, 295, 296, 297, 298; sun sensors in, 289; telemetry of, 292, 293, 294, 297; television camera in, 286, 291, 292, 293
Mariner 5, 281
Mariner R program, 267-269, 270 *ff.*
Mars, 90, 183, 184, 256, 263, 267, 268, 274, 275, 282, 283, 284, 411, 478; Amazonis area on, 296; atmosphere of, 284, 285, 288, 294, 295, 298; "canals" on, 283, 284; craters on, 297-298; day on, length of, 281; diameter of, 282; Elysium region on, 296; gravity on, 282; ionosphere of, 295; life on, question of, 284, 285, 286, 294, 295; moons of, 282; and occultation experiment, 288, 294, 295; photographs of, 291, 292, 293, 295, 296, 297, 298; probes of, 267, 268, 276, 285, 286, 287, 288-298; revolution around sun, 282; rotation of, 293; water vapor on, 285
Mars and Its Canals (Lowell), 283
Mars as the Abode of Life (Lowell), 283
Marshall Space Flight Center, 23, 109, 161, 175, 180, 182, 187, 192, 267, 487
Martin Company, Glenn L., 42, 43, 48, 106, 180, 378, 379, 404, 405
Maskalyne crater, 489, 490
Mason, U.S.S., 358
Massachusetts Institute of Technology, 57, 88, 89, 160, 466
Mathews, Charles W., 108
Mattingly, Thomas K., 461
Max Q, 133, 201
Medaris, John B., 45, 46, 47, 48, 53, 57, 62, 63, 64, 65, 67; quoted, 47, 48, 53
Meisenheimer, John L., 62
Mercury-Atlas 2, 114
Mercury-Atlas 3, 114-115, 117
Mercury-Atlas 4, 117
Mercury-Atlas 5, 117
Mercury-Atlas 6, 100, 102, 103, 117, 120, 126, 127, 129, 130, 131, 132, 155, 156, 157, 420; countdown and liftoff, 132-133; global tracking network for, 129; in orbit, 134-136; and question of landing-bag deployment, 145-147, 148, 149; and retrofire and reentry, 150-154; *see also* Friendship 7; Glenn, John Herschel, Jr.; Mercury Project
Mercury-Atlas 7, and liftoff, 201; *see also* Aurora 7; Carpenter, Malcolm Scott; Mercury Project
Mercury-Atlas 8, 212; countdown and liftoff, 214-215; *see also* Schirra, Walter M., Jr.; Sigma 7
Mercury-Atlas 9, 221, 222; *see also* Cooper, Leroy Gordon, Jr.; Faith 7; Mercury Project
Mercury Control, at Cape Canaveral, 133, 134, 135, 136, 145, 149, 151, 152, 201, 206, 209, 210, 215, 216, 220, 221
Mercury Project, 50, 100-128, 129, 159, 160, 170, 203, 212, 213, 221, 317, 318, 327, 435, 463; and characteristics of spacecraft, 110-112; cost of, 109, 229; end of 229, 327; final report on, 118
Mercury-Redstone, 113, 114, 115
Messier crater, 457, 458, 488
Mesta, Perle, 55
Meteorological satellites, 325-326
Microbarographs, and record of air waves produced by Apollo 4, 420
Micrometeoroids, 61, 251, 269, 288, 305, 335
Microorganisms, space conditions survived by, 364
Microwave radiometer, 270, 271, 277, 278, 279
Microwave transmission, 302, 303
Miller, George P., 163
Millikan, Robert A., 74
Minimum Orbital Unmanned Satellite Experiment (MOUSE), 38

Minners, Howard A., M.D., 198, 200
Minuteman I ICBM, 380
Missile Firing Laboratory, 46
Missile No. 29, 25, 26, 58-61, 64, 65, 68; *see also* Jupiter C
Mission Control, *see* Houston Flight Control
Mitchell, Edgar D., 482
"Molly Brown," *see* Gemini 3
Mondale, Walter F., 406; quoted, 406
Monel metal, 9
Moon, 230, 231, 232, 234, 235, 236, 237, 263, 412, 424, 439, 448, 445; aluminum on, 262; basalt in soil of, 260, 262; and CMR, 303; "cold" vs. "hot" theories of, 255; craters on, 235, 236, 250, 251, 252, 254, 255, 256, 258, 259, 260, 455, 457, 458, 472, 489, 490, 491; diameter of, 234; dimple craters on, 254, 255; and dust hypothesis (Gold), 236, 237, 252-253; gamma radiation on, 239; "graham cracker" surface of, hypothesis of, 252, 255; gravity on, 234; halo craters on, 256; "hot" vs. "cold" theories of, 255; landing sites on, 236, 251, 252, 455, 457, 465, 489, 490, 491, 492, 496, 499; lava flows on (Kuiper), 252, 253; and libration, 234; maria on, 235, 236, 244, 248, 250, 252, 253, 254, 255, 258, 259, 260, 262, 455, 457, 458, 460, 465, 472, 488, 490, 495, 497; mountains on, 236, 250; origin and evolution of, 235; oxygen on, 261-262; photographs of, 250, 251, 252, 254, 255, 256, 259, 260; rayed craters on, 236, 251; revolution around earth, 234, 237, 238; rilles on, 236, 254, 473, 488, 489, 490, 491, 496; round-trip voyage to, within physical capability of astronauts, 347; silicon on, 262; soil of, chemical analysis of, 260-262; surface of, 234, 235, 236, 237, 238, 252, 253, 256, 258, 259, 260, 261, 262, 457, 458, 472-473, 488, 489, 490, 491, 496; temperatures on 235; terrae on, 236, 255; volcanism on, theory of, 253, 255
Morse, Wayne, 316
Moser, Robert, 63, 65, 189, 190, 191, 392
Mrazek, William, 45, 188
Mueller, George E., 333, 375, 385, 386, 389, 416, 432, 487; quoted, 386, 416
Murray, Bruce C., 292, 297
Murrow, Edward R., 171, 310; quoted, 171-172
Mustel, E. R., 87
Myers, Dale D., 401, 402

National Academy of Sciences (NAS), 40, 52, 70, 165, 166, 171, 284
National Advisory Committee for Aeronautics (NACA), 103, 104, 106, 107, 108
National Aeronautics and Space Administration (NASA), 23, 42, 84, 90, 100, 108, 109, 112-120 *passim*, 126, 159, 160, 161, 162, 163, 164, 167-173 *passim*, 182, 192, 193, 194, 195, 196, 231, 232, 238, 267, 268, 303, 326, 328, 333, 334, 373, 374, 375, 376, 424, 425, 426, 427, 435, 438, 439, 440, 443, 452, 454, 473, 482, 487, 490; and AT&T, 307-308; bureaucracy of, 359; cutbacks in budget of, 410, 411, 412; in dispute with JPL, 242, 243, 244, 245, 246; Gemini tapes withheld by, 358-359; and Oversight, House Subcommittee on 243, 246, 400, 402, 412; and Phillips report, 385-388; photographs of Mars displayed by, 297; and Source Evaluation Board, 378, 405; spacecraft 012 certified by, 388-389; spacecraft 012 failure investigated by, 395-401; and spacecraft 017, report on, 405, 406
National Bureau of Standards, 73, 228
National Conference on the Peaceful Uses of Space, 169
National Institute of Health, 166
National Research Council, 284
National Science Foundation, 39
National Security Act, 32
Naval Research Laboratory, 26, 27, 38, 48, 53, 67, 68
Neher, H. V., 275
Neptune P2V, and Carpenter, 211
Nesmeyanov, Alexander N., 39; quoted, 48
Ness, Norman F., 91, 98
Neubert, Erich W., 21
Neugebauer, G., 279*n.*
Neugebauer, Marcia, 90
Neutron monitor, at Deep River,

Canada, 94, 95
Newell, Homer E., 38, 243
Newton, Isaac, 8, 31, 367
Newton, Quigg, 212
Nicks, Oran, 245
Nicolet, Marcel, 84
Nieburg, H. L., 242
Nikolayev, Adrian, 213
Nimbus satellite, 325
Nixon, Richard M., 158, 159; quoted, 54
Noa, U.S.S., 153
North American Aviation, Inc., 29, 36, 178, 180, 182, 191, 373, 378-385 *passim*, 401, 403, 404, 405, 410, 436, 487; and Phillips report, 385-388; selected by NASA to build Apollo 012, 378, 379, 405
Northrop Corporation, 243, 381
Norton Sound, U.S.S., 80
Nose cone, of Jupiter IRBM, 46, 47, 57, 105
Notched zero, in Telstar command-decoder, 313
Nova rocket, 161, 168, 170, 172, 174, 177, 179, 376
Noxon, Victor, 199
Nubium, Mare, 236, 248, 250, 255

Oberth, Hermann, 5, 6
O'Brien, Brian J., 83, 84, 96
Ochs, Larry, 53
Office of Manned Space Flight, 167, 177
Office of Naval Research, 38
Office of Space Flight Programs, 232
Office of Space Science and Applications, 243, 245
O'Keefe, John A., 72
O'Meara, A. P., 48
Oparin, Aleksander Ivanovich, 284
Opel, Fritz von, 6
Operation Backfire, 1, 3, 408
Operation Crossbow, 13
Operation Deep Freeze, 96
Operation Paperclip, 21
"Orange" hydrogen-bomb test, 80
Orbit, distinguished from revolution, 349
Orbit Attitude Maneuvering System (OAMS), 329, 331, 332, 336, 338, 345, 346, 357, 358, 359
Orbiter Project, 39, 41, 44
Orbiting Geophysical Observatory, 441
Orbiting space station, 18
Ordnance Intelligence, United States, 17, 18
Oversight, NASA, 243, 246, 400, 402, 412
Oxygen: liquid, 9, 15, 30, 264, 266, 410, 414; on moon, 261, 262; pure, in Apollo 012 atmosphere, 397, 398, 399, 402; pure, vs. mixed gases in cabin atmosphere, 108, 109, 157, 402-403, 436-437

Paine, Thomas O., 440 and *n.*, 458; quoted, 464
Pan American Airways, 394*n.*
Parachute, drogue, 110, 153, 208, 225, 228
Parker, Eugene N., 88
Parks, Robert J., quoted, 260, 268
Parry-Bonpland crater, 250
Particles, atomic, in space, *see* Cosmic rays
Patrick Air Force Base, 62, 101, 119, 189
Patterson, James H., 261
Paul, Henry C., 63, 66
Pauling, Linus C., 165, 166
Peenemünde rocket development center, 11, 12, 13, 18, 19, 20, 23
Percy, Charles H., 406
Perry, Robert L., quoted, 33-34
Petersen, Norman V., 181, 182
Petrone, Rocco, 189
"Phantom Agena," 344
Phillips, Samuel C., 386, 387, 388, 404, 423, 430, 435, 447, 458, 487; report by, 386-388, 403
Phobos, 282, 283
Pickering, William H., 28, 44, 66, 67, 232, 233, 242, 243, 246, 249, 250, 251, 259, 269, 296, 488; quoted, 294-295
Pickering crater, 457, 458, 488
Piddington, J. H., 98
Pieper, George F., 74
Pierce, John R., 171, 302, 305, 309, 311
Pierce, U.S.S., 201, 212
Pioneer 1, 77, 85
Pioneer 2, 77, 85
Pioneer 3, 77, 79, 85
Pioneer 4, 85
Pioneer 5, 94, 95
Pioneer 6, 90, 91
Pittendrigh, Colin S., 284
Plasma, 91, 92, 275, 281; defined, 88; *see also* Solar wind
Pogo Effect, 424, 431, 434
Pogue, 444
Polaris missile, 18, 80, 119, 122
Poppel, Theodor A., 21
Porkbarrel, in space program, 191-197
Porter, Richard W., 41

Powers, John A., 103, 121, 125, 132, 133, 209, 210, 211; quoted, 209, 210
Pratt and Whitney, 264, 265
"Preliminary Design for an Experimental World-Circling Space Ship" (RAND), 29
"Preliminary Studies of Manned Satellites—Wingless Configuration, Non-Lifting" (NACA), 107
Pressure suit, B. F. Goodrich, 122
Porcellarum, Mare (Ocean of Storms), 236, 258, 491
Program Reader Assembly (PRA), in Apollo, 428, 429

Quarles, Donald A., 40, 41
Quebec, U.S.S., 219, 223

Radiation belts, Van Allen, 70, 71, 73, 76-82 *passim*, 85, 89, 120, 305, 308, 311, 312, 449, 451
Radio Corporation of America, 245, 308, 314, 315, 325, 442
Radiometer: infrared, 272, 279, 325; microwave, 270, 271, 277, 278, 279
RAND Corporation, 29, 30, 31, 32, 33, 104
Randall, Clarence, 54
Ranger 1, 238
Ranger 2, 238, 239
Ranger 3, 121, 122, 126, 239-241
Ranger 4, 241
Ranger 5, 242
Ranger 6, 243-245, 249
Ranger 7, 247-250, 251, 252; and abandonment of sterilization, 247; Inhibiting Command No. 8 for, 248-249; photographs received from, 250, 251, 252, 256, 292
Ranger 8, 253, 254-255; photographs received from, 254-255, 256
Ranger 9, 255-256
Ranger Project, 231, 232-234, 238-256 *passim*
Rankin, Charles, 213
Reaction Control System in Apollo, 429
Rechtin, Eberhardt, 268
Redstone Arsenal, 35, 43, 53, 109, 127, 161
Redstone rockets, 35, 36, 37, 41, 42, 43, 45, 58, 59, 60, 62, 63, 66, 86, 107, 113, 114, 161, 428, 436, 440
Reece, L. D., 393
Reentry control system (RCS), 357, 358
Reese, Eberhard, 21, 24, 188; quoted, 174
Reiner crater, 258
Relay 1, 308, 314-316, 324
Relay 2, 316
Rendezvous, 348; between Agena and Gemini spacecraft, 355, 356, 364, 365, 369; between Apollo 7 and S-4B, 441, 442, between ATDA and Gemini 9, 360; earth-orbit (EOR), 173, 174, 176, 177, 179, 180, 182, 183, 184, 330; on first revolution, 364; between Gemini 6 and Gemini 7, 353, 354, 362; lunar-orbit (LOR), 176, 177, 179, 180, 181, 182, 183, 184, 328, 376
Renzetti, Nicholas, 269
Retrorockets, 110, 134, 136
Rice Institute of Technology, 194
Riphaean Mountains, on moon, 250
Ritland, O. J., 170
Roche, F. E., 83
Rocket airplane, 103
Rocket car, 6
Rocket motor, Wyld, 9
Rocket societies, 6
"Rocket to Outer Space, The" (Oberth), 5
Rocketdyne engines, 36, 161, 180, 265, 381
Rockets: antiaircraft, German plans for, 17; and erosion problem, 8-9; liquid-fuel, 6, 9, 26, 36, 263, 264; multistage, 34, 35, 41, 45, 65-66; posigrade, 110, 115; principle of propulsion of, 8; retrograde, 110, 134, 136; solid-fuel, 5, 6, 42, 45, 257; supersonic antiaircraft, German plans for, 18; used in War of 1812, 6
Rockoon, 68, and *n.*, 69, 70
Rogallo, Francis M., 170*n.*
Rose Knot Victor, U.S.S., 356
Rosser, J. Barkley, 41
Rossi, Bruno, 88
Roudebush, Richard L., 181
Royal Air Force, 12, 13
Ruff, George E., 109
Russell, Richard, 54
Rutherford, Ernest, 260

S-4B hydrogen engine, in Apollo, 416, 421, 430, 432, 441, 448, 450, 467, 474, 475
Sagan, Carl, 279*n.*
Sandler, Carl, 189, 190
Sandys, Duncan, 13, 14
"Sarah" beacon, 211

Satellites, earth, *see* Earth satellites
Saturn 1, 160, 161, 168, 174, 180, 186, 187, 188, 189, 190, 191, 375; launching of, 190; stages of, 188
Saturn 2, 173
Saturn 5, 2, 175-182 *passim*, 187, 191, 376, 381, 389, 405, 425, 426, 430, 433, 435, 436, 437, 439; stages of, 191, 265, 381, 386, 387; *see also* Apollo 4; Apollo 5; Apollo 6; Apollo 8; Apollo 10
Schiaparelli, Giovanni, 283
Schirra, Walter M., Jr., 109, 148, 150, 151, 200, 212-221 *passim*, 347-355 *passim*, 420, 440-447 *passim*; "fireflies" observed by, 216-217; and rendezvous with Gemini 7, 354; *see also* Apollo 7; Gemini 6; Mercury-Atlas 8; Mercury Project; Sigma 7
Schmitt, Harrison H. (Jack), 482, 490, 495
Schneider, William C., 372
Schneiderman, Dan, 268
Schulze, August, 21
Schweickart, Russell L., 466, 467, 468, 469, 470, 471, 482; spacewalk by, 469, 470, 471, 493
Schwidetzky, Walter, 21
Scott, David R., 355, 356, 357, 358, 466, 467, 468, 471
Scott, R. F., 258
Seamans, Robert C., Jr., 163, 180, 191, 378, 396, 397, 405; quoted, 191
Secchi crater, 488
Sedov, Leonid I., 40, 44
See, Elliott M., Jr., 351, 360, 458
Seismometer, lunar, 232, 239
Seitz, Frederick, 166
Senate Appropriations Committee, 167
Serenitatus, Mare, 236
Sergeant rocket, 42, 45, 59, 60, 62, 66, 186
Service Module, on Apollo, 175, 177*n*., 374, 376, 387, 389, 390, 403, 410, 414, 415-416, 421, 422, 426, 447, 448
Service Propulsion System (SPS), in Apollo, 432, 447, 448, 456, 459, 461
Serv-U Company, 380, 381, 382, 384
Shaffer, Philip, 478
Sharp, Robert P., 292
Shea, Joseph F., 177, 181, 375, 388, 402, 405, 406; quoted, 181, 258, 376, 377, 406
Shepard, Alan B., Jr., 109, 114, 115, 117, 127, 130, 133, 134, 135, 144, 151, 152, 153, 167, 170, 204, 205, 206, 223, 225, 436
Shoemaker, Eugene M., 251, 252, 256
Sigma 7, 212, 213, 214, 215; in orbit, 215-221; *see also* Mercury-Atlas 8; Mercury Projects; Schirra, Walter M., Jr.
Signal Communications by Orbiting Relay Equipment (SCORE), 303
Silicon, on moon, 262
Silverstein, Abe, 108, 173, 232, 267; quoted, 173
Simpson, John A., 94, 95, 294
Singer, S. Fred, 38, 75, 79
Sinton, W. M., 284
Sirius, 259
Slayton, Donald K., 109, 123, 130-131, 199, 215, 217, 219, 221, 327, 442, 443, 447, 482
Sloan, Richard K., 292
Small, Ballard B., 63
Smathers, George, 382, 383
Smith, Margaret Chase, 404
Smithson, J. L., 384
Smithsonian Astrophysical Observatory, 51, 321
Smythii, Mare (Smith's Sea), 488, 495
"Snoopy," *see* Apollo 10
Snyder, Conway W., 90
Société Astonomique Française, 6
Society for Space Travel, German (VfR), 6, 7, 8, 9
Society for Studying Interplanetary Communication, in Soviet Union, 6
Solar broom, 93-96
Solar cell, 233, 305, 306, 315
Solar plasma detector, 88
Solar pressure vane, in Mariner 4, 287
Solar wind, 85-92, 96, 97, 98, 99, 269, 275, 281, 282, 287, 288
Sonnett, C. P., 94
Southwest Center for Advanced Studies, 84, 91
Soviet Academy of Sciences, 39, 40
Soviet Union, 2-3, 6, 34, 158, 159, 425, 438, 440, 465; access of, to German rocket development, 22; Astronomical Council of, 88; and Cosmos series, 439; and Leonov's spacewalk, 329, 330, 333, 334, 337, 340; and Luna series, 98, 235, 241, 258; and Lunik series, 86;

Mars probe by, 276, 291; Soyuz series, 465, 468, 493; Sputniks launched by, 51-57, 86, 158, 159; Venus probes by, 86, 88, 268, 270, 280; and Vostok series, 114, 117, 156, 157, 158, 163, 213, 341, 345, 403; Zond series, 439, 448
Spaatz, Carl, 20
Space Act (1958), 164
Space Biology Laboratory, 227
Space Council, 160, 164, 193, 383
Spacecraft-LM Adapter (SLA), in Apollo, 427, 441
"Space Flight Emergencies and Space Flight Safety" (House Subcommittee on NASA Oversight), 401
Space Flight Operations Center, 274
Space Flight Programs Office, 232
Space Science and Applications Office, 243, 245
Space Science Board, of National Academy of Sciences, 284, 285, 286
Space station, orbiting, 18
Space Task Group, NASA, 108, 109, 112, 170
Space Technology Laboratories, 94, 112, 377
Spacewalk: by Aldrin, 369, 370, 371; by Cernan, 361; by Collins, 364; by Gordon, 365-366; by Leonov, 329, 330, 333, 334, 337, 340; by Schweickart, 469, 470, 471, 493; by White, 338-340, 341, 397; *see also* Extravehicular activity (EVA)
Special Committee for the IGY (CSAGI), 38, 39, 52
Sperry Gyroscope Company, 162
"Spider," *see* Apollo 9
Spilhaus, Athelstan, 38
Sputnik I, 28, 32, 47, 51-57, 70, 106
Sputnik II, 56
Sputnik VIII, 268
Squires, R. K., 72
Stafford, Thomas P., 347-355 *passim*, 360-361, 445, 472, 474, 475, 476, 477, 479, 480, 490, 491, 492, 494, 495, 497, 498, 499
Stahl, Charles J., 394
Starfish Project, 308
Stewart, Homer J., 40, 41, 44
Störmer, Carl, 75
Storms, Harrison A., 405
Strang, Charles F., 396
Strategic Air Force, United States, 20
Stuhlinger, Ernst, 63, 65, 66, 188
Submarine V-2, 17
Sun gun, German, 17
Sun sensors, in Mariner probes, 274, 289
Surveyor 1, 259, 264
Surveyor 3, 260
Surveyor 5, 260-261, 490; gold-box chemical analyzer in, 260-262
Surveyor Project, 232, 243, 247, 256-262, 263, 265, 267
Swenson, George W., 52
Swift, Jonathan, 282
Syncom 1, 319-321
Syncom 2, 319-324
Syncom 3, 324
Syncom Project, 315, 316-325

Tannenbaum, Jerome, 52
Target docking adapter (TDA), 365
Taruntius crater, 488
Taylor, Leland R., 384
Teague, Olin E., 183, 194, 246, 400
"Teak" hydrogen-bomb test, 80
Technical Evaluation Group, of Department of Defense Research and Development Board, 32
Telemetry: of Apollo 4, 421, 422; of Mariner, 277-281, 292, 293, 294, 297; of Ranger, 234, 248, 249, 250; of Surveyor, 260-262; of Syncom, 319; of Telstar, 305, 306, 307, 309-311
Telephone, intercontinental, 300
Television, intercontinental, 301
Television camera(s): in Apollo 7, 442, 443, 444; in Apollo 8, 452, 453, 454, 460; in Apollo 10, 475, 476, 477, 480; in Mariner 4, 286, 291, 292, 293; in Tiros satellites, 325
Telstar 1, 81, 171, 215, 304-314; antenna for, 306-307; and British television, 311; and French television, 311; longevity of, 308, 311-312; and notched zero, 313; One Gate in command-decoder circuit of, 313; ruby-crystal maser in, 307; signals from, 305, 306, 307, 309-311; Zero Gate in command-decoder circuit of, 313, 314
Telstar 2, 316
Tessman, Bernhard, 20
Theophilus plain, 490
Thomas, Albert, 193, 196
Thompson, Floyd L., 396, 402, 403

Thor-Able rocket, 77, 85, 94, 120, 303, 325
Thor-Delta rocket, 83
Thor IRBM, 44, 320
Thorneycroft, Peter, 182
Thrust-Augmented Thor (TAD), 324
Tikhonravov, M. K., 6
Tiros satellites, 325
Titan ICBM, 48, 106, 107, 328, 330, 336, 424, 474
Titov, Gherman S., 114, 127, 138, 155, 156, 157
Toftoy, Holger N., 21, 22, 38
Tos I, 325
Traac satellite, 80
Tracking ships, 138, 139, 145, 149, 219, 356, 357
Tracking stations, Mercury, 118, 129
Tranquillitatus, Mare (Sea of Tranquillity), 236, 244, 253, 254, 260, 262, 455, 465, 488, 490, 495, 496
Transistor, deterioration of, caused by radiation, 312
Transit 4B satellite, 81
Trapped radiation, 75, 78, 80; cosmic-ray-neutron theory of, 79; "leaky-bucket" theory of, 78, 79, 82
Truman, Harry S., 34, 159, 162
Tsander, Friedrich A., 6
Tucker, Ernest C., 379
Turkevich, Anthony, 261; quoted, 262
Twigg, John, 189
Tycho crater, 236, 251

United Aircraft Corporation, 264
United States Geological Survey, 251
United States Information Agency (USIA), 159, 171
United States Ordnance Intelligence, 17, 18
United States Strategic Air Force, 20
Unsymmetrical dimethylhydrazine (UDMH), 59, 347
Upper Atmospheric Research Panel, 26
Uprated Saturn 1 booster, 389, 390, 391, 399
Urey, Harold C., 166, 237, 252, 253, 255-256; quoted, 253, 255-256

V-1 flying bomb, 12, 13, 14, 20
V-2 (A-4) rocket, 1, 2, 3, 5, 10, 11, 12, 13, 14, 15, 16, 17, 18, 19, 20, 21, 26, 27, 35
V-3 (A-10) rocket, 18, 24
Van Allen, James A., 38, 52, 61, 68, 69, 70, 75, 76, 77, 78, 79, 83, 84, 96, 275, 281, 294, 295
Van Allen radiation belts, 70, 71, 73, 76-82 *passim,* 85, 89, 120, 305, 308, 311, 312, 449, 451
Van Dolah, Robert W., 397
Vanguard Project, 41, 42, 43, 44, 47, 48-51, 53, 57-58, 69, 71-73, 320
Vehicle Assembly Building (VAB), at KSC, 414-415
Velcro fasteners, in Apollo 012, 397
Venus, 94, 256, 263, 267, 269, 270, 277, 279, 280, 284; American probes of, 90, 94, 244, 267, 268, 269, 270-282; atmosphere of, 270, 280, 282; cloud structure of, 270, 279; diameter of, 270; ionosphere of, 281; and "limb-brightening" effect, 271, 278; radar data on, 279-280; revolution around sun, 270, 273; rotation of, 270, 279, 280, 281; Russian probes of, 86, 88, 268, 270, 280; temperatures on, 270, 271, 278, 279, 280
Verein für Raumschiffahrt (VfR), 6, 7, 8, 9
Vernov, Sergei N., 79
Viking rocket, 27, 38, 41, 42, 50, 51
Von Braun, Wernher, 6, 7, 19, 20, 21, 23, 24, 35, 36, 38, 39, 41, 44, 45, 46, 56, 62, 66, 67, 108, 109, 177, 182, 183, 188, 189, 419, 487; quoted, 19, 20, 53, 183, 184
Voskhod I, 355
Voskhod II, 330, 333-334, 337
Vostok series, 114, 117, 156, 157, 158, 163, 213, 341, 345, 403
Voyager Project, 285, 287, 288, 411

Wac Corporal, 34, 35
War of the Worlds (Wells), 283
Wasp, U.S.S., 341, 354, 355, 372
Waterman, Alan T., 39
Weather Satellites, *see* Meteorological Satellites
Webb, James E., 161, 162, 163, 164, 167, 169, 170, 180, 182, 192, 246, 378, 379, 381, 383, 385, 396, 397, 401, 403, 404, 405, 410, 411, 425, 439-440, 458; quoted, 169-170, 246, 401, 404, 416
Weightlessness, 31, 127, 139, 147, 201, 202, 224, 333, 351, 355
Welles, Orson, 298
Wells, H. G., 283

Westinghouse Electric Company, 475
Wexler, Harry, 38
White II, Edward H., 335, 336, 337, 338, 427, 458; and spacecraft 012, 373, 375, 377, 389, 390; spacewalk by, 337-340, 341, 397; as victim of spacecraft fire, 395
White, George C., Jr., 396
White Sands Proving Ground, 21, 26, 34, 35, 49, 50
Whiteside, John, 189
Wiesner, Jerome B., 160, 166, 167, 182, 196
Wiesner report, 160, 163
Williams, John, 396
Williams, John J., 380
Williams, Walter C., 118, 120, 125, 144, 146, 157, 216
Winkler, Johann, 6
"World Reaction to the United States and Soviet Space Programs" (USIA), 159
Wyld, James, 9

X-15 rocket airplane, 181, 378
X-258 rocket motor, 324
XS-1 rocket airplane, 103

Yates, Don, 60, 63
Yaw orientation, 202-203
Yorktown, U.S.S., 448, 463
Young, John W., 327, 330, 362, 365, 366, 367, 368, 436, 472, 474, 475, 479, 480, 481, 494 495, 497; on Gemini 3 ("Molly Brown"), 330-333; on Gemini 10, 362, 363, 364

Zeiler, Albert, 188, 189
Zero Gate, in Telstar command-decoder circuit, 307, 308
Ziolkovsky, Konstantin, 5, 6, 27, 263
Zot gun (HHMU), 336, 338, 364